Structural Nanocrystalline Materials

Nanocrystalline materials exhibit exceptional mechanical properties, representing an exciting new class of structural materials for technological applications. The advancement of this important field depends on the development of new fabrication methods, and an appreciation of the underlying nanoscale and interface effects. This authored book addresses these essential issues, presenting a fundamental, coherent, and current account at the theoretical and practical level of nanocrystalline and nanocomposite bulk materials and coatings. The subject is approached systematically, covering processing methods, key structural and mechanical properties, and a wealth of applications. This is a valuable resource for graduate students studying nanomaterials science and nanotechnologies, as well as researchers and practitioners in materials science and engineering.

CARL C. KOCH is Professor and Associate Department Head of Materials Science and Engineering at North Carolina State University. His principal areas of research include nanocrystalline materials; amorphization by mechanical attrition; mechanical alloying; rapid solidification; high-temperature intermetallics; oxide superconductors; and metastable materials. He is a Fellow of the American Physical Society, the Minerals, Metals and Materials Society (TMS), ASM International and the American Association for the Advancement of Science.

ILYA A. OVID'KO is Professor and Head of Laboratory for Nanomaterials Mechanics and Theory of Defects at the Institute of Problems of Mechanical Engineering (Russian Academy of Sciences) in St. Petersburg. His main research interests include nanostructured bulk materials and films; grain boundaries; and advanced materials (nanocrystalline, composite, and non-crystalline).

SUDIPTA SEAL is a Professor for the Advanced Materials Processing and Analysis Center at the University of Central Florida (UCF). He is also the Nanocoordinator for the Office of Research at UCF, and Director of the Surface Engineering and Nanotechnology Facility. His key research interests include surface engineering, nanoscience and nanotechnology.

STAN VEPREK is Professor Emeritus and former Director of the Institute for Chemistry of Inorganic Materials, Technical University Munich, and Visiting Principal Scientist at the Singapore Institute of Manufacturing Technology (SIMTech), Singapore. His areas of research interest include materials and surface science; plasma CVC and PVD; thin film coatings; and light-emitting Si and microcrystalline Si. He also teaches courses at the Technical University München and the National University of Singapore.

Structural Nanocrystalline Materials

Fundamentals and Applications

CARL C. KOCH, ILYA A. OVID'KO, SUDIPTA SEAL,
AND STAN VEPREK

CAMBRIDGE UNIVERSITY PRESS
Cambridge, New York, Melbourne, Madrid, Cape Town, Singapore, São Paulo

Cambridge University Press
The Edinburgh Building, Cambridge CB2 8RU, UK

Published in the United States of America by Cambridge University Press, New York

www.cambridge.org
Information on this title: www.cambridge.org/9780521855655

© C. C. Koch, I. A. Ovid'ko, S. Seal, S. Veprek 2007

This publication is in copyright. Subject to statutory exception
and to the provisions of relevant collective licensing agreements,
no reproduction of any part may take place without
the written permission of Cambridge University Press.

First published 2007

Printed in the United Kingdom at the University Press, Cambridge

A catalog record for this publication is available from the British Library

ISBN-13 978-0-521-85565-5 hardback
ISBN-10 0-521-85565-9 hardback

Cambridge University Press has no responsibility for the persistence or accuracy of URLs for external or third-party internet websites referred to in this publication, and does not guarantee that any content on such websites is, or will remain, accurate or appropriate.

Contents

Preface		*page* ix
Acknowledgments		xii
1	**Introduction**	1
	1.1 Coatings and thin films	2
	1.2 Bulk parts	5
	1.3 Other nanostructured materials for structural applications	8
	1.4 Topics to be covered	21
	References	21
2	**Processing of structural nanocrystalline materials**	25
	2.1 Introduction	25
	2.2 Methods for preparation of bulk structural nanocrystalline materials	27
	2.3 Preparation of nanostructured, hard, and superhard coatings	72
	References	85
3	**Stability of structural nanocrystalline materials – grain growth**	93
	3.1 Introduction	93
	3.2 Experimental methods for measuring grain growth	94
	3.3 Grain-growth theories for conventional grain size materials	99
	3.4 Grain growth at ambient temperature in nanocrystalline materials	108
	3.5 Inhibition of grain growth in nanocrystalline materials	109
	3.6 Experimental studies of isothermal grain-growth kinetics in nanocrystalline materials	114
	3.7 Thin films and coatings	118
	References	128
4	**Mechanical properties of structural nanocrystalline materials – experimental observations**	134
	4.1 Elastic properties of nanostructured materials	134
	4.2 Anelastic properties	136
	4.3 Hardness and strength	138
	4.4 Ductility of nanocrystalline materials: optimization of strength and ductility	151

4.5	Superplasticity of nanocrystalline materials	169
4.6	Creep of nanocrystalline materials	173
4.7	Fatigue of nanocrystalline materials	177
4.8	Fracture and fracture toughness of nanocrystalline materials	180
4.9	Mechanical properties of superhard nanostructured coatings	182
4.10	Summary	194
	References	196

5 Mechanical properties of structural nanocrystalline materials – theory and simulations
204

5.1	Introduction	204
5.2	Specific structural features of nanocrystalline materials	208
5.3	Basic concepts on plastic deformation processes in nanocrystalline materials	217
5.4	Nanoscale and interface effects on lattice dislocation slip	221
5.5	Deformation modes associated with enhanced diffusion along grain boundaries and their triple junctions. Competition between deformation mechanisms. Effect of a distribution of grain size	224
5.6	Grain-boundary sliding in nanocrystalline materials	229
5.7	Interaction between deformation modes in nanocrystalline materials. Emission of lattice dislocations from grain boundaries. Twin deformation mode	245
5.8	Interaction between deformation modes in nanocrystalline materials. Rotational deformation mode	258
5.9	Fracture mechanisms in nanocrystalline materials. Generation, growth and convergence of nanocracks	268
5.10	Strain-rate sensitivity, ductility and superplasticity of nanocrystalline materials	291
5.11	Diffusion in nanocrystalline materials	303
5.12	Concluding remarks	306
	References	308

6 Corrosion of structural nanomaterials
317

6.1	Introduction	317
6.2	Effect of defects and grain size	318
6.3	Corrosion of metallic and alloyed nanostructured materials	319
6.4	Nanocrystalline nickel	319
6.5	Nanocrystalline cobalt and its alloys	320
6.6	Zirconium and its alloys	326
6.7	304 Austenitic steels: wet corrosion	331
6.8	304 Austenitic steels: dry corrosion	333
6.9	Magnetic nanocomposites	336
	References	338

7	**Applications of structural nanomaterials**	341
	7.1 Introduction	341
	7.2 Ceramic nanocomposites for load-bearing applications	343
	7.3 Bulk nanoribbons can tie up photonic circuits	343
	7.4 Functionally gradient nanoparticles	344
	7.5 Nanotechnology in automotive applications	344
	7.6 Nanoclay–polymer composites for structural applications	346
	7.7 Nanotechnology in the consumer world	347
	7.8 Nanobelts for actuator applications	350
	7.9 Nanosteel for high wear, toughness and hardness applications	350
	7.10 Copper–carbon nanotube composite for high heat applications	351
	7.11 Metal matrix nanocomposites for structural applications	352
	7.12 Application of ferrofluids with magnetic nanoparticles	352
	7.13 Industrial applications of nanocomposite coatings	355
	7.14 Applications of electrodeposited nanostructures	358
	7.15 Potential military applications	359
	7.16 Concluding remarks	360
	References	360
	Index	362

Preface

Nanoscience and nanotechnology has become an identifiable, if very large, diverse, and multidisciplinary field of research and emerging applications. It is one of the most visible and growing research areas in science and technology. Government research funding agencies throughout the world have recognized its potential importance with substantial special initiatives to support its growth and development. The realms of nanotechnology applications are being explored not only of chemistry, materials, and engineering, but the frontiers of medicine as well. This on-going research is what enables the continuing expansions in nanotechnology. Over the past two decades there has been a revolution in the material-science field that has sparked great interest and research in all areas of science and engineering. Nanotechnology and nanoscience are leading this revolution fueled by the industrial progress, the scientific ability to fabricate, model and manipulate objects (things) with a small numbers of atoms, and the almost daily discovery of new phenomena of the nanoscale.

Nanostructured materials include atomic clusters, layered (lamellar) films, filamentary structures, and bulk nanostructured materials. The common thread to these various material forms is the nanoscale dimensionality, i.e. at least one dimension less than 100 nm (more typically less than 50 nm), often less than 10 nm.[1] While this dimension requirement may appear to be arbitrary, it is usually at these length scales that the "physics" often changes and leads to very different properties, often superior, than those of conventional materials. Typical manifestations of nanoscale effects in solids are the dislocation instability in nanoscale crystallites and the quantum confinement of charge carriers in semiconductor quantum dots. Nanoparticles can be considered to be of "zero" dimensionality and examples include a large range of nanoscale powders of interest for diverse applications such as dispersions in cosmetics and pharmaceuticals. Quantum dots for optoelectronic applications may also fall into this category. A layered or lamellar structure is a one-dimensional nanostructure in which the magnitudes of length and width are much greater than the thickness that is nanoscale. Thin films with quantum well structures for electronic device applications are examples of this category. Two-dimensional nanostructures have the length much larger than the width or diameter and nanowires or nanotubes may fit this division. The nanostructures that contain the "bulk" definition relevant to this book are three-dimensional and consist of crystallites, or in certain cases quasicrystals and/or amorphous material, that are

---

[1] This applies for optoelectronic materials, catalysts, superhard nanocomposites, etc.

nanoscale in dimension. Structural geometry of nanostructures becomes more complicated in the case of composite materials. Categories of nanocomposites consisting of at least two phases, with at least one phase having characteristic sizes less than 100 nm, are specified by size scales and geometry of constituent phases.

Surfaces and interfaces are also important in explaining nanomaterial behavior. In bulk materials, only a relatively small percentage of atoms will be at or near a surface or interface. In nanomaterials, the small feature size ensures that many atoms, perhaps half or more in some cases, will be near interfaces. Surface properties such as energy levels, electronic structure, and reactivity can be quite different from interior states, and give rise to quite different material properties.

Structural nanomaterial can exist in various forms, such as, nanometals, nano-oxides and nanocomposites. Each type exhibits different properties, thus having an immense potential for many applications in every field of science and engineering. These materials can find applications in mechanical, aerospace and electronics engineering. For example, aerospace is in constant search of lighter, more flexible and stronger materials; higher efficiency fuels and more powerful engines among others. Electronics are ruled by Moore's law,[2] and the need for smaller yet more powerful integrated circuits makes, nanomaterials the holy grail of the electronics field. Other fields that are worth mentioning owing to their large interest and research into the materials are medical science, forensic science, civil engineering, and cosmetology.

As the field of nanoscience/nanotechnology has developed, new journals, books, and topical conferences have been devoted to the dissemination of the research results that are increasing at a rapid rate. However, often "nano" refers to the size of a component rather than the dimensions of the internal microstructure of a bulk material. In many cases, the specialized nanomaterials journals and books leave out bulk nanostructured materials.

This book takes for its theme *Structural Nanocrystalline Materials: Fundamentals and Applications*. Since structural materials are typically "bulk" or coatings the book will focus mainly on these categories although the two-dimensional nanotubes are also briefly addressed. Since many books on nanostructured materials have already been published we must justify another book in this area. First, since the field is moving so rapidly owing to its novelty and the huge number of researchers working in the area, many concepts only a few years old have become obsolete and another "snap-shot" of the state-of-the-art of the field is worthwhile. Secondly, the area of structural nanomaterials has been relatively neglected although several excellent reviews of the mechanical behavior of nanocrystalline materials have been published. This book aims to cover the complete cycle of structural nanomaterials from their processing, characterization, stability, mechanical behavior, corrosion, and applications.

The introductory chapter briefly covers the need for structural nanocrystalline materials. Chapter 2 covers the variety of processing methods that have been used to make nanomaterials for bulk or thin-film structural applications. Chapter 3 describes the stability of nanocrystalline microstructures and also covers the various methods used to characterize the

---

[2] Moore's law means that the number of transistors on a chip doubles, their size shrinks and speed increases by a factor of 2 every 18–24 months.

nanocrystalline structure. Chapter 4 reviews the experimental evidence for the mechanical behavior of bulk and thin film nanocrystalline materials. Chapter 5 covers the theoretical understanding of the mechanical properties of nanomaterials. Chapter 6 reviews the limited knowledge regarding the corrosion behavior of nanocrystalline materials that is important for structural applications. Chapter 7 presents existing and potential applications for structural materials.

Acknowledgments

The research on nanocrystalline materials by Carl C. Koch has been supported for over 15 years by the US National Science Foundation, most recently under grant number DMR-0201474. His recent work is also supported by the US Department of Energy under grant number DE-FG02-02ER46003. He also wishes to thank his colleagues at North Carolina State University, Professors Ronald Scattergood and K. L. Murty, and post-doctoral fellow Dr. Khaled Youssef. The many graduate and undergraduate students who have contributed to his research in this area also should be acknowledged.

Ilya Ovid'ko gratefully acknowledges the support from the Office of US Naval Research (grant N00014-05-1-0217), INTAS (grant 03-51-3779), INTAS-AIRBUS (grant 04-80-7339), Russian Academy of Sciences Program "Structural Mechanics of Materials and Construction Elements," Ministry of Education and Science of Russian Federation and St. Petersburg Scientific Center. Also, Ilya Ovid'ko would like to express his warmest thanks to Dr. S. V. Bobylev, Dr. M. Yu. Gutkin, Dr. R. A. Masumura, Dr. N. F. Morozov, Dr. C. S. Pande, Dr. A. G. Sheinerman, and Dr. N. V. Skiba for fruitful discussions and collaboration.

Sudipta Seal would like to thank NASA Glenn (NAG 32751), NASA-KSC, FSEC-UCF, US-Russia CRDF grant, Florida Space Grant Consortium (FSGC), Office of Naval Research – Young Investigator Award (ONR YIP) (ONR: DURIP: N000140310858, ONR-YIP N000140210591), National Science Foundation (NSF – BES: 0552438, CMS: 0548815, BES: 0541516, ECE: 0521497, DMII: 0500268, EEC: 0453436, DMR-0421253, CTS 0350572, DMI: 0334260, CTS 0404174, EEC: 0304525, EEC: 0136710, EEC: 0139614, EEC: 0086639, EEC: 0086639, EEC: 9907794), National Institute of Health (R01AG022617), Missile Defense Agency (MSTAR BAA 05-012), and DOD, DOE (DOE: DE-FG02-04ER83994), NASA SBIR Ph I and II, Plasma Process Inc., Materials Interface, Praxair, Lockheed Martin, State of Florida, Siemens Westinghouse and University of Central Florida for funding and supporting his current research in material engineering, functional coatings and nanotechnology. Professor Seal also thanks his research group for assistance in this work. Finally he acknowledges his wife and parents for continued encouragement in his scientific quest.

Stan Veprek would like to thank for financial support the NATO Science for Peace Project SfP 972379, German Science Foundation (DFG), European Commission Growth Programme in the frame of the 5th RTD Framework Programme under the contract number G5RD-CT-2000-0222 Project "NACODRY," the 6th RTD Framework Programme under the Project "MACHERENA," and Contract No. CA 505549-1 – "DESHNAF." He further thanks the German Space Agency (DLR), Volkswagen Stiftung under Grant No. I/77 192, and Alexander von Humboldt Stiftung, for supporting several visiting scientists who were participating in the research on nanocrystalline materials in his Institute. Thanks are also extended to Professor Li Shizhi, Professor A. S. Argon, Professor D. M. Parks, Professor Q. F. Fang, Dr. Maritza Veprek-Heijman, Dr. P. Holubar, Dr. M. Jilek, Dr. M. Sima, and Dr. T. Cselle for fruitful discussions, and to all co-workers who participated in the work on superhard nanocomposites at the Technical University Munich.

1 Introduction

With the explosion of research interest in nanocrystalline materials in recent years, one sub-area that has received significant attention is the mechanical behavior of materials with grain sizes less than 100 nm. The great interest in the mechanical behavior of nanocrystalline materials originates from the unique mechanical properties first observed and/or predicted by the pioneers of this field, Gleiter and co-workers, in materials prepared by the gas condensation method (Gleiter, 1989). Among these early observations or predictions were:

- lower elastic moduli than for conventional grain size materials – by as much as 30%–50%;
- very high hardness and strength – hardness values for nanocrystalline pure metals ($\sim$10 nm grain size) that are 2–10 or more times higher than those of larger grained ($>$1 μm) metals;
- increased ductility – perhaps even superplastic behavior – at low homologous temperatures in even normally brittle ceramics or intermetallics with nanoscale grain sizes, believed to be caused by grain boundary, diffusional deformation mechanisms.

While some of these early observations and predictions have been verified by subsequent studies, in particular the high hardness and strength values, some have been found to be caused by high porosity in the early bulk samples (for example, elastic constant behavior) or to other artifacts introduced by the processing procedures. The ductility issue remains a subject of present research and while most nanocrystalline materials don't exhibit the high predicted ductilities (Koch *et al.*, 1999), there are recent examples of good ductility along with high strength in a limited number of cases (Karimpoor *et al.*, 2003; Li and Ebrahimi, 2004; Youssef *et al.*, 2004). Similarly, while nanocrystalline ceramics or intermetallics have not been found to be ductile at ambient temperature, ductility, and in some cases even superplasticity (McFadden *et al.*, 1999), has been found at lower homologous temperature than for the conventional grain size counterparts.

An important question in this field is "what are the needs for structural nanocrystalline materials?" Are there applications for which the superior properties and/or the grain size are required for feasible performance? Are there cases in conventional structural applications where nanostructured materials could be substituted in a cost effective way? This chapter will attempt to address these questions for structural nanomaterials in the form of coatings and thin films as well as bulk parts.

1.1 Coatings and thin films

Coatings and thin films are applied to structural bulk materials in order to improve the desired properties of the surface, such as wear resistance, friction, corrosion resistance and others, yet keeping the bulk properties of the material unchanged. A typical example is nitriding and carbonitriding of steel parts for engines and other machines at relatively low temperatures of about 500 °C in order to increase the hardness of the surface and reduce wear.

Modern nanostructured coatings and thin films for structural and functional applications, which were developed during the past 10–15 years, are used mainly for wear protection of machining tools and for the reduction of friction in sliding parts. One distinguishes between nanolayered coatings, where a few nanometers thin layers of two different materials are deposited subsequently, and nanocomposites, which are, in the optimum case, isotropic. The superhard nanocomposites, such as nc-$(Ti_{1-x}Al_x)$N/a-Si_3N_4 (nc- and a- stand for nanocrystalline and X-ray amorphous, respectively), show superior cutting performance as compared with conventional, state-of-the art hard coatings $(Ti_{1-x}Al_x)$N that presently dominate the applications for dry machining. The costs of their large-scale industrial production are comparable with those of the conventional coatings (Jilek *et al.*, 2004). Also, the heterostructures and multilayer coatings are successfully applied on industrial scale (Münz *et al.*, 2001; Münz, 2003). Low-friction nanostructured coatings consisting of a hard transition-metal carbide or nitride in combination with a solid lubricant, such as diamond-like carbon (DLC), MoS_2, WS_2 and others that combine with a high hardness and low friction (Voevodin *et al.*, 1996a; 1996b; 1997a; 1997b; Voevodin and Zabinski, 1998; 2000; Voevodin *et al.*, 2002). They are applied in a variety of bearings and sliding parts operating without liquid lubricants, which is an important advantage particularly in a hostile environment, and when the movable parts have to stop and go very frequently, e.g. in the textile industry.

The recent development of nanocomposites consisting of a hard transition-metal nitride or carbide in combination with soft and ductile metal (Baker *et al.*, 2005) is likely to find numerous applications in a variety of machine parts. The hardness of these coatings varies between about 13 and 30 GPa depending on the composition. When deposited under energetic ion bombardment and temperatures below about 350 °C, an enhancement of the hardness up to about 50 GPa was found (Zeman *et al.*, 2000), in a similar way as for hard transition-metal nitrides (e.g. 100 GPa for TiAlVN and 80 GPa for TiN) (Musil *et al.*, 1988; for further examples see Veprek *et al.*, 2005). However, this hardness enhancement is of a little use because, upon annealing to ≥ 500 °C, these coatings soften (Karvankova *et al.*, 2001). Unfortunately, these nanocomposites were often confused (see Musil 2000; and personal communication) with the thermally highly stable superhard nanocomposites prepared according to the generic design principle (Veprek and Reiprich, 1995).

The nanolayered coatings should be subdivided into multilayers and superlattices. When a 3–4 µm thick monolytic layer of a hard ceramic material, such as TiN, is replaced by a stack of 20–100 nm thin multilayers of TiN and another hard nitride, boride or carbide (see Holleck *et al.*, 1990; Zhitomirsky *et al.*, 1999; Soe and Yamamoto, 1999), the resistance against brittle failure strongly increases because the crack cannot propagate through the

1.1 Coatings and thin films

whole layer. Usually, also an increase of the hardness above that of the rule-of-mixtures is found. Similar enhancement was found also in metallic multilayers, for example Fe/Cu and Ni/Cu (Bunshah et al., 1980) and Ni$_3$Al/Ni (Tixier et al., 1999), and in metal/nitride multilayers, for example Ti/TiN (Kusano et al., 1998). The majority of hard protecting coatings applied to machining tools nowadays are such multilayers.

The concept of superlattices goes back to a theoretical paper of Koehler who suggested depositing alternate epitaxial crystal layers of two different materials A and B with similar lattice parameters but with elastic moduli as different as possible (Koehler, 1970). The thickness of the layer should be small so that dislocation multiplication sources cannot operate. If, under an applied stress, a dislocation moves within the layer with the smaller modulus, it will be hindered in crossing the interface A/B owing to the repelling image force induced within the stronger layer by elastic deformation, when the dislocation approaches that interface. Koehler's concept was experimentally confirmed by Lehocky, who prepared heterostructures consisting of different metals (Lehocky, 1978a; 1978b). Later, several research groups applied this concept to hard transition-metal nitrides and obtained superhard heterostructures with hardnesses of 50–56 GPa for TiN/NbN and TiN/VN systems (Shinn et al., 1992). In a note added in proof, Koehler stated that the ideas described in his paper are also valid if one of the materials is amorphous (Koehler, 1970). Indeed, later work by a number of researchers confirmed that the hardness enhancement can also be achieved in polycrystalline heterostructures and in laminates consisting of amorphous and crystalline materials. For further information the reader should consult some of the numerous reviews (see, for example, Barnett, 1993; Veprek, 1999; Chung and Sproul, 2003; Barnett et al., 2003; Münz et al., 2001; Münz, 2003).

Examples of nanostructured, low-friction hard coatings are nanocomposites consisting of a hard transition-metal carbide (MeC) in combination with amorphous, "diamond-like" carbon (DLC) (Voevodin et al., 1997a), which were originally developed in order to reduce the high-compressive stress in DLC that causes delamination. Alternatively, coatings consisting of multilayers and combination of multilayers with nanocomposites (Voevodin et al., 1996a; Voevodin et al., 1997b) or functionally gradient nanolayered nanocomposites (Voevodin and Zabinski, 1998) can be used. The high degree of hardness of these coatings (exceeding 30 GPa) allows their use under the conditions of a high contact load, and the high toughness provides them with a long lifetime also under impact load. The low friction of these coatings is the result of the formation of a thin hydrated, graphitic surface layer. Such coatings find many applications in sliding gears operating either under a high contact load or in a dusty environment, where liquid lubricants cannot be used. Another advantage of such a "solid lubricant" is the low friction when starting a machine, because when two surfaces lubricated by oil are at rest, the lubricant is pressed out of the surface asperities and, therefore, upon starting, the friction is high. The disadvantage of the DLC-based nanocomposites is the fact that the low coefficient of friction is achieved only in humid atmosphere. In order to solve this problem, Voevodin and co-workers included into the nc-MeC/DLC nanocomposites a second lubricant phase, such as MoS$_2$ or WS$_2$, that provides a low coefficient of friction in a dry environment. These coatings show an excellent adaptation of surface properties upon changing the environment (Voevodin and Zabinski, 2000; Voevodin et al., 2002).

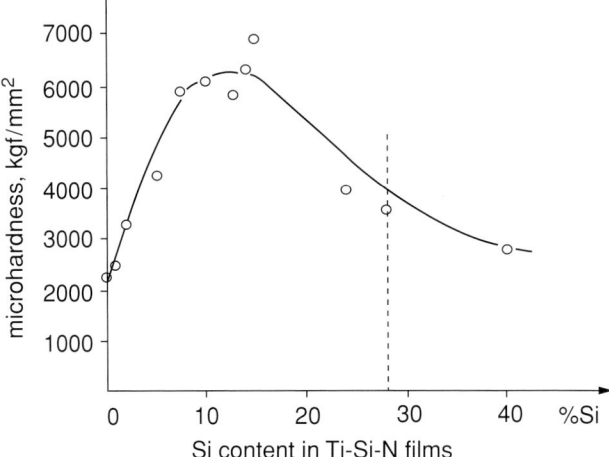

Figure 1.1. Hardness of Ti-Si-N films as a function of Si content (from Li et al., 1992, with permission).

Superhard nanocomposites, such as nc-TiN/a-Si$_3$N$_4$ (here a- stands for X-ray amorphous) were pioneered by Li and co-workers who used plasma CVD in order to deposit ≥5 μm thick "Ti–Si–N" coatings at a temperature of 560 °C. A strong hardness enhancement up to ≥60 GPa (Figure 1.1) was found at a silicon content of about 12 at.% (Li et al., 1992). Originally, the researchers attributed the large increase in the hardness to dispersion hardening. However, when, later on, these coating were studied in more detail and the two-phase nature of immiscible stoichiometric nitrides consisting of 3.0–3.5 nm TiN nanocrystals imbedded in X-ray-amorphous Si$_3$N$_4$ was found (Veprek et al., 1995a), it became clear that the dispersion hardening cannot operate in such systems. Therefore, a generic design concept for the preparation of such nanocomposites was elaborated (Veprek and Reiprich, 1995) and its validity was verified on several nc-(Me$_n$N)a-Si$_3$N$_4$ (Me = W, V, . . .) systems in which superhardnesses in excess of 40–50 GPa were achieved (Veprek et al., 1995b; Veprek et al., 1996). Compared with conventional hard nitride and carbide coatings the advantage of the superhard nanocomposites is the significantly higher hardness which strongly decreases abrasive wear.

Li and co-workers had noticed that, when the hardness reaches its maximum value, the columnar structure, which is typical for refractory ceramic coatings deposited at a relatively low homologous temperature, vanishes (Li et al., 1992). This was further confirmed for nc-TiN/a-Si$_3$N$_4$ (Veprek et al., 1995a) and extended to other systems (Veprek et al., 1995b; Veprek et al., 1996). Furthermore, it was found, that the Si$_3$N$_4$ layer provides the nanocomposites with a much higher oxidation resistance than TiN (Veprek et al., 1995a; Veprek and Reiprich, 1995; Veprek et al., 1996). This is because of the formation of a dense interfacial Si$_3$N$_4$ layer, the thickness of which was estimated to be about 0.3–0.5 nm (Veprek and Reiprich, 1995). We shall see later in this book that the thickness of the interfacial layer of about one monolayer of Si$_3$N$_4$, BN and possibly other covalent nitrides, is crucial for the optimum properties of these materials. More-recent work has suggested that the formation of the stable nanostructure probably occurs during the deposition by spinodal mechanism (Zhang and Veprek, 2006). The high immiscibility of the stoichiometric TiN and Si_3N_4 (Rogl and Schuster, 1992) provides the nanocomposites with a high thermal stability up to $\geq 1100\,°C$. In the case of the nc-$(Ti_{1-x}Al_x)$N/a-Si_3N_4 nanocomposites, the decomposition and softening, which for the conventional $(Ti_{1-x}Al_x)$N coatings occurs at 800–900 °C, is suppressed up to about 1250 °C (Veprek *et al.*, 2004). Let us emphasize that the high thermal stability and oxidation resistance of the nanocomposites is a very important advantage for their application in dry machining when the temperature of the tool and coatings reaches $\geq$ 800 °C. For these reasons, the superhard nanocomposites prepared according to the design principle represent a new class of advanced nanocrystalline materials that are already used in large-scale industrial applications.

1.2 Bulk parts

Bulk nanostructured materials are defined as bulk solids with nanoscale or partly nanoscale microstructures. This category of nanostructured materials has historical roots going back many decades, but for nanostructured matrices the relatively recent focus is the result of new discoveries of unique properties of some nanoscale materials.

Early in the twentieth century, when "microstructures" were revealed primarily with the optical microscope, it was recognized that refined microstructures, for example, small grain sizes, often provided attractive properties such as increased strength and toughness in structural materials. A classic example of property enhancement due to a refined microstructure – with features too small to resolve with the optical microscope – was age hardening of aluminum alloys (Mehl and Cahn, 1983). The phenomenon, discovered by Alfred Wilm (1906) was essentially explained by Merica *et al.* (1919) and the microstructural features responsible were first inferred by the X-ray studies of Guinier (1938) and Preston (1938). With the advent of transmission electron microscopy (TEM) and sophisticated X-ray diffraction methods, it is now known that the fine precipitates responsible for age hardening, in Al–4% Cu, for example, are clusters of Cu atoms (Guinier–Preston (GP) zones) and the metastable partially coherent $\theta^{'}$ precipitate (Porter and Easterling, 1983). Maximum hardness is observed with a mixture of GPII (or $\theta^{''}$) (coarsened GP zones) and $\theta^{'}$ with the dimensions of the $\theta^{'}$ plates, typically about 10 nm in thickness by 100 nm in diameter. Therefore, the important microstructural feature of age-hardened aluminum alloys is nanoscale. Carbide precipitates in certain steels can be nanoscale and affect the mechanical behavior in advantageous ways. A recently developed cast stainless steel with nanoscale alloy carbides has been found to have exceptional strength and creep resistance at operating temperatures up to 800 °C (Maziasz and Pollard, 2003). The many examples of nanoscale second phases on the strength of structural materials are confined to matrices wherein the grain size is "conventional," that is typically tens of microns. The focus of recent work, and of this book, is on structural materials where the matrix has a nanoscale grain size.

If nanocrystalline bulk structural materials could be processed in a cost-effective manner and on a scale sufficient for some present structural applications, it is clear that with their special, often superior, properties they might supplant existing conventional grain-size materials. However, at this stage of development this is not possible. Most methods for preparing truly nanocrystalline materials (grain sizes <100 nm) do not allow for significant quantities or sizes of artifact-free bulk materials at the present time. Major breakthroughs in processing will be needed to accomplish this. However, some processing methods, to be discussed in more depth in Chapter 2, are capable of producing limited sizes of artifact-free nanostructured materials. What are the potential uses for such small sample sizes? One potential structural application for nanostructured materials comes from the requirements for miniaturization of components for engineering components for biotechnology, sensors, electro-optics, and micro-nanoscale integration. Material grain sizes must be a factor of 50–100 times smaller than component/feature sizes. Otherwise, mechanical properties will be highly variable and unsuitable for fabrication or end-use. Miniaturization scales <5–10 μm thus require grain sizes <100 nm, that is, nanocrystalline materials. Grain sizes in conventional engineering metals and alloys are typically in the range of 5–50 μm, which would mean that essentially single-crystal behavior would be operative in components of this size. This would lead to large variations in elastic and plastic properties owing to the inherent crystallographic differences in behavior. In conventional parts the grain anisotropies are averaged over millions of grains and such variation is not observed. A general guideline for obtaining uniform and predictable properties using polycrystalline materials is that there should be at least 50 grains on the span of a component or feature cross-section (Ehmann *et al.*, 2000). When this is the case, the resultant mechanical properties will be the uniform polycrystalline averages expected by the manufacturing engineer.

Once the force components, deflections, and other functional requirements are specified for miniaturized components, and material properties are available, the design protocols follow well-established rules. For structural integrity, yield stresses determine the maximum stresses and deflections that are allowed before the onset of plastic yielding failures. Fracture toughness determines the onset of fracture failure, either in static loading or cyclic fatigue loading. Fracture toughness and fatigue-crack growth correlations are used to determine fatigue life. Miniaturization can affect failure modes through changes in the properties owing to grain size reduction or through reduction in component cross-section sizes. With all else constant, this would increase the resistance to fracture and fatigue failures for nanostructured materials. However, fatigue life depends on the number of cycles needed for an initial crack to grow to a critical size. For higher-toughness materials, this length can typically be in the range of centimeters. Consequently, with all else constant, miniaturized components with cross-section sizes on the order of 1 mm or less would have reduced fatigue life because sub-critical cracks grow completely through the cross-section causing premature failure. The existing limited studies on fatigue of nanostructured materials will be discussed in Chapter 4.

Miniaturized parts can be made in relatively high-volume production using the so-called Swiss machining techniques developed decades ago for precision mechanical watch movements. For the current state-of-the-art, multi-axis precision machine tools are commercially available that incorporate high-speed spindles and very small tooling. Figure 1.2 (Scattergood and Koch, 2003) shows an example of miniaturized parts made using these techniques. Figure 1.3 (Scattergood and Koch, 2003) shows a machining tool with

Figure 1.2. Miniaturized parts produced by micromachining. (www.remmele.com)

Figure 1.3. Miniature machine tool. Reprinted from Y. Okazaki, N. Mishima, and K. Ashida "Microfactory-Concept, History, and Developments" in *Journal of Manufacturing Science and Engineering*, Nov 2004, vol 126, with permission.

dimensions of the order of a few centimeters. These advances will undoubtedly change the protocols for designing, manufacturing, and handling very small components for high-volume production applications. The mechanical properties of the nanostructured materials that will be used for such miniaturized parts will have a critical effect on the machining parameters. Delicate tooling requires low tool forces, therefore machining feed rates must be reduced. However, very-high-speed machining appears to be a necessary requirement for volume production using these materials. Burr formation is another important consideration for machining miniaturized parts. Even small burrs become unacceptable when feature sizes are reduced. Depending on the application, removing burrs after machining can be difficult and costly. Since burr formation involves complex deformation paths with a final fracture process to separate the chip from the work-piece, the interplay of the material properties that control burring is not easily identified. Experience shows that high-strength materials, with lower ductility, perform better in this regard. This may be an advantage for nanostructured materials.

1.3 Other nanostructured materials for structural applications

As discussed earlier, nanocrystalline materials have an average grain size less than 100 nm. The small grain size results in positioning the atoms in grain boundaries that are not part of the original crystalline lattice (Sun and Murray, 1999). This structure leads to new material properties and it is possible to utilize this phenomenon to create new structural materials with hardness comparable to that of diamond (Berger, 1996). To obtain materials with combinations of properties, multiphase structures are used. The nanocomposites can be used as hard surfaces with excellent properties and great corrosion resistance. There is a current need for high-performance magnetic composites for a wide variety of structural applications. It is possible to fabricate nanocomposites with new magnetic properties leading to the development of large structural magnets.

1.3.1 *Magnetic nanomaterials and composites*

Magnetism in small scales has created a lot of interest in potential technological applications (Heinrich and Bland, 1994). Magnetic nanomaterials are the most popular basis for data storage in modern computers, but they are also used in many other fields of science and technology including engineering, medicine, and biotechnology. In magnetic nanoparticles, the magnetization direction often fluctuates and is not necessarily fixed as in larger crystals. Furthermore, if nanoparticles are in close proximity, the magnetic properties are often dominated by phenomena that take place at the interface (Berger, 1996). Magnetic nanoparticles based on transition metals show a variety of unusual magnetic behaviors when compared with the bulk materials. This is mostly because of the surface and the interface effects of symmetry breaking, electronic environment, charge transfer, and magnetic interactions. Reducing the size or dimensions of the magnetic systems changes drastically the electronic properties by reducing the symmetry of the system. This further introduces a quantum limitation in the material; this is often an intriguing aspect of nanoscale magnets.

1.3 Other nanostructured materials for structural applications

These magnets are regarded as molecularly complex with a net magnetic moment considered as a single domain (Slonczewski, 1996). However, in many technological applications such molecular magnets may suffer from poor magnetic densities, but research efforts are on-going to develop magnetic network materials where magnetic centers are combined with organic linker molecules in three-dimensional nanoporous structures. These materials have great structural flexibility due to the variable coordination chemistry of the transition-metal centers. Through changes in the ligands it is possible to control the magnetic properties of the material.

Through the use of appropriate interstitial material, the assembly of these magnetic building blocks into ordered two-dimensional arrays would allow for tunable and externally controllable inter-particle interactions that modify the macroscopic material properties for future application as superior performance magnetic memory, sensors, and ultra-high-speed large-scale device architectures (Urazhdin et al., 2003). The control of structure on the nanoscale has been used to improve the performance of magnetic materials, and this progress in turn will contribute to improvements in performance of electric motors and generators; this is a key component in many structural applications.

The past ten years have seen great interest in the magnetism of reduced dimensionality structures. Researchers around the world have recently started to study one-dimensional and zero-dimensional magnets by laterally structuring thin magnetic films into magnetic quantum wires and dots. As well as being an excellent experimental means of studying fundamental magnetic phenomena, these low-dimensionality magnets may in the future form the basis of new data storage and computing technologies.

One of the biggest challenges of these nanomagnets lies in the complex characterization mode. This is why a new type of nanomagnetic probe uses the Magneto Optical Kerr Effect (MOKE), a phenomenon by which the polarization of light is rotated when it is reflected from the surface of a magnetic material. The NanoMOKE probe can determine quantitative hysteresis loops from small to large arrays of nanomagnets or from individual nanomagnets (Krivorotov et al., 2005). A hybrid magneto-optical magnetometer has been designed and constructed for probing the magnetic properties of submicron nanomagnets. For example, 10 nm-thick square nanomagnets have been fabricated individually and in small arrays from $Ni_{80}Fe_{14}Mo_5$ ("supermalloy") by electron-beam lithography. Hysteresis loops with a good signal-to-noise ratio have been obtained from individual nanomagnets as small as 400 nm and from arrays of nanomagnets ranging in size from 500 nm to 75 nm (Cowburn et al., 1998).

The strength of a magnet used in structural applications is measured in terms of coercivity and saturation magnetisation values. These values increase with a decrease in the grain size and an increase in the specific surface area (surface area per unit volume) of the grains. It has been shown that magnets made of nanocrystalline yttrium–samarium–cobalt grains possess very unusual magnetic properties owing to their extremely large surface area (Cowburn et al., 1998). Typical applications for these high-power rare-earth nanocrystalline magnets include quieter submarines, automobile alternators, land-based power generators, motors for ships, ultra-sensitive analytical instruments, linear motors using Hallback arrays, and in magnetic resonance imaging (MRI) in medical diagnostics (Buschow, 1988).

One can also create nanocomposites with soft and hard magnetic attributes via exchange reaction. The system has excellent hard-magnetic properties: the hard-magnetic phase gives rise to a high anisotropy and coercivity, whereas the soft magnetic phase enhances the saturation magnetization (Jianu et al., 2004). Because of their unusual high remanence, large energy product and low cost, the exchange spring magnets provide attractive potential applications as permanent magnets. Micromagnetic calculations (Kneller and Hawing, 1991) and experimental results have shown that the exchange coupling is achieved in nanocomposite materials in which the hard-magnetic ($RE_2Fe_{14}B$, where RE = rare earth) and additional soft (a-Fe or Fe_3B) nanophases are crystallographic coherent. Nanocomposite exchange spring magnets may be obtained from amorphous ribbons with RE–Fe–B defined composition, using suitable annealing treatments. By reducing the rare-earth content we alter the crystallization process. For example, in nanocomposite $Nd_2Fe_{14}B/Fe_3B$ or a-Fe the magnetic properties are strongly related to the microstructure. The optimum structure consists of uniformly distributed soft- and hard-magnetic nanophases.

1.3.2 Tougher and harder cutting tools

Large-scale cutting tools made of nanocrystalline materials, such as cemented tungsten carbide, tantalum carbide, and titanium carbide, are much harder, much more wear- and erosion-resistant, and last longer than their conventional (large-grained) counterparts. They also enable the manufacturer to machine various materials much faster, thereby increasing productivity and significantly reducing manufacturing costs. Also, for the miniaturization of microelectronic circuits, the industry needs microdrills, drill bits with diameters much less than the thickness of an average human hair, with enhanced edge retention and far better wear resistance. Since nanocrystalline carbides, nitrides and silicides are much stronger, harder, and wear-resistant, they are currently being used in these microdrills (Singh et al., 2003, Mohanan et al., 2005).

1.3.3 Better insulation materials

New insulation materials have been created by utilizing the nanoscale properties of materials. A new foam-like structure called "aerogels" are in fact nanocrystalline oxide materials synthesized by the sol–gel technique discussed later in Chapter 2. These aerogels are porous and extremely lightweight and they can withstand a load equivalent to 100 times their weight. Aerogels are composed of three-dimensional, continuous networks of nano/microcrevices with air or gases trapped at their interstices. Since they are porous and air is trapped at the interstices, aerogels are currently being used for insulation in offices, homes, etc. By using aerogels for insulation, heating and cooling bills are drastically reduced, thereby saving power and reducing the attendant environmental pollution. They are also being used as materials for "smart" windows, which darken when the Sun is too bright, just as in changeable lenses in prescription and sunglasses. They lighten themselves when the Sun is not shining too brightly.

Table 1.1. *Mechanical properties of the nanocomposites in the dry state*

Filler content (%)	E-modulus (MPa)		Tensile strength (MPa)		Elongation (%)	
	Nap	PAA–Nap	Nap	PAA–Nap	Nap	PAA–Nap
0	30.5 ± 2.1		7.0 ± 0.2		375 ± 100	
10	49.1 ± 1.7	56 ± 6.3	6.8 ± 0.5	6.5 ± 0.3	343 ± 73	354 ± 29
25	82.1 ± 6.3	79.2 ± 3.3	5.8 ± 0.2	6.0 ± 0.3	270 ± 16	137 ± 60
50	242 ± 27.9	n.d.*	4.8 ± 0.9	n.d.	8.7 ± 3	n.d.

Reprinted from Liu *et al.*, 1997, with permission from Elsevier.
* n.d., not determined

Table 1.2. *Mechanical properties of the nanocomposites after immersion in phosphate-buffered saline for 24 h*

Filler content (%)	E-modulus (MPa)		Tensile strength (MPa)		Elongation (%)	
	Nap	PAA–Nap	Nap	PAA–Nap	Nap	PAA–Nap
0	7.1 ± 0.4		4.4 ± 0.3		87.2 ± 9.1	
10	17.7 ± 1.7	16.7 ± 1.9	3.9 ± 0.2	3.8 ± 0.2	91 ± 16	80 ± 11
25	15.5 ± 0.6	18.5 ± 0.7	2.9 ± 0.3	2.8 ± 0.2	51 ± 9	51 ± 8
50	11.4 ± 1.0	n.d.*	0.6 ± 0.1	n.d.	4.8 ± 0.5	n.d.

Reprinted from Liu *et al.*, 1997, with permission from Elsevier.
* n.d., not determined

1.3.4 High-functional polymer nanocomposites

Addition of nanoparticles not only improves the mechanical properties, but also has been shown to improve thermal stability, in some cases allowing use of polymer-matrix nanocomposites using a wide variety of fillers including carbon nanotubes. An additional 100 °C above the normal service conditions is achieved. Decrease in material flammability has also been studied (Kashiwagi *et al.*, 2002), an especially important property for transportation applications where choice of material is influenced by safety concerns. Clay/polymer structural nanocomposites have also been considered as matrix materials for fiber-based composites destined for aerospace components.

Nano-apatite (Nap) is used as a filler to make composites with polyethylene glycol/poly(butylene terephthalate) (PEG/PBT) block copolymer created by hydrothermal synthesis. Also poly(acrylic acid) (PAA) coating of Nap has been tested. The resulting improvement in the mechanical properties as a function of nanodomains in wet and dry conditions is shown in Table 1.1 and Table 1.2 (Liu *et al.*, 1997).

It can be seen from Table 1.1 that the increase in the percentage of the nano-apatite filler in the dry state causes the elastic modulus to increase greatly and the addition of the PAA doesn't have a dramatic effect on the values. It is also observed that with the use of 50% nanofiller in the first case, the elastic modulus increases from 30.5 MPa to 242 MPa which is an increase in magnitude of approximately 8 times the original value. With this size of an

Table 1.3. *Effect of functional PP on the mechanical properties of nano-SiO$_2$/PP composites at nano-SiO$_2$ content of 0.84 vol. %*

Sample	Content of grafted PP (vol. %)	Young's modulus (GPa)	Tensile strength (MPa)	Elongation to break (%)	Area under tensile curve (MPa)	Notched impact strength (kJ/m^2)
Neat PP	0	1.22 ± 0.017	41.8 ± 0.76	74.2 ± 6.63	24.0 ± 2.35	2.36 ± 0.217
Untreated nano-SiO$_2$/ PP	0	1.42 ± 0.019	43.5 ± 0.45	105 ± 20.2	33.2 ± 2.93	2.55 ± 0.124
SiO$_2$-c-PS/PP-c-PS/PP	0	1.32 ± 0.016	43.8 ± 0.13	141 ± 11.9	42.6 ± 3.7	2.63 ± 0.248
	5.1	1.33 ± 0.022	43.6 ± 0.30	161 ± 19.2	48.3 ± 5.43	2.74 ± 0.700
	10.2	1.30 ± 0.011	43.9 ± 0.38	70.7 ± 10.7	22.0 ± 3.2	2.50 ± 0.124
SiO$_2$-c-PS/PP-g-PS/PP	5.1	1.34 ± 0.024	43.4 ± 0.25	62.9 ± 7.10	19.5 ± 2.06	2.71 ± 0.182
SiO$_2$-c-PEA/PP-c-PEA/PP	0	1.38 ± 0.008	42.8 ± 0.30	141 ± 0.70	41.8 ± 0.23	2.78 ± 0.104
	5.1	1.30 ± 0.004	43.5 ± 0.20	55.0 ± 4.40	17.2 ± 1.28	3.09 ± 0.313
	10.2	1.17 ± 0.005	41.5 ± 0.63	47.5 ± 3.50	13.7 ± 0.86	3.21 ± 0.200
SiO$_2$-c-PEA/PP-g-Pea/PP	5.1	1.31 ± 0.012	40.2 ± 0.27	124 ± 8.10	31.8 ± 1.30	3.18 ± 0.401

From Wu *et al.*, 2004, with permission.
Percent grafting of SiO$_2$-c-PS and SiO$_2$-c-PEA are 11.4% and 15.3% respectively.

1.3 Other nanostructured materials for structural applications

Table 1.4. *Mechanical properties of PP (MI = 8.5 g/10 min) based nanocomposites filled with different polymers grafted SiO$_2$**

Grafted polymers	Nanocomposites						Neat PP
	PS	PBA	PVA	PEA	PMMA	PMA	
Tensile strength (MPa)	34.1	33.3	34.1	34.1	34.1	34.1	32.0
Young's Modulus (GPa)	0.92	0.86	0.81	0.88	0.89	0.85	0.75
Elongation to break (%)	9.3	12.6	10.0	4.6	12.0	11.9	11.7
Area under tensile stress–strain curve (MPa)	2.4	3.3	2.3	0.8	3.2	2.9	2.2
Unnotched Charpy impact strength (kJ/m^2)	19.8	19.4	22.9	14.6	20.5	4.7	8

From Zhang and Rong, 2003, with permission.
*Content of SiO$_2$ = 3.31 vol%
PBA – Polybutyl acrylate   PVA– Polyvinyl acetate
PEA – Polyethyl acrylate   PMMA– Polymethyl methacrylate
PMA – Polymethyl acrylate PS – Polystyrene

increase the material becomes stiffer and it will take a larger force to cause the material to stretch. Tensile strength was found to decrease, which means that the filler is not reinforcing the structure and also explains why the elongation percentage decreased. From this study it can be determined that the use of Nap filler is beneficial only when there is a need for a high elastic modulus under dry conditions. Otherwise it would be better to use the composite without nano-apatite filler.

Another composite is a grafted nanosilica/polystyrene and nanosilica/polyethyl acrylate (Wu *et al.*, 2004). These composites were created by placing the ingredients in a high-speed blender and then extruding them at 210–230 °C. The nanosilica composites in all the tests were fixed to 0.84 vol. %. Table 1.3 shows the results from testing of the components created with this composition. SiO$_2$-c-PS and SiO$_2$-c-PEA represent nanosilica/polystyrene and nanosilica/ polyethyl acrylate respectively. The "-c-" means that the homopolymerized PS or PEA generated during the graft reaction is associated with the graft copolymer and the "-g-" means that the PS or PEA homopolymer has been isolated from the graft copolymer. Table 1.3 shows how the addition of nanoparticles increases all of the properties of the polypropylene. When the product is grafted with the polystyrene or the polyethyl acrylate the properties change, and depending on what is needed for an application a specific combination can be found that can increase one property but change another one. Even if they are grafted with the other polymers there is still an increase in all of the properties compared to the use of only polypropylene.

Other research laboratories have also performed the tests with nanosilica and a polymer graft composition (Zhang and Rong, 2003). Table 1.4 shows the results that have been found by using other polymers in the grafting process of composite polymers. From the data presented in Table 1.4, it can be seen that the overall mechanical properties of the polymer increase when the nanosilica is incorporated into the system. The impact strength (fracture toughness) increases the most in these experiments. This occurs because the

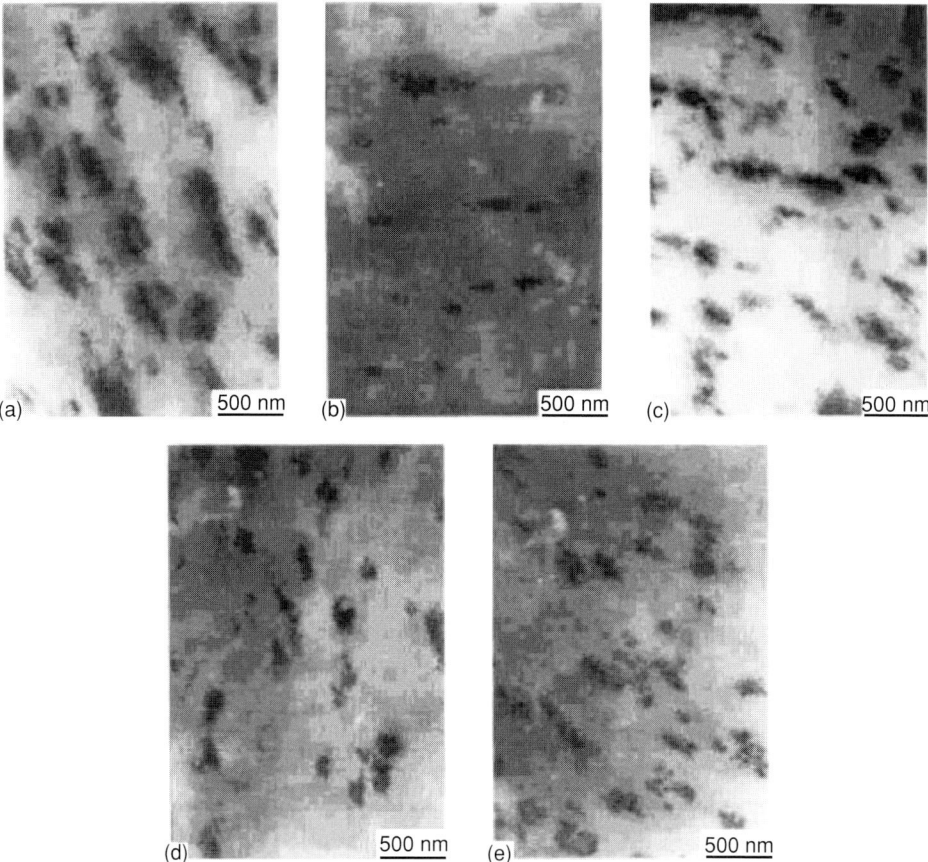

Figure 1.4. TEM micrograph of PP-based nanocomposites filled with (a) SiO_2 as received (content of $SiO_2 = 1.96$ vol.%), (b) SiO_2-g-PS (content of $SiO_2 = 1.96$ vol.%), (c) SiO_2-g-PS (content of $SiO_2 = 6.38$ vol.%), (d) SiO_2-g-PMMA (content of $SiO_2 = 1.96$ vol.%), and (e) SiO_2-g-PMMA (content of $SiO_2 = 6.38$ vol.%). (Reprinted with permission, Zhang and Rong, 2003.)

nanoparticles deflect the cracks as they propagate through the material. This is of great advantage because it can prevent an object from breaking when impacted by blunt force. These data in the table also show that a desired property can be achieved from the combination of different polymers in the grafting process. See Figure 1.4 (Zhang and Rong, 2003), for TEM images of nanopolymer composites, where ultrafine particulates are dispersed in the matrix.

Another study investigated the composites based on polyacrylate polymer matrices with the use of a kaolin (kaolinite is a mineral with a chemical composition of $Al_2Si_2O_5(OH)_4$, a layered silicate mineral also known as china clay) and $CaCO_3$ nanofillers. Composites of styrene-butyl acrylate (S-BA) and methyl methacrylate-butyl acrylate (MMA-BA) were filled with the kaolin filler and then their mechanical properties were determined (Kovacevic et al., 2002). Also composites of calcium carbonate and calcium carbonate treated with stearic acid with nanofiller were used in poly(vinyl acetate) (PVAC) matrix. The modified

$CaCO_3$ was used because the addition of sodium stearate decreases the surface activity of the filler material. The kaolin filler is expected to have a layered structure, because it is a clay material, and with this structure the PA matrix is allowed to intercalate between the filler layers and that creates a nanocomposite structure. The $CaCO_3$ composites were created by using a PVAC emulsion and two mixtures were created and then blended. The mixture was dried so the composite film would contain the desired nanofiller content. It is observed that the Young's modulus, yield strength, and ultimate strength increase as the volume fraction of the filler increases; this is because the smaller particles reinforce the material, leading to the increase in the mechanical properties. As for the elongation at fracture, in this experiment the value decreased as the filler volume increased. This occurs because the material is becoming stiffer and starting to mimic the properties of other materials that do not stretch as much before they break. When comparing the mechanical properties of the kaolin and the carbonate it can be seen that the carbonate has a higher Young's modulus. This is because the calcium carbonate is in the nanodimension prior to the creation of the composite material. Because of the smaller particles of $CaCO_3$ there is a higher adhesion level, which increases the number of molecules with restricted mobility and this explains why the $CaCO_3$ composites have higher mechanical properties than the kaolin composites. Kovacevic *et al.* (2002) also mentioned that it is important to recognize that the polymer–surface interactions are essential for understanding the behavior of the polymers containing the nanofiller material and also in the determination of the structure and properties of the solid polymer films.

Nanostructure-filled polymer ceramic composites can be used in various structural applications including microsystem technology. Hanemann *et al.* (2004) discussed how these composites can be used in various structural systems. They determined that, by using nanosized ceramics instead of microsized ceramics, the physical properties change and the nanoscale results in an improved sinter activity because of the large surface area provided by the nanoparticles. Nanoscaled ceramics (e.g. alumina and zirconia) with sizes less than 40 nm can be used as thickeners in coatings and also as flame retardants in plastics. The use of the nanoceramic also helps to create dense structures which then make the parts more durable and less likely to break when in use in a specific application. Hanemann *et al.* (2004) determined that the refractive index of polymers can be modified by dispersed nanoscaled ceramics (alumina and zirconia) without an increase in the optical damping due to scattering. This brings up the concern of agglomerated nanoparticles and the difficulty in deagglomeration of the ceramic nanoparticles. This is difficult because of the large surface area and the large surface energy the nanoparticles have, leading to a tendency to produce agglomerated particles.

As a general comment, these properties are very useful in creating plastics. With the integration of nanoparticles it is possible to create structural components which can replace metals that are currently being used for a specific application. Also, because these materials are polymer based, they have a lower electrical conductivity and this feature is useful where having a durable non-conductive material is required. Polymers are also less corrosive then metals and with the increased mechanical properties these polymers can be used in aquatic vehicles that need to travel deep under the ocean, and because of the corrosion resistance these vehicles would be able to last longer without maintenance.

1.3.5 Carbon nanotubes

Carbon nanotubes, long thin cylinders of carbon, were discovered in 1991 by Iijima (1991). These are large macromolecules that are unique for their size, shape, and remarkable physical properties. They can be thought of as a sheet of grapheme (a hexagonal lattice of carbon) rolled into a cylinder. These intriguing structures have sparked much excitement in recent years and a large amount of research has been dedicated to their understanding. Currently, the physical properties are still being discovered and disputed.

Simply put, carbon nanotubes (Figure 1.5) exist as a macromolecule of carbon, analogous to a sheet of graphite (the pure, brittle form of carbon in your pencil lead) rolled into a cylinder. Graphite looks like a sheet of chicken wire, a tessellation of hexagonal rings of carbon. Sheets of graphite in your pencil lay stacked on top on one another, but they slide past each other and can be separated easily, which is how the graphite is used for writing. However, when coiled, the carbon arrangement becomes very strong. In fact, nanotubes have been known to be up to 100 times as strong as steel and almost 2 mm long! These nanotubes have a hemispherical "cap" at each end of the cylinder. They are light, flexible, thermally stable, and are chemically inert. They have the ability to be either metallic or semi-conducting depending on the "twist" of the tube. Carbon nanotubes can be found in both single-walled (SWCNT) and multiwalled forms (MWCNT). Determining the elastic properties of SWNTs has been one of the most hotly disputed areas of nanotube study in recent years. On the whole, SWNTs are stronger than metallic glasses and have a much larger elastic limit than steels, and are resistant to damage from physical forces. Pressing on the tip of the nanotube will cause it to bend without damage to the tip or the whole CNT. When the force is removed, the tip of the nanotube will recover to its original state (Dekker, 1999). Quantizing these effects, however, is rather difficult and an exact numerical value cannot be agreed upon. The value of Young's modulus (elastic modulus) for SWNTs lies close to 1 TPa. The maximum tensile strength is close to 30 GPa (Yu *et al.*, 2000).

The results of various studies over the years have shown a large variation in the value reported. In 1996, researchers at NEC in Princeton and the University of Illinois measured the average modulus to be 1.8 TPa (Treacy *et al.*, 1996). This was measured by first allowing a tube to stand freely and then taking a microscopic image of its tip. The modulus is calculated from the amount of blur seen in the photograph at different temperatures. In 1997, Gao *et al.* presented a talk at the Fifth Foresight Conference on Molecular Nanotechnology where they reported three variations of Young's modulus to five decimal places that were dependent on the chiral vector. They concluded that a (10,10) armchair tube had a modulus of 640.30 GPa, a (17,0) zigzag tube had a modulus of 648.43 GPa, and a (12,6) tube had a value of 673.94 GPa.

Using tight-binding calculations, Young's modulus was dependent on the size and chirality of the SWNT, ranging from 1.22 TPa for the (10,0) and (6,6) tubes to 1.26 TPa for the large (20,0) SWNT. However, using first principle calculations, the calculated value for a generic tube was found to be 1.09 TPa.

The previous evidence would lead us to assume that the diameter and shape of the nanotube was the determining factor for its elastic modulus. However, when working with

1.3 Other nanostructured materials for structural applications

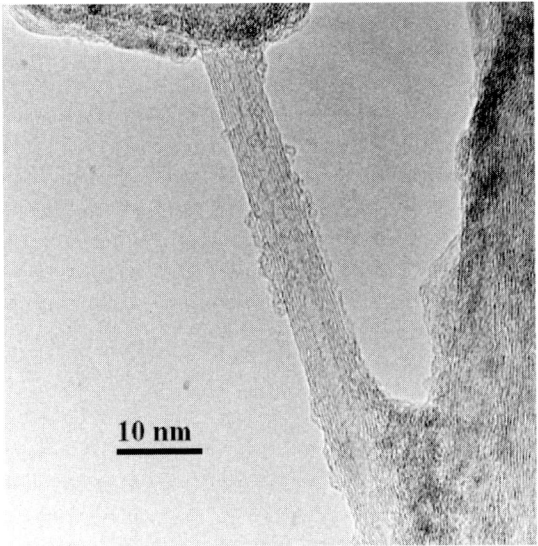

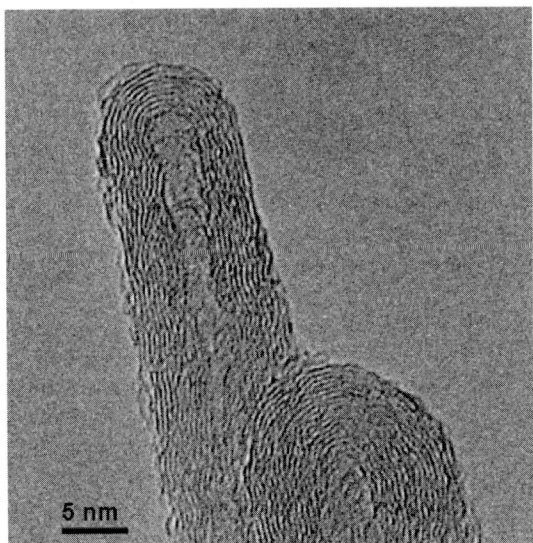

Figure 1.5. Carbon nanotubes. (Sudipta Seal, personal research photographs.)

different MWNTs, Forró *et al.* (2000) noted that their modulus measurements of MWNTs in 1999 using AFM did not strongly depend on the diameter, as had been recently suggested. Instead, they argued that the modulus of MWNTs correlates to the amount of disorder in the nanotube walls. However, evidence showed that the value for SWNTs does in fact depend on diameter; an individual tube had a modulus of about 1 TPa while bundles (or ropes) of 15–20 nm in diameter had a modulus of about 100 GPa (Saito *et al.*, 1998).

It has been suggested that the controversy into the value of the modulus is the result of the author's interpretation of the thickness of the walls of the nanotube. If the tube is considered

1 Introduction

Table 1.5. *Mechanical properties of carbon nanotubes and other related materials*

	Young's modulus (GPa)	Tensile strength (GPa)	Density (kg/m^3)
SWNT	1054	75	1300
MWNT	1200	~150	2600
SWNT bundle	563	~150	1300
Graphite (in plane)	350	2.5	2.6
	686[a]	19.6	
Diamond	1050–1150	130	3.5
	Bulk modulus (450–500)		
Kevlar	186	3.6	7.8
Steel	208	0.2–1.5	7.8
Wood	16	0.008	0.6

[a] From Kelly and Macmillan (1986).

Table 1.6. *Mechanical properties of nanocomposites used as matrix for glass fiber-reinforcement*

Filler type/content (wt.%)	Young's modulus (GPa)	Ultimate tensile strength (MPa)	Fracture toughness (MPa m$^{1/2}$)
Epoxy	2.59 ± 0.08	63 ± 1.03	0.65 ± 0.062
Epoxy/0.1% CB	2.75 ± 0.14	63.3 ± 0.85	0.77 ± 0.030
Epoxy/0.3% CB	279 ± 0.03	63.1 ± 0.59	0.86 ± 0.063
Epoxy/0.1% CB DWCNT–NH$_2$	2.76 ± 0.09	63.6 ± 0.96	0.77 ± 0.058
Epoxy/0.3% CB DWCNT–NH$_2$	2.94 ± 0.05	67 ± 0.48	0.92 ± 0.042

Reprinted from Gojnya *et al.* (2005), with permission of Elsevier.

to be a solid cylinder, then it would have a lower Young's modulus. If the tube is considered to be hollow, the modulus gets higher; and the thinner we treat the walls of the nanotube, the higher the modulus will become. Table 1.5 lists the mechanical properties of carbon nanotubes compared to other related materials. Since carbon nanotubes possess high strength, they can act as perfect fillers for high-strength nanocomposites. In a recent study, superior physical properties of epoxy-based double-wall carbon nanotubes (DWCNTs) composites were observed as compared with nanocomposites containing highly conductive carbon black (CB) (Gojnya *et al.*, 2005). The average outer diameter for the DWCNTs is 2.8 nm (Figure 1.6a), and the CB consists of spherical particles with a diameter of 30 nm (Figure 1.6b). The mechanical properties are listed in Table 1.6. From the information in Table 1.6 it can be seen that the addition of the carbon to the material improves all of the mechanical properties. The largest increase is with the double-walled carbon nanotubes. This shows

1.3 Other nanostructured materials for structural applications

Figure 1.6. SEM images of (a) DWCNT=NH$_2$ and (b) carbon black used as nanoscaled fillers in the study. (Reprinted from Gojnya *et al.* (2005), with permission from Elsevier.)

that the use of carbon nanotubes is useful in reinforcing materials. Figure 1.7 shows a TEM image of DWCNT-polymer nanocomposites, and Figure 1.8 shows how the carbon nanotubes prevent the crack from going through the entire part. This is very important because this image shows how the carbon nanotubes effect the epoxy material in making CNT-based polymer composites for high-strength applications.

In general, understanding the atomic and molecular processes occurring at the interface of two materials when they are brought together, separated, or moved with respect to

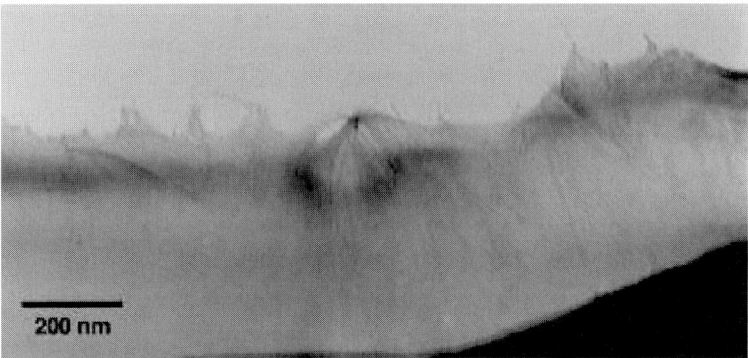

Figure 1.7. TEM image of a DWCNT-NH$_2$/epoxy nanocomposite. (Reprinted from Gojnya *et al.* (2005), with permission from Elsevier.)

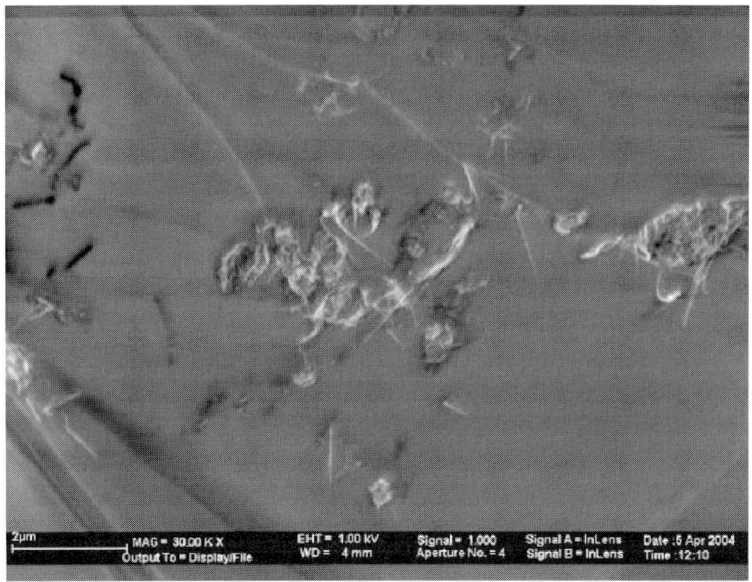

Figure 1.8. SEM micrograph from a fracture surface of a tested GFRP specimen. Some small aggregates can be observed in the matrix-rich areas. The observed loose structure is caused by the re-aglomeration of the CNTs. (Reprinted from Gojnya *et al.* (2005), with permission from Elsevier.)

each other, is central to many technological problems, including adhesion, contact formation, friction, wear, and lubrication, all essential properties in structural nanomaterials. For example, molecular dynamics computer simulations have predicted the occurrence of an interfacial "jump-to-contact" instability when two material surfaces are brought to close proximity (about 0.5 nm), leading to formation of nanoscale adhesive junctions whose resistance to shear is the cause of energy dissipation, that is, nanoscale friction and wear. Furthermore, preventing the formation of such junctions requires that the narrow spacing between the approaching surfaces be lubricated (as, for example, in the case of a read/write head and the surface of an ultra-high-density information storage device) by a very thin lubricating film. It has been found through both theoretical simulations and experiments that the energetic, structural, dynamic, thermodynamic, and rheological properties differ greatly under nanoscale confinements from their bulk behavior. Reorientation of crystal structure in nanoscale dimension is capable of supporting load; may undergo dynamic phase changes; and show high visco-elastohydrodynamic response. This further depends on the degree of confinement and the nature of the interface bonding. These properties are key parameters in many structural materials.

1.4 Topics to be covered

Given the importance of detailed understanding of the mechanical behavior of nanostructured materials for the applications described above, this book will focus upon these properties, the processing of nanostructured materials, and the stability of nanostructured microstructures. Chapter 2 will cover the various processing methods pertinent to structural nanocrystalline materials. Chapter 3 will discuss the stability of nanocrystalline microstructures. Chapters 4 and 5 will cover the experimental and theoretical aspects of the mechanical behavior of nanocrystalline materials respectively. Chapter 6 will describe other properties of structural nanocrystalline materials which are important for their performance, such as corrosion and oxidation resistance. The final chapter, Chapter 7, will summarize present and potential applications of structural nanocrystalline materials with emphasis on specific material systems.

References

Baker, M., Kench, P. J., Tsotsos, C., Gibson, P. N., Leyland A., and Matthews, A. (2005). *J. Vac. Sci. Technol. A*, **23**, 423.
Barnett, S. A. (1993). In *Physics of Thin Films Vol. 17: Mechanic and Dielectric Properties*, ed. Francombe, M. H., and Vossen, J. L. Boston: Academic Press, p. 2.
Barnett, S. A., Madan, A., Kim, I., and Martin K. (2003). *MRS Bulletin*, **28**, 169.
Berger, L. (1996). *Phys. Rev. B*, **54**, 9353.
Bunshah, R. F., Nimmagadda, R., Doerr, H. J., Movchan, B. A., Grechanuk, N. I., and Dabidzha, E. V. (1980). *Thin Solid Films*, **72**, 261.
Buschow, K. H. J. (1988). In *Ferromagnetic Materials Vol 4*, ed. Wohlfarth, E. P., and Buschow, K. H. J. Amsterdam: Elsevier, p. 1.
Cowburn, R. P., Koltsov, D. K., Adeyeye A. O., and Welland, M. E. (1998). *Appl. Phys. Lett.*, **73**, 3947.
Chung, Y.-W., and Sproul, W. D. (2003). *MRS Bulletin*, **28**, 164.
Dekker, C. (1999). *Physics Today*, **52**(5), 22.
Ehmann, K. F., DeVor, R. E., Kapoor, S. G., and Ni, J. (2000). *Micro/Meso-Mechanical Manufacturing M^4, NSF Workshop*, May 16–17.
Forró, L., Salvetat, J.-P., Bonard, J.-M., Bacsa, R., Thompson, N. H., Garaj, S., Thier-Nga, L., Gaál, R., Kulik, A., Ruzicka, B., Degiorgi, L., Bachtold, A., Schönenberger, C.,

Pekker, S., and Hernandi, K. (2000). *Science and Application of Nanotubes*, ed. Tomànek, D., and Enbody R. J. New York: Kluwer Academic / Plenum Publishers, p. 297.

Gao, G., Cagin, T., and Goddard III, W. A. (1997). *Energetics, Structure, Mechanical and Vibrational Properties of Single Walled Carbon Nanotubes (SWNT)* (http://www.wag.caltech.edu/foresight/foresight_2.html). *Nanotechnology*, **9** (1998), 1984.

Gleiter, H. (1989). *Progress in Mater. Sci.*, **33**, 223.

Gojnya, F., Wichmann, M., Fiedler, B., Bauhofer, W., and Schulte, K. (2005). *Composites A*, **15**, 1.

Guinier, A. (1938). *Nature*, **142**, 569.

Hanemann, T., Boehm, J., Henzi, P., Honnef, K., Litfin, K., Ritzhaupt-Kleissl, E., and Hausselt, J. (2004). *IEEE Proc.-Nanobiotechnol.*, **151**, 167.

Heinrich, B., and Bland, J. A. C. (1994). *Ultrathin Magnetic Structures*. Berlin: Springer.

Holleck, H., Lahres, M., and Woll, P. (1990). *Surf. Coat. Technol.*, **41**, 179.

Iijima, S. (1991). *Nature*, **354**, 56.

Jianu, A., Valeanu, M., Lazar, D. P., Lifei, M., Tomut, M., Pop, V., and Alexandru, S. (2004). *Romanian Rep. Phys.*, **56**, 385.

Jilek, M., Cselle, T., Holubar, P., Marstei, M., Veprek-Heijman, M. G. J., and Veprek, S. (2004). *Plasma Chem. Plasma Process.*, **24**, 493.

Karimpoor, A. A., Erb, U., Aust, K. T., and Palumbo, G. (2003). *Scripta Materialia*, **49**, 651.

Karvankova, P., Männling, H.-D., Eggs, C., and Veprek, S. (2001). *Surf. Coat. Technol.*, **146–147**, 280.

Kashiwagi, T., Grulke, E., Hilding, J., Harris, R., Awad, W., and Douglas, J. (2002). *Macromolecules Rapid Commun.*, **23**, 761.

Kelly, A., and Macmillan, N. H. (1986). *Strong Solids*, 3rd edition. Oxford: Clarendon Press.

Kneller, E. F., and Hawing, R. (1991). *IEEE Trans. Mag.*, **27**, 358.

Koch, C. C., Morris, D. G., Lu, K., and Inoue, A. (1999). *MRS Bulletin*, **24**, 54.

Koehler, J. S. (1970). *Phys. Rev. B*, **2**, 547.

Kovacevic, V., Lucic S., and Leskovac, M. (2002). *J. Adhesion Sci. Technol.*, **16**, 1343.

Krivorotov I. N., Emley, N. C., Sankey, J. C., Kiseiev, S. I., Ralph, D. C., and Buhrman, R. A. (2005). *Science*, **307**, 228.

Kusano, E., Kitagawa, M., Nanto, H., and Kinbara, A. (1998). *J. Vac. Sci. Technol. A*, **16**, 1272.

Lehocky, S. L. (1978a). *J. Appl. Phys.*, **49**, 5479.

Lehocky, S. L. (1978b). *Phys. Rev. Lett.*, **41**, 1814.

Li, H., and Ebrahimi, F. (2004). *Appl. Phys. Lett.*, **84**, 4307.

Li, S. H., Shi, Y. L., and Peng, H. R. (1992). *Plasma Chem. Plasma Process.*, **12**, 287.

Liu, Q., Wijn J., and Blitterswijk, C. (1997). *Biomaterials*, **18**, 1263.

Maziasz, P. J., and Pollard, M. (2003). *Adv. Mater. Processes*, **161**, 57.

McFadden, S. X., Mishra, R. S., Valiev, R. Z., Zhilyaev, A. P., and Mukherjee, A. K. (1999). *Nature*, **398**, 684.

Mehl, R. F., and Cahn, R. W. (1983). In: *Physical Metallurgy*, 3rd edition. Amesterdam: North Holland, p. 1.

Merica, P. D., Waltenburg, R. G., and Scott, H. (1919). *Bulletin, AIME* (June), 913.

References

Mohanan, J. L., Arachchige I. U. and Brock, S. L. (2005). *Science*, **307**(5708), 397.
Münz, W.-D. (2003). *MRS Bulletin*, **28**, 173.
Münz, W.-D., Lewis D. B., Hovsepian, P. Eh., Schönjahn, C., Ehiasarian, A., and Smith, I. J. (2001). *Surf. Engineering*, **17**, 15.
Musil, J., Kadlec, S., Vyskocil, J., and Valvoda, V. (1988). *Thin Solid Films*, **167**, 107.
Musil, J. (2000). *Surf. Coat. Technol.*, **125**, 322.
Porter, D. A., and Easterling, K. E. (1983). *Phase Transformations in Metals and Alloys*. New York: Van Nostrand Reinhold, p. 291.
Preston G. D. (1938). *Nature*, **142**, 570.
Rogl, P., and Schuster, J. C. (1992). *Phase Diagrams of Ternary Boron Nitride and Silicon Nitride Systems*. Materials Park, Ohio: ASM The Materials Society.
Saito, R., Dresselhaus, G., and Dresselhaus, M. S. (1998). *Physical Properties of Carbon Nanotubes*. Singapore: Imperial College Press, Utopia Publications.
Scattergood, R. O., and Koch, C. C. (2003). Unpublished, North Carolina State University.
Shinn, M., Hulman, L., and Barnett, S. A. (1992). *J. Mater. Res.*, **7**, 901.
Singh, S., Godhkindi, M. M., Krishnarao, R. V., Murthy, B. S., and Mukunda P. G. (2003). *Rev. Adv. Mat. Sci.*, **5**, 337.
Slonczewski, J. C. (1996). *J. Magn. Mater.*, **159**, L1.
Soe, W.-H., and Yamamoto, R. (1999). *Radiat. Effects Defects Solids*, **148**, 213.
Sun, S. H., and Murray, C. B. (1999). *J. Appl. Phys.*, **85**, 4325.
Tixier, S., Böni, P., and Van Swygenhoven, H. (1999). *Thin Solid Films*, **342**, 188.
Treacy, M. M. J., Ebbesen, T. W., and Gibson, J. M. (1996). *Nature*, **381**, 678.
Urazhdin, S., Birge, N. O., Pratt, W. P., and Bass, J. (2003). *Phys. Rev. Lett.*, **91**, 146 803.
Veprek, S. (1999). *J. Vac. Sci. Technol. A*, **17**, 2401.
Veprek, S., and Reiprich, S. (1995). *Thin Solid Films*, **268**, 64.
Veprek, S., Reiprich, S., and Li, S. H. (1995a). *Appl. Phys. Lett.*, **66**, 2640.
Veprek, S., Haussmann, M., and Reiprich, S. (1995b). *J. Vac. Sci. Technol. A*, **14**, 46.
Veprek, S., Haussmann, M., Reiprich, S., Li, S. H., and Dian, J. (1996). *Surf. Coat. Technol.*, **86–87**, 394.
Veprek, S., Männling, H.-D., Jilek, M., and Holubar., P. (2004). *Mater. Sci. Eng. A*, **366**, 202.
Veprek, S., Veprek-Heijman, G. M. J., Karvankova, P., and Prochazka, J. (2005). *Thin Solid Films*, **476**, 1.
Voevodin, A. A., and Zabinski, J. S. (1998). *Diamond and Related Materials*, **7**, 463.
Voevodin. A. A., and Zabinski, J. S. (2000). *Thin Solid Films*, **370**, 223.
Voevodin, A. A., Schneider, J. M., Rebholz, C., and Matthews, A. (1996a). *Tribology Int.*, **29**, 559.
Voevodin, A. A., Capano, M. A., Safriet, A. J., Donley, M. S., and Zabinski, J. S. (1996b). *Appl. Phys. Lett.*, **69**, 188.
Voevodin, A. A., Prasad, S. V., and Zabinski, J. S. (1997a). *J. Appl. Phys.*, **82**, 855.
Voevodin, A. A., Walck, S. D., and Zabinski J. S. (1997b). *Wear*, **203–204**, 516.
Voevodin, A. A., Fitz, T. A., Hu, J. J., and Zabinski, J. S. (2002). *J. Vac. Sci. Technol. A*, **20**, 1434.
Wilm, A. (1906). German patent DRP244554.

Wu, C. L., Zhang M. Q., Rong M. Z., Lehmann, B., and Friedrich, K., (2004). *Plastics, Rubbers and Composites*, **33** (2/3), 71.

Youssef K. M., Scattergood, R. O., Murty, K. L., and Koch, C. C. (2004). *Appl. Phys. Lett.*, **85**, 929.

Yu, M. F., Files, B. S., Arepalli, S., and Ruoff, R. S. (2000). *Phys. Rev. Lett.*, **84**, 5552.

Zeman, P., Cerstvy, R., Mayrhofer, P. H., Mitterer, C., and Musil, J. (2000). *Mater. Sci. Eng. A*, **289**, 189.

Zhang, M., and Rong, M. (2003). *Chinese J. Polymer Sci.*, **21**(6), 587.

Zhang, R. F., and Veprek, S. (2006). *Mater. Sci. Eng. A*, **424** (1–2), 128.

Zhitomirsky, V. N., Grimber, I., Rapoport, L., Travitzky, N. A., Bocman, R. L., Goldsmith, S., and Weiss, B. Z. (1999). *Surf. Coat. Technol.*, **120–121**, 219.

2 Processing of structural nanocrystalline materials

2.1 Introduction

Structural nanomaterials are finding applications in bulk materials, films, coatings, and composites. Applications vary from wear-resistance coatings to load-bearing structures. Nanophase or nanocrystalline materials are also being used in electronics, refractory, biological, and catalytic applications. Progress in a wide range of structural applications for nanomaterials crucially depends on the development of new fabrication and processing technologies, along with a fundamental understanding of the relationship between the structure and properties of the feedstock powders and consolidated parts. Among the most important issues discussed here are experimental data, and theoretical and computer models concerning mechanical properties in nanostructured materials, which, in general, are different from the conventional coarse-grained counterparts. The competition between conventional and unusual deformation modes is believed to cause the unique mechanical properties of nanomaterials, serving as a basis for their structural applications. Fabrication of nanomaterials with bimodal (nano- and sub-micro-particles) composites, that exhibit both very high strength and reasonable ductility, represents a promising strategy in the synthesis of nanomaterials with enhanced properties for various structural applications. High strain rate and low-temperature superplasticity of some nanocrystalline materials are the subjects of growing fundamental research efforts motivated by a range of new applications of these super strong and super plastic materials in net shaping technologies.

Structural materials and composites containing at least one phase that is less than 100 nm are often termed as *structural nanomaterials/composites* (Roy *et al.*, 1986). These materials exhibit enhanced mechanical (Sekino *et al.*, 1996,), magnetic (Sternitzke, 1997), elevated temperature (Riedel *et al.*, 1995), optical (Niihara, 1991; Lambeth *et al.*, 1996; Meldrum *et al.*, 2001), and excellent catalytic properties (Özkar *et al.*, 1992). The commercial applications of nanomaterials beyond the boundaries of *laboratories* lie on the successful production and consolidation of these materials into components preserving the nanostructures. The traditional consolidation techniques have the strong limitation of not being able to retain the nanograin size owing to the problem of grain growth. This chapter summarizes results of numerous studies towards nanopowder production and manufacturing of structural nanomaterials.

Most of the nanomaterial consolidation and fabrication techniques use nanopowders as the feedstock materials. Although synthesis and production of nanopowders have developed by leaps and bounds in the past decade, it is not yet possible to produce a large amount of nanopowders at an affordable cost for their successful consolidation into nanomaterials used in structural applications. There exists a need to develop techniques which are not only economical but also capable of producing high volumes of nanoparticles.

The *handling of nanopowders* will be an issue as well. The ultra-fine powders have relatively high surface area. Therefore these powders are susceptible to contamination due to their high surface activity. The presence of such surface contaminants would definitely compromise the physical, chemical, and mechanical properties of the final products. Therefore nanopowders should be prepared, stored and handled in a novel manner to avoid any contamination so that the benefit of the "nanosize effect" can be harnessed to the fullest extent. Developing means of proper handling and storage of nanoparticles prior to their processing to make engineering components is another challenging aspect of structural nanocomposite manufacturing.

Consolidation of nanopowders to composites with retained nanostructures is another challenging engineering problem the nanotechnology community will face. The density of the green compact depends primarily on the frictional forces of the powder particles. These forces are often originated from electrostatic, van der Waals and various surface adsorption forces. These interaction forces are significantly high in nanoparticles forming hard agglomerates and inter-agglomerates that are relatively large. Based on the thermodynamic treatment of the shrinkage of the pore, Mayo (1996) suggested that the finest pore size usually yields the highest densification rate. Large pores require not only higher temperature but also prolonged sintering times for their successful elimination; consequently, it becomes difficult to retain the grain size in the nanometer domain. Large pores undergo pore-boundary separation that restricts attaining the full density in the consolidated nanoparticles (Koch, 1993; Mayo, 1996). During sintering of nanoparticles, pores smaller than the critical size shrink (Kingery *et al.*, 1976; Groza, 1999), while larger pores undergo the pore-boundary separation. The fraction of the grain boundaries in nanomaterials is rather large compared with that for coarse-grained materials. The density of the grain-boundary regions is less than the grain interior. This is mainly the result of the relaxation of atoms in the grain boundaries, and the fact that they also contain many lattice defects (Groza and Dowding, 1996).

There are numerous conflicting views on the sintering behavior of nanoparticles. Nanoparticles showed depressed onset of sintering temperature (Chen and Mayo, 1993; Skandan *et al.*, 1994; Risbud *et al.*, 1995) to the range of $0.2T_m$–$0.3T_m$ compared with that of conventional powders that normally exhibit a range of $0.5T_m$–$0.8T_m$, where T_m is the melting point in Kelvin (K). These results are possibly attributed to the structural instability of these particles owing to the presence of high surface area. Synthesis of nanoparticles of ceramics, metallics, and their mixtures, has made substantial progress in the past decade; however, consolidation of such nanoparticles into fully densified large structural components remains a difficult problem.

Thus consolidation and fabrication technology can strongly affect the morphology and the materials properties in the nanometer domain; thus innovations for manufacturing of functional structural nanocomposites with retained nanostructure remain a challenge. Nanocomposite materials are expected to have extensive applications in cars, ships, airplanes, and even in space vehicles. Although substantial progress has been made in the understanding of the structure–property relationships in nanoceramics, further progress is needed in the areas of nanocomposite manufacturing using various cost-effective consolidation techniques that will be discussed here. A number of techniques are reviewed in this chapter in order to provide a snapshot of the recent advances in manufacturing of structural nanomaterials and nanocomposites. The role that the consolidation technologies might play in the future development of novel and unique three-dimensional multifunctional structures, which cannot be realized by using traditional manufacturing techniques, is also emphasized. A detailed description of each technique is beyond the scope of this book, and interested readers are encouraged to consult the publications presented in the References. A brief review of the various powder production and consolidation processes is also presented here.

2.2 Methods for preparation of bulk structural nanocrystalline materials

2.2.1 Inert-gas-condensation methods

The inert gas condensation method for the processing of nanocrystalline materials has as its precursor the studies in Japan on the formation of ultrafine particles. Much of this work has been summarized by Uyeda in his review (Uyeda, 1991). It was pointed out that the interest in this area in Japan was stimulated by a paper published by Kubo (1962) in which he predicted that ultrafine metal particles should exhibit physical properties that would be very different from those of their bulk counterpart. Uyeda discussed a variety of gas or vacuum evaporation methods that have been used to obtain ultrafine "metal smoke" particles. These include various methods for heating metals to evaporation such as resistance, plasma flame, laser and electron-beam heating, arc-discharge heating, or high-frequency induction heating. Subsequent work by Granqvist and Buhrman (1976) and Tholen (1979) concentrated on the gas-condensation methods and defined process parameters such as the type of gas used, gas pressure, and evaporation rate. These methods were used by Gleiter and co-workers (Gleiter, 1989) whose studies of the structure and properties of the consolidated nanocrystalline materials made by this method really stimulated the interest in the field as it now exists. The steps believed to result in nanocrystalline materials by the inert gas condensation method involve: (1) the aggregation of evaporated atoms into clusters by first establishing a population of evaporated atoms by appropriate heating of metal targets; (2) "cooling" of the atoms by collisions with "cold" inert gas atoms; and (3) growth of clusters by addition of atoms to individual clusters and by agglomeration of clusters by collisions between them. A schematic drawing of a gas-condensation chamber for the synthesis of nanocrystalline materials is given in Figure 2.1, after Siegel and Eastman (1989). The method consists of evaporating a metal by heating it inside a chamber that was evacuated to a pressure of about 10^{-7} Torr and then backfilled with a partial pressure of an inert gas, typically a few hundred pascals of helium. The evaporated atoms collide with the gas atoms inside the chamber, lose their kinetic energy, and condense in the form of small, discrete crystals of loose powder. Convection currents, generated by the heating of the inert gas by the evaporation source and cooled by the liquid-nitrogen-filled collection device (cold finger),

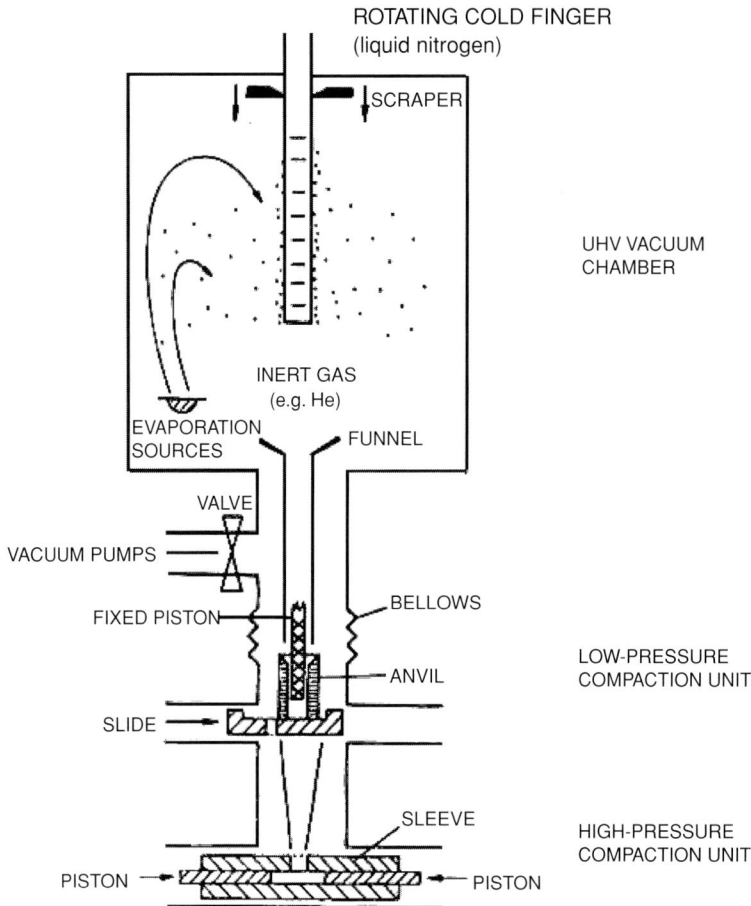

Figure 2.1. Schematic of inert gas-condensation method (Siegel and Eastman, 1989). (Reproduced with permission from VCH.)

carry the condensed fine powders to the collector device, from where they can be stripped off by moving an annular Teflon ring down the length of the tube into a compaction device. Compaction is carried out in a two-stage piston-and-anvil device initially at low pressures in the upper chamber to produce a loosely compacted pellet, which is then transferred in the vacuum system to a high-pressure unit where final compaction takes place. The scraping and compaction processes are carried out under ultra-high-vacuum conditions to maintain cleanliness of the particle surfaces (and subsequent interfaces) and also to minimize the amount of any trapped gases. Besides pure metals, nanocrystalline alloys could be produced by evaporating the different metals from more than one evaporation source. Oxides, nitrides, carbides, etc., of the metals can be synthesized by filling the chamber with oxygen or nitrogen gases or by maintaining a carbonaceous atmosphere. Densities of as-compacted samples in the early studies typically had measured values of about 75%–90% of the bulk density for metal samples. The resultant significant porosity in these materials had a major

2.2 Methods for preparation of bulk

effect on the mechanical properties measured, as will be discussed at length in Chapters 4 and 5. The details of the conventional inert-gas-condensation process can be found in the review by Siegel (1991). Because of the evident extrinsic defects such as porosity and oxygen contamination in nanocrystalline metals made by this technique, several advances were made in the process to minimize the deleterious effects. An improved synthesis chamber and compaction unit was developed at Argonne National Laboratory (Sanders *et al.*, 1997). The new synthesis chamber provided a much cleaner processing environment, with a base vacuum better by an order of magnitude and a dynamically pumped evaporating atmosphere. The new compaction unit is cleaner owing to an actively pumped chamber, and has the capability to do warm compactions. The production of cleaner powders and the use of warm compaction resulted in significant increases in the measured densities of the compacted powders to values about 98% of theoretical density. These improvements led to the measurement of mechanical properties that were not as subject to the masking by the extrinsic defects such as porosity. Other advances have been made in the gas-condensation method (Siegel, 1991) that involve the use of forced gas to improve the transport of the condensed atom clusters. This can lead to narrower particle-size distributions and to the scale-up of cluster production to levels more practical for industrial production of nanocrystalline materials.

2.2.2 *Mechanical attrition methods*

The ball milling of powders – mechanical attrition – was first developed as a powder metallurgy method to produce dispersion-strengthened alloys with fine, uniform dispersoid distributions (Benjamin, 1970). Subsequently, it has been used as a powerful nonequilibrium processing method that can synthesize a variety of metastable structured materials (Koch, 1991). The ball milling of powders can be divided into two categories: (1) the milling of elemental or compound powders – "mechanical milling," and (2) the milling of dissimilar powders – "mechanical alloying," in which material transfer occurs. This subject has been reviewed by a number of authors (see Koch, 1991; Suryanarayana, 2001; Suryanarayana, 2004). Nanostructured materials are one class of metastable materials that can be made by ball milling. Besides being discussed in the examples of general reviews of ball milling, this specific topic has also been reviewed by itself (see Koch, 1993). The details of the mechanical attrition processes, equipment used, etc., have been covered in a number of reviews (see Koch, 1991), and will not be repeated here.

Mechanical attrition has been found to refine the grain size to the nanoscale of all solid elements studied. The minimum grain size achieved is, however, dependent upon a number of variables as well as on properties of the element, alloy, or compound being milled. The minimum grain size obtainable by milling, d_{min}, has been attributed to a balance between the defect/dislocation structure introduced by the plastic deformation of milling and its recovery by thermal processes (Eckert *et al.*, 1992). It has been found that the minimum grain size induced by milling scales inversely with the melting temperature of a group of face-centered cubic (fcc) structure metals studied (Eckert *et al.*, 1992). These data are plotted in Figure 2.2 along with data for other metallic elements and carbon (graphite) (Koch, 1993). For these data, only the lower-melting-point metals show a clear inverse dependence

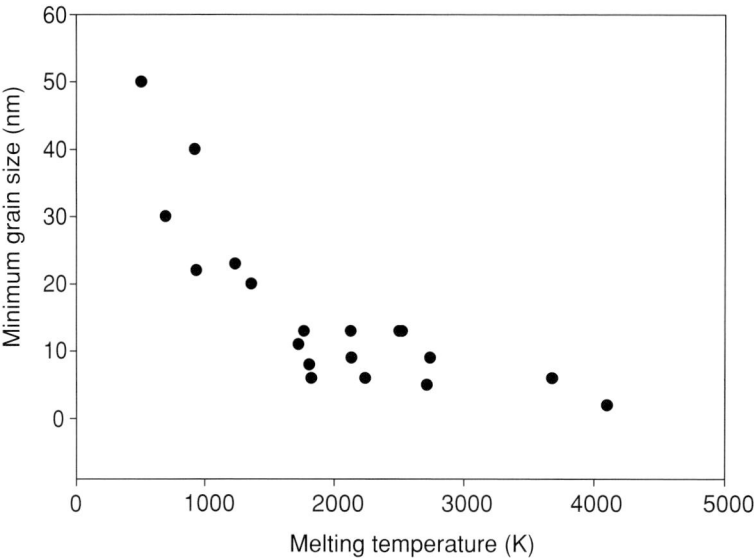

Figure 2.2. Minimum grain size vs. melting temperature.

of minimum grain size on melting temperature. The minimum grain size for elements with higher melting temperatures ($>T_m$ for Ni), exhibit essentially constant values with melting temperature for given crystal structure classes. For these elements it appears that d_{min} is in the order: fcc < bcc < hcp. However, these data must be considered with some skepticism since they were obtained by several different research groups using different milling equipment, and the measurements of grain size were mainly by the analysis of X-ray diffraction line broadening, which can be questionable in terms of absolute values. In spite of these potential problems, however, the values for d_{min} are in remarkable agreement between the data of Fecht et al. (1990) who used a high-energy shaker mill (Spex 8000) and the data of Oleszak and Shingu (1996) who used a conventional horizontal low-energy ball mill. The milling time for Oleszak and Shingu to obtain d_{min} was, however, orders of magnitude longer than in the high-energy mill. These results suggest that total strain, rather than milling energy or ball–powder–ball collision frequency, is responsible for determining the minimum nanocrystalline grain size. This is different from ball-milling-induced amorphization or disordering, where it appears the energy and frequency of ball–powder–ball collisions determine the final structures formed in "driven systems." It is, however, consistent with observations of nanocrystallites formed at high strain values using other non-cyclic deformation methods, as will be discussed later in this chapter. These results suggest that mill energy per se is not critical to the final microstructure, but, of course, the kinetics of the process are dependent on the energy, and times for attaining the same microstructure can be several orders of magnitude longer in the low-energy mills than in high-energy mills, as described above.

Milling temperature has been observed to affect the rate at which the nanocrystalline structure develops. The milling time at which a given grain size was attained in a TiNi intermetallic compound was a function of milling temperature, with smaller grains formed at

2.2 Methods for preparation of bulk

lower milling temperatures (Yamada and Koch, 1993). Shen and Koch (1995) also observed smaller nanocrystalline grain sizes in both Cu and Ni milled at $-85\ °C$ compared with samples milled at room temperature. For example, for Cu, $d = 26 \pm 3$ nm for room temperature milling and $d = 17 \pm 2$ nm for milling at $-85\ °C$. Evidence for smaller nanocrystalline grain sizes formed by milling at low temperatures has now been observed in a number of materials including the intermetallic compound CoZr (Pathak, 1995) and elemental Zn (Zhang and Koch, 2000). Since milling at temperatures lower than ambient can bias the defect accumulation induced by plastic deformation with respect to thermal recovery, higher dislocation densities, and therefore, as observed, finer grain sizes, can be obtained. The first use of milling at cryogenic temperatures, however, has been for the purpose of introducing fine nanoscale nitrides or oxynitrides into Al (Luton et al., 1989). This method has subsequently been used similarly to dispersion harden other materials including NiAl (Whittenberger et al., 1990) and Inconel 625 (He and Lavernia, 2001). In this case, liquid nitrogen is introduced into the milling vial along with the powders and milling balls. The chemical reaction of the nitrogen with the metal matrix powders produces the fine nitride particles which can help stabilize the nanoscale grain size during subsequent thermal/mechanical powder consolidation, as well as act as dispersion hardening agents.

A recent study of the milling of Zn at cryogenic temperatures (CM), albeit with the Zn milled under an argon atmosphere, has revealed a modulated cyclic variation in hardness/strength with milling time which has implications for the formation mechanisms for nanocrystalline grain size formation by plastic deformation (Zhang et al., 2002a). The measured density and hardness values for cryomilled Zn are shown in Figure 2.3 as a function of milling time. The hardness values for Zn milled at room temperature are also shown. A relative density greater than 98% of theoretical density was obtained for all of the Zn samples after milling and powder compaction. The magnitude of the hardness peaks decreases with milling time, thus exhibiting a modulated oscillatory hardening manifest as damped oscillations of the hardness. At longer milling times the hardness reaches a steady-state value about twice that of unmilled Zn. Transmission electron microscopy showed that large variations in the dislocation density and grain-size distribution occurred during cryomilling. The observations suggest that dynamic recrystallization takes place in larger grains when the dislocation density due to strain-hardening reaches a critical level. A reaction-rate model was developed which accounts for the dynamic recrystallization effect and the observed oscillations in hardness.

The behavior of Zn milled at room temperature (RT) showed very different behavior (Zhang et al., 2003). It did not exhibit the oscillatory variations in hardness with milling time, but showed another phenomenon of interest – namely the *in situ* consolidation of the powder during milling into solid spheres. Small spheres were observed after about 1 h of milling and the sphere size increased with milling time such that after 25 h of milling sphere sizes of about 4–10 mm diameter were obtained. The dense spheres could be pressed into disks to obtain material suitable for a variety of mechanical tests including tensile tests of sub-size samples (Zhang et al., 2002b). Harris et al. (1993) had shown previously that milling Cu powders for long times at ambient temperature produced spherical balls, which in most cases were hollow, and often contained smaller spheres within the cavities.

2 Processing of structural nanocrystalline materials

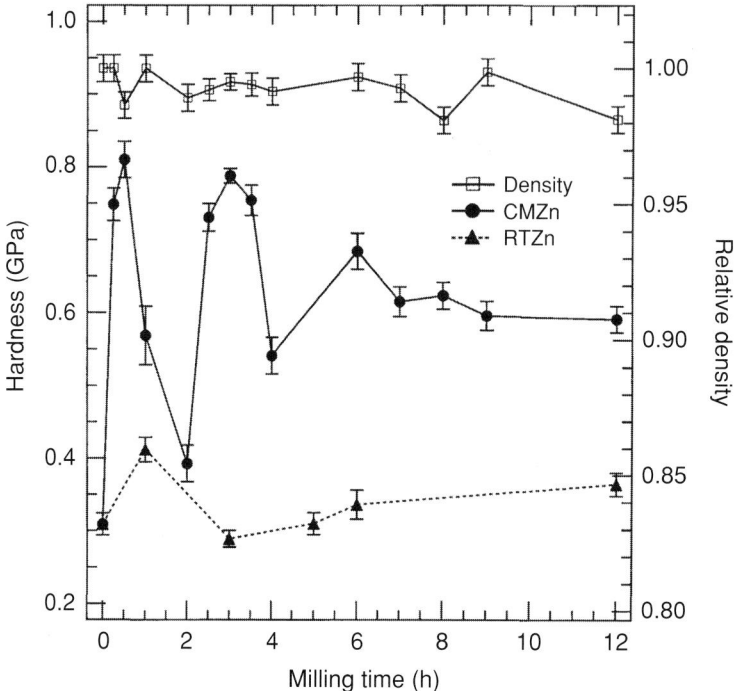

Figure 2.3. Hardness vs. milling time for CM Zn (solid circles) and RT Zn (solid triangles). The relative density for CM Zn samples is shown by the open squares. (From Zhang *et al.* (2002a), reproduced with permission from Elsevier.)

Subsequent to the Zn studies it was found that several other metals such as Al and Al alloys could also be produced by *in situ* powder consolidation during milling (unpublished results, North Carolina State University, 2003), but in many cases the spheres so produced were hollow or contained voids and cracks. It has since been shown that a combination of milling procedures involving milling at cryogenic temperatures (in argon atmosphere) and at room temperature along with variations in milling ball size can lead to solid spheres up to or greater than 10 mm in diameter in Cu, Al, and Al–Mg alloys. These spheres are ideal for forming disks on which samples for many mechanical property tests can be obtained. Results of such tests will be described in detail later in Chapter 4.

The mechanisms for the formation of nanocrystalline microstructures by mechanical attrition were first proposed by Fecht *et al.* (1990). This explanation was based upon the TEM studies of ball-milled Ru and AlRu powders. The observed phenomenology of nanocrystallization by mechanical attrition was summarized as occurring in three stages, as follows.

- Stage 1. Deformation localization in shear bands containing a high dislocation density.
- Stage 2. Dislocation annihilation/recombination/rearrangement to form cells/subgrain structures with nanoscale dimensions – further milling extends this structure throughout the sample.

2.2 Methods for preparation of bulk

- Stage 3. The orientation of the grains becomes random, that is, low-angle grain boundaries disappear as high-angle grain boundaries replace them, presumably by grain-boundary rotation/sliding.

Stage 2 might be considered to be a form of self-assembly since the dense dislocation arrays form into subgrain boundaries in order to lower the energy of the system. This mechanism proposed by Fecht and co-workers appears to be a logical description, and may in most cases be the process of nanocrystallization by mechanical milling. However, from the results on cryomilled Zn discussed above, in some special cases, high-angle nanocrystalline grains can also be formed by a "discontinuous" dynamic recrystallization process in contrast to the continuous rearrangement of dislocation structures during deformation which leads to subgrains and then grains with high-angle boundaries.

Nanocrystalline grains are observed during the mechanical alloying of dissimilar component powders. Klassen *et al.* (1994) followed the phase formation and microstructural development during mechanical alloying of Ti and Al powder blends of overall composition $Ti_{25}Al_{75}$. TEM revealed nanocrystalline grains of partially ordered $L1_2$ phase with a crystallite size of 10–30 nm in the alloy layers at the interface between the pure Ti and Al lamellae at very early stages of the milling process. The alloy phase that develops between the pure powder components consists of nanocrystalline grains presumably because of the multiple nucleation events and the slow growth which occur at the relatively low temperatures (100–200 °C above ambient) during milling. Trudeau *et al.* (1992) prepared nanocrystalline FeTi by both low-milling-energy mechanical alloying of elemental Fe and Ti powders, and by mechanical milling of FeTi compound powders. Higher mill energies resulted in amorphization. The grain size of the mechanically milled FeTi steadily decreased with milling time while that for the mechanically alloyed Fe/Ti first increased and then decreased to values essentially identical to those for the FeTi compound milled powders. This effect is illustrated in Figure 2.4.

It might be expected that brittle components would simply fracture during milling and be reduced in size with little or no changes in the internal microstructure of the powders. However, it has been found that ball milling of some nominally brittle materials can lead to alloying of the brittle components, e.g. Si and Ge (Davis and Koch, 1987), and the introduction of significant plastic deformation and high dislocation densities in brittle compounds, e.g. Nb_3Sn (Cho and Koch, 1991). Si and Ge are completely brittle at room temperature and yet complete solid solutions of Si–Ge alloys were obtained across the binary phase diagram. Thus alloying on the atomic scale was observed by mechanical alloying of the brittle components. Nb_3Sn is an extremely brittle intermetallic compound which fractures elastically until tested at temperatures above 1400 °C. However, ball milling Nb_3Sn can produce large amounts of plastic deformation as observed by TEM of the milled powder. The dislocations so produced then presumably induce a nanocrystalline grain structure similar to that formed in milled ductile metals. It is not yet clear how ball milling can produce large plastic deformation in materials that are very brittle under uniaxial stress conditions. It is suggested that the high hydrostatic stress component that may exist in the powders during milling can favor plastic deformation over fracture and allow a large dislocation density to be generated (Cho and Koch, 1991). Mechanical attrition has also been found to

34   2 Processing of structural nanocrystalline materials

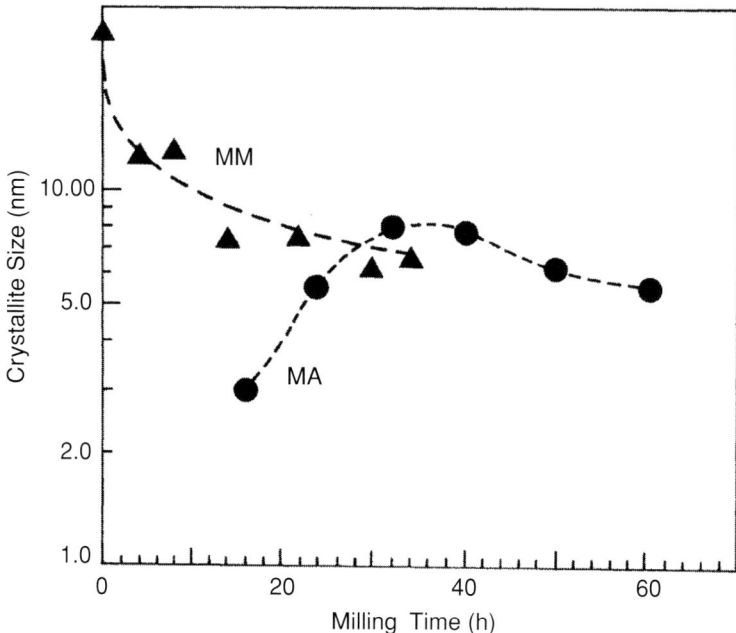

Figure 2.4. Crystallite size for FeTi intermetallic phase for low-energy mechanical alloying of Fe and Ti powders (solid circles), and mechanical milling of the FeTi compound powders (solid triangles) (From Trudeau *et al.* (1992), reproduced with permission from Trans Tech.)

induce nanocrystalline microstructures in brittle ceramics such as ZrO_2 and ceramic powder mixtures such as Fe_2O_3/Cr_2O_3 and ZrO_2/Y_2O_3 (Fecht, 2002).

The application of mechanical attrition to polymeric materials was initiated by Shaw and co-workers (Shaw, 1998). In order to fracture the polymer particulates and, on the microscopic level, the polymer chains, the milling was conducted at temperatures below the glass-transition temperature of the given polymer. Shaw's group has studied a number of homopolymers such as polyamide, polyethylene, acrylonitrile-butadiene-styrene, polypropylene, and polystyrene. Refinement of the microstructure typically occurred and milling-induced structural and property changes were noted that were very material specific. Subsequently others have studied milling-induced changes in the structure of several semi-crystalline and amorphous homopolymers (Koch *et al.*, 2000). Milling of poly(methyl methacrylate) (PMMA) resulted in monotonic decreases in molecular weight and glass-transition temperature, reflecting the milling-induced scission of the polymer chains. Polyisoprene (PI) exhibited much different behavior in that the decrease in glass transition temperature, T_g, given by $\Delta T_g = T_{g,0} - T_g(t_m)$, where $T_{g,0}$ is the glass transition temperature of the unmilled polymer and t_m is the milling time, first increased and then decreased as illustrated in Figure 2.5. In this case cryomilled PI does not exhibit a monotonic increase in ΔT_g but instead shows a sharp maximum at relatively short milling times (2 h) followed by a drop to almost zero before again increasing slightly for longer milling times. This unusual, but reproducible, behavior strongly suggests that the PI chains undergo chemical

2.2 Methods for preparation of bulk

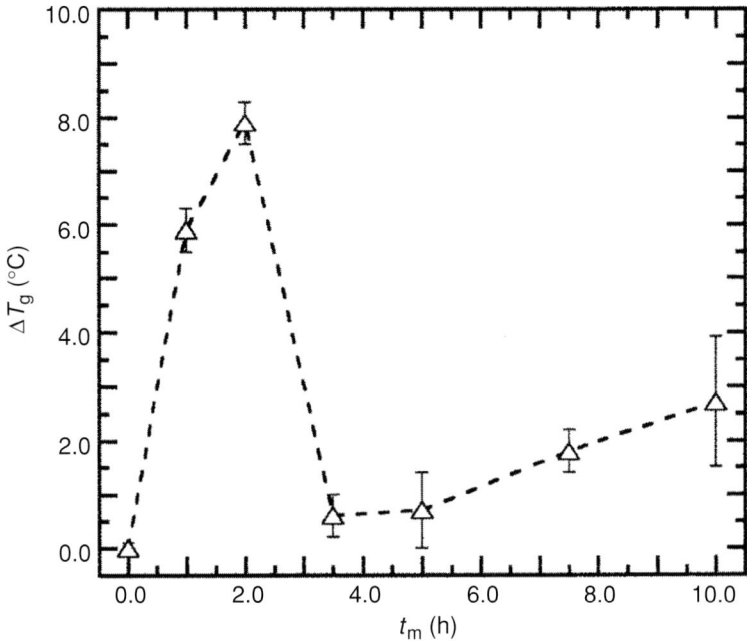

Figure 2.5. Dependence of ΔT_g on milling time, t_m, for cryomilled polyisoprene. (From Koch et al. (2000), reproduced with permission from Trans Tech.)

crosslinking during cryomilling. In this case we imagine a dynamic competition between chains breaking (causing a decrease in molecular weight) and crosslinking (promoting an increase in molecular weight) under the nonequilibrium conditions of milling. Sol–gel analysis and FTIR spectroscopy gave further evidence for the milling-induced crosslinking in PI. Mechanical attrition was also performed at cryogenic temperatures to incorporate PI into PMMA (Smith et al., 2000). TEM clearly showed that the solid-state blending by mechanical attrition of polymeric materials can yield nanoscale dispersions of immiscible polymers. The strong evidence for free-radical formation and crosslinking induced by milling suggests many possibilities for the design of novel new polymeric materials with nanoscale microstructures.

A serious problem with the milling of fine powders is the potential for significant contamination from the milling media (balls and vial) and/or atmosphere. If steel balls and containers are used, iron contamination can be a problem. It is most serious for the highly energetic mills, for example, the Spex shaker mill, and depends upon the mechanical behavior of the powder being milled as well as its chemical affinity for the milling media. For example, milling Ni to attain the minimum grain size in a Spex mill resulted in Fe contamination of 13 at.%, while the Fe contamination in nanocrystalline Cu similarly milled was only <1 at.% (Shen and Koch, 1996). Lower-energy mills result in much less, often negligible, Fe contamination. Other milling media, such as tungsten carbide or ceramics, can be used but contamination from such media is also possible. Interstitial element (oxygen, nitrogen, etc.) contamination can be controlled by milling and subsequent powder handling in a pure inert-gas atmosphere, with care taken that the milling vial is leak-free during processing.

As a "two-step" processing method, in general the powders formed by mechanical attrition with a nanocrystalline internal microstructure require a compaction step to obtain bulk materials for mechanical testing and for structural applications. The description of particulate compaction will be given in Section 2.2.4.

2.2.3 *Nanocrystalline powders by chemical reactions*

Chemistry has always played an important role in the development of new materials and technologies. The advantages of chemical synthesis lie in the versatility in fabricating and designing new materials with novel properties. Since chemical synthesis works at molecular-level mixing, the final product will have a homogenous chemical composition, required for the tailoring of the properties. A very fundamental understanding of the thermodynamics, crystal chemistry, phase diagrams, and reaction kinetics will tailor the many potential benefits chemical synthesis offers (Rao, 1993).

Chemical synthesis can be scaled up for bulk production. There are a variety of chemical methods available for making nanoparticles of metals, oxides, intermetallics, semiconductors, quantum dots, and glasses. However, we will limit the description of chemical methods only to the development of the structural nanomaterials, the main theme of this book. Several reducing agents can be used to make nanoparticles in a simple, cost-effective manner. For example, nanocrystals of Mo, one of many main components for multifunctional structural nanomaterials, can be obtained by reduction of $MoCl_3$ in an organic solution (toluene) with $NaBEt_3H$ as reducing agent at room temperature. This reaction provides a high yield of Mo nanoparticles with a size of 1–5 nm. The reaction equation is as follows:

$$MoCl_3 + 3\,NaBEt_3H \rightarrow Mo + 3\,NaCl + 3\,BEt_3 + (3/2)\,H_2. \tag{2.1}$$

Another industrially important structural material is aluminum. Al nanoparticles can be made by decomposing $Me_2EtNAlH_3$ (Me = methyl, CH_3) in toluene and subsequent heating the solution to 110 °C for 2 h. However, in order to speed up the reaction, Ti in the form of titanium isopropoxide is used as catalyst. Al nanoparticles 50–80 nm in size have been produced. So it is clear that, in chemical synthesis methods, not only does one need reducing or oxidizing agents, but a proper choice of catalysts and surfactants will play a role in controlling the size, shape, and agglomeration of the ultrafine particles.

In fact, chemically synthesized nanoparticles (also nanomaterials) are already playing a role in numerous industries. Carbon black (nanocarbon) makes rubber tires wear resistant; nanophosphors are used in flat panel displays and cathode ray tubes (CRT) to display colors; nanocrystalline materials are used for insulation; nanoalumina and nanosilica are used in slurries used for chemo-mechanical polishing of silicon wafers; nanoiron oxide creates the magnetic material used in disk drives of audio/video tapes; nanozinc oxide or nanotitania are used in many sunscreens to block harmful UV rays; and nanoplatinum is critical to the operation of catalytic converters. The list of the applications of nanomaterials appears to be limitless.

2.2 Methods for preparation of bulk

There are certain challenges in the chemical fabrication of nanostructures. While contamination is a problem, severe agglomeration takes place in the chemical baths. This can significantly reduce the properties of nanomaterials. Also agglomeration often causes trouble in consolidating nanoparticles in fully dense components; this is discussed in subsequent sections. The problem of agglomeration may be avoided in liquid-phase reactions by common methods such as spray drying or freeze drying (Neilson, 1982; Real, 1986). Although chemical processes are easy to scale up, the process may not be straightforward all the time.

In chemical synthesis, we often use precursors that are finally converted to nanoparticles either by precipitation or by decomposition. Both homogeneous and heterogeneous nucleation take place in the solution bath depending on the saturation state of the precursor molecules. Subsequent to nucleation, growth of the particles is evident. Growth usually proceeds by diffusion, which is dependent on the concentration gradients, reaction temperatures, and pH. This will control the monodispersity of the nanoparticles. Both particle-size distribution and shape control the physical properties of the nanostructures. It is important to control the chemical homogeneity and stoichiometry. This is achieved via careful control of the reaction chemistry.

Metals, ceramics, and intermetallics are key to the development of structural nanomaterials. These fine powders can find applications in composites, magnets, catalytic supports, and various aspects of powder metallurgy. Since this book deals with structural nanomaterials, we will describe in detail two chemical-synthesis methods: (1) sol–gel, and (2) microemulsion in the production of metals, ceramics and alloys. However, other chemical routes of fabricating structural materials will be described briefly.

2.2.3.1 Sol–gel process

The basic steps involved in the sol–gel synthesis of oxide nanoparticles are shown in the block diagram presented in Figure 2.6. The synthesis of oxide nanoparticles by this procedure is based upon the hydrolysis and condensation of metal alkoxide $M(OR')_n$, where M is the metal, O is the oxygen and R' is the organic group. Since the metal alkoxide and water are not soluble in each other, they are required to be dissolved in a common alcoholic solvent in order to carry out their reaction. The very high reactivity of water with the metal alkoxide is explained by the partial charge model proposed by Livage *et al.* (1988). According to this model, the alkoxide group, being highly electronegative, creates a positive charge on the central metal atom. On the other hand, in the water molecule, there exists a partial negative charge on the oxygen atom. As a result, the water molecule attacks the metal atom from the alkoxide and results in the hydrolysis of the alkoxide.

During the synthesis, two equal parts of alcoholic solutions are made. In one part, water and an organic polymer as a surfactant are dissolved in appropriate proportions in the alcoholic solvent. The second part of the alcoholic solution is prepared by dissolving completely the metal alkoxide in the common solvent. After homogenizing both the solutions, they are mixed rapidly. The order of mixing the two solutions does not affect the properties of the reaction product (Fegely *et al.*, 1985). The process is carried out in a dry-nitrogen glove box to avoid any premature reaction of atmospheric moisture with the

38   2 Processing of structural nanocrystalline materials

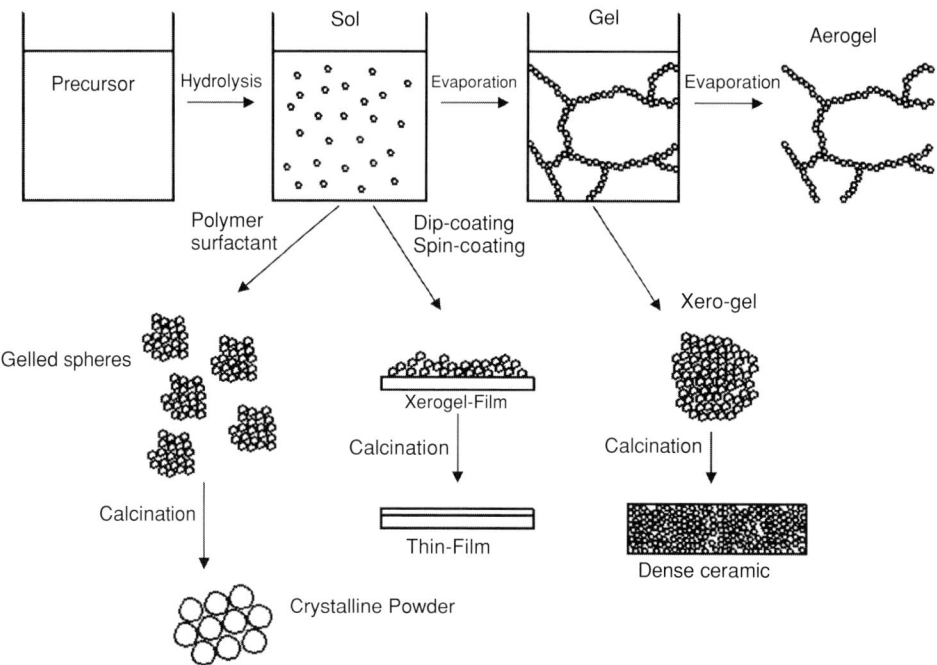

Figure 2.6. A schematic diagram of a sol–gel process.

alkoxide. The following reactions immediately take place upon mixing the two parts of the solutions:

$$\text{hydrolysis:} \quad M(OR')_4 + H_2O \rightarrow M(OH)_4 \, 4R'OH, \quad (2.2)$$
$$\text{condensation:} \quad M(OR)_4 + M(OR)_4 \rightarrow (OH)_3M-O-M(OH)_3 \, H_2O. \quad (2.3)$$

Here M represents the metal atom, R' is the alkoxide group and O is the oxygen atom. The result is the formation of the M–O–M bond within the solution. The kinetics of hydrolysis and condensation reactions are governed mainly by the ratio of molar concentrations water to alkoxide (R). In general, low R values (<3) are suitable for fiber and thin-film formation while large R values (>3) generate powder particles (Nagpal et al., 1992). The oxide nanoparticle size and its distribution and particle surface morphology are hence affected by this R value.

2.2.3.2   The growth mechanism and the particle shape

The shape of the oxide particles obtained by using the present sol–gel technique is modeled to depend on the growth mechanisms (Pierre, 1991). The growth of the oxide particles occurs either by mononuclear or polynuclear growth mechanisms. In the mononuclear growth mechanism, the particles grow by formation of successive layers. During the operation of this growth mechanism, the first step is created on the nucleated particle surface. A layer has the time to achieve its completion before a new step appears. Growth proceeds via a layer-by-layer mechanism, and these layers correspond to the densest packing of polynuclear

2.2 Methods for preparation of bulk

complexes. The particle surfaces are locally smooth at the molecular level, but faceted at the microscopic scale. These particles take well-defined shapes, corresponding to particular crystal structures. In the mononuclear growth region, the particle growth rate is proportional to the particle surface area

$$dr/dt = k_m r^2, \quad (2.4)$$

where k_m is a proportionality constant. For two particles of initial radii r and r_0 (with $r_0 < r$), the relative particle size difference increases with growth, according to the relationship

$$dr/r = (dr/r_0) \cdot (r_0/r). \quad (2.5)$$

As the difference in the particle size increases when the growth proceeds, the mononuclear growth mechanism does not aid the formation of monodisperse powders.

As the particles grow initially by a mononuclear growth mechanism, the area of the layer increases and the growth mechanism progressively changes to a polynuclear growth mechanism. During this event, the formation of new steps is fast enough to nucleate a new layer before the previous one has reached its completion. The particle surface becomes less smooth at the molecular scale, but less faceted at the macroscopic scale. The particles hence lose their particular crystallographic shape and become more spherical. A shape transition thus occurs owing to the change in the growth mechanism. During polynuclear growth mechanisms, the particle growth rate becomes constant and the relative size difference between the two particles, nucleated at different instants, attenuates slightly according to the relationship (Pierre, 1991)

$$dr/r = (r_0/r)(dr_0/r_0). \quad (2.6)$$

As the particles keep growing by a polynuclear growth mechanism, at certain times the supply of chemical complexes becomes insufficient since they are consumed by the growing particles. Under these conditions, the supply of the polynuclear complex becomes insufficient with time and the transport of chemical complexes by diffusion become rate controlling. The growth mechanism again changes and enters a "diffusion growth regime." The growth rate of the particle in this regime behaves according to the relationship (Pierre, 1991)

$$dr/dt = [D(C - C_S) V_M]/r, \quad (2.7)$$

where D is the diffusion coefficient, V_M is the solid material molar volume, C_S is the saturation concentration of the polynuclear complexes, and C is the actual solute concentration. During the operation of this growth mechanism, the relative size difference between two particles attenuates according to

$$dr/r = (r_0/r)^2 (dr/r_0). \quad (2.8)$$

The attenuation in the size difference is much more efficient during the operation of the diffusion growth mechanism than that during the polynuclear growth mechanism. Hence, the diffusion growth mechanism significantly enhances the formation of monodispersed powders. As in polynuclear growth, the diffusion regime corresponds to the particles with a rough surface at the molecular level, but smooth at the macroscopic level. From a comparison of the three growth mechanisms, it appears that the particle plays an important role in deciding the growth mechanism. The mononuclear growth process dominates when the particle size is small, polynuclear growth dominates when the particles are bigger, and diffusion dominates in the last stage. However, the solute supersaturation also plays an important role at the very end of the process. The growth mechanism reverts to the polynuclear regime and even to the mononuclear regime. Consequently, the particles shape changes from spherical to faceted. This explains the evolution of varying particle shapes and their transitions depending on the experimental conditions.

2.2.3.3 Colloidal stability of an oxide nanoparticle sol

When the particles grow in the solution, they are likely to collide with each other owing to Brownian motion. This may cause excessive aggregation of the particles. Thus, forming a stable colloidal suspension of oxide nanoparticles is essential for producing a nanocrystalline monodispersed ceramic-oxide powder. The mechanism of the stability of the sol hence plays an important role.

(a) Dispersion forces

In a colloid, the dispersion forces (van der Waals forces of attraction) exist between the particles, which essentially aggregate the particles (Brinker and Scherer, 1990). For a colloidal particle, the dispersion force is the addition of forces of all the atoms within the particle itself. Hence, for colloidal particles in the nanometer range, the attractive dispersion force becomes quite significant, resulting in flocculation of colloids. These forces also depend on the shape of the particles. According to theory proposed by Mahanty and Ninham (1976), for particles having a plate-like shape separated by distance h, the attractive potential is given by

$$V_A = -A/12\pi h^2, \tag{2.9}$$

where A is the Hamaker constant and is the material property, and h is the distance away from the particle surface. It can be noted that for plate-shaped particles, the attractive force decays in proportion to $1/h^2$ while for atoms it decays as $1/h^6$.

Hence, the decay of dispersion force with h is slower for the plate-shaped particles than that for atoms. Moreover, for spherical particles, the dependence of the attractive dispersion force on h is logarithmic; hence the dispersion force decays more slowly than that for plate-shaped particles. The attractive dispersion force between two spherical particles (of size, say, in the nanometer scale) can extend over the distances of nanometers, and hence can cause coagulation of particles. Hence, the closest approach of the particles must be avoided to prevent the coagulation. This can be done by two different mechanisms: by creating electrostatic repulsion between two particles, or by adsorbing a thick organic layer (called a steric barrier) on the particles.

(b) DLVO theory of electrostatic stabilization

In a colloid, the repulsive electrostatic force is created by the electrical double layer associated with the colloidal particle. The hydrous oxides generally have OH^- groups on their

2.2 Methods for preparation of bulk

surface. The protonation and deprotonation of the M–OH bonds result in the creation of a charge on the particle surface:

$$M-OH + H^+ \rightarrow M-OH_2^+, \quad (2.10)$$
$$M-OH + OH^- \rightarrow M-O^- + H_2O. \quad (2.11)$$

These H^+ and OH^- ions are thus the charge-determining ions. Whether the particle is positively or negatively charged or is neutral depends upon the pH of the solution. The pH at which the charge on the particle is zero is called the "point of zero charge (PZC)." At pH < PZC, the particle is positively charged while for pH > PZC, the charge on the particle is negative (Brinker and Scherer, 1990).

This surface charge of the colloidal particle then attracts ions (known as "counter-ions") of opposite charge in the solution. These counter-ions bind to the particles via van der Waals forces as well as electrostatic potential of charge determining ions. The water molecules are also attracted towards the surface charge and are held by van der Waals forces as well as by hydrogen bonds. Hence, in a colloid, the particle surface charge is screened by the counter-ions. The charge-determining ions and counter-ions create an electrical double layer. This electrical double layer, associated with every colloidal particle, is responsible for generating a repulsive force between two neighboring particles, thus avoiding flocculation of particles.

(c) *Steric stabilization*

Although aqueous systems are most commonly stabilized by an electrostatic double layer, it is also possible to prevent coagulation by using a thick adsorbed layer, which constitutes a steric barrier. To provide an effective barrier, the adsorbed polymeric layer must satisfy the following requirements:

(i) the surface of the polymer should be completely covered with the polymer (to prevent polymer chains from attaching both the particles);
(ii) the polymer should be firmly anchored to the surface so that it is not displaced during Brownian collisions;
(iii) the layer must be thick enough (typically >3 nm) to keep the point of closest approach outside the range of the attractive van der Waals forces; and
(iv) the nonanchored portion of the polymer must be well solvated by the liquid.

The adsorbed layer of the polymer prevents coagulation of the particles in two ways: entropicly and enthalpicly. When the two particles approach each other, the adsorbed layers overlap, resulting in a decrease in the degree of motion of the polymer chains. This is the entropic contribution of the adsorbed polymer against the particle coagulation. As the particles approach each other, the solvent molecules, which surround the particles, are squeezed out. However, an osmotic pressure is created that tends to suck the liquid back into the space between the particles. This increases the energy required for the particles' approach and it is an enthalpic contribution. The block polymers are particularly effective steric barriers in this respect, because one end can be made to be strongly adsorbed at the particle surface, while the rest of the chain can be made to have a high affinity for the solvent. The higher

Table 2.1. *Polymer surfactants that may be used in the solgel process*

Surfactant name	Type	Molecular weight
Klucel E	Hydroxypropyl (HPC)	60 000
Mazon 210–141	Poly(propylene glycol) PPG w/carbox. Acid graft	2200
PVP K-15	Poly(vinylpyrrolidone)	10 000
K-30	(PVP)	40 000
K-60		160 000
Butvar B-73	Poly(vinylbutyral)	50–80 K
B-74	(PVB)	100–150 K
Pluronic P123	Poly(propylene glycol)/polyethylene glycol)	4025/1725
L42	PPG/PEG block polymer	1300/325
Kessco 1000DS	Poly(ethylene glycol)	1600
Esters 1000 MO	PEG w/ester linkage to hydrocarbon	1300
200DS		800
200 MS		500
200 MO		500
PPG Custom MO	PPG w/ester linkage to ol. or laur.	1300
ML		2300
PPG 150	Poly(propylene glycol)	10 000
Ethylcellulose N50	Ethylcellulose	50 000
N100		100 000
N200		200 000
Myrj 52	PEG stearate	2020
45		580
Brij 58	PEG cetyl ether	1080
76	PEG stearyl ether	660
78	PEG stearyl ether	1100
Gelva TS-30	PVA emulsion in water	40–80 K
S-52		∼300 000

Reprinted from Mates and Ring (1987), with permission from Elsevier.

the affinity of the solvent for the polymer, the greater the enthalpic contribution to the stabilization (Brinker and Scherer, 1990).

2.2.3.4 Selection of a polymer as a surfactant

During the synthesis of monodispersed powders by the sol–gel process (i.e. by hydrolysis of alkoxides) a large amount of agglomeration occurs when the reactant concentration is above 0.1 wt% solids (Mates and Ring, 1987). As mentioned in the previous section the agglomeration can be effectively prevented by electrostatic as well as by steric means. A stericly stabilizing surfactant for the growing lyophilic particles should have several other unique properties in addition to adsorption and stabilization (Jean and Ring, 1986, 1988). First, the surfactant should allow the polymeric reaction products to diffuse through the adsorbed layer. Secondly, the surfactant must not be incorporated into the structure of the growing particle. For this reason, a stericly stabilizing surfactant must be physically adsorbing and not chemically binding at the surface. In search for such a surfactant, twenty-seven surfactants were tested (Mates and Ring, 1987) as listed in Table 2.1.

The surfactants were from two different categories:

(1) homopolymers with sufficiently high molecular weight to concentrate at the solid–liquid interface and
(2) block polymers having one segment that stabilizes against agglomeration.

All the surfactants were evaluated based on the following criteria.

Solubility in ethanol: Any surfactant which could not form a clear, homogeneous solution with ethanol at a concentration of 20 g/l after gentle, overnight agitation at room temperature was eliminated. Surfactants that were eliminated at this stage were ethylcellulose, a poly(ethylene glycol)–stearate, and poly(vinyl alcohols).

Stabilizing effect: LiCl salt was added, at three times its critical coagulation concentration (CCC), to coagulate the sol (product of 0.8 M (water)/0.2 M (alkoxide) reaction). The stabilizing effect of the polymer was tested by considering its ability to avoid the coagulation under these conditions.

Repeptization: The ability of the surfactant was tested to repeptize the previously unprotected sol (product of 0.8 M/0.2 M reaction), which was coagulated by adding LiCl at three times its CCC.

Critical polymer concentration (CPC): This is defined as the minimum amount of polymer required to protect the sol (product of 0.8 M/0.2 M reaction) from coagulation by 0.01 M LiCl.

Reversible adsorption: The polymer, one and half times the CPC, was added to the sol (product of 0.8 M/0.2 M reaction) for adsorption. Four successive ethanol washes were then given to desorb the polymer. LiCl salt (0.01 M) was then added to the sol to cause coagulation. Any polymer that prevented coagulation under these conditions, was considered to be irreversibly adsorbed.

Stability during nucleation and growth: All the tests mentioned above were performed on the previously precipitated particles. Hence, in this test, the oxide particles were synthesized in the presence of a polymer surfactant. The effective surfactant should not affect the water–alkoxide reaction and should be adsorbed on the particle surface before they collide with each other resulting in the formation of monodisperse particles.

It is found that, among the twenty-seven polymers listed in Table 2.1, only the HPC polymer satisfied all the test conditions. It was only the HPC polymer that stericly stabilized the particle spheres both during and after the growth, allowing stable suspension of mostly single particles. Hence, the HPC polymer appears to be the most effective polymer for providing steric stabilization against particle coagulation within the sol. This polymer has been successfully utilized by many investigators (Shukla *et al.*, 2002; Fegely *et al.*, 1985) in the synthesis of sub-micron sized as well as nanocrystalline oxide particles. Figure 2.7 shows a TEM micrograph of zirconia nanoparticles.

Because of the adsorption of the HPC polymer on the surface of the oxide nanoparticles and its dependence on the sol–gel synthesis parameters, it is possible to control the oxide nanocrystallite's size and its distribution during the processing. Figure 2.8 shows the variation in the TiO_2 and ZrO_2 nanocrystallite's size as a function of [HPC] and R (Nagpal *et al.*, 1992; Shukla *et al.*, 2002). It is to be noted that increase in both R and [HPC] results in a decrease in the nanocrystallite's size. There are two reasons for the decrease in nanocrystallite's size with increase in R. First, high nucleation rates are associated with high R values

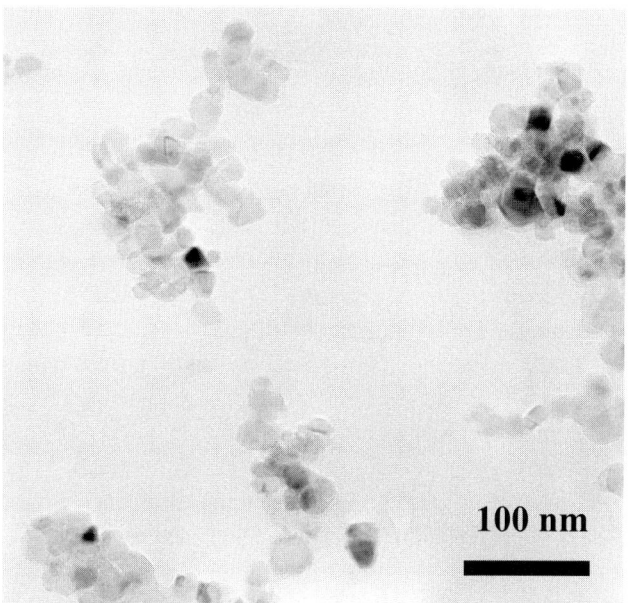

Figure 2.7. Transmission electron micrograph (TEM) of sol–gel-derived zirconia nanoparticles in the size range 20–40 nm. (Courtesy of Seal Research Group.)

and the HPC adsorption increases with increase in R. The decrease in the nanocrystallite's size with increase in [HPC] is attributed to the increased steric hindrance caused by the increased adsorption of HPC polymer with increase in the initial concentration of the HPC polymer.

Sol–gel methods have been extensively used to synthesize nanocrystalline oxides and ceramics, such as tin oxide (SnO_2) in the past decade by several groups (Chatelon et al., 1997; Licciulli and Mazzarelli, 2001; Ando et al., 1994); thin films of alumina (Al_2O_3) (Lalauze et al., 1993), molybdenum oxide (MoO_3) (Li et al., 1999; Galatsis et al., 2001), titania (TiO_2) (Traversa, 1995), barium strontium titanate ((Ba,Sr)TiO_3) (Zhu et al., 2002), and zirconia/yttria stabilized zirconia (ZrO_2/YSZ) (Tan and Wu, 1998) have also been synthesized using the sol–gel technique. Table 2.2 shows typical precursors used for sol–gel-derived nanocrystalline powders (Shukla and Seal, 2004).

Composite Fe/silica xerogel hybrids are formed by reacting Fe(III) nitrate in a conventational silica sol–gel formulation using tetraethylorthosilicate (TEOS) as a source point of silica xerogel. Ferric ions are reduced in the gel matrix by passing hydrogen gas at 400–700 °C for 4 h of exposure. Fe particles in the range 3–30 nm are formed in the gel matrix (Basumallick et al., 1997; Roy et al., 1993; Chatterjee and Chakrabarty, 1992; Wang et al., 1995).

Other metal/oxide nanohybrids, such as Ru/silica xerogel, have been fabricated by conventional metal-ion doping in the silica xerogels. The addition of Ru(III) chloride to sol–gel containing TEOS at the levels of 0.5 wt%, 1.0 wt%, and 2.0 wt% Ru leads to silica xerogel nanocomposites doped with Ru(III) ions. The average diameter of Ru is 3–5 nm (Lopez et al., 1993).

2.2 Methods for preparation of bulk

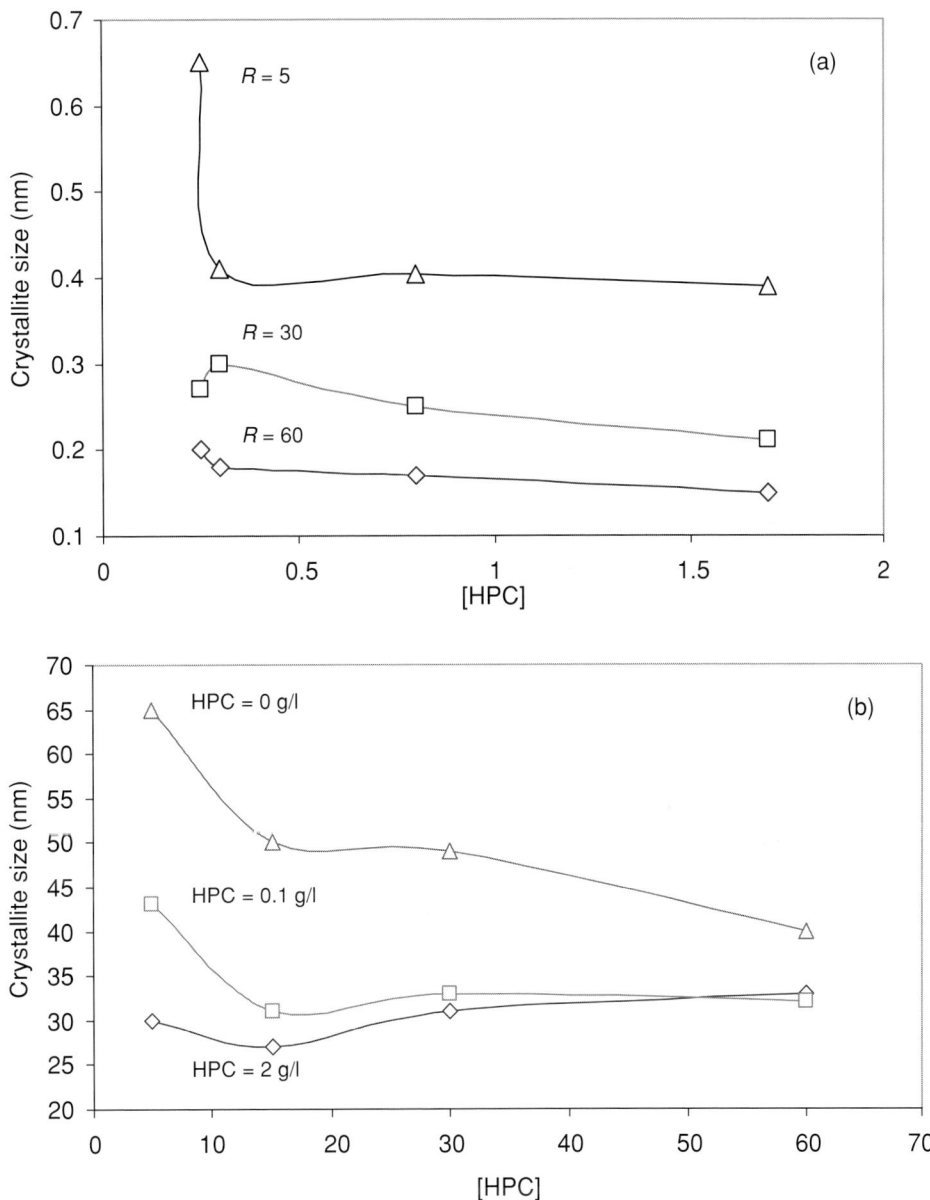

Figure 2.8. The effect of HPC polymer surfactant and R on the (a) TiO_2 (Nagpal *et al.*, 1992) and (b) ZrO_2 (Shukla *et al.*, 2003, with kind permission of Springer Scientific and Business Media, and the author) nanoparticles via the sol–gel route. Molecular weight of HPC in (a) is 1 150 000 g/mol and in (b) 80 000 g/mol.

Sol–gel processing of nanomaterials has attracted considerable attention for the simplicity of the method. Complex powder composition can be created by controlling the hydrolysis and condensation steps in the sol–gel process. The flexibility in the process can be utilized to control the particle size, pore size, surface area, density and other physico-chemical properties. Both inorganic and organic hybrid powders can be tailored according to the

Table 2.2. *Typical sol–gel precursors used to make nanocrystalline oxides*

Oxides	Precursors	Crystallite Size (nm)
Zinc oxide	Zinc acetate	300–600
Zinc oxide–Pd^{2+} catalyst		
Lead zirconate titanate	Lead acetate trihydrate, zirconium acetylacetone, Titanium isopropoxide	–
Titania	Titanium butanol alkoxide	50
	Titanium butoxide	80
Titania–Niobium oxide	Titanium butoxide, Niobium ethoxide	55–60
Titania–Niobium oxide	Titanium sulfate, Niobium chloride	12
Titania	Tetraethyl ortho-titanate	3–30 nm
	Titanium isopropyl alkoxide	<100
Indium oxide	Indium nitrate	8
Molybdenum Oxide–Titania	Molybdenum(V) isopropoxide – Titanium butoxide	20–100
Tungsten oxide	Tungsten(VI) ethoxide	30–100

From Shukla and Seal, 2004. Reprinted with permission from the Encyclopedia of Nanoscience and Nanotechnology, edited by H.S. Nalwa, American Scientific Publishers, Los Angeles (2004) Vol. 10, p37. Copyright American Scientific Publishers, http://www.aspbs.com.

nature of the applications. Control of nanoparticle size remains a challenge in the sol–gel process. By controlling the host matrix pore size, one can control the particle size; however, convenient and cost-effective routes are yet to be developed to create monodispersed nanoparticles.

2.2.3.5 Microemulsion technology

Microemulsions, another wet chemical synthetic route of production of nanomaterials, have been known for more than a century. Floor cleaning, lubrication for machining and metal working processes, foods, agrochemicals, detergents, paints, cosmetics, pharmaceuticals, and tertiary oil recovery are a few examples of the widespread industrial and domestic applications of microemulsions. Microemulsions were described as special colloidal dispersions by Hoar and Schulman (1943).

Microemulsions are defined as systems of single-phase and thermodynamically stable isotropic solutions composed of water, oil, and one or more amphiphiles. Such a definition was put forward by Danielson and Lindman (1981), and is widely accepted, as reported by Solans and co-workers (1997). The microemulsion can form large numbers of reactors, which can be successfully used to synthesize nanoparticles mainly because these nanoreactors are separated from one another and thereby the growth of the synthesized nanoparticles is restricted. Therefore it becomes a valuable process tool to synthesize nanoparticles for designing and manufacturing metals, ceramics, and composite components with improved mechanical, electrical, magnetic, optical and electronic properties. The use of microemulsions to synthesize monodispersed nanoparticles of the platinum group of metals was first reported by Boutonnet *et al.* (1982). The process of microemulsion to produce nanoparticles is unique in the sense that it can produce monodispersed particles and has capability to control not only the size but also the shape of the nanoparticles. Almost all types of nanoparticles can be synthesized with the help of the microemulsion method, which has drawn the attention of many researchers for the synthesis of nanoparticles in the past decade. Microemulsion fundamentals, thermodynamics, kinetics, and the physicochemical aspects of microemulsions and nanoparticles synthesis are summarized in the subsequent sections.

(a) *Microemulsion fundamentals*

The surface-tension forces at the boundary between the two immiscible liquids are free to exert their influence equally in all directions (Prince, 1977), and eventually the dispersed phase assumes the form of spheres in order to have a minimum surface area per unit volume. These spheres may consist of either water or oil. When they are water, the emulsions are called water-in-oil or simply "w/o." On the other hand, the emulsions are described as oil-in-water or "o/w," when the spheres consist of oil. The formation of an emulsion yields an enormous increase in the surface area of contact between the two immiscible liquids, which in turn requires work to be performed on the system as per the following relation:

$$work = -\gamma \, dA, \qquad (2.12)$$

where γ is the interfacial tension between the two liquids when emulsifying agent is present in the system, and dA is the increase in the area of contact between the two liquids.

Equation (2.12) indicates that the emulsification is a strong function of the interfacial energy. The smaller the value of γ, the smaller is the droplet size. Moreover, a decrease in the value of interfacial energy decreases the amount of work required for emulsification. When the value of interfacial tension approaches zero, the work term becomes almost zero, which implies that the emulsification process is spontaneous, and such a system of emulsion is termed as a microemulsion. However, the work done in a macroemulsion is of mechanical origin, unlike in the case of microemulsions where the work performed for dispersion is mostly chemical in nature. Since changing the chemical composition of the immiscible liquids is neither preferred nor practicable, a chemical substance is introduced into the system that has the ability to reduce the interfacial energy to a great extent at low concentration level. Such a material is called the emulsifier or surfactant.

Molecules of the surfactant can spontaneously arrange themselves as a monomolecular film or monolayer between the two immiscible liquids owing to the strong attraction of one portion by the oil and another portion by the water. The hydrophile–lipophile balance (HLB) is the convenient measure of such attractions and counter-attractions by the two liquids. Such attraction forces are directly responsible for orienting these molecules perpendicular to the interface and parallel to one another. The surfactant molecule has an aliphatic tail and a polar head, which are normally represented by a zigzag line and a circle, respectively. The polar head is oriented towards the aqueous phase as it is water soluble. The tail will be dissolved in the oil phase. A circle with a positive sign indicates a cationic surfactant. The polar head and aliphatic tail of the oriented monolayer molecules interact with one another. The forces of attraction and repulsion among the heads are different than among the tails.

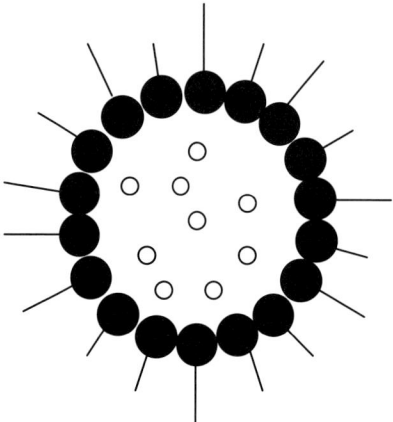

Figure 2.9. A schematic representation of single water-in-oil microemulsion droplet.

Among the tails, the London dispersion forces play an important role; they are inversely proportional to the seventh power of the lateral distance between them. On the other hand, the forces operating among the heads mainly depend on the hydrogen bonding with each other and water molecules. As the concentration of the surfactant molecules on the interface increases, the net repulsion forces among the heads and the tails increases, which helps in controlling the curvature of the film and consequently the geometrical shape of the droplets. A co-surfactant, which is either an alcohol or an amine, is used along with the surfactant in order to make the emulsification process more effective. It reduces not only the critical micelle concentration (CMC) but also the interfacial tension above the CMC. Usually the CMC is defined as the concentration of micelles at which a sharp change occurs in any of the wide variety of properties of the surfactant solution such as electrical conductance, transference number, surface tension, etc. Co-surfactants help to reduce the rigidity of the film, thereby making the transition process of film to the microemulsion droplets easy (Di Meglio et al., 1989). The surfactant and the co-surfactant interact in such a way as to greatly enhance the interfacial stability by reducing droplet size. A spherical water-in-oil microemulsion droplet is shown schematically in Figure 2.9. The heads with long and short tails denote the surfactant and co-surfactant, respectively. The water molecules are shown inside the droplet. The diameter of such a spherical droplet in actual practice may vary in the range of 10–200 Å. The continuous medium outside the microemulsion droplet is the oil phase.

The surfactants, which are employed for synthesis of materials through the microemulsion route, can be classified into three groups called cationic, anionic, and nonionic (Osseo-Asare, 1999). Cetyldimethylbenzylammonium bromide (CDBA), cetyltrimethylammonium bromide (CTAB), and didodecyldimethylammonium bromide (DDAB) are examples of cationic surfactants. The anionic surfactants usually used are sodium bis(2-ethylhexyl)sulfosuccinate, which is also known as Aerosol OT or simply AOT, dioleyl phosphoric acid (DOLPA), sodium dodecyl sulfate (SDS), and potassium oleate. Ionic surfactants with two hydrocarbon chains do not need a co-surfactant to enable them to form microemulsions. AOT and DDAB can form microemulsions without the help of a co-surfactant.

Polyoxyethylene-(n)-dodecyl ether (e.g. tetraoxyethylene dodecyl ether ($C_{12}EO_4$)), polyoxyethylene-(n)-nonylphenyl ether (e.g. NP-4, $n = 4$ (Triton N-42); NP-5, $n = 5$ (Triton N-57); NP-6, $n = 6$ (Triton N-60); NP-9, $n = 9$–10 (Triton N-101)), polyoxyethylene-(n)-octylphenyl ether (e.g. OP-1, $n = 1$ (Triton X-15); OP-3, $n = 3$ (Triton X-35); OP-5, $n = 5$ (Triton X-45); OP-10, $n = 9$–10 (Triton X-100)), polyoxyethylene-(n)-dodecylphenyl ether (e.g. DP-6, $n = 6$), and sorbitan monooleate (SPAN 80) are a few examples of non-ionic surfactants used for materials synthesis. Temperature plays an important role on the behavior of the nonionic surfactants because these are water soluble at low temperatures and oil soluble at elevated temperatures. The narrow temperature range where the solubility changes is called the phase inversion temperature (PIT) at which the interfacial tension becomes extremely low (Saito and Shinoda, 1970). The types of microemulsions formed below and above the PIT are oil-in-water and water-in-oil, respectively.

(b)   Thermodynamics

The treatment of the surface chemical behavior of a microemulsion system by using thermodynamic principles in order to understand the molecular interactions taking place at the interface is quite complex. However, these data are important to tailor a particle in the nanometer domain. A few initial studies on the concept of thermodynamical stability and the interfacial free energy of microemulsions were initiated by Adamson (1969), Ruckenstein, and Chi (1975), and Wagner (1976). Wagner postulated that a thermodynamically stable emulsion could be obtained only if the concentration of surfactant required for zero interfacial tension is lower than its critical micelle concentration. Microemulsions are thermodynamically stable unlike emulsions, which are stable owing to kinetic restrictions. Ruckenstein (1989, 1999) and Kegel *et al.* (1999) have recently attempted to provide the phase behavior of microemulsions with respect to the excess dispersed phases in equilibrium using various thermodynamic variables. A microemulsion is a mixture of oil, water, surfactant, co-surfactant, and electrolyte in which water or oil are dispersed in oil or water as globules of 10–100 nm diameter. The surfactant and co-surfactant are mostly located at the interface between the oil and water phases. However, these species are also at equilibrium between the two phases. Hence, mixing is not at the molecular level in a microemulsion, and this restricts the application of the principles of solution thermodynamics to a substantial extent. A brief outline of the thermodynamic treatment of the microemulsion will be described mostly from the work of Ruckenstein (1989) in this section, and readers may refer to the other reports (Ruckenstein, 1999, and Kegel *et al.*, 1999) for details.

Thermodynamics of microemulsions should adequately consider the entropy of the dispersion of the globules in the medium, the free energy of formation of the interface between the two media, the mutual interactions among the globules, and the equilibrium distribution of the surfactant and co-surfactant not only at the interface but also between the two media of the microemulsion. Ruckenstein (1989) conveniently decomposed the Helmholtz free energy (f) per unit volume of the entire system into two parts: the free energy (f_0) of dispersion of fixed globules without interactions among them is one, and the free energy (Δf) due to the dispersion of the globules in the continuous phase and to their interactions is the other. Such splitting of the free energy allows one to apply Gibbs thermodynamics and statistical mechanics for the first and second part, respectively. Now, in order to develop the basic thermodynamic equations for a single-phase microemulsion that contains spherical globules of uniform size, Ruckenstein (1989) has represented the Helmholtz free energy per unit volume of the microemulsion in the following form:

$$f = f_0 + \Delta f. \tag{2.13}$$

Gibbs thermodynamics provides the following expression for the infinitesimal change of f_0 as follows:

$$df_0 = \gamma dA + C_1 dc_1 + C_2 dc_2 + \sum \mu_i dn_i - p_2 d\phi - p_1 d(1-\phi), \tag{2.14}$$

where C_1 and C_2 are the bending stresses associated with the curvatures c_1 and c_2, respectively; A is the interfacial area per unit volume between the two media of the microemulsion; μ_i and n_i are the chemical potential and the number of molecules of species i per unit volume, respectively; and ϕ is the volume fraction of the surfactant and co-surfactant molecules present at the interface of the globules. The chemical potentials were taken at pressure p_1 for the species present in the continuous phase and at pressure p_2 for those present in the dispersed phase. For spherical shape globules of radius r, $c_1 = c_2 = 1/r$, and $C_1 = C_2 = C/2$. Now using Eqs. (2.13) and (2.14), the following relation can be obtained for microemulsions containing spherical globules:

$$df = \gamma dA + C d(1/r) + \sum \mu_i dn_i - p_2 d\phi - p_1 d(1-\phi) + d\Delta f. \tag{2.15}$$

The equilibrium state of the microemulsion is completely determined by n_i, the temperature T, and pressure p. Therefore, the values of r and ϕ emerge from the condition that f be a minimum with respect to r and ϕ. For the spherical globules:

$$A = \frac{3\phi}{r}. \tag{2.16}$$

Therefore, one obtains

$$\gamma = \frac{r^2}{3\phi} \left(\frac{\partial \Delta f}{\partial r} \right)_\phi - \frac{C}{3\phi} \tag{2.17}$$

and

$$p_2 - p_1 = \left(\frac{\partial \Delta f}{\partial \phi} \right)_r + \frac{r}{\phi} \left(\frac{\partial \Delta f}{\partial r} \right)_\phi - \frac{C}{r\phi}. \tag{2.18}$$

An additional relationship between p_2 and p_1 is provided by the mechanical equilibrium condition between microemulsion and environment. Considering a variation of dV of the volume V of the microemulsion at constant N_i ($= n_i V$) and T, one can write

$$\gamma d(AV) + VC d\left(\frac{1}{r}\right) - p_2 d(V\phi) - p_1 d[V(1-\phi)] + d(V\Delta f) - p dV_e = 0, \tag{2.19}$$

2.2 Methods for preparation of bulk

where dV_e is the variation in the volume of the environment that is equal to $-dV$. Combining Eqs. (2.17) and (2.18), the following relation can be written:

$$\left(\frac{3\phi}{r}\gamma\right) - (p_2 - p_1)\phi + (p - p_1) + \Delta f = 0. \tag{2.20}$$

Equations (2.17) and (2.19) yield the following relation:

$$p_2 = p + \Delta f + (1 - \phi)\left(\frac{\partial \Delta f}{\partial \phi}\right)_r + \frac{r}{\phi}\left(\frac{\partial \Delta f}{\partial r}\right)_\phi - \frac{C}{r\phi} \tag{2.21}$$

and

$$p_1 = p + \Delta f - \phi\left(\frac{\partial \Delta f}{\partial \phi}\right)_r. \tag{2.22}$$

Equation (2.22) demonstrates that the pressure p_1 in the continuous medium of the non-interacting system with free energy f_0 differs from the pressure p, and that the pressure p is equal to the sum of the pressure p_1 and the osmotic contribution due to the free energy Δf. The pressure p acts both in the microemulsion and on the continuous medium of the microemulsion. A globule senses in its immediate vicinity the pressure p_1 of the non-interacting system plus the osmotic pressure due to the free energy Δf. Using Eqs. (2.17) and (2.21), the modified Laplace equation, which is valid for the microemulsion, can be written in the following form:

$$p_2 - p = \frac{2\gamma}{r} - \frac{C}{3\phi r} + \Delta f + (1 - \phi)\left(\frac{\partial \Delta f}{\partial \phi}\right)_r + \frac{r}{3\phi}\left(\frac{\partial \Delta f}{\partial r}\right)_\phi. \tag{2.23}$$

In addition to the first two terms on the right-hand side, which are present in the case of a single droplet in a continuum, Eq. (2.23) contains terms that result from the entropy of dispersion of the globules in the continuous medium and from the energy of interactions among globules.

In the case of the microemulsion that is in equilibrium with an excess dispersed phase, the composition of the excess dispersed phase is expected to be the same as that of the dispersed phase. Therefore, the chemical potentials of the excess dispersed phase and the dispersed phase are the same. The chemical potential of any species of the microemulsion is equal to its chemical potential in the non-interacting system. The chemical potential in the dispersed phase is expressed in Eq. (2.21) at pressure p_2, while the pressure in the excess dispersed phase is equal to the external pressure p to which both the microemulsion and the excess phase are subjected. Therefore, at equilibrium, p_2 should be equal to p. Consequently, Eq. (2.21) will take the following form:

$$\Delta f + (1 - \phi)\left(\frac{\partial \Delta f}{\partial \phi}\right)_r + \frac{r}{\phi}\left(\frac{\partial \Delta f}{\partial r}\right)_\phi - \frac{C}{r\phi} = 0. \tag{2.24}$$

The values of r and ϕ for a microemulsion in equilibrium with an excess dispersed phase can be obtained by using Eqs. (2.17) and (2.23) with known expressions for γ, C and Δf. From the empirical formula of Carnaham and Starling (1969) for hard spheres using molecular dynamic simulations as shown in Eq. (2.24), and combining with Eqs. (2.17) and

(2.23), the expressions for γ and C may be obtained:

$$\Delta f = -\frac{3\phi kT}{4\pi r^3}\left[1 - \ln\phi - \phi\frac{4-3\phi}{(1-\phi)^2} + \ln\left(\frac{4\pi r^3}{3v_c}\right)\right], \quad (2.25)$$

$$\gamma = \frac{kT}{4\pi r^2}\left[\ln\left(\frac{4\pi r^3}{3v_c\phi}\right) - \frac{8\phi - 5\phi^2}{(1-\phi)^2} + \phi\right], \quad (2.26)$$

$$C = \frac{3\phi kT}{4\pi r^2}\left[2\ln\left(\frac{4\pi r^3}{3v_c\phi}\right) + \frac{6\phi^2 - 5\phi - \phi^3}{(1-\phi)^2}\right], \quad (2.27)$$

where v_c is the volume of a molecule in the continuous phase. The values of the γ and C were estimated and plotted as functions of r by Ruckenstein (Ruckenstein, 1989; 1999). The values of γ and C were found to be very small with increasing values of r. For values of ϕ greater than 0.5, the interfacial tension exhibited negative values, and hence the interface between the microemulsion and the excess dispersed phase became unstable to thermal perturbations. It can be mentioned here that surfactant, co-surfactant, and hydrophobic phases play an important role on the values of γ and C; however, Eqs. (2.25) and (2.26) do not reflect the effect of these process variables. In reality, development of any mathematical relation that incorporates all these aspects is a remote possibility considering the degree of complexity of the microemulsion system. Any change in the thermodynamic variables results in the change in the particle size and shape of the system.

(c) *Reaction kinetics and mechanisms of particle formation in microemulsions*
The nanoparticles are prepared by using the microemulsion method in various ways such as by mixing of microemulsions containing the reactants necessary for the precipitation or reduction reaction. However, extra care should be taken to use an identical system of microemulsions for each precursor in order to avoid the complexities further. The two precursors come into contact with each other upon mixing owing to collision and coalescence of the droplets. This causes the reaction to take place. When the chemical reaction rate is high, the overall rate of the formation of the particles is expected to be controlled by the rate of coalescence of the emulsion droplets that contain the reactants. In such cases, the interfacial rigidity plays an important role in controlling the particle size. A relatively rigid interface decreases the rate of coalescence, hence leading to a slow reaction rate.

However, a less-rigid interface in the microemulsion enhances the rate of precipitation. It is quite likely that one can control the kinetics of the reaction by controlling the structure of the interface. The structure of the oil, surfactant, and co-surfactant, and the ionic strength of the aqueous phase can significantly affect the interfacial rigidity and the reaction kinetics (Hou et al., 1988; Chew et al., 1990). Figure 2.10 depicts the formation of particles by using two microemulsion droplets formed from an identical system, one carrying the precursors (R1) and the other containing the precipitation or reducing agent (R2). Surfactants play an important role in modifying the interfacial rigidity. The figure further indicates that the exchange of the R1 and R2 takes place after these two droplets come close together forming the nuclei. Such nuclei are denoted by the filled circles inside the droplets after their separation at the end of the exchange event as shown in Figure 2.10. Under certain

2.2 Methods for preparation of bulk 53

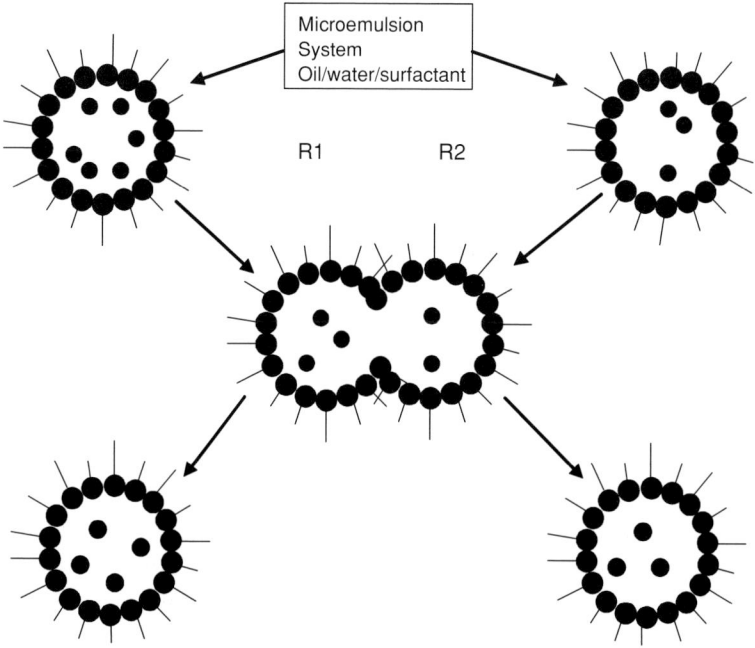

Figure 2.10. A schematic diagram showing nanoparticle synthesis using a water-in-oil microemulsion from the identical system containing the precursor.

circumstances, when the size of the nuclei is above the critical one, growth of such nuclei takes place during the next exchange event. The process repeats and the nanoparticles form. However, in reality the situation is more complex because of the influence of various factors like viscosity, interface rigidity, mobility of the droplets, etc. Several kinetic models have been proposed in order to explain the particle formation process in microemulsion systems (Fletcher *et al.*, 1987; Fletcher and Horsup, 1992; Hirai *et al.*, 1993; Natarajan *et al.*, 1996, and Tojo *et al.*, 1997).

Although a microemulsion is thermodynamically stable, the droplets are kinetically labile (Fletcher *et al.*, 1987). The droplet-size distribution is in a state of dynamic equilibrium. Fletcher *et al.* (1987) proposed that the mechanism of such a dynamic equilibrium involves coalescence of the droplets to form short-lived droplet dimers followed by their re-separation. The phenomena of intermicellar exchange and the formation of ultrafine particles have been studied through simulation and stochastic modeling by Natarajan *et al.* (1996). The study revealed an increase in the particle size with aqueous core size and total aqueous phase volume, and an increase in the particle number density with surfactant concentration at fixed aqueous phase to surfactant molar ratio. Bandyopadhyaya *et al.* (1997) put forward a model to calculate the size of fine particles synthesized by the precipitation process in reverse micelles.

Tojo and coworkers (Tojo *et al.*, 1998) studied the formation of particles in microemulsions by Monte Carlo computer simulation. They reported that uni-modal and bi-modal particle-size distributions might be obtained based on the surfactant film flexibility and concentration. The particle-size study on the reaction time suggested the occurrence of well-defined nucleation and growth process. Ingelsten *et al.* (2001) investigated the effects of surfactant and temperature on the kinetics of the formation of platinum nanoparticles in water-in-oil microemulsions. The surfactants used were poly(ethylene glycol) monododecyl ethers ($C_{12}E_4$, $C_{12}E_5$, $C_{12}E_6$), sodium bis(2-ethylhexyl) sulfosuccinate (AOT), and mixtures of the alcohol ethoxylates and AOT. The oil domain was *n*-heptane. The reaction rate for platinum particle formation was approximately the same in microemulsions based on either of the alcohol ethoxylates but considerably lower for microemulsions based on AOT. The average diameter of the platinum particles was found to be about 5 nm, which was consistent with the microemulsion droplet size.

Nanoparticles can also be synthesized by directly adding the reducing or precipitating agent in the form of a liquid solution or a gas to the primary reactant that is present in the microemulsion droplet. In this method, only one type of microemulsion is used instead of two. Moreover, the chances of particle growth in this system due to the collision of the microemulsion droplets are also minimized in this case because of the increased interface stability. Initially, the percolation of the precipitation or reducing agent will be accessible to the water core of the microemulsion. Subsequently, nuclei of the precipitate are formed inside the water core; eventually these grow till the precipitation or reduction reaction is almost completed inside the microemulsion droplet. Therefore the size of the grown particle will be limited by the size of the water droplet that acts as the nanoreactor in such cases. Often, the diameter of the particle formed inside the water droplet is almost equal to the size of the water droplet. The reaction scheme is shown in Figure 2.11. Such percolation of the precipitating or reducing agent has been reported for the synthesis of various metal-boride and metal-oxide nanoparticles (Nagy, 1999; Wu *et al.*, 2001). Recently Patil *et al.* (2002) synthesized 3–5 nm size ceria nanoparticles by using a microemulsion system consisting of sodium bis(2-ethylhexyl)sulfosuccinate (AOT), toluene and water. AOT was dissolved in toluene and an appropriate portion of aqueous cerium nitrate solution was added. The mixture was stirred for 30–40 min and ammonium hydroxide aqueous solution was then added drop by drop. The reaction was carried out for 1 h and then the reaction mixture was allowed to separate into two layers. The upper layer was toluene containing non-agglomerated ceria nanoparticles and the lower layer was an aqueous phase. The particle morphology was studied using high-resolution transmission electron microscopy (HRTEM) while the chemical-phase analysis was done using an energy dispersive spectroscopy (EDS) system. An HRTEM image of the nanoparticles of ceria is shown in Figure 2.12.

López-Quintela and Rivas (1993) have reported that ultrafine particles could be obtained with the help of a controlled chemical reaction in microemulsions. Controlling the size of the microemulsion droplet can control the size of such ultrafine particles. By using such a procedure, metallic Ni and Ni_3Fe alloy particles were successfully synthesized.

Porta *et al.* (2002) have prepared gold metallic nanoparticles in the micelle core of the clear w/o microemulsions of $NaAuCl_4$ and $NaBH_4$. The microemulsions were produced by using the four components cetyltrimethyl-ammonium bromide, 1-butanol, *n*-octane, and aqueous solutions of inorganic salts. Arcoleo and Liveri (1996) and Aliotta *et al.* (1995) have successfully synthesized gold nanoparticle aggregates using 0.0103 M tetrachloroauric acid ($HAuCl_4$), sodium bis(2-ethylhexyl)sulfosuccinate (AOT), and 0.0624 M hydrazine

2.2 Methods for preparation of bulk

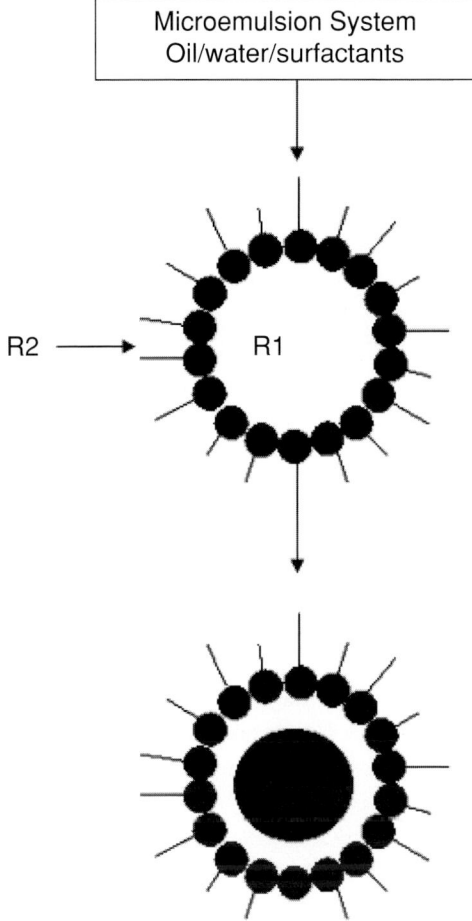

Figure 2.11. Schematic diagram showing nanoparticle synthesis using the single type of water-in-oil microemulsion droplets.

sulfate (N_2H_4, H_2SO_4), in an AOT/n-heptane solution (0.295 M). Gold nanoparticles were characterized with atomic force microscopy and it was found that the aggregates' sizes were within the range of 30–100 nm. Calorimetric measurements confirmed (Aliotta *et al.*, 1995) the formation of nuclei of gold nanoparticles and their subsequent aggregation in the microemulsions.

Qiu and co-workers (Qiu *et al.*, 1999) have synthesized spherical copper nanoparticles in a microemuslion system consisting of sodium dodecyl sulfate as a surfactant, isopentanol as a co-surfactant, cyclohexane as the oil phase, and aqueous $CuSO_4$ solution as the reactant solution, using $NaBH_4$ as a reducing agent. Copper particles with size of 3 nm were obtained with a molar ratio of water to surfactant of 2, whereas the size increased to 30 nm at the ratio value of 39.

Approximately 20 nm size crystalline Bi nanoparticles (Figure 2.13) were synthesized by Fang *et al.* (2000; 2001) using a reverse microemulsion technique. Bismuth(II) nitrate, sodium borohydrate ($NaBH_4$), and cyclohexane were used as the precursors. To prevent

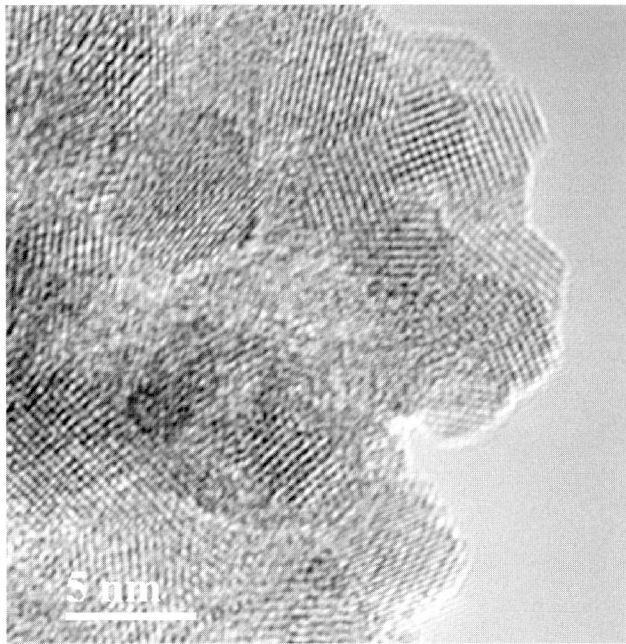

Figure 2.12. High-resolution transmission electron micrograph (HRTEM) image of the nanoparticles of ceria. The 5–7 nm ceria crystallites, shown as lattice fringes, are made through a microemulsion process. (Courtesy of Seal Research Group.)

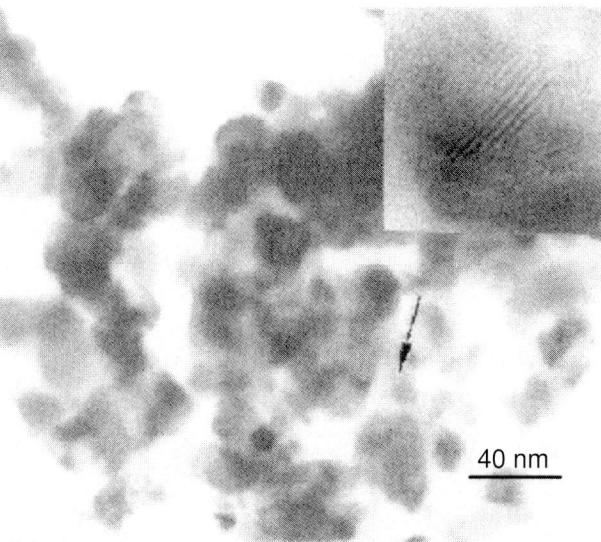

Figure 2.13. TEM image of Bi nanoparticles produced by microemulsion. Reprinted from (Fang et al., 2001), with permission from Elsevier.

oxidation of the Bi metallic particles, an *in situ* polymerization technique using methyl methacrylate as the monomer and 2-hydroxyethyl methacrylate as the co-monomer with crosslinking by ethylene glycol dimethacrylate. Such polymerization coating protects the bismuth nanoparticles from oxidation during post-synthesis annealing.

2.2 Methods for preparation of bulk

Qi et al. (1997) have synthesized well-dispersed copper nanoparticles by reduction of aqueous copper chloride solution using $NaBH_4$ in the nonionic water-in-oil microemulsion formed by Triton X100, n-hexanol, cyclohexane, and water. However, absorption spectra of such synthesized particles revealed the presence of a surface monolayer of the copper chloride on copper nanoparticles.

Iron particles with diameters of less than 100 nm were prepared by Rivas et al. (1993) using microemulsions consisting of an aqueous solution of $FeCl_2$, n-heptane, and AOT. The R values ($R = [H_2O]/[AOT]$) of 5, 10, and 15, along with various percentage values of the aqueous phase were used in the study to control the size of the Fe nanoparticles. The reducing agent employed in the synthesis was $NaBH_4$.

Arcoleo et al. (1998) characterized the Pd nanoparticles that were synthesized by using the microemulsion method in AOT and n-heptane along with an aqueous solution of $Pd(NO_3)_2$ and a hydrazine monohydrate as a reducing agent. Nanoparticles with a radius of 3–5 nm were obtained from this method with an excellent long-term stability of the size. Researchers have concluded that formation of such highly stable Pd nanoparticles was possibly attributed to the formation of a novel dimeric surfactant with the Pd nanoparticles.

In another study, Berkovich and Garti (1997) prepared colloidal Pd particles in the size range 1–4 nm in a water-organic microemulsion in the presence of the cationic surfactant aliquat 336 (trioctylmethylammonium chloride) in order to study the transfer hydrogenolysis of bromotoluene with sodium formate. Papp and Dékány (2001) synthesized nanoparticles of Pd by reduction of palladium acetate with ethanol in systems containing tetrahydrofuran (THF) and tetradodecylammonium bromide (TDAB) surfactant as stabilizer.

Pillai and Shah (1996) have synthesized cobalt ferrite ($CoFe_2O_4$) using water-in-oil microemulsions consisting of water, CTAB as a surfactant, n-butanol as co-surfactant, and n-octane as the oil phase. Hydroxides were precipitated in the aqueous cores of water-in-oil microemulsions; they were subsequently separated and calcined to give the magnetic oxide. XRD studies confirmed the cobalt ferrite phase with crystallite size less than 50 nm. These ferrite nanoparticles exhibited a very high intrinsic magnetic coercivity and saturation values of 1440 Oe and 65 emu/g, respectively. Pillai et al. (1993) synthesized barium ferrite ($BaFe_{12}O_{19}$) nanoparticles of less than 100 nm size. The process carbonates of Ba and iron were formed in the aqueous cores of the microemulsions consisting of CTAB, n-butanol, and the octane system, which are separated, dried and calcined to form nanoparticles of barium ferrite.

Boakye et al. (1994) reported the synthesis of molybdenum sulfide nanoparticles (high-temperature structural material) based on a microemulsion system that consisted of poly-oxyethylene(5)nonylphenyl ether (NP-5), cyclohexane, and an aqueous solution of acidified ammonium tetrathiomolybdate. The nanoparticles were found to be in the size range 10–80 nm. Nanoparticles of Fe_3O_4, $MnFe_2O_4$, $\gamma\text{-}Fe_2O_3$, and $CoFe_2O_4$ in the size range 10–20 nm were successfully synthesized by O'Connor et al. (1999), using the microemulsion procedure for lightweight high-strength magnetic applications.

In conclusion, the microemulsion process is an essential tool for the synthesis of nanoparticles of metals, alloys, and other compounds like metals, oxides, sulfides, halides and mixed oxides. Nanoparticles in the size range less than 5 nm with a controlled shape and size distribution have a wide range of applications, such as catalysis, nanoporous membranes, nanocomposites, and as precursor powders for functional alloys and ceramics. However, the problem of improving the yield remains to be solved. Moreover the particles are coated with surfactant, and it is not only difficult to remove the surfactant, but this also entails the problems of agglomeration and grain growth after the surfactant is removed. Therefore development of a viable approach for successful removal of the surfactant from the surface of the nanoparticles formed by the microemulsion process is needed in order to scale up the process to the industrial scale for production of the nanoparticles.

2.2.3.6 Other chemical methods for nanoparticle production

Iron nanoparticles have been made by ultrasonic irradiation of iron pentacarbonyl. Cao *et al.* (1995) developed a dull powder using sonochemical decomposition of $Fe(CO)_5$ in decane with varying solution concentrations. The powder size varies from 59 nm to 200 nm.

Several types of reducing agents have been used to produce fine metal powders from inorganic salts. Reducing agents such as $NaBEt_3H$, $LiBEt_3H$, and $NaBH_4$ are commonly used to make metal nanopowders. Mostly metal chloride salts of $CrCl_3$, $MoCl_3$, WCl_4 are commonly used. These salts are reduced in toluene solution with $NaBEt_3H$ at room temperature. This will form metal colloids in high yield (Zeng and Hampden-Smith, 1993). When the same metal chlorides are reduced in tetrahydrofuran (THF) solution with $LiBEt_3H$ and $NaBEt_3H$, metal carbides (M_2C) are formed with 95% yield. Both metal and metal colloids were shown to be composed of 1–5 nm sized particles by transmission electron microscopy. The proposed equation for the reduction is as follows, for $M = Cr$, Mo, $x = 3$; and $M = W$, $x = 4$:

$$MCl_x + x\,NaBEt_3H \rightarrow M + x\,NaCl + x\,BEt_3 + (x/2)\,H_2. \quad (2.28)$$

The chemical reaction may result from direct hydride transfer to form a metal-hydride intermediate, which subsequently will eliminate H_2, or may be reduced directly by an electron-transfer mechanism (Bönnemann *et al.*, 1990). Reduction of $MoCl_4(THF)_2$ or $MoCl_3(THF)_3$ in THF at room temperature will form a black colloid with H_2 gas evolution. The black colloid consists of 2–4 nm sized Mo_2C crystals. On heating the powder to 500 °C, the XRD peak sharpened and corresponded to that of Mo_2C (Figure 2.14).

Intermetallics are also fabricated by wet chemical synthesis similar to the production of metal powders. Mostly, reduction methods are adapted in preparing intermetallic powders. For example, Buhro *et al.* (1995) synthesized nanopowders of TiAl, $TiAl_3$, NiAl, and Ni_3Al by the reductions of $TiCl_3$ or $NiCl_2$ with $LiAlH_4$ in a mesitylene slurry followed by heating in the solid state (<500 °C). During the reaction process, Al and Ti, or Al and Ni are initially precipitated. Subsequent heating gives an exothermic reactive sintering process to produce nanocrystalline intermetallic powders of 25–40 nm.

Buhro *et al.* (1995) further produced nanointermetallics of Ti_2B, Ni_2B, WC–CO, Fe–B, and Ni–B using reduction reactions with $NaBH_4$ as a reducing agent. The byproducts were either dried at elevated temperatures or washed away by organic solvents. Various particle sizes and shapes were obtained by controlling the reaction conditions.

Bönnemann *et al.* (1994) prepared a whole array of intermetallic colloids via chemical reduction, shown in Table 2.3. Further, a wide variety of various intermetallic alloy synthesis procedures is listed in Table 2.4.

2.2 Methods for preparation of bulk

Table 2.3. *Preparation of nanocrystalline alloys using co-reduction of metal salts in THF*

Metal salt	Reducing agent	Condition t (h)	T (°C)	Metal content (%)	DIF D (Å)	Boron content (%)
RhCl$_3$ IrCl$_3$	LiBEt$_3$H	5	65	Rh: 33.5 Ir: 62.5	2.14	0.15
PdCl$_2$ PtCl$_2$	LiBEt$_3$H	5	65	Pd: 33.6 Pt: 63.4	2.25 1.96	0.04
PtCl$_2$ IrCl$_2$	NaBEt$_3$H	12	65	Pt: 50.2 Ir: 48.7	2.25 1.95	0.15
CuCl$_2$ SnCl$_2$	LiBEt$_3$H	4	65	Cu: 49.6 Sn: 47.6	2.96 1.80	0.00
FeCl$_3$ CoCl$_2$	LiBEt$_3$H	1.5	23	Fe: 30.1 Co: 31.4	2.02 1.77	0.00
Co(OH)$_2$ Ni(OH)$_2$	NaBEt$_3$H	7	65	Co: 48.3 Ni: 45.9	2.05	0.25
FeCl$_3$ CoCl$_2$	LiH+10%BEt$_3$	6	65	Fe: 47.0 Co: 47.1	2.02	0.00
FeCl$_3$ CoCl$_2$	LiBEt$_3$H	5	23	Fe: 54.8 Co: 24.5	2.02 1.17	0.00
CoCl$_2$ PtCl$_2$	LiBEt$_3$H	7	65	Co: 21.6 Pt: 76.3	1.93 2.23	0.00
RhCl$_3$ PtCl$_2$	LiBEt$_3$H	5	65	Rh: 26.5 Pt: 65.5	2.24 1.96	0.04

Reprinted from Bönnemann *et al.* (1994), with permission of Elsevier.

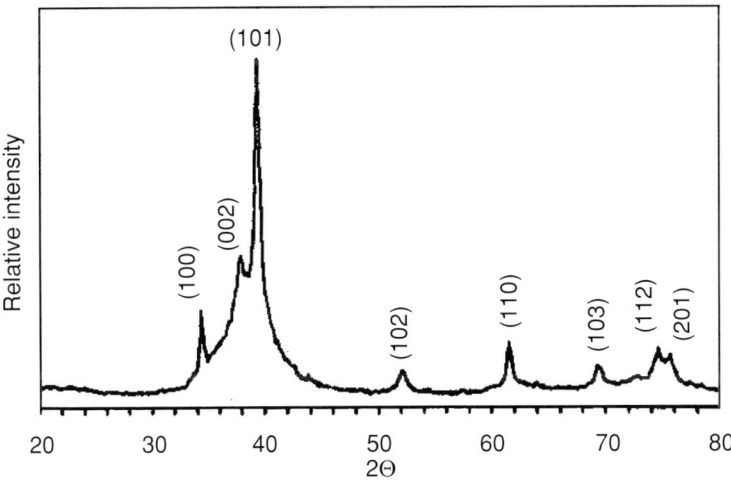

Figure 2.14. X-ray powder diffraction pattern of powders separated from reduction of MoCl$_3$(THF)$_3$ in THF after heating to 500 °C for 4 h in vacuum. (From Zeng and Hampden-Smith (1993), reproduced with permission.)

Steels are often considered as structural materials and have a wide range of applications. M50 is a class of steel important for the aircraft industry. It is used in the main shaft bearings in gas turbine engines, since M50 has excellent resistance to tempering and wear. The composition of M50 steel is 4% Cr, 4.5% Mo, 1%V, and Fe the rest. Conventional

Table 2.4. *Synthesis of nanointermetallics*

Number	Metal salt	Reducing agent	Condition t (h)	Condition T (°C)	Colloidal alloy solution color	Workup solvent	Solvent added for precipitation	Metal content in isolated colloid (%)	Mean particle size (nm)
1	$FeBr_2$ $CoBr_2$	$N(octyl)_4BEt_3H$	18	50	Dark brown to black. Fe and Co completely dissolved	Ethanol	Ether	Fe:13.36 Pt:14.39	2.3
2	$NiBr_2$ $CoBr_2$	$N(octyl)_4BEt_3H$	16	25	Dark red to black. Ni and Co completely dissolved	Ethanol	Ether	Ni:23.80 Co:23.80	2.8
3	$PtCl_2$ $CoBr_2$	$N(octyl)_4BEt_3H$	18	23	Deep reddish brown to black. Pt and Co completely dissolved	Toluene	Pentane/ethanol (25:1)	Pt:25.40 Co: 6.47	–
4	$PbCl_2$ $PtCl_2$	$N(octyl)_4BEt_3H$	16	23	Deep brown to black. Pd and Pt completely dissolved	Ether	Ethanol	Pd: 25.90 Pt: 33.60	2.8
5	$RhCl_2$ $PtCl_2$	$N(octyl)_4BEt_3H$	18	50	Deep red to black. Rh and Pt completely dissolved	Ether	Ethanol	Pd: 25.90 Pt: 33.60	2.3
6	$CuCl_2$ $PtCl_2$	$N(octyl)_4BEt_3H$	16	23	Deep red to black. Cu and Pt completely dissolved	Ether	Ethanol	Cu:15.60 Pt: 55.40	2.3

Reprinted from Bönnemann *et al.*, (1994), with permission of Elsevier.

2.2 Methods for preparation of bulk

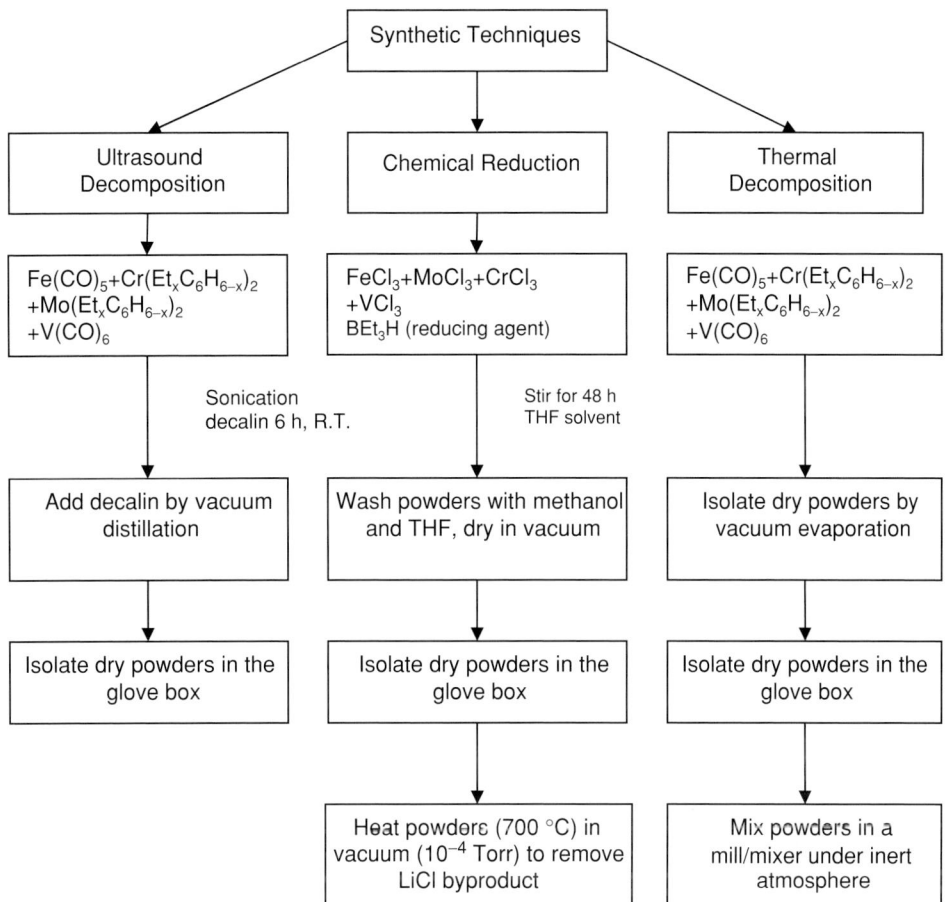

Figure 2.15. A schematic outline for the chemical synthesis of nanosteel powders. (From Gonsalves, et al. (1997), reproduced with permission.)

steels have grain sizes in micrometers. The presence of carbide particles act as fatigue crack initiation sites in the bearing materials (Kayser et al., 1952). If one can produce M50 steels with nanophase grains, the mechanical properties of these steels can be further improved.

Gonsalves et al. (1997) have utilized three different methods of producing the M50 steel nanopowders. Figure 2.15 illustrates the detailed synthesis outline. Following chemical synthesis, the powders are heated at 420 °C is an H_2 atmosphere. This will reduce the carbon and oxygen content prior to compaction.

Using a sonochemical reactor, a dispersion of $Fe(CO)_5$, $Cr(Et_xC_6H_{6-x})_2$, $Mo(Et_xC_6H_{6-x})_2$ and $V(CO)_6$ is allowed to react in decalin for 6 h at room temperature. The color of the solution turns black and it is sonicated till shiny metal particles are observed. The sonication is stopped and the decalin solvent removed via vacuum distillation. Fine black powder from the bottom of the reactor is isolated and stored under inert gas.

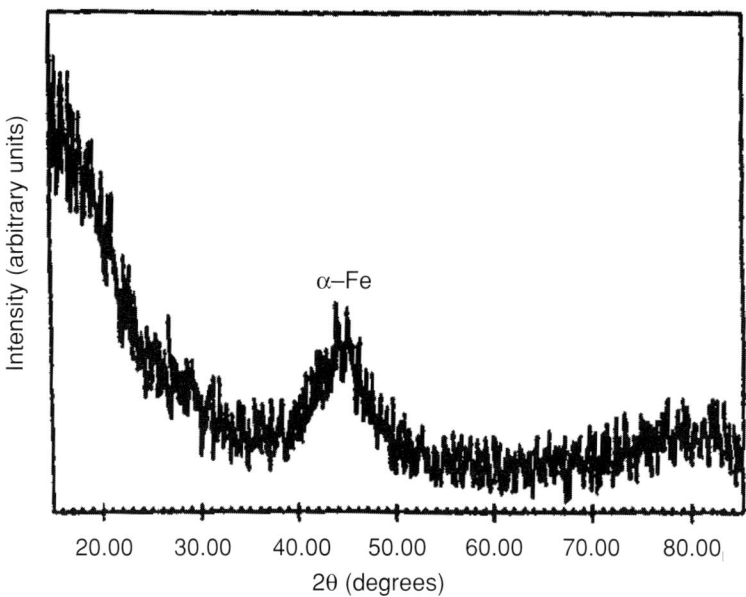

Figure 2.16. XRD pattern of a sonochemically prepared M50 steel powder. (From Gonsalves et al. (1997), reproduced with permission.)

The nanosteel powders are also prepared by a co-reduction method. A suspension of $FeCl_3$, $MoCl_3$, $CrCl_3$, and VCl_3 in THF is prepared in appropriate portions. Then 1 M lithium triethyl borohydride in THF is added followed by stirring at room temperature. The whole reaction is carried out in a glove box. After stirring the reaction for 48 h, a black suspension is obtained. The THF solvent is removed via vacuum distillation and the black powders are washed in degassed methanol and dried in vacuum.

Using thermal chemical decomposition, nano-M50 steel powders are created using a dispersion of $Fe(CO)_5$, $Cr(Et_xC_6H_{6-x})_2$, $Mo(Et_xC_6H_{6-x})_2$, and $V(CO)_6$ in a dry decalin. (Et denotes the ethyl ($\cdot C_2H_5$) radical.) The dispersion is refluxed for 6 h. The solution turns black within 3 h and the mixture is left to reflux until the formation of shiny metallic particles. The powders are dried in inert atmosphere and mixed in a mill for 8 h under an inert atmosphere.

Figure 2.16 shows the XRD pattern of a sonochemically prepared M50 steel powder showing amorphous and agglomerated phases. The peak at 44–45° is the reflection of α-Fe. The low-temperature chemical synthesis prevented grain growth and full recrystallization. A TEM micrograph shows ultrafine steel powders of 4 nm particles (Figure 2.17).

The morphology of the powders produced by thermal decomposition was examined by SEM and EDAX analysis. These powders show porous coral-like features as shown in Figure 2.18. EDAX analysis (Figure 2.19) from the powders shows a homogenous composition of the expected stoichiometry for conventional M50 steel.

In this section, we have outlined various wet chemical synthetic methods for the production of nanometals, oxides, intermetallics, alloys, and steels related to the application

2.2 Methods for preparation of bulk

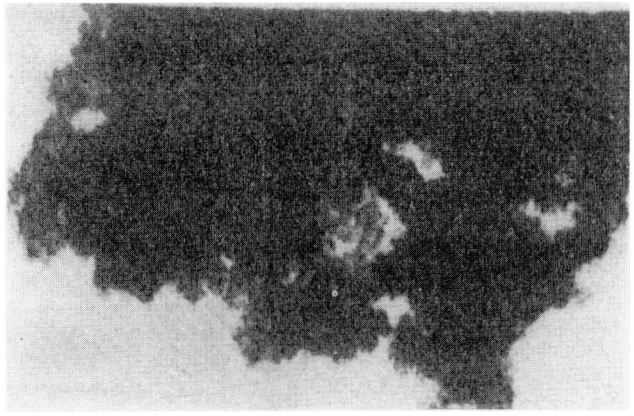

96 nm

Figure 2.17. A TEM micrograph shows ultrafine steel powders of 4 nm particles. (From Gonsalves *et al.* (1997), reproduced with permission.)

Figure 2.18. SEM picture of M50 steel powders produced via thermal decomposition. (From Gonsalves *et al.* (1997), reproduced with permission.)

of structural nanomaterials. Solution chemistry can be tailored to produce select particle size and chemistry via various reaction methods to obtain the final desired product. While yield and contamination remain a problem, agglomeration at the nanoscale is hard to avoid. Thus a new modification is needed to produce monodipersed nanoparticles. Fast-developing membrane-related research can be employed for size control and stabilization of nanopowders. Chemical synthesis of nanomaterials is a fast-growing area of research. This has great potential for fabricating technologically advanced and useful materials.

64 2 Processing of structural nanocrystalline materials

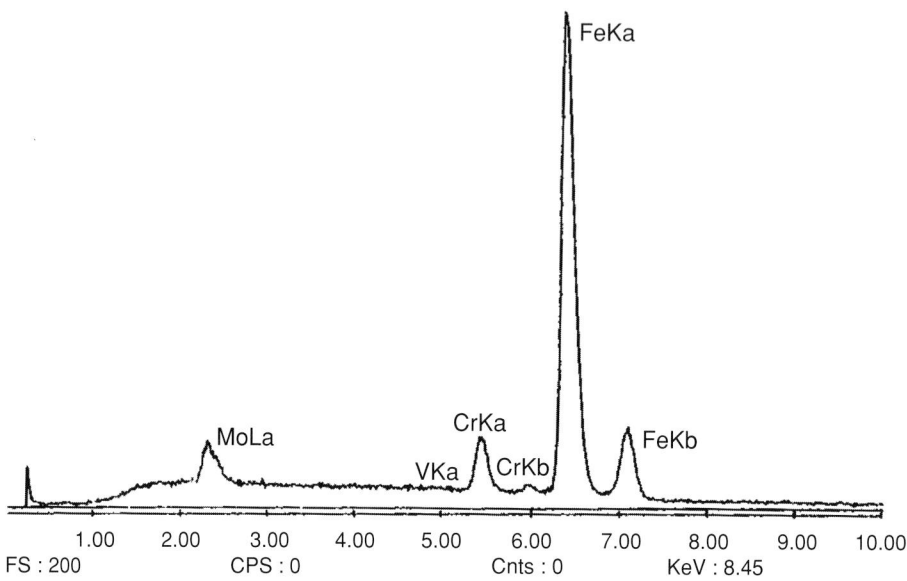

Figure 2.19. EDAX analysis of M50 steel nanopowders produced by thermal decomposition of chemical precursors. (From Gonsalves *et al.* (1997), reproduced with permission.)

2.2.4 Powder consolidation methods

For all the "two-step" processes for the formation of nanostructured materials, the first step provides either a nanoscale particulate, as in the inert-gas-condensation method, or many chemical reaction products, or a powder particle, often tens of microns in size, with a nanoscale microstructure such as in mechanical attrition. These particulates must then be consolidated into the bulk form in order to have bulk samples for mechanical testing and eventually for bulk parts for the applications of structural nanocrystalline materials. The consolidation problem remains an active area for more research and development and has not been adequately resolved to date. The problem is to form, typically by a combination of pressure and temperature, good atomic bonding between the particulates such that theoretical densities are reached along with the complete inter-particle bonding. This should be done without significant coarsening of the nanoscale microstructure or by introduction of any structural defects or unwanted phases. This topic has been reviewed, for example, by Groza (2002) and Mayo (1996). The densification of conventional powders is a well-studied area of materials science and engineering and comprises the processing methods of pressure compaction, sintering, and full-density processing methods (e.g. hot isostatic pressing, extrusion, etc.). The different physical behavior associated with the densification of nanoscale powders must be considered. The behavior of micron-scale powder with nanoscale microstructures also has the same problem of maintaining a nanocrystalline microstructure during consolidation. A special question regarding densification of nanoscale particulates is whether the physics of the sintering process is size-scale dependent. This has

2.2 Methods for preparation of bulk

particular importance for the densification of ceramic or other brittle particulates where the sintering process is most important and deformation by pressure plays a lesser role. The effect of surface impurities on the bonding/sintering of nanoscale particulates must also be considered because of the much higher surface area compared with conventional size powders.

The driving force for the densification of nanoscale particulates can be high because of the high surface energy due to the high surface area of the nanoparticles. This suggests that sintering can occur at much lower temperatures than in conventional (micron or larger) particle sizes. The enhanced kinetics for sintering of nanoparticulates is verified by experiments that show sintering starts in nanoparticles at temperatures of about 0.2–0.4 T_m (melting temperature) compared to about 0.5–0.8 T_m for conventional powders. Rapid surface diffusion may be responsible for the enhanced sintering kinetics but cannot explain the low-temperature densification alone. There is still uncertainty about the precise mechanisms for sintering in nanoscale powders and specific additional mechanisms such as grain-boundary sliding may also play a role. The activation energies determined for the early stages of sintering of nanoparticles are lower than those for conventional diffusion, even grain-boundary diffusion, so new mechanisms may be operative.

Surface impurities play a critical part in the densification of nanoparticles. In most cases, ultraclean surfaces allow for rapid sintering of nanoparticles, even at room temperature. Surface contamination usually suppresses sintering. There are, however, some examples where surface impurity layers can enhance sintering. An example of the latter effect is a layer of SiO_2 on nanocrystalline Si_3N_4 (Groza, 2002).

Particulate compaction at room temperature, which provides a sample with "green density," is the first step in conventional sintering practice and is important for success in compaction of nanoscale particulates. Homogeneous powder compaction with a small and uniform distribution of porosity is desired for subsequent sintering and good final densities. Agglomeration of the powders, which makes for non-uniform porosity and large pores, should be avoided. Because of frictional forces, green density decreases as particle size decreases, as illustrated in Figure 2.20. Therefore, micron-size powders with nanoscale internal grain structures, as produced by mechanical attrition, should be easier to densify than nanoscale particulates. Cold compaction of metal powders requires plastic yielding, therefore high pressures for strong nanocrystalline metals of the order of several gigapascals (GPa) are needed. Oxidation of the surfaces of fine powders can limit densification.

Warm compaction assists densification by removal of volatile contaminants and also, for metals, allows for more plastic deformation. Sanders *et al.* (1997) obtained densities >95 % in inert gas condensation (IGC) of metal nanocrystalline powders compacted at elevated temperatures, in the region of 375–575 K.

Certain ceramic nanopowders have been fully densified by conventional (pressureless) sintering; in some cases without excessive grain growth (Skandan, 1995). Conventional sintering has been less successful with nanocrystalline metals and intermetallics with substantial grain growth observed when good densities are obtained (Groza, 2002).

Pressure-assisted sintering aids densification while restricting grain growth. Shear stresses are effective in collapsing pores and also disrupt surface oxide layers. Since

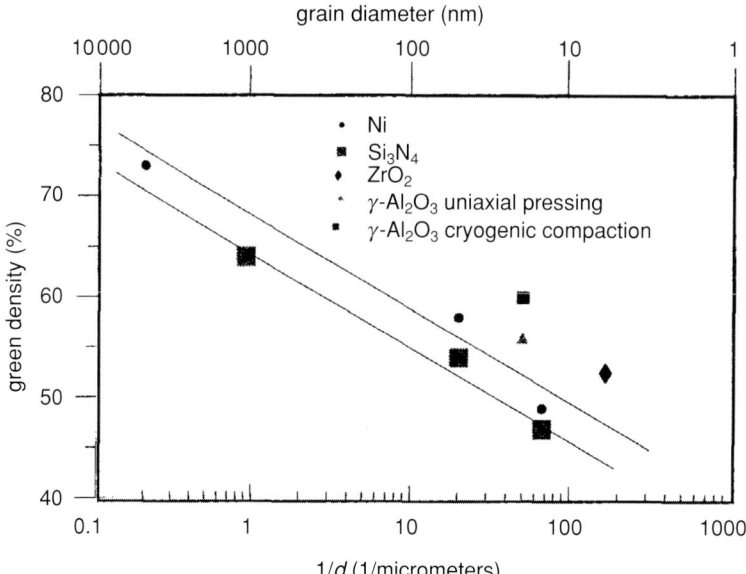

Figure 2.20. The effect of particle size on green density upon conventional dry compaction. (From Groza (2002), reproduced with permission from William Andrew Publishing.)

deformation processes that have significant shear stress components are desired, we can list the processes in order of decreasing effectiveness as follows: extrusion – sinterforging – uniaxial hot pressing – hot isostatic pressing (HIP). In nanocrystalline powders the effect of the applied stress is noticeable only if it exceeds the intrinsic curvature-driven sintering stress. Theoretical or near-theoretical densities have been obtained by hot pressing powders produced by mechanical attrition and with grain sizes maintained below 100 nm. One example of this is a 16 nm grain size in mechanically alloyed Fe–10%Al hot pressed at 823 K and heat treated for 1 h at 1223 K (Perez et al., 1996). The stabilization of the grain size was attributed to grain-boundary pinning by Al_2O_3 and AlN particles.

Non-conventional consolidation methods for densification of nanocrystalline powders include microwave sintering, field-assisted sintering methods, and shockwave consolidation. Microwave sintering of nanocrystalline ceramic powders reduces the processing time. However, densification of, for example, Al_2O_3 and TiO_2 was not improved over that obtained by conventional consolidation methods (Freim et al., 1994; Lewis et al., 1997). Field-assisted sintering has been carried out in several ways using electric or magnetic fields (Misra et al., 1996). The pulsed electric fields in plasma-activated sintering (PAS) enhances densification, presumably by promoting removal of surface oxides and adsorbates, thus enhancing particle bonding. Shockwave-consolidation methods have also been used such that peak pressures of tens of gigapascals can lead to densification by plastic yielding along with localized heating. Both metallic and ceramic powders have been densified in this way (Korth and Williamson, 1995).

2.2 Methods for preparation of bulk

2.2.5 Bulk processing methods for severe plastic deformation

The possibility of producing very fine grain structures by severe plastic deformation was suggested by research using conventional deformation methods taken to very high degrees of strain. It has been known for many decades, going back to the 1950s, that the structure of deformed metals can change with increasing plastic deformation such that random dislocation arrays can lower the energy of the system by "self-assembly" into "cells" or "subgrains" such that there is a high dislocation density in the cell walls and a lower dislocation density within the cells. The cells are typically the result of plastic deformation, the cell boundaries are somewhat diffuse, and strengthening due to cell structures gives an $m = 1$ dependence in the equation $\tau = \tau_0 + kGbd^{-m}$, where τ is the shear flow stress, G is the shear modulus, b is Burgers' vector, d is the cell size, and k and m are constants (Thompson, 1977). Subgrains, like cells, show small misorientations with their neighbors, but have sharper boundaries, and are formed by plastic deformation and thermal recovery processes. The strengthening gives an m value of 0.5, the same as for high-angle grain boundaries in the Hall–Petch equation. In most cases, the early studies of microstructures produced by severe plastic deformation gave cell or subgrain sizes in the micron down to sub-micron size scale, but not into the nanoscale. However, there have been some reports of severe plastic deformation by multiple pass rolling or drawing that produced nanoscale microstructures (Pavlov, 1985). For example, Pavlov has given evidence for the decomposition of the microstructure of Ni, Pt, and several Pt-based alloys first into "fragments" of about 15 nm diameter. The fragment size decreased further with continued deformation, and the misorientation between crystallites increased. Eventually, partial amorphization was reported. These results have not been reproduced to the present authors' knowledge. Rack and Cohen (1970) reported the cell structure developed in a series of Fe–Ti alloys by wire drawing to large values of true strain up to about 7. The size of the cells decreased with increasing deformation and reached values of about 50 nm at the highest strains. However, these were all cells with very low angle misorientations. In recent years special methods of mechanical deformation have been developed for producing sub-micron and even nanoscale grains with high-angle grain boundaries. These methods, the microstructures developed, and the properties of the materials with the refined grains so produced have been reviewed by Valiev et al. (2000).

The major methods reviewed (Valiev et al., 2000), and those that have received the most attention from researchers in the field, are severe plastic torsion straining under high pressure (HPT) and equal-channel angular pressing (ECAP). In the case of HPT a disk-shaped sample is compressed to pressures of about 2–6 GPa and then one of the dies is moved with respect to the other. With enough rotation, significant shear strains can be achieved. Even though it would be predicted that the strain distribution might be radial from the disk center, electron microscopy studies have shown relatively uniform microstructures across the disk samples in some cases. This is consistent with more-detailed calculations that take into account both the compressive and torsional stress states. While in most studies sub-micron grain sizes have been produced by HPT, in some cases nanostructured materials have been prepared. This method has also been successfully used for the consolidation of powders.

The ECAP method which allows for the deformation of bulk samples by pure shear was first developed by Segal (Segal et al., 1981). In this method a billet is pressed through a die with two channels at angles of intersection typically 90–120°. The billet is subjected to severe deformations without changing its dimensions. Multiple passes through the die provide accumulative strain. The grain sizes developed by this method are typically in the sub-micron (200–300 nm) range. A large body of experimental research and modeling studies has been developed for this technique, as reviewed by Valiev et al. (2000) and in subsequent journal articles and conference proceedings. There are examples of sub-micron-size grain structures induced by the severe strain of ECAP in several metals that provide an excellent combination of both increased strength along with good ductility.

Another bulk deformation method that has been applied to billet deformation is multiple forging. This was developed by Salischev and co-workers (Valiahmetov et al., 1990). It involves repeated free forging of a billet with changes of the axis of the applied stress and has been done at elevated temperatures in a variety of Ti, Ni, and Mg-based alloys. The mechanism of grain refinement has been attributed to dynamic recrystallization at the elevated temperatures of processing (Valiev et al., 2000). For example, Belyakov et al. (2000) have deformed 304 stainless steel by multiple compressions with the deformation axis changed after every pass to total strains of over 6 at 873 K. The formation of the sub-micron grains observed was attributed to a continuous dynamic recrystallization phenomenon.

Accumulative roll-bonding is a process that can be applied to the grain refinement of sheet samples and, in principle, scaled up for commercial application to large production quantities of sheet material. The process was first proposed by Saito et al. (1999), although similar methods had previously been used to prepare nanostructured or amorphous materials by cyclical rolling and stacking of sheets of dissimilar metals (Atzmon et al., 1984; Shingu et al., 1990). Accumulative roll-bonding involves the rolling of a sheet, cutting, cleaning the surfaces, stacking, and rolling the stacked pair to sufficient deformation (typically 50%) to attain good bonding. The process is then repeated and the strain so accumulated can reach large values, dependent only on the number of cycles used. Whereas the earlier studies using dissimilar metal sheets and foils resulted in nanocrystalline or often amorphous structures, the more-recent work applied to single-phase sheets has usually resulted in sub-micron rather than nanoscale grain sizes. However, recent work on dissimilar foils also exhibits nanostructured grain formation (see Battezzati et al., 1999). The number of rolling/stacking cycles in these studies of dissimilar metal composites was large, typically 60–75 times, while the reports of accumulative roll-bonding on single-composition sheets were only 7–8 times. Since the total strain is proportional to the number of rolling/stacking cycles, this may explain why nanostructured grains have been obtained in the dissimilar stacked foil experiments and only sub-micron grains in the single-composition studies. However, more recently, Dinda et al. (2005) have carried out rolling experiments on elemental metal foils that differed from the accumulative roll-bonding in that no lubricating agents or intermediate cleaning were used. However, very large total strains were achieved by up to 100 folding and rolling cycles. In this case nanostructured materials were formed, consistent

2.2 Methods for preparation of bulk

with the contention that high total strain is necessary to develop the nanocrystalline grain structure.

There have been several attempts to design and use alternative techniques for producing refined grain structures by severe plastic deformation. Among these are repetitive corrugation and straightening (Huang et al., 2001), friction stir welding (Benavides et al., 2000), multipass coin-forging (Ghosh and Huang, 2000), constrained groove pressing (Park and Shin, 2002), and twist extrusion (Beygelzimer et al., 2002). Repetitive corrugation and straightening involves the bending and straightening of sheet samples repetitively to build up significant plastic strain. The advantages cited for this method, which can be adapted to rolling-mill technology, are the creation of bulk sheet material with fine grains and free from contamination and porosity (Huang et al., 2001). A range of fine grains was observed in Cu subjected to repetitive corrugation and straightening, but the average grain size was sub-micron, not nanocrystalline. Friction stir welding carried out with the sample cooled to liquid-nitrogen temperature has resulted in sub-micron grain sizes in aluminum alloys (Benavides et al., 2000). The mechanism has been attributed to dynamic recrystallization. In multipass coin-forge processing a metal sheet surface is coined between two sine-wave-shaped dies, with successive rotation of the workpiece, followed by flat forging or rolling. The process is repeated until the deformed surface zones meet. Sub-micron grain sizes have been obtained. Constrained groove pressing (Park and Shin, 2002) is very similar in principle and results from multipass coin-forging. The concept of twist extrusion – extruding a prismatic billet through a die with a twist channel – has been proposed (Beygelzimer et al., 2002) but no experimental results have been presented.

It has been known for some time that the microstructures of surfaces subjected to wear processes can be similar to those observed by mechanical attrition; that is, nanostructured or sub-micron grain sizes (Fecht, 2002). Recently, K. Lu and co-workers have developed a method to produce controlled severe surface deformation on metallic surfaces (Tao et al., 1999). This method is ultrasonic shot peening. Steel balls (3 mm diameter) are vibrated at ultrasonic frequencies and impact the sample surface when resonated by a large number of balls over a short period of time. The surface layers of the sample are severely deformed and nanoscale microstructures are developed in the near surface regions. The refined grains become coarser as the distance from the surface increases. By using cross-sectional TEM the progress of the deformation-induced grain refinement can be followed. This surface mechanical attrition (SMA) technique promises to provide a surface treatment to enhance mechanical and possibly corrosion properties of metals so treated.

For practical production of structural nanostructured materials in industrial operations it would be desirable to use conventional deformation processing methods such as rolling, extrusion, or wire drawing followed by appropriate annealing treatments to give microstructures for optimized mechanical behavior. Although it has been demonstrated in the past (see Rack and Cohen, 1970) that severe conventional deformation processes can produce refined microstructures, these microstructures on the nanoscale have been cell or subgrain boundaries. High-angle grain boundaries are needed as a substantial percentage of the boundaries to take advantage of the nanocrystalline properties. There have been some recent results that indicate such hoped-for processes may be possible.

Wang et al. (2002) have reported the preparation and tensile testing of nanostructured/sub-micron grain-size Cu. They rolled the Cu to 93% deformation at liquid-nitrogen temperatures and then annealed it at low temperatures up to 200 °C. The original heavily cold-worked Cu had a high dislocation density along with some resolvable grains less than 200 nm in size. Annealing resulted in the development of well-defined grains with high angle boundaries. The annealing treatment (3 min at 200 °C) that optimized strength and ductility produced a mixture of nanoscale/ultrafine grains (80–200 nm) along with about 25% volume fraction of coarser grains (1–3 μm). The coarser grains were the result of secondary recrystallization. Rolling the Cu at room temperature did not produce the above effects, suggesting that biasing the accumulation of defects (dislocations) versus recovery by thermal processes by rolling at liquid-nitrogen temperatures was needed to obtain the required deformed structure as a precursor to the development of the fine-grained structures on subsequent annealing.

Another example of using conventional thermomechanical processing to obtain a mixture of nanostructured and sub-micron grain-size microstructures was given by Ueji et al. (2002). The authors used plain low-carbon steels that were austenitized and quenched to produce a martensitic structure. The as-quenched sheets were rolled to a modest strain of 0.8 and then annealed at temperatures of 673–973 K. The steel annealed at about 823 K exhibited a microstructure consisting of fine-grained ferrite with grain sizes of 50–300 nm along with uniformly distributed carbide particles. The as-quenched lath martensite in these low-carbon steels had a high dislocation density, which on subsequent modest plastic deformation by rolling apparently allowed a sufficiently high dislocation density to produce the fine-grained microstructure observed. These examples suggest that it should be possible to obtain nanostructured/sub-micron grain sizes in materials using conventional deformation processing methods combined with annealing in carefully chosen procedures for given materials. An initial high dislocation density appears to be a prerequisite for such behavior.

2.2.6 Electrodeposition

Electrodeposition is an old industrial process for providing mostly surface coatings such as protective galvanic zinc coatings for steel. However, it is a complex process and can produce, in the past probably inadvertently, nonequilibrium structures such as amorphous or nanocrystalline coatings. Since the late 1980s electrodeposition has been studied as a method for the preparation of nanocrystalline materials and it has moved into the commercial production of such materials. Much of this work was pioneered by Erb and co-workers and this subject has been reviewed by Erb et al. (2002). Using special processing methods (to be outlined below), a large number of metals, alloys, composites, and ceramics have been electrodeposited with nanocrystalline microstructures. Electrodeposition is another "one-step" processing method in that the coatings or free-standing foils are nanocrystalline and consolidation of particulates is not required. This eliminates one category of artifact, which is porosity or incomplete particulate bonding, that is common to "two-step" processes. However, other artifacts that can mask inherent behavior of nanocrystalline materials may still be present and will be discussed below.

2.2 Methods for preparation of bulk

Electrodeposition occurs by the nucleation of crystallites on the substrate surface and their subsequent growth along with nucleation of new crystallites. In order to have a nanocrystalline grain size, nucleation events should be favored over growth. The variables in electrodeposition include bath composition, bath pH, temperature, overpotential, bath additives, and direct current vs. pulse electrodeposition, etc. It has been stated (Erb *et al.*, 2002) that the two most important mechanisms that are rate-determining steps for formation of nanoscale grains in electrodeposition are charge transfer at the electrode surface and surface diffusion of adions on the crystal surface (Bockris and Razumney, 1967). One method for inhibiting growth of nucleated crystallites is by reducing the surface diffusion of adions by adsorption of foreign species (which may be referred to as "grain refiners" or simply "additives") on the growing surface. A large number of additives have been identified and used as grain refiners. These are typically organic materials such as saccharin, coumarin, thiourea, and polyacrylamide. They have been developed for specific systems and depend upon surface-adsorption characteristics, compatibility with the electrolyte, the temperature stability, etc. The problem with many of these organic additives is that they become trapped at the grain boundaries and are probably responsible for the brittle mechanical behavior observed in nanocrystalline materials processed by electrodeposition with additives. In the co-electrodeposition of alloys, the solute ion can act like an organic additive but without the deleterious embrittlement effects. The control of other deposition parameters such as bath pH, substrate rotation speed, current density, and bath temperature can also result in nanocrystalline deposits in some cases.

The other important process parameter that is critical in controlling the grain size is the overpotential. The overpotential, or "overvoltage," is the deviation of an electrode potential from its equilibrium value required to produce a net flow of current across an electrode/solution interface. Grain growth is favored at low overpotential and high surface diffusion rates while high overpotential and low surface diffusion favor extensive nucleation. A powerful method to achieve high overpotentials is by the use of pulse plating. In this case the peak current density can be much higher than the limiting current density attained for the same electrolyte during direct current plating. In direct current plating only one parameter can be varied, namely, the current density, whereas in pulse electrodeposition there are three parameters which can be varied independently: the peak current density (J_p), the pulse on-time (T_{on}) and the pulse off-time (T_{off}). Figure 2.21 is a schematic illustration of a pulse electrodeposition waveform showing the pulse parameters. In this figure, T_{on} and T_{off} are defined as the times during which the plating current is passing or interrupted, respectively. J_p is the maximum or peak current density during the pulse duration T_{on}, and J_m is the average current density. These additional plating parameters make possible the creation of a range of mass transport, electrocrystallization, and adsorption/desorption situations that are not otherwise possible in direct current plating (Puippe, 1986). In general it has been found that a high peak current density results in smaller grain sizes since the high electrode overpotential greatly increases the free energy available for the formation of new nuclei, resulting in higher nucleation rates and thus smaller crystal size. The influence of pulse on-time and off-time can not predict whether nucleation will be favored or not, and contradictory results have been observed in different systems. This is presumably

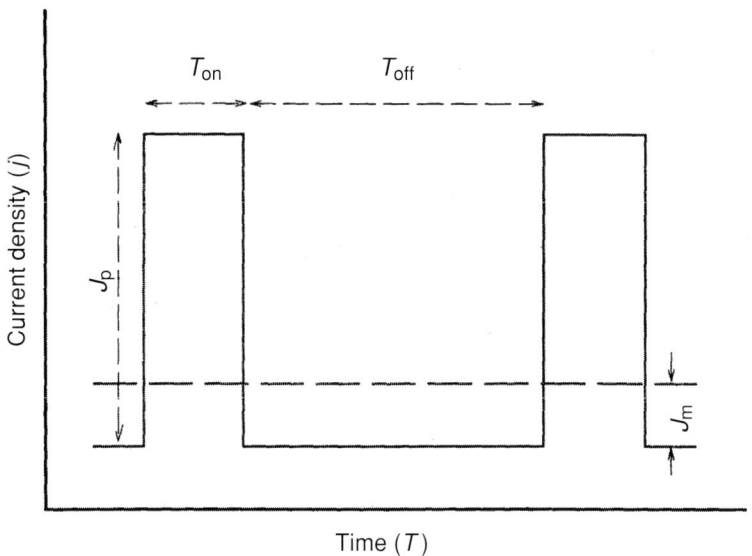

Figure 2.21. Schematic representation of pulse electrodeposition waveform showing the important pulse parameters. (Youssef (2003), reproduced with permission of the author.)

because the electrocrystallization process is strongly affected by adsorption and desorption processes, which are system-dependent.

While electrodepositon is a one-step process for producing nanocrystalline microstructures, artifacts are possible. Trapped impurities at the grain boundaries from bath additives, if used, can be a source of embrittlement. In some systems high internal stresses can be generated during electrodeposition which can cause cracking (Li and Ebrahimi, 2003). Therefore, while electrodeposition is a powerful method for producing nanocrystalline materials, the processing variables must be optimized to prevent artifacts.

2.3 Preparation of nanostructured, hard, and superhard coatings

Thin films and coatings of nanostructured hard and superhard materials have been prepared by a variety of deposition techniques, such as thermally or plasma-induced chemical vapor deposition (TCVD, PCVD), reactive sputtering, vacuum arc evaporation, laser ablation, and hybrid techniques consisting of a combination of these. Other techniques, such as electrodeposition, have also been used but mainly for the deposition of functional materials (Zhitomirsky *et al.*, 2002). Here we briefly discuss different methods for the preparation of hard and superhard nanostructured coatings, and, afterwards, the principle of the most frequently used deposition techniques together with their merits and demerits for the deposition of structural superhard nanocomposites.

Heterostructures and multilayers are prepared by sequential deposition of two materials with different elastic moduli. Because a sharp interface between the layers consisting of the different materials is necessary in order to avoid decrease of the hardness at a very

2.3 Preparation of nanostructured, hard, and superhard coatings

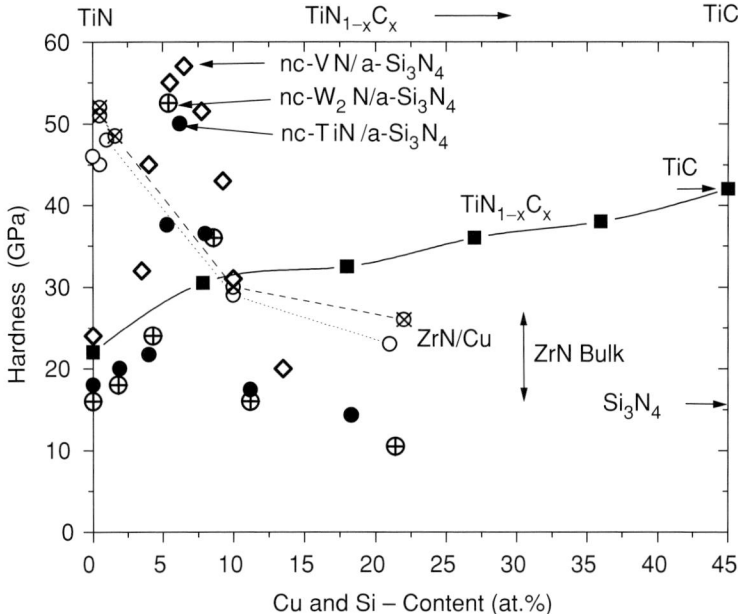

Figure 2.22. Dependence of the hardness of superhard nanocomposites nc-M_nN/a-Si_3N_4 on silicon content in comparison with the ZrN/Cu coatings in which the hardness enhancement was achieved by energetic ion bombardment during their deposition at a relatively low temperature. The hardness of the substitutionally miscible titanium carbonitride follows the rule-of-mixtures (Veprek et al., 2005a).

small period of <3–5 nm (Chu and Barnett, 1995; Barnett and Madan, 1999; Barnett et al., 2003), two components with a high immiscibility have to be chosen. The best control of the interface provides reactive sputtering from two separate cathodes with shutters (Barnett, 1993). This technique is, however, not suitable for large-scale industrial coating of tools, where substrate holders with a planetary motion are rotated in a chamber which is equipped with four or more cathodes consisting of the two different materials (Münz et al., 2001; Münz, 2003). Because the transition of the substrates being coated from one cathode to the other one is rather smooth in this case, the interface is not sharp.

As mentioned in Chapter 1, superhardness can be achieved in ceramic coatings either by energetic ion bombardment during the deposition or by the formation of a stable nanostructure by spinodal phase segregation. We are primarily interested in the second class of nanocomposites because they have a high thermal stability.

Figure 2.22 shows how one can distinguish between the different mechanisms of hardness enhancement. In the case of the stable nanocomposites nc-M_nN/a-Si_3N_4 (M = Ti, W and V), the hardness shows a pronounced maximum with values significantly higher than that of the individual nitrides, at a silicon content of about 6–8 at.%. In contrast, for M(1)N/M(2) coatings consisting of a hard transition-metal nitride and ductile metal (e.g. ZrN/Cu), the maximum hardness enhancement is achieved for the smallest content of that metal. A similar or even higher hardness enhancement is achieved for pure transition-metal nitrides, borides, and carbides; thus, the hardness enhancement reported by Musil (2000) is not related to any nanostructure (Veprek, 1999; Veprek *et al.*, 2005a, and references therein). With increasing content of the metal the hardness enhancement decreases. The hardness of titanium carbonitride, TiC_xN_{1-x}, which forms a stable, substitutional solid solution, follows the rule-of-mixtures. Another difference between these two types of nanocomposites is their temperature stability, as will be discussed in Chapter 3. For achieving the hardness enhancement in the M(1)N/M(2) coatings a sufficiently low deposition temperature of ≤ 350 °C has to be used. In contrast, the formation of the stable nanostructure in the superhard nanocomposites nc-M_nN/a-Si_3N_4, nc-M_nN/a-BN and others during the deposition requires a sufficiently high activity (partial pressure) of nitrogen and fast diffusion, because the spinodal decomposition is thermodynamically driven and kinetically controlled (Veprek and Reiprich, 1995). More recent calculations have shown that this is fulfilled for a nitrogen pressure of $\geq 2 \cdot 10^{-3}$ mbar and deposition temperature of 550–600 °C, and that this system is indeed spinodal (Zhang and Veprek, 2006). The spinodal nature of the phase segregation and resulting high thermal stability will be discussed in more detail in Chapter 3. Here we just emphasize that, if either a too low nitrogen pressure or deposition temperature is used, the formation of the fully developed nanostructure will not be completed during the deposition. In spite of that, one may achieve a hardness enhancement due to the energetic ion bombardment. However, this will be lost whenever such a coating is exposed to a somewhat higher temperature of >450 °C. With this in mind, we shall now discuss the deposition techniques in view of their suitability for the preparation of the stable, superhard nanocomposites.

2.3.1 Thermal CVD

The majority of thermal CVD of hard coatings uses as reactants halides in combination with hydrogen and additional reactive gases, such as nitrogen, for the deposition of nitrides. Let us illustrate the problems associated by the example of the deposition of TiN, which is an important hard coating for machining tools as well as a scratch-resistant, decorative coating because of its gold-like color and hardness of 20–22 GPa. We consider the chemical reaction

$$TiCl_4 + 2H_2 + 0.5N_2 = TiN + 4HCl. \qquad (2.29)$$

Because at low temperatures the chemical equilibrium is on the side of the chlorides, high temperatures of 1000 °C are needed to shift the equilibrium towards TiN and HCl. Similar conditions apply to the deposition of TiC and TiC_xN_{1-x} (see Veprek, 1983). Using ammonia, lower temperatures are sufficient, but the ammonia complicates the process and there is a large concentration of chlorine incorporated into the deposited film.

Because at such high temperatures the diffusion in the solid is fast and miscibility increases, nanocomposites can be prepared only in refractory systems which display a large immiscibility. An example is the TiN/Si_3N_4 system in which the stoichiometric nitrides are immiscible. Because each of these nitrides has, at a given temperature, a well defined saturation pressure of nitrogen (Chase *et al.*, 1985; Rogl and Schuster, 1992), the stoichiometry

2.3 Preparation of nanostructured, hard, and superhard coatings

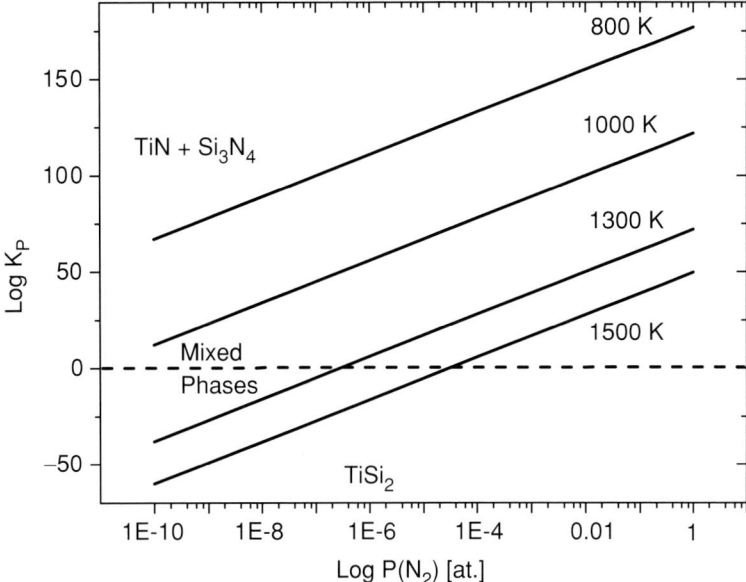

Figure 2.23. Dependence of the logarithms of the equilibrium constant of reaction (2.30) on the nitrogen pressure for different temperatures (Veprek *et al.*, 2005b). The free energies of formation were taken from Chase *et al.* (1985) and Barin (1993).

and concomitant immiscibility always requires the system to be under at least that equilibrium pressure. Otherwise, the nitrides will become substoichiometric and at least partially miscible. This is illustrated in Figure 2.23 for the reaction (2.30):

$$6\,TiSi_2 + 11\,N_2 = 6\,TiN + 4\,Si_3N_4. \tag{2.30}$$

Hirai and Hayashi reported on the deposition of TiN/Si_3N_4 composites from $TiCl_4$, NH_3, and H_2 in which TiN was always crystalline and Si_3N_4 either amorphous or crystalline, depending on the deposition temperature which was varied between 1050 and 1450 °C (Hirai and Hayashi 1982; 1983). These authors did not report about the mechanical properties of the deposited films. Judging from the micrographs published in their papers one estimates that the crystallite size was in the range of 10–100 μm. This shows that even in such a system which is immiscible at high temperature, superhard nanocomposites cannot be formed owing to rapid diffusion and growth of the nuclei.

An alternative possibility of reducing the deposition temperature in order to obtain nanocomposites in these systems would be the use of organometallic reactants. However, this will limit the range of possible deposits to carbides and carbonitrides because carbon can substitute nitrogen in many hard transition-metal nitrides, and carbon is miscible with silicon, forming SiC and $Si_xC_yN_z$ mixed compounds. Although the latter are very interesting hard and strong materials with a high thermal stability, they will not be discussed here because they are not nanocomposites but amorphous solid solutions.

2.3.2 *Low-pressure, glow discharge plasma-induced CVD*

The low-pressure, non-isothermal plasmas of glow discharges will be considered here, and the applications of thermal plasmas for the preparation of the nanocomposites will be discussed in the following section. The effect of a non-isothermal glow discharge plasma on a chemical system can be threefold (Veprek *et al.*, 1971; Veprek, 1972; 1982; 1983).

(a) A weak discharge with a low degree of dissociation but a sufficient concentration of energetic electrons and ions can catalyze reactions which are possible thermodynamically but hindered due to a high activation energy. An example is the deposition of silicon from silane, whose free energy of formation is positive (Wagner and Veprek, 1982; Veprek, 1982).

(b) The high internal energy of an intense glow discharge plasma with a high degree of dissociation can significantly shift the chemical equilibrium thus enabling strongly endothermic reactions to occur at relatively low temperature. AlN and TiN single crystals were deposited in a chloride/nitrogen system even without using hydrogen, i.e. the equilibrium constant was shifted by many orders of magnitude (Veprek *et al.*, 1971; Veprek, 1980).

(c) When the growing film is exposed to energetic ion bombardment, i.e. a negative bias of -50 to a few hundred volts is applied, the crystallite size is decreased, the surface of the film is smoothed, the stoichiometry can be altered owing to preferential sputtering, and compressive stress can be built in the growing film (see Veprek, 1980).

In the majority of plasma CVD experiments, most of these effects play a combined role, and their individual contribution to the final properties of the films is difficult to exactly separate. The silicon/hydrogen system is a case example where these effects of the plasma were shown in illustrative separate experiments. The reader may consult the few selected references for further information regarding the possible effects of plasmas in CVD (Veprek *et al.*, 1971; Veprek, 1980; 1983). Here we concentrate on systems which are relevant for the preparation of superhard nanocomposites by means of plasma CVD.

When hydrogen is used as in reaction (2.29), stoichiometric TiN coatings with chlorine impurity content below 0.7 at.% can be obtained at sufficiently high deposition rates when an intense plasma and deposition temperature of 500–550 °C is used (Patscheider *et al.*, 1996). In a weak discharge, however, the chlorine content exceeds 2 at.%, which is too high for a long-term stability of such coatings (see Patscheider *et al.*, 1996; Li *et al.*, 1992). Therefore, a sufficiently high deposition temperature and intense plasma are necessary in order to obtain stoichiometric coatings with chlorine content below 0.7 at.% (Veprek and Reiprich, 1995; Veprek *et al.*, 2000), which, as known from industrial practices, guarantees a long-term stability.

Similar conditions apply for the deposition of the nanocomposites. However, as mentioned above, one has, in addition to the thermodynamic conditions, also to assure the kinetic conditions for the phase formation and segregation. The role of a sufficiently high deposition temperature was already discussed. Moreover, a sufficiently high plasma density is a prerequisite for fast chemical kinetics to form the stoichiometric nitrides. This is illustrated

2.3 Preparation of nanostructured, hard, and superhard coatings

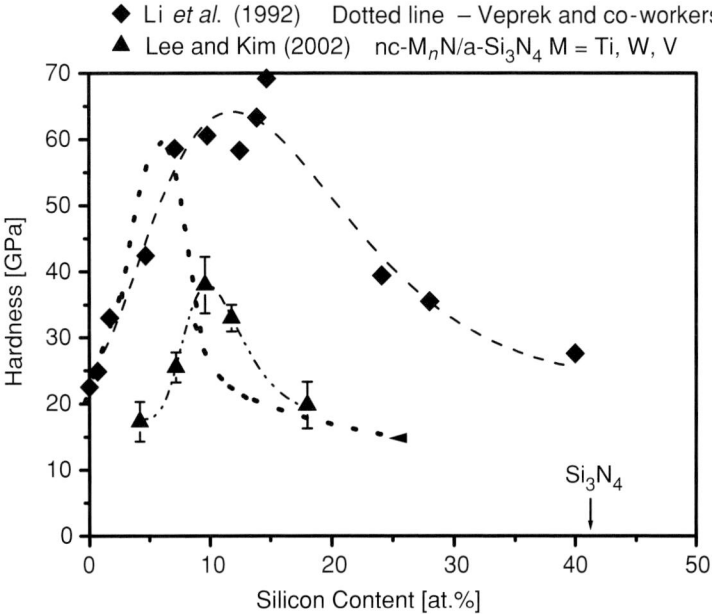

Figure 2.24. Dependence of hardness on silicon content in superhard nc-TiN/a-Si$_3$N$_4$ nanocomposites and in "Ti–Si–N" coatings deposited by plasma CVD (Veprek et al., 2005a).

by Figure 2.24, which shows a comparison of the dependence of the hardness on silicon content in the coatings deposited by means of plasma CVD, as reported in three different papers. The superhard quasibinary nc-M$_n$N/a-Si$_3$N$_4$ nanocomposites show a well-defined maximum of hardness at about 7–8 at.% of silicon whereas the maximum is broader and shifted to a higher Si content in the "Ti–Si–N" coatings of Li et al. (1992). This is because Li et al. used a weaker discharge (electric current density at the substrate of <1 mA/cm^2) than Veprek et al. (2.5–3 mA/cm^2 in a d.c. discharge or a high power density in an r.f. discharge). These parameters are not specified in the work of Lee and Kim (2002), but judging from the overall description of the deposition conditions specified there, the plasma density of their inductively coupled radio frequency (r.f.) discharge was also low. At such a low plasma density the kinetics of the reaction are too slow and the system does not reach the equilibrium composition of stoichiometric TiN and Si$_3$N$_4$ because TiSi$_x$ is also formed. Because the maximum hardness is achieved when the surface of the TiN nanocrystals is covered by about one monolayer of Si$_3$N$_4$ (see Veprek and Reiprich (1995) and Chapter 4 in this book), more silicon is needed to achieve this coverage if there are more silicon containing phases present in the coating.

The lower value of the maximum hardness found by Lee and Kim and shown in Figure 2.24 is most probably caused by impurities. This issue will be discussed in more detail in Chapter 4. Therefore here we shall emphasize only that oxygen impurities of more than 0.05–0.1 at.% strongly degrade the hardness and other properties of these nanocomposites (Veprek et al., 2004; 2005a; 2005b). The high purity that is required for the superhardness is difficult to achieve in many of the deposition techniques and equipments.

2.3.3 Deposition by means of thermal plasmas

As a matter of fact, thermal plasmas, which operate at gas temperatures of many thousand degrees Celsuis, suffer problems similar to thermal CVD. However, since the gas-flow velocity in typical plasma jets reaches several hundred meters per second, solids can be treated via the control of the dwell time of the reactants and particles in that plasma (see Sayce, 1975; Pfender, 1985; Akashi, 1985; Heberlein, 2002). Because the resultant nano-sized particles have a rather large size distribution, particle size filters were used in order to obtain an almost monodispersed deposit (Girshick *et al.*, 2002; Hafiz *et al.*, 2005). The characteristic of thermal plasma jets is a high deposition rate at a relatively small lateral area and high quenching rates. At a reduced pressure, this area is somewhat larger, but scanning of the plasma jet over the substrate is necessary if a larger area should be coated uniformly.

For the deposition of "Si–Ti–N" coatings, Hafiz *et al.* used $TiCl_4$, $SiCl_4$, and NH_3 as reactants which are dissociated in the hot plasma of a direct current (d.c.) arc and rapidly quenched in a fast nozzle expansion (Hafiz *et al.*, 2005). Solid particles of the size of about 20 nm nucleate upon this quenching, and they are accelerated in the hypersonic gas flow. The deposition on a substrate placed downstream of the nozzle occurs by ballistic impaction at a velocity of about 1700 m/s. This makes the process different from the other deposition techniques, such as thermal and plasma-induced CVD, sputtering and vacuum arc evaporation, where care is taken to avoid nucleation in the gas phase.

Using the thermal plasma hypersonic plasma particle deposition operating with a 45 kW d.c. plasma torch and gas flow rates of 30–37.5 slm (standard liters per minute), Hafiz *et al.* deposited coatings of thickness 10–25 μm at deposition rates of 2–10 μm/min. The highest hardness of 24 GPa was achieved at a deposition temperature of 820 °C, but only 15 GPa was achieved at 210 °C. XPS and XRD analyses revealed the presence of nanocrystalline TiN, $TiSi_2$ and Si, and amorphous Si_3N_4. Because the as-deposited coatings had some degree of porosity, plasma sintering of the deposit was applied after the deposition. After 5 min sintering in argon plasma at a temperature of 800 °C the density increased by about 13% to $9.82 \cdot 10^{22}$ atoms/cm^3 (Hafiz *et al.*, 2005), which compares with the theoretical density of stoichiometric TiN of $10.15 \cdot 10^{22}$ atoms/cm^3.

The researchers did not report on any large-area coatings, nor about any other mechanical properties. The relatively large surface roughness as seen in the SEM micrographs (see Figure 2.2 in Hafiz *et al.* (2005)) and small hardness call for further improvement of this deposition technique if it is to be compatible with the others currently used ones.

2.3.4 Sputtering

In the sputter-deposition process a solid material is vaporized by momentum transfer from ions impinging on that target with kinetic energy of several hundred to a few thousand electronvolts (Behrisch, 1981; 1983; Behrisch and Wittmark, 1991; Thornton and Greene, 1994; Musil *et al.*, 1993). The atoms and small clusters sputtered from the target are allowed to deposit at a substrate whose temperature and negative electrical bias are controlled. One distinguishes between simple physical sputtering in an inert gas, such as argon, and reactive

2.3 Preparation of nanostructured, hard, and superhard coatings

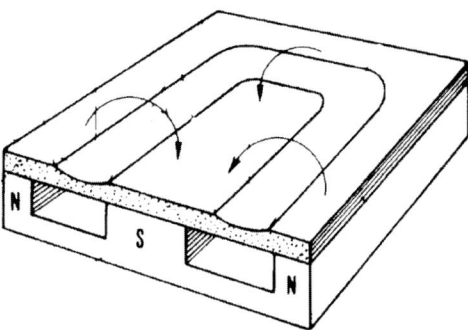

Figure 2.25. Schematics of a magnetron target which is used for sputtering. The arrows show the magnetic field lines.

sputtering where the depositing atoms react with a reactive gas, such as oxygen or nitrogen, in order to form oxides or nitrides. The electric power delivered to the target can be a d.c., pulsed d.c., bipolar pulsed, or r.f. The advantage of the r.f. sputtering is the possibility also of sputtering electrically insulating materials because the displacement r.f. current with frequency between about 1 and 20 MHz that flows through the insulator can build up a sufficient d.c. self-bias (Chapman, 1980). Also bipolar pulsed d.c. discharges can be used for that purpose because the positive and negative charges that build up at the surface of the target within the respective single pulse are compensated for by the following pulse of the opposite polarity.

The most common arrangement of the target which is used in research and industrial production is the magnetron shown schematically in Figure 2.25. The electrons are trapped into trajectories circulating around the magnetic field lines. This makes their effective length of motion much longer than the dimension of the target. Furthermore, because the magnetic field is stronger at the surface of the target than elsewhere, it acts as a magnetic mirror which repels a part of the electrons back. These two effects in combination increase significantly the effective length of the electron movement and, consequently, the probability of electron impact ionization of the gas. As a result, the plasma density and the resultant current of the energetic ions to the target are much larger and the maintenance voltage of the glow discharge is much lower than in the case of a simple, planar cathode.

Because, for a given material, the sputter yield increases with increasing mass of the ions (Behrisch, 1981; 1983; Behrisch and Wittmark, 1991), argon is typically used as the sputtering gas at an operating pressure in the range of about $5 \cdot 10^{-4}$ to $2 \cdot 10^{-3}$ mbar, and the reactive gas is just added to it.

A typical example of the application of reactive magnetron sputtering for the preparation of the "Ti–Si–N" nanocomposite coatings is the work of Vaz and co-workers (Vaz et al., 1998; 1999; 2000; 2001; 2002), whereas the preparation of the "Ti–Al–Si–N" coatings by magnetron sputtering was reported by, among others, Carvalho et al. (2004); this technique are already used in large-scale industrial production. Another example is the work of Diserens et al. (Diserens et al., 1998; 1999; Patscheider et al., 2001). Both d.c. and r.f. discharges were used in these and other papers. In the majority of the papers the deposition was done from Ti and Si targets (in the work of Carvalho et al. on the "Ti–Al–Si–N" an

80   2 Processing of structural nanocrystalline materials

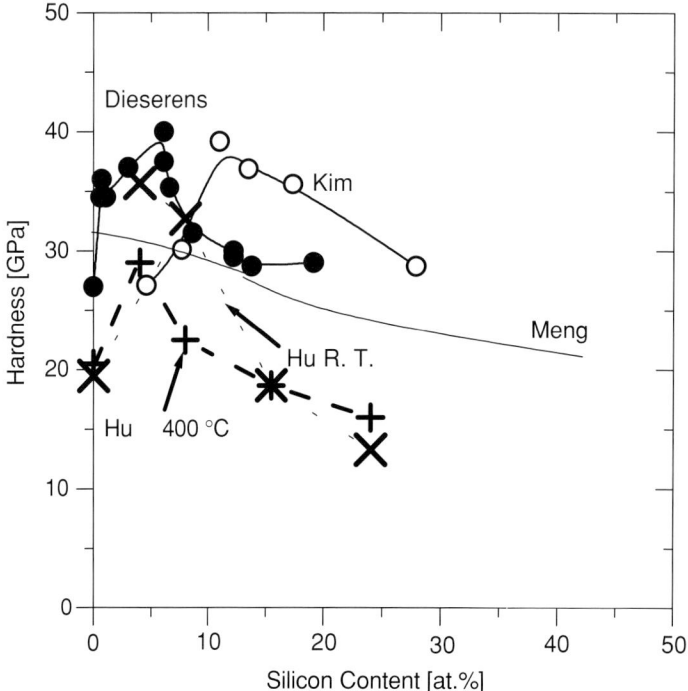

Figure 2.26. Example of differences in the properties of the "Ti–Si–N" coatings prepared by different groups by means of reactive sputtering (Veprek et al., 2005a). (For the original data see: Diserens et al., 1999; Kim et al., 2002; Hu et al., 2002; Meng et al., 2002.)

Al target was also used). Because of the relatively low electric conductivity of pure Si, r.f. sputtering was preferably used on that target.

The results obtained by different groups show relatively large scattering as illustrated in Figure 2.26 (Veprek et al., 2005a; 2006). The maximum of the hardness appears at different silicon content, the maximum achievable hardness is different in different papers and there is no hardness enhancement found in the work of Meng et al. (2002). Instead, the reported hardness decreases from about 32 GPa for pure TiN to about 21 GPa for Si_3N_4. One notices that the reported hardness for pure TiN is much higher than the correct value of the bulk hardness of about 20–22 GPa (Holleck, 1986). This suggests that at least a part of the hardness enhancement is caused by energetic ion bombardment. A similar conclusion applies to the results of Diserens and co-workers, who reported hardness of TiN of 27 GPa. This possible explanation is further supported by the results of Hu and co-workers, who found a higher hardness in coatings deposited at room temperature (see "Hu R.T." in Figure 2.26) than in those deposited at 400 °C. Considering the contribution to the hardness enhancement by the ion bombardment, the actual hardness enhancement due to the nanocomposite formation should be only between about 30 and 35 GPa in the work of Diserens and co-workers. We shall see in Chapter 4 that such a corrected value is in the range expected for such coatings with oxygen impurity content of ≥ 0.5 at.%. The reason for the differences in these and many other papers is either a too low deposition temperature, or

2.3 Preparation of nanostructured, hard, and superhard coatings 81

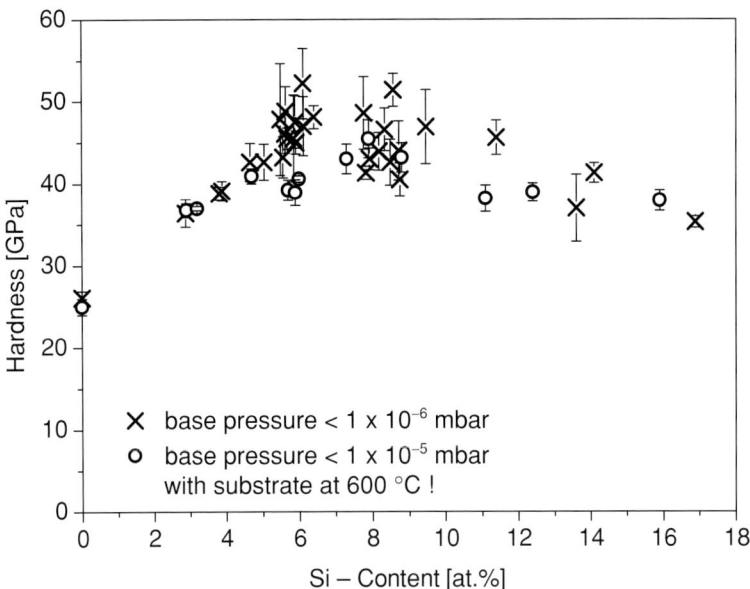

Figure 2.27. Hardness vs. silicon content in nc-TiN/a-Si$_3$N$_4$ superhard nanocomposites deposited by means of reactive sputtering (Prochazka *et al.*, 2004; Veprek *et al.*, 2005a).

too low pressure of nitrogen used during the deposition, and impurities, in contradiction to the data and explanation regarding the necessity of sufficiently high nitrogen pressure and deposition temperature, which were published 10 years ago (Veprek and Reiprich, 1995). We refer the reader to recent papers for further details (Prochazka *et al.*, 2004; Veprek *et al.* 2005a; 2005b; 2006).

When, following the generic design concept (Veprek and Reiprich, 1995) a sufficiently high deposition temperature of 550–600 °C and nitrogen pressure of ≥ 0.002 mbar were used under clean conditions that allowed a decrease in the oxygen impurity content, a hardness of 52 GPa was achieved in nc-TiN/a-Si$_3$N$_4$ (Prochazka *et al.*, 2004) as shown in Figure 2.27 (Prochazka *et al.*, 2004; Veprek *et al.*, 2005a). This figure also illustrates the importance of low background pressure in the deposition system. With a background pressure in the range of 10^{-6} mbar, the maximum achievable hardness was limited to about 46 GPa by about 0.2 at.% of oxygen impurities. A prolonged outgassing of the deposition system allowed a background pressure of less than $1 \cdot 10^{-6}$ mbar to be reached, and a decrease in the oxygen impurities to about 0.1 at.%, which resulted in a maximum achievable hardness of 52 GPa. Let us emphasize that such a low impurity content was achieved in sputter-deposited coatings only as a result of a combination of the low background pressure and high deposition rates of 1.6–1.9 nm/s, which are about one order of magnitude higher than the usually reported ones. Also the background pressure of $(2–6) \cdot 10^{-6}$ mbar reported in the literature (Diserens *et al.*, 1998; 1999; Patscheider *et al.*, 2001; Vaz *et al.*, 1998; 1999) was in many cases measured with the substrate at room temperature, and it increased up to $>1 \cdot 10^{-5}$ mbar when the substrate was heated to the chosen deposition temperature of ≤ 350 °C (Patscheider, personal communication, 2004). A simple estimate shows that several atomic percent (at .%) of oxygen impurities are expected in coatings deposited in such a way (Veprek *et al.* 2005a; 2005b). The detrimental role of oxygen will be discussed in more detail in Chapter 4. Here we emphasize that it is crucial to be considered when depositing coatings by means of sputtering, which usually suffers from low deposition rates and relatively large background pressure of mostly water vapor. This problem is relatively easier to manage in the case of plasma CVD, which operates at higher total pressure of several mbar, high gas flow rates and deposition rates of 0.6–1.2 nm/s. In such a way, coatings with oxygen impurity content of 0.01–0.05 at.% were routinely prepared and characterized (Veprek *et al.*, 2000).

2.3.5 Vacuum-arc evaporation

The electron emission from the cathode of an arc discharge occurs via a thermionic mechanism. Therefore, the temperature of the cathodic spot reaches several thousand degrees Celsius. As a result, the material within the spot is molten and evaporates. The vacuum arc burns within the evaporating atoms (which are fully ionized) between a cathode with the localized, hot spot, and an anode, where it is diffuse and, therefore, the temperature remains low. The advantage of deposition of coatings by means of vacuum arc are the high deposition rate of several micrometers per hour and the fully ionized plasma of the evaporated atoms that are than depositing on the substrate. A disadvantage is the uncontrolled movement of the cathodic spot over the surface and emission of droplets of molten metal from the cathodic spot. These droplets may be removed by using magnetic filters (see Musil *et al.*, 1993), but this results in low deposition rates and a small area where the substrates can be placed.

In order to make the vacuum-arc technique applicable to industrial processing, variety of the control of the movement of the cathodic spot has been developed. The real breakthrough was achieved by Curtins who developed the "PLATIT" technique of the cathode movement control as shown in Figure 2.28a (Curtins, 1995). The permanent magnet 1 together with coils 2 and 3 (symbols $+$ and $-$ mean the opposite direction of the electric current in the coils) induce a magnetic field 6 parallel to the surface of the planar cathode 5 in a similar manner as in the case of the magnetron shown in Figure 2.25. The Lorenz force due to the magnetic field and the electric current cause the cathodic spot to move in a closed trajectory. The electric current in the coils is provided by a power supply 7, which is controlled by a microprocessor 8 in such a way that the cathodic spot undergoes an oscillatory movement to left and right. As a result, the eroded area of the target, which is a measure of the utilization of the target material, is much larger. The problem associated with a long-term operation is the variation of the magnetic field at the surface of the cathode, which requires a periodic adjustment. A solution of this problem was recently achieved in a modified system (Curtins, 2004).

Another approach was developed by Jilek and co-workers, who used a central cathode with a magnetic field parallel to the axis (Holubar *et al.*, 2000). This allowed a much better utilization of the target material and the deposition of the nc-$(Ti_{1-x}Al_x)$N/a-Si_3N_4 superhard nanocomposites on an industrial scale. The advantage of the central cathode is a higher plasma density at the coated substrates and tools; this is needed for the

2.3 Preparation of nanostructured, hard, and superhard coatings

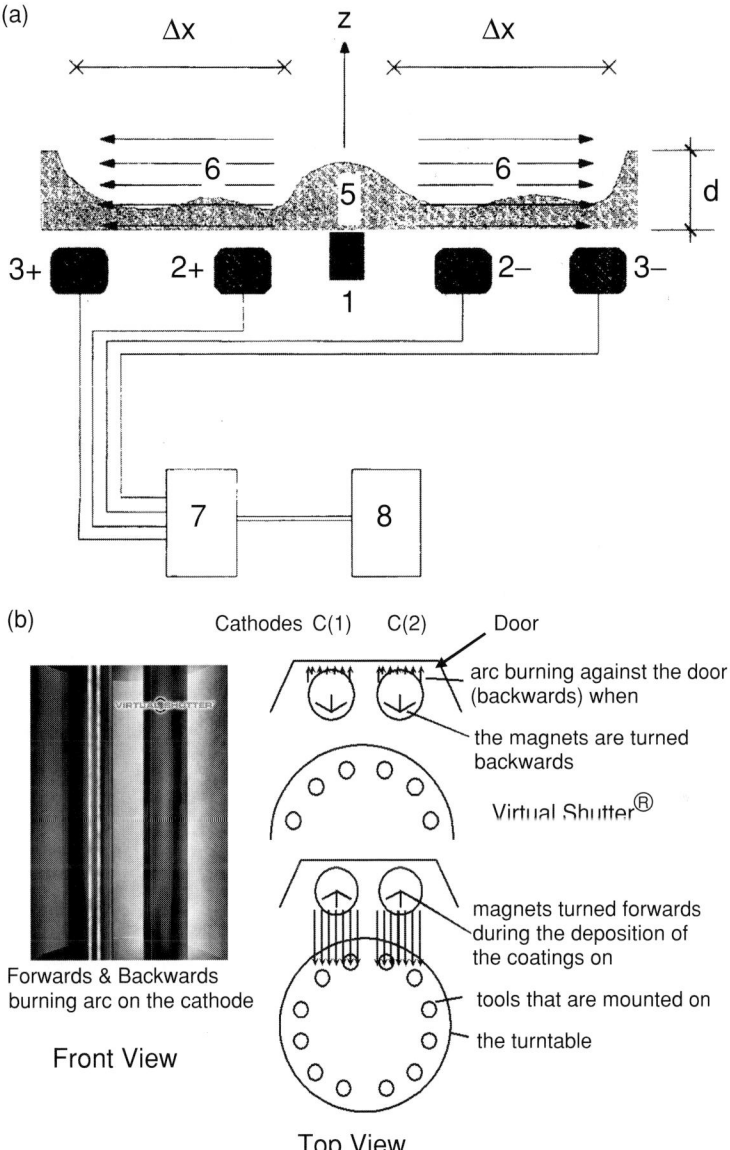

Figure 2.28. (a) Schematics of the control of the movement of the cathodic spot in the system PLATIT: 1, permanent magnet; 2 and 3, coils; 5, target; 6, horizontal magnetic field lines generated by 1, 2, 3; 7, power supply for the coils; 8, microprocessor control. (From Curtins (1995), with the permission of the author and with permission from Elsevier.) (b) The new coating unit π^{80}, which was developed by Jilek et al. (2004).

formation of a stable nanostructure. In the more-recent development of the Lateral Rotating Arc Cathodes (LARC®), vacuum-arc technology became ready for large-scale industrial production, as illustrated in Figure 2.28b (Jilek et al., 2004). A strong and concentrated magnetic field of a special geometry enables very fast movement of the arc spot which significantly reduces the emission of droplets resulting in a very smooth surface. In such

84 2 Processing of structural nanocrystalline materials

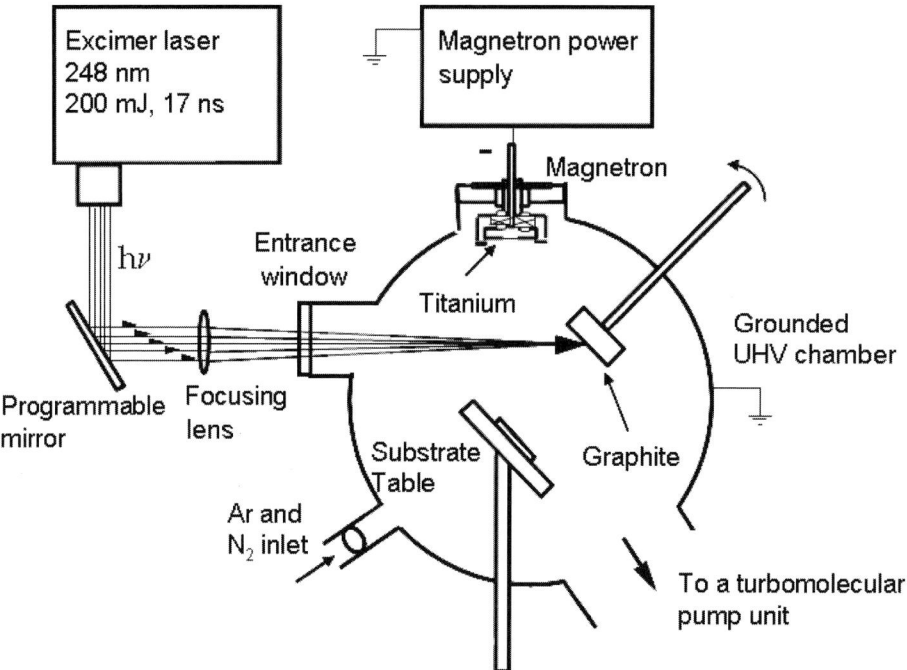

Figure 2.29. Schematic arrangement of the laser-ablation and magnetron sputtering hybrid deposition technique used by Voevodin et al. (1996a; 1996b; Voevodin and Zabinski, 2000). (With the permission of the authors, and permission of Elsevier.)

a way, the LARC® technology removes the disadvantages of the vacuum arc without sacrificing the high deposition rates. Furthermore, during the cleaning of the substrates, the magnetic field is turned by 180° so that the arc is burning towards the wall of the chamber (see Figure 2.28b), and the material being evaporated from the cathode, which was contaminated as a result of the exposure to air during loading of the chamber, is deposited on that wall. After the cleaning is finished, the magnets are turned back so that the deposition of pure material from the cleaned cathodes occurs directly on the cleaned substrates (Virtual Shutter®).

2.3.6 Hybrid techniques

Hybride deposition techniques are used when different materials (elements) have to be deposited simultaneously and, for some fundamental reasons, they cannot be deposited by the same technique. A typical example is the deposition of the "Ti–Si–N" nanostructured coatings by means of vacuum-arc evaporation of titanium and sputtering of silicon, which cannot be evaporated by vacuum arc because of cracking of the Si-cathode under the high thermal load of the arc (Choi et al., 2004). Low-melting-point metals, such as aluminum or copper, can also not be evaporated by conventional vacuum-arc equipment. Therefore, sputtering of copper and vacuum-arc evaporation of titanium were used for the deposition of TiN/Cu nanocomposites (Myung et al., 2003).

Voevodin and co-workers used a laser ablation of carbon in combination with magnetron sputtering of titanium in order to prepare TiC/DLC nanocomposites because the sputtering of carbon is a relatively inefficient process due to the high vaporization temperature and concomitant low sputter yield. The schematic of the apparatus is shown in Figure 2.29. More recently, these researchers have used this technique to deposit hard and low friction adaptable WC/DLC/WS$_2$ coatings (Voevodin *et al.*, 1996a; 1996b; Voevodin and Zabinski, 2000). The laser ablation provides a high deposition rate of the majority element carbon whereas the somewhat lower and easily adjustable sputtering rate of titanium is used in order to obtain the desirable concentration.

Another alternative is a combination of PVD techniques such as reactive sputtering and plasma CVD. In such a way, superhard nc-TiN/a-Si$_3$N$_4$ nanocomposites with hardness of 48 GPa were prepared by Brunell and co-workers, when Ti was sputtered and Si$_3$N$_4$ was deposited from silane in a nitrogen discharge (Brunell *et al.*, 2005).

References

Adamson, A. W. (1969). *J. Colloid Interface Sci.*, **29**, 744.
Akashi, K. (1985). *Pure & Appl. Chem.*, **57**, 1197.
Aliotta, F., Arcoleo, V., Buccoleri, S., La Manna, G., and Liveri, V. T. (1995). *Thermochim. Acta*, **265**, 15.
Ando, M., Suto, S., Suzuki, T., Tsuchida, T., Nakayama, C., Miura, N., and Yamazoe, N. (1994). *J. Mater. Chem.*, **4**, 631.
Arcoleo, V., and Liveri, V. T. (1996). *Chem. Phys. Lett.*, **258**, 223.
Arcoleo, V., Goffredi, M., Liveri, V. T., and Longo A. (1998). *Mater. Sci. Eng.*, **C6**, 7.
Atzmon, M., Verhoeven, J. D., Gibson, E. D., and Johnson, W. L. (1984) *Appl. Phys. Lett.*, **45**, 1052.
Bandyopadhyaya, R., Kumar, R., Gandhi, K. S., and Ramkrishna, D. (1997). *Langmuir*, **13**, 3610.
Barin, I. (1993). *Thermochemical Data of Pure Substances*. Weinnheim: VCH – Verlag.
Barnett, S. A. (1993). In *Physics of Thin Films Vol. 17: Mechanic and Dielectric Properties*, ed. Francombe, M. H., and Vossen, J. L. Boston: Academic Press, p. 2.
Barnett, S., and Madan, A. (1999). *Phys. World*, **11**, 45.
Barnett, S. A., Madan, A., Kim, I., and Martin, K. (2003). *MRS Bulletin*, **28**, 169.
Basumallick, A., Biswas, K., Mukherjee, S., and Das, G. C. (1997). *Mater. Letters*, **30**, 363.
Battezzati, L., Pappalepore, P., Durbiano, F., and Gallino, I. (1999). *Acta Mater.*, **47**, 1901.
Behrisch, R. (ed.) (1981). *Sputtering by Particle Bombardment I*. Berlin: Springer-Verlag.
Behrisch, R. (ed.) (1983). *Sputtering by Particle Bombardment II*. Berlin: Springer-Verlag.
Behrisch, R., and Wittmark K. (eds.) (1991). *Sputtering by Particle Bombardment III*. Berlin: Springer-Verlag.
Belyakov, A., Sakai, T., Miura, H., and Kaibyshev, R. (2000). *Phil. Mag. Lett.*, **80**, 711.
Benavides, S., Li, Y., and Murr, L. E. (2000). In *Ultrafine Grained Materials*, ed. Misra, R. S., Semiatin, S. L., Suryanarayana, C., Thadhani, N. N., and Lowe, T. C. Warrendale, PA: TMS, pp. 155–163.
Benjamin, J. S. (1970). *Metall. Trans.*, **1**, 2943.

Berkovich, Y., and Garti, N. (1997). *Colloids and Surfaces* A: *Physicochem. Eng. Aspects*, **128**, 91.

Beygelzimer, Y., Orlov, D., and Varyukhin, V. (2002) In *Ultrafine Grained Materials* II, ed. Zhu, Y. T., Langdon, T. G., Misra, R. S., Semiatin, S. L., Saran, M. J., and Lowe, T. C. Warrendale, PA: TMS, p. 297.

Boakye, E., Radovic, L. R., and Osseo-Asare, K. (1994). *J. Colloid Interface Sci.*, **163**, 120.

Bockris, J. O. M., and Razumney, G. A. (1967). *Fundamental Aspects of Electrocrystallization*. NY: Plenum Press, p. 27.

Bönnemann, H., Brijoux, W., and Joussen, T. (1990). *Angew. Chem. Int. Ed. Engl.*, **29**, 273.

Bönnemann, H., Brijoux, W., Brinkmann, R., *et al*. (1994). *J. Mol. Catal.*, **86**, 129.

Boutonnet, M., Kizling, J., Stenius P., and Maire, G. (1982). *Colloids Surfaces*, **5**, 209.

Brinker, C. J., and Scherer, G. W. (1990). *Sol–Gel Science: The Physics and Chemistry of Sol-Gel Processing*. San Diego: Academic Press Inc.

Brunell, I., Prochazka, J., and Veprek, S. (2005). Unpublished results.

Buhro, W. E., Haber, J. A., Waller, B. E., and Trentler, T. J. (1995). *Am. Chem. Soc. Symp.*, **210**, 20.

Cao, X., Koltypin, Y., Kataby, G., Prozorov, R., and Gedanken, A. (1995). *J. Mater. Res.*, **10**, 2952.

Carnaham, N. F., and Starling, K. E. (1969). *J. Chem. Phys.*, **51**, 635.

Carvalho, S., Rebouta, L., Ribeiro, E., Vaz, F., Denannot, M. F., Pacaud, J., Riviere, J. P., Paumier, F., Gaboriaud, R. J., and Alves, E. (2004). *Surf. Sci. Technol.*, **177–178**, 369.

Chapman, B. (1980). *Glow Discharge Processes*. New York: John Wiley & Sons.

Chase, M. W., Davies, C. A., Downey, J. R., Frurip, D. J., McDonald, R. A., and Syverud, A. N. (1985). *J. Phys. Chem. Data*, **14** (Supplement No. 1).

Chatelon, J. P., Terrier C., and Roger, J. A. (1997). *J. Sol–Gel Sci. Technol.*, **10**, 55.

Chatterjee, A., and Chakrabarty, D. (1992). *J. Mat. Sci.*, **27**, 4115.

Chen, D. J., and Mayo, M. J. (1993). *NanoStructured Materials*, **2**, 469.

Chew, C. H., Gan, L. M., and Shah, J. D. O. (1990). *Disp. Sci. Tech.*, **11**, 593.

Cho, Y. S., and Koch, C. C. (1991). *Mater. Sci. Eng. A*, **141**, 139.

Choi, S. R., Park, I.-W., Kim, A. H., and Kim, K. H. (2004). *Thin Solid Films*, **447–448**, 371.

Chu, X., and Barnett, S. A. (1995). *J. Appl. Phys.*, **77**, 4403.

Curtins, H. (1995). *Surf. Coat. Technol.*, **76–77**, 632.

Curtins, H. (2004). *European Patent* EP 1 392 879 B1.

Danielson, I., and Lindman, B. (1981). *Colloids Surfaces*, **3**, 391.

Davis, R. M., and Koch, C. C. (1987). *Scripta Metall.*, **21**, 305.

Di Meglio, J. M., Dvolaitzky, M., and Taupin, C. (1989). In *Progress in Microemulsions*, ed. Martellucci, S., and Chester, A. N. New York: Plenum Press, p. 263.

Dinda, G. P., Rosner, H., and Wilde, G. (2005). *Scripta Mater.*, **52**, 577.

Diserens, M., Patscheider, J., and Lévy, F. (1998). *Surf. Coat. Technol.*, **108–109**, 241.

Diserens, M., Patscheider, J., and Lévy, F. (1999). *Surf. Coat. Technol.*, **120–121**, 158.

Eckert, J., Holzer, J. C., Krill, C. E., and Johnson, W. L. (1992). *J. Mater. Res.*, **7**, 1751.

Erb, U., Aust, K. T. and Palumbo, G. (2002). In *Nanostructured Materials: Processing, Properties, and Applications*, ed. Koch, C. C. Norwich, NY: William Andrew Publishing, p. 179.

References

Fang, J., Stokes, K. L., Wiemann, J., and Zhou, W. (2000). *Mater. Lett.*, **42**, 113.
Fang J., Stokes K. L., Wiemann J. A., Zhou, W. L., Dai, J., Chen, F., and O'Connor, C. J. (2001). *Mater. Sci. Eng.*, **B83**, 254.
Fecht, H. J. (2002). In *Nanostructured Materials, Processing, Properties, and Applications*, ed. Koch, C. C., Norwich, NY: William Andrew Publishing, p. 73.
Fecht, H. J., Hellstern, E., Fu, Z., and Johnson, W. L. (1990). *Metall. Trans. A*, **21A**, 2333.
Fegely Jr., B., White, P., and Bowen, H. K. (1985). *J. Am. Ceram. Soc. Bull.*, **64**, 1115.
Fletcher, P. D. I., and Horsup, D. I., (1992). *J. Chem. Soc. Faraday Trans.*, **88**, 865.
Fletcher, P. D. I., Howe, A. M., and Robinson, B. H. (1987). *J. Chem. Soc. Faraday Trans.*, **83**, 185.
Freim, J., McKittrick, J., Katz, J., and Sickafus, K., (1994). NanoStructured Mater., **4**, 371–385.
Galatsis, K., Li, Y. X., Wlodarski, W., Conini, E., Faglia, G., and Sberveglieri, G. (2001). *Sensors and Actuators B*, **77**, 472.
Ghosh, A. K., and Huang, W. (2000). In *Ultrafine Grained Materials*, ed. Misra, R. S., Semiatin, S. L., Suryanarayana, C., Thadhani, N. N., and Lowe, T. C. Warrendale, PA: TMS, p. 173.
Girshick, S. L., Heberlein, J. V. R., McMurry, P. H., Geberich, W. W., Irdonoglou, D. I., Rao, N. P., Godwani, A., Tymiak, N., Di Fonzo, F., Fan, M. H., and Neumann, D. (2002). In *Innovative Processing of Thin Films and Nanocrystalline Powders*, ed. Choy, K. L. London: Imperial College Press, p. 165.
Gleiter, H. (1989). *Progress in Materials Science*, **33**, 233.
Gonsalves, K. E., and Rangarajan, S. P. (1997). In *Chemistry and Physics of Nanostructures and Related Non-Equilibrium Material*, ed. Ma, E., Fultz, B., Shull, R., Morall, J., and Nash, P., p. 149. TMS Meeting, Orlando, Fl.
Granqvist, C. G., and Buhrman, R. A. (1976). *J. Appl. Phys.*, **47**, 2200.
Groza, J. R. (1999). *NanoStructured Materials*, **12**, 987.
Groza, J. R. (2002). In *Nanostructured Materials, Processing, Properties, and Applications*, ed. Koch, C. C. Norwich, NY: William Andrew Publishing, p. 115.
Groza J. R., and Dowding, R. J. (1996). *NanoStructured Materials*, **7**, 749.
Hafiz, J., Wang, X., Mukherjee, R., Mook, W., Perrey, C. R., Deneen, J., Herberlein, J. V. R., McMurry, P. H., Gerberich, W. W., Carter, C. B., and Girshick, S. L. (2005). *Surf. Coat. Technol.*, **188–189**, 364.
Harris, A. M., Schaffer, G. B., and Page, N. W. (1993). In *2nd Inter. Conf. Mech. Alloying for Structural Applications*, ed. de Barbadillo, J. J., Froes, F. H., and Schwarz R., ASM Materials Park, OH, p. 15.
He, J., and Lavernia, E. J. (2001). *J. Mater. Res.*, **16**, 2724.
Heberlein, J. (2002). *Pure & Appl. Chem.*, **74**, 327.
Hirai, T., and Hayashi, S. (1982). *J. Mater. Sci.*, **17**, 1320.
Hirai, T., and Hayashi, S. (1983). *J. Mater. Sci.*, **18**, 2401.
Hirai, T., Sato, H., and Komasawa, I. (1993). *Ind. Eng. Chem. Res*, **32**, 3014.
Hoar, T. P., and Schulman, J. H. (1943). *Nature*, **152**, 102.
Holleck, H. (1986). *J. Vac. Sci. Technol.*, **A 4**, 2661.
Holubar, P., Jilek, M., and Sima, M. (2000). *Surf. Coat. Technol.*, **133–134**, 145.
Hou, M. J., Kim, M., and Shah, D. O. (1988). *J. Colloid Interface Sci.*, **123**, 398.

Hu, X., Han, Z., Li, G., and Gu, M. (2002). *J. Vac. Sci. Technol.*, **A 20**, 1921.
Huang, Y. J., Zhu, Y. T., Jiang, H., and Lowe, T. C. (2001). *Acta Mater.*, **49**, 1497.
Ingelsten, H. H., Bagwe, R., Palmqvist, A., Skoglundh, M., Svonberg, C., Holmberg, K., and Shah, D. O. (2001). *J. Colloid Interface Sci.*, **241**, 104.
Jean J. H., and Ring, T. A. (1986). *J. Am. Ceram. Soc. Bull.*, **65**, 1574.
Jean J. H., and Ring, T. A. (1988). *Colloids and Surfaces*, **29**, 273.
Jilek, M., Cselle, T., Holubar, P., Morstein, M., Veprek-Heijman, M. G. J., and Veprek, S. (2004). *Plasma Chem. Plasma Process*, **24**, 493.
Kayser, F., and Cohen, M. (1952). *Metal. Progr.*, **61**, 79.
Kegel, W. K., Theo, J., Overbeek, G., and Lekkerkerker, N. W. (1999). In *Handbook of Microemulsion Science and Technology*, ed. Kumar, P., and Mittal, K. L., New York: Marcel Dekker, p. 13.
Kim, S. H., Kim, J. K., and Kim, K. H. (2002). *Thin Solid Films*, **420–421**, 360.
Kingery, W. D., Bowen, H. K., and Uhlmann, D. R. (1976). *Introduction to Ceramics*, 2nd edition. New York: Wiley, p. 486.
Klassen, T., Oehring, M., and Bormann, R. (1994). *J. Mater. Res.*, **9**, 47.
Koch, C. C. (1991). In *Materials Science and Technology*, ed. Cahn, R. W., Haasen, P., and Kramer, E. J., vol. 15. Weinheim: VCH, p. 193.
Koch, C. C. (1993). *NanoStructured Mater.*, **2**, 109.
Koch, C. C., Smith, A. P., Bai, C., Spontak, R. J., and Balik, C. M. (2000). *Mater. Sci. Forum*, **343–346**, 49.
Korth, G. E., and Williamson, R. L. (1995). *Metall. Mater. Trans.*, **26A**, 2571.
Kubo, R. (1962). *J. Phys. Soc. Japan*, **17**, 975.
Lalauze, R., Visconte, E., Montanaro, L., and Pijolat, L. (1993). *Sensors and Actuators B*, **13**, 241.
Lambeth, D. N., Velu, E. M. T., Bellesis, G. H., Lee, L. L., and Laughlin, D. E. (1996). *J. Appl. Physics*, **79**, 4496.
Lee, E.-A., and Kim, K. H. (2002). *Thin Solid Films*, **420–421**, 371.
Lewis, D., Rayne, R. J., Bender, B. A., Kurihara, L. K., Chow, G. M., Flijlet, A., Kincaid, A., and Bruce, R. (1997). *NanoStructured Mater.*, **9**, 97.
Li, H., and Ebrahimi, F. (2003). *Mater. Sci. Engr. A*, **347**, 93.
Li, S. H., Shi, Y. L., and Peng, H. R. (1992). *Plasma Chem. Plasma Process*, **12**, 287.
Li, Y. X, Ghantasala, M. K., Galatsis K., and Wlodarski, W. (1999). In *Proceedings of SPIE – The International Society for Optical Engineering*, **3892**, 364.
Licciulli, A., and Mazzarelli, S., (2001). *J. Sol–Gel Sci. Technol.*, **21**, 195.
Livage, J., Henry, M., and Sanchez, C. (1988). In *Progress in Solid State Chemistry*, **18**, 259.
Lopez, T., Gomez, R., Noverao, O., Solis, A. R., Mora, E. S., Castillo, S., Poulain, E., and Magaden, J. M. (1993). *J. Catalysis*, **141**, 114.
López-Quintela M. A., and Rivas, J. (1993). *J. Colloid Interface Sci.*, **158**, 446.
Luton, M. J., Jayanth, C. S., Disko, M. M., Matras, S., and Vallone, J. (1989). *Mat. Res. Soc. Symp. Proc.*, **132**, 79.
Mahanty, J., and Ninham, B. W. (1976). *Dispersion Forces*. New York: Academic Press.
Mates, T. E., and. Ring, T. A. (1987). *Colloids and Surfaces*, **24**, 299.

References

Mayo, M. J. (1996). *Int. Mater. Rev.*, **41**, 85.

Meldrum, A., Boatner, L. A., and White, C. W. (2001). *Nuclear Instruments Methods Phys. Res. B*, **178**, 7.

Meng, W. J., Zhang, X. D., Shi, B., Tittsworth, R. C., Rehn, L. E., and Baldo, P. M. (2002). *J. Mater. Res.*, **17**, 2628.

Misra, R. S., Mukherjee, A. M., and Yamazaki, K. (1996). *J. Mater. Res.*, **11**, 1144.

Münz, W.-D. (2003). *MRS Bulletin*, **28**, 173.

Münz, W.-D., Lewis D. B., Hovsepian, P. Eh., Schönjahn, C., Ehiasarian, A., and Smith, I. J. (2001). *Surf. Eng.*, **17**, 15.

Musil, J., Vyskocil, J., and Kadlec, S. (1993). In *Physics of Thin Films, Vol. 17, Mechanic and Dielectric Properties*, ed. Francombe, M. H., and Vossen, J. L. Boston: Academic Press, p. 80.

Musil, J. (2000). *Surf. Coat. Technol.*, **125**, 322.

Myung, H. S., Lee, H. M., Shaginyan, L. R., and Han, J. G. (2003). *Surf. Coat. Technol.*, **163–164**, 591.

Nagy, J. B. (1999). In *Handbook of Microemulsion Science and Technology*, ed. Kumar, P., and Mittal, K. L. New York: Marcel Dekker, p. 499.

Nagpal, V., Davis, R., and Riffle, J. (1992). In *Polymeric Materials Science and Engineering*, vol. 67. Washington DC: ACS, Books and Journal Division, p. 235.

Natarajan, U., Handique, K., Mehra, A., Bellare, J. R., and Khilar, K. C. (1996). *Langmuir*, **12**, 2670.

Neilson, F. (1982). *Manufacturing Chemist*, **53**, 38.

Niihara, K. (1991). *J. Ceram. Soc. Japan*, **99**, 974.

O'Connor, C. J., Seip, C. T., Carpenter, E. E., Li, S., and John, V. T. (1999). *NanoStructured Mater.*, **12**, 65.

Oleszak, D., and Shingu, P. H. (1996). *J. Appl. Phys.*, **79**, 2975.

Osseo-Asare, K. (1999). In *Handbook of Microemulsion Science and Technology*, ed. Kumar, P., and Mittal, K. L. New York: Marcel Dekker, p. 549.

Özkar, S., Ozin, G. A., and Prokopowicz, R. A. (1992). *Chem. Mater.*, **4**, 1380.

Papp, S., and Dékány, I. (2001). *Colloid Polymer Sci.*, **279**, 449.

Park, J.-J. and Shin, D. H. (2002). In *Ultrafine Grained Materials II*, ed. Zhu, Y. T., Langdon, T. G., Misra, R. S., Semiatin, S. L., Saran, M. J., and Lowe, T. C. Warrendale, PA: TMS, p. 253.

Pathak, D. K. (1995). *Ph.D. Thesis*, North Carolina State University.

Patil, S., Kuiry, S. C., Seal, S., and Vanfleet, R. (2002). *J. Nanoparticles Res.*, **4**, 433.

Patscheider, J., Li S. Z., and Veprek, S. (1996). *Plasma Chem. Plasma Process.*, **16**, 341.

Patscheider, J., Zehnder, T, and Diserens, M. (2001). *Surf. Coat. Technol.*, **146–147**, 201.

Pavlov, V. A. (1985). *Phys. Met. Metall.*, **59**, 1.

Perez, R. J., Huang, B., and Lavernia, E. J. (1996). *NanoStructured Mater.*, **7**, 565.

Pfender, E. (1985). *Pure & Appl. Chem.*, **57**, 1179.

Pierre, A. C. (1991). *Ceramic Bull.*, **70**, 1281.

Pillai, V., and Shah, D. O. (1996). *J. Mag. Magn. Mater.*, **163**, 243.

Pillai, V., Kumar, P., Multani, M. S., and Shah, D. O. (1993). *Colloids and Surfaces A: Physicochemical Eng. Aspects*, **80**, 69.

Porta, F., Prati, L., Rossi M., and Scarý, G. (2002). *Colloids and Surfaces A*: *Physicochem. Eng. Aspects*, **211**, 43.
Prince, L. M. (1977). In *Microemulsions Theory and Practice*, ed. Prince, L. M. New York: Academic Press.
Prochazka, J., Karvankova, P., Veprek-Heijman, M. G. J., and Veprek, S. (2004). *Mater. Sci. Eng. A*, **384**, 102.
Puippe, J. Cl. (1986). In *Theory and Practice of Pulse Plating*, ed. Puippe, J. Cl., and Leaman, F. Orlando, FL: AESF, p. 1.
Qi, L., Ma, J., and Shen, J. (1997). *J. Colloid Interface Sci.*, **186**, 498.
Qiu, S., Dong, J., and Chen, G. (1999). *J. Colloid Interface Sci.*, **216**, 230.
Rack, H. J., and Cohen, M. (1970). *Mater. Sci. and Eng.*, **6**, 320.
Rao, C. N. R. (1993). *Mat. Sci. Eng. B*, **18**, 1.
Real, M. W. (1986). *Proc. Br. Ceram. Soc.*, **38**, 59.
Riedel, R., Kleebe, H. J., Schonfelder, H., and Aldinger, F. (1995). *Nature*, **374**, 526.
Risbud, S. H., Shan, C.-H., and Mukherjee, A. K. (1995). *J. Mater. Res.*, **10**, 237.
Rivas, J., Lopez-Quintela, M. A., Lopez, J. A., Liz, L., and Duro, R. J. (1993). *IEEE Trans. Magnetics*, **29**, 2655.
Rogl, P., and Schuster, J. C. (1992). *Phase Diagrams of Ternary Boron Nitride and Silicon Nitride Systems*. Materials Park, Ohio: ASM The Materials Society.
Roy, R., Roy, R. A., and Roy, D. M. (1986). *Mater. Letters*, **4**, 323.
Roy, S., Das, D., Chakrabarty, D., and Agarwal, D. C. (1993). *J. Appl. Physics*, **74**, 4746.
Ruckenstein, E. (1989). In *Progress in Microemulsions*, ed. Martellucci, S., and Chester, A. N. New York: Plenum Press, p. 3.
Ruckenstein, E. (1999). In *Handbook of Microemulsion Science and Technology*, ed. Kumar, P., and Mittal, K. L. New York: Marcel Dekker, p. 45.
Ruckenstein E., and Chi, J. C. (1975). *J. Chem. Soc. Faraday Trans. II*, **71**, 1690.
Saito, H., and Shinoda, K. (1970). *J. Colloid Interface Sci.*, **32**, 647.
Saito, Y., Utsunomiya, H., Tsuji, N. and Sakai, T. (1999). *Acta Mater.*, **47**, 579.
Sanders, P. G., Youngdahl, C. J., and Weertman, J. R. (1997). *Mater. Sci. Eng. A*, **234–236**, 77.
Sayce, I. G. (1975). *Pure & Appl. Chem.*, **48**, 215.
Segal, V. M., Reznikov, V. I., Drobyshevkij, A. E., and Kopylov, V. I. (1981). *Metally*, **1**, 115.
Sekino, T., Nakajima, T., and Niihara, K. (1996). *Mater. Letters*, **29**, 165.
Shaw, W. J. D. (1998). *Mater. Sci. Forum*, **269–272**, 19.
Shen, T. D., and Koch, C. C. (1995). *Mater. Sci. Forum*, **179–181**, 17.
Shen, T. D., and Koch, C. C. (1996). *Acta Mater.*, **44**, 753.
Shingu, P. H., Ishihara, K. N., Uenishi, K., Kuyama, J., Huang, B., and Nasu, S. (1990). In *Solid State Powder Processing*, ed. Clauer, A. H., and deBarbadillo, J. J. Warrendale, PA: TMS, p. 21.
Shukla, S., Seal. S., and Mishra, S. (2002). *J. Sol–Gel Sci. Technol.*, **23**, 151.
Shukla S. V., and Seal, S. (2004). In *Encyclopedia of Nanoscience and Nanotechnology*, vol. 10, ed. Nalwa, H. S. San Diego, CA: Academic Press, p. 27.
Siegel, R. W. (1991). In *Materials Science and Technology*, ed. Cahn, R. W., Haasen, P., and Kramer, E. J., vol. 15. Weinheim: VCH, p. 583.

References

Siegel, R. W., and Eastman, J. A. (1989). *Mater. Res. Soc. Symp. Proc.*, **209**, 3.

Skandan, G., Hahn, H., Kear, B. H., Reddy, M., and Cannon, W. R. (1994). *Mater. Letters*, **20**, 305.

Skandan, G. (1995). *NanoStructured Mater.*, **5**, 111.

Smith, A. P., Ade, H., Balik, C. M., Koch, C. C., Smith, S. D., and Spontak, R. J. (2000). *Macromolecules*, **33**, 2595.

Solans, C., Pons, R., and Kunieda, H. (1997). In *Industrial Applications of Microemulsions*, ed. Solans, C., and Kunieda, H. New York: Marcel Dekker Inc.

Sternitzke, M. (1997). *J. Euro. Ceramic Society*, **17**, 1061.

Suryanarayana, C. (2001). *Progress in Mater. Sci.*, **46**, 1.

Suryanarayana, C. (2004). *Mechanical Alloying and Milling*. New York: Marcel Dekker, Inc.

Tan G. L., and Wu, X. (1998). *Thin Solid Films*, **330**, 59.

Tao, N. R., Sui, M. L., Lu, J., and Lu, K. (1999). *NanoStructured Mater.*, **11**, 433.

Tholen, A. R. (1979). *Acta Metall.*, **27**, 1765.

Thompson, A. W. (1977). *Metall. Trans. A*, **8A**, 833.

Thornton, J. A., and Greene, J. E. (1994). In *Handbook of Deposition Technologies for Films and Coatings*, 2nd edition, ed. Bunshah, R. F. Park Ridge: Noyes Publ., p. 249.

Tojo, C., Blanco M. C., and López-Quintela, M. A. (1997). *Langmuir*, **13**, 4527.

Tojo, C., Blanco M. C., and López-Quintela, M. A. (1998). *J. Non-Crystalline Solids*, **235**, 688.

Traversa, E. (1995). *J. Intelligent Materials Systems and Structures*, **6**, 860.

Trudeau, M. L., Schultz, R., Zaluski, L., Hosatte, S., Ryan, D. H., Doner, C. B. Tessier, P., Strom-Olsen, J. O., and Van Neste, A. (1992). *Mater. Sci. Forum*, **88–90**, 537.

Ueji, R., Tsuji, N., Minamino, Y., Koizumi, Y., and Saito, Y. (2002). In *Ultrafine Grained Materials II*, ed. Zhu, Y. T., Langdon, T. G., Misra, R. S., Semiatin, S. L., Saran, M. J., and Lowe, T. C. Warrendale, PA: TMS, pp. 399–408.

Uyeda, R. (1991). *Progress in Mater. Sci.*, **35**, 1–96.

Valiahmetov, O. R., Galeyev, R. M., and Salishchev, G. A. (1990). *Fiz. Metall. Metalloved*, **10**, 204.

Valiev, R. Z., Islamgaliev, R. K., and Alexandrov, I. V. (2000). *Prog. Mater. Sci.*, **45**, 103.

Vaz, F., Rebouta, L., Ramos, S., da Silva, M. F., and Soares, J. C. (1998). *Surf. Coat. Technol.*, **108–109**, 236.

Vaz, F., Rebouta, L., Almeida, B., Goudeau, P., Pacaud, J., Riviere, J. P., and Bessa e Sousa, J. (1999). *Surf. Coat. Technol.*, **120–121**, 166.

Vaz, F., Rebouta, L., Goudeau, P., Pacaud, J., Garem, H., Riviere, J. P., Cavaleiro, A., and Alves, E. (2000). *Surf. Coat. Technol.*, **133–134**, 307.

Vaz, F., Rebouta, L., Goudeau, Ph., Giraadeau, T., Pacaud, J., Riviere, J. P., and Traverse, A. (2001). *Surf. Coat. Technol.*, **146–147**, 274.

Vaz, F., Carvalho, S., Rebouta, L., Silva, M. Z., Paúl, A., and Schneider, D. (2002). *Surf. Coat. Technol.*, **408**, 160.

Veprek, S. (1972). *J. Crystal Growth*, **17**, 101.

Veprek, S. (1980). In *Current Topics in Materials Science*, Vol. 4, ed. Kaldis, E. Amsterdam: North-Holland, p. 151.

Veprek, S. (1982). *Pure & Appl. Chem.*, **54**, 1197.

Veprek, S. (1983). *Thin Solid Films*, **130**, 135.
Veprek, S. (1999). *J. Vac. Sci. Technol.*, **A 17**, 2401.
Veprek, S., and Reiprich, S. (1995). *Thin Solid Films*, **268**, 64.
Veprek, S., Brendel, C., and Schäfer, H. (1971). *J. Crystal Growth*, **9**, 266.
Veprek, S., Niederhofer, A., Moto, K., Bolon, T., Männling, H.-D., Nesladek, P., Dollinger, G., and Bergmaier, A. (2000). *Surf. Coat. Technol.*, **133–134**, 152.
Veprek, S., Männling, H.-D., Niederhofer, A., Ma, D., and Mukherjee, S. (2004). *J. Vac. Sci. Technol.*, **B 22**, L5.
Veprek, S., Veprek-Heijman, G. M. J., Karvankova, P., and Prochazka, J. (2005a). *Thin Solid Films*, **476**, 1.
Veprek, S., Karvankova, P., and Veprek-Heijman, M. G. J. (2005b). *J. Vac. Sci. Technol.*, **B 23**, L 17.
Veprek, S., Männling, H.-D., Karvankova, P., and Prochazka, J. (2006). *Surf. Coat. Technol.*, **200** (12–13), 3876.
Voevodin, A. A., Schneider, J. M., Rebholz, C., and Matthews, A. (1996a). *Tribology Int.*, **29**, 559.
Voevodin, A. A., Capano, M. A., Safriet, A. J., Donley, M. S., and Zabinski, J. S. (1996b). *Appl. Phys. Lett.*, **69**, 188.
Voevodin. A. A., and Zabinski, J. S. (2000). *Thin Solid Films*, **370**, 223.
Wagner, C. (1976). *Colloid Polym. Sci.*, **254**, 400.
Wagner, J. J., and Veprek, S. (1982). *Plasma Chem. Plasma Process.*, **2**, 95.
Wang, J. P., Han, D., Luo, H. L., Lu, Q. X., and Sun, Y. W. (1995). *Appl. Phys.*, *A*, **61**, 407.
Wang, Y., Chen, M., Zhou, F., and Ma, E. (2002). *Nature*, **419**, 912.
Whittenberger, J. D., Arzt, E., and Luton, M. J. (1990). *J. Mater. Res.*, **5**, 271.
Wu, Z., Benfield, R. E., Guo L., Li, H., Yang, Q., Grandjean, D., Li, Q., and Zhu, H. (2001). *J. Phys. Condens. Matter*, **13**, 5269.
Yamada, K., and Koch, C. C. (1993). *J. Mater. Res.*, **8**, 1317.
Youssef, K. M. (2003). Ph.D. thesis, North Carolina State University.
Zeng, D., and Hampden-Smith, M. J. (1993). *Chem. Mater.*, **5**, 681.
Zhang, R. F., and Veprek, S. (2006). *Mater. Sci. Eng. A*, **424** (1–2), 128.
Zhang, X., and Koch, C. C. (2000). In *Ultrafine Grained Materials*, ed. Mishra, R. S., Semiatin, S. L., Suryanarayana, C., Thadhani, N. N., and Lowe, T. C. Warrendale, PA: TMS, p. 289.
Zhang, X., Wang, H., Scattergood, R. O., Narayan, J., and Koch, C. C. (2002a). *Acta Mater.*, **50**, 3995.
Zhang, X., Wang, H., Scattergood, R. O., Narayan, J., Koch, C. C., Sergueeva, A. V., and Mukherjee, A. K. (2002b). *Acta Mater.*, **50**, 4823.
Zhang, X., Wang, H., Scattergood, R. O., Narayan, J., and Koch, C. C. (2003). *Mater. Sci. Eng. A*, **A, 344**, 175.
Zhitomirsky, I., Petric, A., and Niewczas, M. (2002). *JOM-J. of the Minerals Metals & Materials Soc.*, **54**, 31.
Zhu, W., Deng, J., Tan O. K., and Chen, X. (2002). *Key Engineering Materials*, **214–215**, 183.

3 Stability of structural nanocrystalline materials – grain growth

3.1 Introduction

Knowledge of the thermal stability of nanocrystalline materials is important for both technological and scientific reasons. From a technological point of view, the thermal stability is important for consolidation of nanocrystalline particulates without coarsening the microstructure. That is, many methods, as described in Chapter 2, for synthesis of nanocrystalline materials result in particulate products which must be consolidated into bulk form. Since most consolidation processes involve both heat and pressure, the thermal stability of the nanoscale microstructure is always at risk. The goal of particulate consolidation is to attain essentially 100% theoretical density and good particulate bonding while preventing or minimizing grain growth of the nanocrystalline grains.

Understanding the scientific nature of stability, grain growth of nanocrystalline microstructures is a criterion for allowing strategies for minimizing grain growth to be developed. A basic scientific question with regard to nanocrystalline materials is whether their behavior involves "new physics" or is simply the expected grain-size-dependent behavior extrapolated to nanocrystalline grain sizes. Thermal stability is an important phenomenon to be addressed in this regard. The thermal stability in a broader sense involves not only the stability of the grain structure, that is the microstructure, but also the stability of the structure of the grain boundaries in nanocrystalline materials. A number of investigations on the thermal stability of nanocrystalline materials have been conducted. Grain growth in nanocrystalline materials has been reviewed by Suryanarayana (1995), Weissmuller (1996), and Malow and Koch (1996a,b). In this chapter we will discuss the thermal stability, grain growth of nanocrystalline materials with reference to experimental methods for measuring grain growth, grain-growth theories for conventional grain-size materials which may be applicable, grain growth (secondary recrystallization) at ambient temperatures in nanocrystalline metals, strategies for inhibition of grain growth in nanocrystalline materials, and examples of experimental studies of grain-growth kinetics in nanocrystalline materials.

3.2 Experimental methods for measuring grain growth

3.2.1 X-ray diffraction line broadening analysis

In most studies of the grain size of nanocrystalline materials either X-ray line-broadening analysis is used or direct measurements by transmission electron microscopy (TEM) are performed. First, we discuss the X-ray line-broadening methods. There are a number of factors that can cause broadening of a diffraction peak. All those factors that are associated with the instrumental arrangement of the diffractometer are in combined form commonly referred to as the "instrumental broadening." This includes such factors as the quality of the alignment (X-ray beam to sample, sample to detector, θ to 2θ, etc.), and the receiving slit. The material characteristics of interest that cause line broadening include small crystal size, lattice strain, stacking faults, and dislocations. In order to determine the experimental broadening effects due to material parameters, β, the instrumental broadening, b, must be taken away from the total experimental broadening function, B. There are various approaches to first determining the peak shapes of the sample and instrumental broadening (Klug and Alexander, 1974). Depending on the assumptions made regarding line shape, different mathematical calculations have to be made, which depend upon the mathematical functions that describe the respective line shapes of the instrumental and sample broadening. The most commonly used functions are the Gaussian and the Cauchy function or combinations of them. If the instrumental, the sample, and the experimental combination of these diffraction lines are assumed to have a Gaussian line shape, the broadening effects are related through the relationship

$$B^2 = b^2 + \beta^2. \quad (3.1)$$

For all profiles taken as a Cauchy function, the line breadth is convoluted as a simple sum: $B = b + \beta$. Equation (3.1) is the more widely used, however, more realistic are cases of mixed line functions. Klug and Alexander (1974), state that the shape of the intrinsic broadening profile is approximately a Cauchy function, and that of the instrumental broadening is approximately the Gaussian function. In this case the relation becomes

$$\beta/B = \sqrt{1 - b^2/B^2}. \quad (3.2)$$

The instrumental peak broadening is typically determined with a large grained, strain-free standard sample with the same absorption coefficient as the sample to be measured (Delhez et al., 1980). It is not trivial to obtain such a perfect reference sample and some error results from the instrumental broadening corrections (Krill and Birringer, 1998). Once the diffraction line and peak shapes of the reflections have been adequately corrected for instrumental broadening and $K_{\alpha 2}$ contributions, the latter by using the Rachinger correction (Klug and Alexander, 1974), a number of different analyses can be performed to deduce the crystallite size and the internal strain present in the sample. The most used of these methods are the Scherrer, the Williamson–Hall, the Warren–Averbach, and modifications of these. The oldest and simplest of the methods is the use of the expression proposed by Scherrer in 1918 which assumes that the crystallite size is directly related to the width of the X-ray

3.2 Experimental methods for measuring grain growth

diffraction peaks (Guinier, 1963). The Scherrer equation is

$$\Delta(2\theta) = 0.9\lambda/d\cos(\theta_0), \qquad (3.3)$$

where $\Delta(2\theta)$ is the peak broadening, λ is the wavelength of the X-rays, θ_0 is the position of the peak maximum, and d is the volume average of the crystal dimension normal to the diffracting plane. Bertaut (1950) subsequently showed that it is the unit-cell column length, L, that is measured by the Scherrer equation and that grain size must be estimated from this value. The Scherrer equation does not take the broadening of the diffraction line by internal strains into consideration. If it is known that internal strains are negligible then the Scherrer equation can be used to estimate the grain size. If internal strain calculations are needed one approach is the method of Williamson and Hall (1953). By the use of this approach an estimation of the grain size and the separation of the broadening effect due to internal strain can be achieved. Bragg's law, written in the vectorial form (magnitudes) (Guinier, 1963) is

$$s = 2\sin\theta/\lambda. \qquad (3.4)$$

The magnitude of the scattering vector $k = 2\pi s$ will exhibit broadening in reciprocal space according to the real-space broadening of the diffraction peaks. The broadening in reciprocal space owing to grain refinement is represented by the Scherrer formula as

$$\Delta k = 0.9(2\pi/L), \qquad (3.5)$$

while the strain contribution (root mean square strain $\langle e^2 \rangle^{0.5}$) can be expressed as

$$\Delta k = A\sqrt{\langle e^2 \rangle}, \qquad (3.6)$$

where A is a constant depending upon the spatial distribution of the strains (close to unity for dislocations). Combination of the sum of both contributions with Bragg's law gives

$$\frac{\beta \cos\theta}{\lambda} = \frac{0.9}{d + \sqrt{\langle e^2 \rangle}}. \qquad (3.7)$$

In a plot of the broadening versus the scattering vector, the strain can be obtained from the slope of the resulting curve and the grain size can be deduced from the intercept with the y-axis. However, often inconsistent results are obtained by this method partly because of anisotropy in internal strains. Therefore, this method is used to obtain only the strains, and the Scherrer equation is used to estimate the grain size.

The method of Warren and Averbach (Warren, 1990) requires the study of the shapes of the diffraction lines rather than looking at the breadth of the diffraction peak. This method is more rigorous than the above methods and it requires the accurate knowledge of the peak shapes. The power of the total reflection (hkl) can be represented by a Fourier series, of which the corrected cosine coefficients ($A(hkl)$) are related to the area averaged length, L, normal to the diffracting plane of the coherent domain, as well as the distortion in the domain. The experimentally obtained coefficient, $A_L^s(hkl)$ and the distortion coefficient, $A_L^d(hkl)$ are related as follows:

$$A_L(hkl) = A_L^s(hkl) A_L^d(hkl). \qquad (3.8)$$

Since only the distortion coefficient is dependent on the order of the reflections, the knowledge of the Fourier cosine coefficients of more than one order of a reflection makes it possible to separate strain and size effects. When plotting $\ln(A_L)$ against $h_0^2 = h^2 + k^2 + l^2$, the strain is related to the slope of the curve and the intercept at $h_0 = 0$ gives the size coefficient A_L^s. The initial slope of the curve in a plot of A_L^s as a function of the grain size intercepts the grain size axis at the average domain size L. In order to perform the analysis in the manner described, peaks at least two orders of reflection from the same plane (e.g. (111), (222)) have to be known. Often two orders of reflection are not obtainable experimentally. In this case, there are single-peak methods available in which assumptions about the strains have to be made (Enzo *et al.*, 1988).

Ungar and co-workers (Ungar and Borbely, 1996; Ungar *et al.*, 1998) further modified the peak-profile analysis methods of Williamson–Hall and Warren–Averbach to estimate grain-size distributions as well as dislocation density and dislocation arrangements.

Generally, grain-size determination by the use of X-ray diffraction line-broadening analysis is advised only if the average grain-size does not exceed 100 nm and the grain-size distribution is narrow. Correction for instrumental broadening becomes critical for grain sizes greater than about 30 nm. In cases where an inhomogeneous grain-size distribution exists, with some grains >100 nm, other methods, in particular, transmission electron microscopy, must be used.

3.2.2 Transmission electron microscopy, TEM

Transmission electron microscopy, TEM, is one of the most direct methods for the determination of the grain size. High-resolution TEM (HRTEM) can reveal grains at the smallest sizes as well as information on the nature of the grain boundaries. More sophisticated Z-contrast imaging can provide chemical information regarding grain-boundary segregation. The advantages of TEM and HRTEM are obvious in terms of providing direct images of the grain size, shape, and size distribution. The analysis of the micrographs involves the application of straightforward stereological relationships that provide various possibilities for characterizing the grains. TEM can distinguish between normal and abnormal grain growth and, with varying magnifications, can determine the distribution of inhomogeneous grain structures. In order to get accurate statistical information on nanocrystalline grain size by TEM, dark-field TEM must be carried out on many fields of view in many samples. The usual problems of specimen preparation for TEM are exacerbated for some nanocrystalline materials, which are often prepared in powder form. However, bulk structural nanocrystalline materials should not present unusual difficulties regarding TEM specimen preparation. It has become clear that TEM must be used with/instead of X-ray diffraction line-broadening analysis to determine the grain size. It is absolutely necessary to reveal the grain-size distribution.

3.2.3 High-resolution scanning electron microscopy, SEM

The resolution of modern scanning electron microscopes is typically <10 nm (Verhoeven, 1986) so in principle SEM could be used to measure nanocrystalline grain size. The problem

3.2 Experimental methods for measuring grain growth

is obtaining appropriate surface contrast for grain sizes in the nanoscale regime. This can be achieved for fracture surfaces of materials that fracture intergranularly. Examples of using SEM on such materials have been reported for ceramic nanocrystalline materials (Hofler and Averback, 1990) and for the intermetallic TiAl (Chang *et al.*, 1992). For TiAl, good agreement between the SEM results and those from the analysis of X-ray line broadening was obtained. In materials which do not fail intergranularly some other contrast method, such as chemical etching, must be found to reveal the grain boundaries. This is not a trivial problem for nanostructured materials. No reports of the successful use of SEM to determine grain sizes in nanocrystalline materials that do not fracture intergranularly have been given.

3.2.4 Atomic force microscopy and scanning tunneling microscopy, AFM and STM

Nanocrystalline Ag and Pd samples prepared by the inert-gas condensation and compaction method have been imaged by the use of STM and AFM (Sattler *et al.*, 1994). The STM probes the local electron density of states at the Fermi level, while the AFM is sensitive to the total local density of electron states. The STM imaging method was found to be very sensitive to inhomogeneous oxidation during the scanning procedure, which created an unstable image. The STM was able to distinguish between low-angle grain-boundary crystallites, while the AFM imaged clusters of crystallites without the possibility to differentiate between some of the individual crystallites. Since these two methods image specimen surfaces they suffer from the same problems as described above for high-resolution SEM in terms of requiring appropriate surface contrast to image the grains. In these cases, the efficacy is dependent upon the existence of some kind of surface (height) contrast, which may be naturally occurring with some processing methods but otherwise require some artificial procedure such as etching. In Figure 3.1 an example of an AFM image of the surface of electrodeposited nanocrystalline Zn is presented. The grain size determined from this surface contrast was in good agreement with other measurements such as X-ray diffraction line broadening analysis and TEM (Youssef, 2003).

3.2.5 Raman spectroscopy

Models have been developed to measure the nanocrystalline grain size in semiconductors from the experimentally measured parameters of Raman scattering lines; that is, line broadening, frequency shift, and coefficient of asymmetry (Richter *et al.*, 1981; Campbell and Fauchet, 1986). This analysis has been used to estimate the nanocrystalline grain size in ball-milled Si (Shen *et al.*, 1995) and nanocrystalline Si and C prepared by several methods (Obraztsova, 1994). However, as with X-ray diffraction, Raman line broadening can have other sources than small crystallite size. Melendres *et al.* (1989) observed significant Raman line broadening in nanocrystalline TiO_2 that was uncorrelated with the initial grain size. The broadening decreased with annealing at elevated temperatures and was attributed to intra-grain defects. An example of how much care has to be exercised in order to avoid misinterpretation of Raman spectra is the example of Raman scattering from nc-Si and

98   3 Stability of structural nanocrystalline materials – grain growth

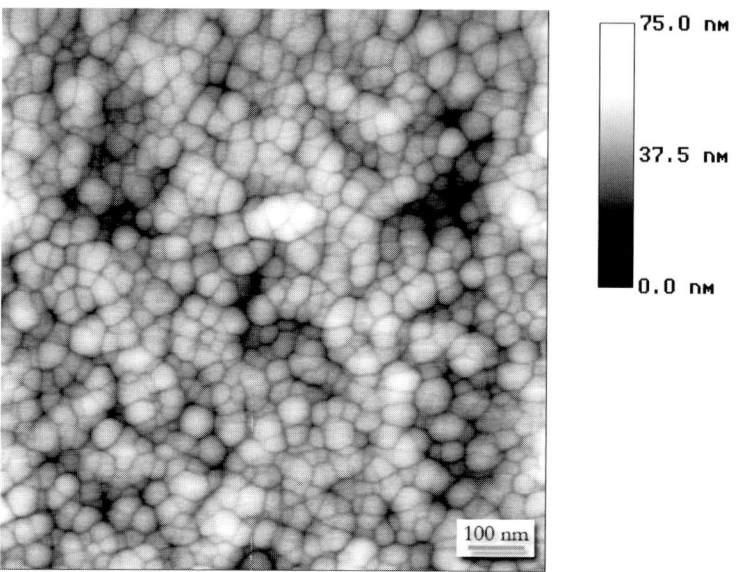

Figure 3.1. AFM image of surface morphology of electrodeposited nanocrystalline Zn. (From Youssef (2003), reproduced with permission of the author.)

nc-Si/a-Si:H nanocomposites. After the first report on the effect of the crystallite size on the Raman spectra of nc-Si (Iqbal *et al.*, 1981; Iqbal and Veprek, 1982), the phonon confinement model was developed (Richter *et al.* 1981; Campbell and Fauchet, 1986) and widely used to determine the crystallite size and fraction of nc-Si in the nanocomposites. However, these models cannot account for the effect of stress in thin coatings, which significantly changes both the Raman shift and the relative intensity of the Raman signal from nc-Si and a-Si (Sarott *et al.*, 1982; Veprek *et al.*, 1987). It was shown by Ossadnik and co-workers that using Raman scattering without additional characterization of the nc-Si and nc-Si/a-Si:H coatings in order to determine the crystalline fraction and crystallite size leads to incorrect results (Ossadnik *et al.*, 1999).

3.2.6 Differential scanning calorimetry (DSC)

During heating of a polycrystalline sample in a differential scanning calorimeter, the reduction of the interfacial enthalpy of the grain ensemble due to grain growth will be released. For growth of conventional grain-size (micron and above) grains, these enthalpies are typically too small to be resolved in a differential scanning calorimeter. However, as pointed out by Chen and Spaepen (1988; 1991; 1992) the grain growth of nanoscale grains ($\sim$10–20 nm) will allow for measurable enthalpy release because of the high interface-to-volume ratio in these materials. A well-known analysis method which serves to determine the activation energy of a process causing heat release in a linear heating experiment is the Kissinger method (Kissinger, 1957). If the heat release of the thermal event causes a peak in the linear heating experiment in a DSC, the shift of the peak with the change in the heating rate can be used to determine the activation energy of the event that caused the peak. The Kissinger method was originally developed for the analysis of nucleation and growth rate processes from a number of scans at different heating rates, b. The activation energy, Q, of the rate process can be determined by using the peak temperature, T_p, in a plot of $\ln(b/T_p^2)$ vs. $1/T_p$. Chen and Spaepen (1988; 1991; 1992) have shown that this type of analysis is equally valid for the grain-growth process. Activation energies so determined can be compared with the activation energies for lattice or grain-boundary diffusion to gain information on possible mechanisms for the observed grain growth.

Isothermal DSC measurements can be a powerful, if little used, tool to extract the important parameters for analysis of grain growth. Chen and Spaepen (1991) have presented a detailed analysis of the methods for extracting the parameters that describe the process of grain growth from DSC measurements. In addition to obtaining activation energies and grain growth exponents, isothermal DSC can distinguish between normal and abnormal grain growth. Qualitatively, normal grain growth gives a monotonically decreasing DSC signal with time, while abnormal grain growth, similar to nucleation and growth reactions, exhibits a peak in signal with time at temperature. TEM measurements are needed to guide the analysis from DSC in terms of providing accurate grain-size data.

3.3 Grain-growth theories for conventional grain size materials

Grain growth is the process by which either the mean grain size in a single-phase material or the mean matrix grain size in a composite material with second-phase particles increases. This process results in essential changes of the microstructure and thereby strongly influences mechanical and other properties of materials. In particular, grain growth can eliminate the nanocrystalline state with its remarkable properties by transforming it into the conventional coarse-grained state.

Grain growth is a cooperative process involving simultaneously many individual grains of various sizes and shapes (see Atkinson, 1988; Pande and Masumura, 2003). One distinguishes between "normal" and "abnormal" grain-growth processes. During normal grain growth, the range of grain sizes and shapes is relatively narrow (Figure 3.2). During abnormal grain growth (sometimes called secondary recrystallization), a few large grains develop consuming their small neighbors (Figure 3.2).

Grain growth in coarse-grained polycrystals commonly occurs through the standard mechanism involving migration (movement) and coalescence of grain boundaries. These processes result in an increase of the mean grain size and a decrease of the number of grains in a polycrystalline aggregate. Grain growth through migration and coalescence of grain boundaries is commonly viewed as the sole mechanism for grain growth in coarse-grained polycrystals (see Atkinson, 1988). However, there is an alternative mechanism related to crystal-lattice rotation in a grain, resulting in both disappearance of a grain boundary and coalescence of its adjacent grains (Li, 1962). Though this alternative mechanism does not effectively operate in coarse-grained polycrystals, it is capable of essentially contributing to grain growth in nanocrystalline materials where crystal-lattice rotations at the nanoscale level are enhanced compared with those in coarse grains (see discussion below).

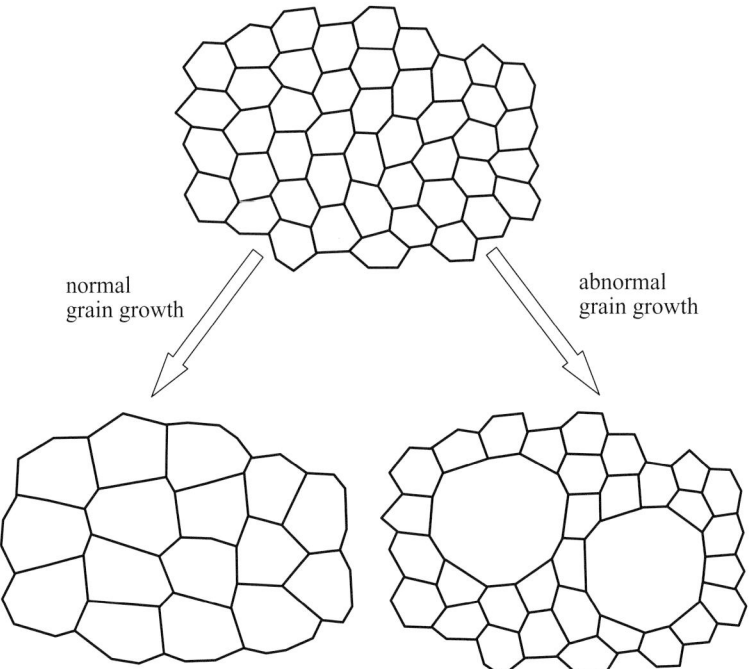

Figure 3.2. Normal and abnormal grain-growth regimes.

The dominant driving force of the grain growth is the reduction of the free energy of grain boundaries whose total area decreases during this process. The driving force tends to eliminate grain boundaries and, thus, transform initially poly- and nanocrystalline samples into single crystalline samples. However, besides the driving force, there are hampering forces for grain growth. The dominant hampering force in conventional coarse-grained polycrystals is related to a finite mobility of migrating grain boundaries (see Atkinson, 1988). The competition between the driving and hampering forces for grain growth is crucial for stability of nano- and polycrystalline structures.

An adequate theoretical description of grain growth and structural stability in solids is a formidable challenge because it requires detailed knowledge of many structural and geometrical parameters of grains, grain boundaries and their ensembles, as well as how this information is coupled to mobility and energy characteristics of grain boundaries at a given temperature. At present, although there are many theoretical approaches to a description of grain growth in coarse-grained polycrystals (see Atkinson, 1988; Pande and Masumura, 2003), laws governing this process with its main experimentally detected features are still under discussion. In the case of nanocrystalline materials, grain-growth theory is in its infancy. Its progress in the future is expected to be based both on achievements of grain-growth theories for conventional grain size materials and on involvement of the specific structural and behavioral features of nanocrystalline materials in examination. Below we briefly consider classical theories of normal grain growth in single-phase materials with conventional grain sizes and then discuss grain-growth peculiarities in nanocrystalline materials.

3.3 Grain-growth theories for conventional grain size materials

Theories of normal grain growth in conventional coarse-grained polycrystals attempt to account for its main attributes experimentally observed in a wide variety of materials. These attributes are briefly as follows (see Kurtz and Carpay, 1980; Pande and Masumura 2003).

(a) During grain growth in a polycrystal, there is a relatively narrow range of grain sizes and shapes.
(b) During grain growth, after sufficient time, the distribution of grain sizes scaled to the mean grain size remains self-similar.
(c) The final grain-size distribution resulted from grain growth is generally insensitive to the initial distribution.
(d) During grain growth, the mean grain size (radius), R, increases with time t as follows:

$$R^2 - R_0^2 = Kt^m \text{ or, after sufficient time } \left(\text{for } R^2 \gg R_0^2\right), R = Kt^m. \tag{3.9}$$

Here K and m are constant, and m is termed the grain-growth exponent.

One of the early theoretical models of grain growth in conventional-grain-size materials was suggested by Burke and Turnbull (1952). They considered migration of an isolated grain boundary and described the driving force as the pressure due to grain-boundary surface curvature. In these circumstances, the area of grain boundary and therefore its energy decrease when the grain boundary migrates towards its center of curvature. Burke and Turnbull (1952) deduced the parabolic law $R = Kt^{1/2}$ (given by formula (3.9) with grain-growth exponent $m = 0.5$) for the mean grain size R dependence on time t. However, as noted by Atkinson (1988), besides experiments indicative of the parabolic grain growth kinetics $R = Kt^{1/2}$ in single-phase materials, there are also experiments showing the grain-growth exponent $m < 0.5$. In addition, theoretical representations (Burke and Turnbull, 1952) on curvature-driven migration of isolated grain boundaries are not intrinsically relevant in a description of grain growth as a cooperative process involving simultaneously many individual grains of various sizes and shapes.

Cooperative character of grain growth is taken into consideration in the theories operating with topology of arrays of grains. Following pioneering papers by Smith (1948; 1953; 1964), grain growth is regulated by the topological requirements of space filling by grains, coupled with the minimum energy factor that requires the grain-boundary surface tension to be in equilibrium. This idea lies in the basis of most theories of grain growth in coarse-grained polycrystals.

A theoretical analysis of grain-growth topology and energy characteristics in three-dimensional (3D) polycrystals is very difficult. Its simplification is possible by considering grain growth in a two-dimensional (2D) plane. Along with simplification of 3D theories, 2D theories are important and self-sufficient in a description of 2D grain growth at surfaces/sections of polycrystals, which is experimentally monitored by standard mean grain-intercept measurements. Besides, all the basic ideas and mathematical approaches of 3D and 2D theories are the same, which could be useful in a development of the theory of grain growth in nanocrystalline materials. In the context discussed, hereinafter, we will focus our analysis on mostly theories of 2D grain growth.

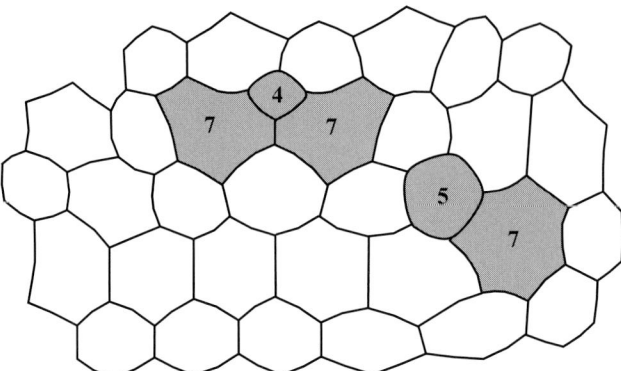

Figure 3.3. Array of grains with convex and concave grain boundaries. Grains with number n of grain boundaries different from the mean number $= 6$ (inherent to dominant grains) are shown as grey polygons.

Following Smith (1948; 1953; 1964), let us consider a 2D cell (polygon) structure whose cells, edges and vortices serve as geometric models of 2D grains, grain boundaries and their junctions at a polycrystal surface/section, respectively. Topological invariance of 2D plane filling by grains is given by the Euler equation:

$$C - E + V = 1. \qquad (3.10)$$

Here C, E, and V are the numbers of 2D grains, grain boundaries, and grain boundary junctions, respectively. Topology of a 2D grain structure stable relative to small deformations is specified by both the mean number $n = 6$ of grain boundaries per one grain and the presence of only triple junctions of grain boundaries. Grain growth is accompanied by local topological transformations characterized by local changes in n, keeping the combination (3.10) constant.

As noted by Smith (1953, 1964), the above topological requirements on 2D grain structure should be coupled with the minimum energy factor that requires the grain-boundary surface tension to be at equilibrium. With the assumption that all the grain boundaries of a 2D polycrystal have the same tension, the equilibrium angles formed by grain boundaries at triple junctions are equal to 120°. Since a polycrystal is characterized by distributions in the grain size and its shape parameter n, equilibrium angles (120°) exist because of grain-boundary curvature. Grains with $n > 6$ and relatively large grains with $n = 6$ have concave grain boundaries (Figure 3.3). Grains with $n < 6$ and relatively small grains with $n = 6$ have convex grain-boundaries (Figure 3.3). The grain-boundary curvature causes the main effective driving force for grain boundary migration carrying grain growth. With the assumption that all triple junctions of grain boundaries have infinite mobility, 120° angles are unchanged during grain growth. In this situation, grain growth is specified by evolution of grain size (R) and shape (parameterized by n) distributions in time, which is driven by grain-boundary curvature distributed over grains and dependent on both the grain size and shape. Grains with concave grain boundaries tend to grow, because grain boundaries tend to move towards their centers of curvature. Conversely, grains with convex grain boundaries tend to shrink owing to the curvature effect.

The general statements discussed above are exploited by almost all the theories of grain growth in coarse-grained polycrystals. The theories under consideration are distinguished by their specific simplifying assumptions and analytical methods.

For instance, one of the first attempts to combine topological requirements and grain-growth kinetics is reflected in the von Neumann–Mullins relationship (Mullins, 1956):

$$dA_n(t)/dt = M(n - 6), \qquad (3.11)$$

where $A_n(t)$ is the area of the grain, with n grain boundaries at time instant t, and M is a constant ($M > 0$). Formula (3.11) describes grain growth in array of hexagon-like grains (with $n = 6$) and other grains as the process occurring via growth of grains with $n > 6$ and shrinkage of grains with $n < 6$. In doing so, grain-boundary curvature is associated with the grain shape specified by n and drives the grain-boundary migration when n is different from the number $n = 6$ of grain boundaries that characterizes "equilibrium" hexagon-like grains.

Hillert (1965) described 2D grain growth as a process carried by topological transformations of 5–7 pairs each consisting of two neighboring grains with 5 and 7 grain boundaries (Figure 3.4). Such 5–7 pairs can be defined as perfect lattice dislocations (called cellular dislocations) in a regular array of hexagon-like grains, in which case topological transformations of 5–7 pairs (Figure 3.4) are treated as cellular dislocation climb. As a result of an elemental series of topological transformations shown in Figure 3.4, one grain disappears, and a 5–7 pair moves (cellular dislocation climbs) to a new position. Moral and Ashby (1974) generalized the idea of cellular dislocation climb on the situation with 3D grain growth. In doing so, cellular dislocations have line cores threading 13–15 pairs, that is, pairs of two neighboring 3D grains (polyhedra) with 13 and 15 grain boundaries (facets) in a packing of dominant "equilibrium" grains (polyhedra) with 14 grain boundaries (facets). Moral and Ashby (1974) derived the parabolic law $R = Kt^{1/2}$ in the situation where the cellular dislocation density is constant, and the grain growth kinetics $R = Kt^m$ with $m < 0.5$ in the situation where the cellular dislocation density decreases during grain growth.

Mean-field theories of grain growth (see Feltham, 1957; Hillert, 1965; Louat, 1974; Mullins, 1998) consider growth of an isolated grain in an average or "mean" field of all other grains of a polycrystalline aggregate. It is a statistical approach generalizing theoretical representations (Burke and Turnbull, 1952) on grain growth associated with curvature-driven migration of isolated grain boundaries. More precisely, mean-field theories operate with evolution of grain-size distribution $f(R)$ in time owing to both a boundary-curvature-driven increase of grain sizes and a decrease in the number of grains during grain growth. In doing so, these theories involve in consideration the boundary-curvature-driven tendency of growth of comparatively large grains and shrinkage of comparatively small grains. For illustration, let us briefly discuss the key aspects of mean-field theories (Feltham 1957; Hillert 1965), using the unified formulation given by Hunderi and Ruym (1980). These classical mean-field theories deal with time-dependent grain-size distribution $f(R,t)$ and its continuity equation in one-dimensional R-space:

$$\partial f/\partial t + \partial (fv)/\partial R = 0. \qquad (3.12)$$

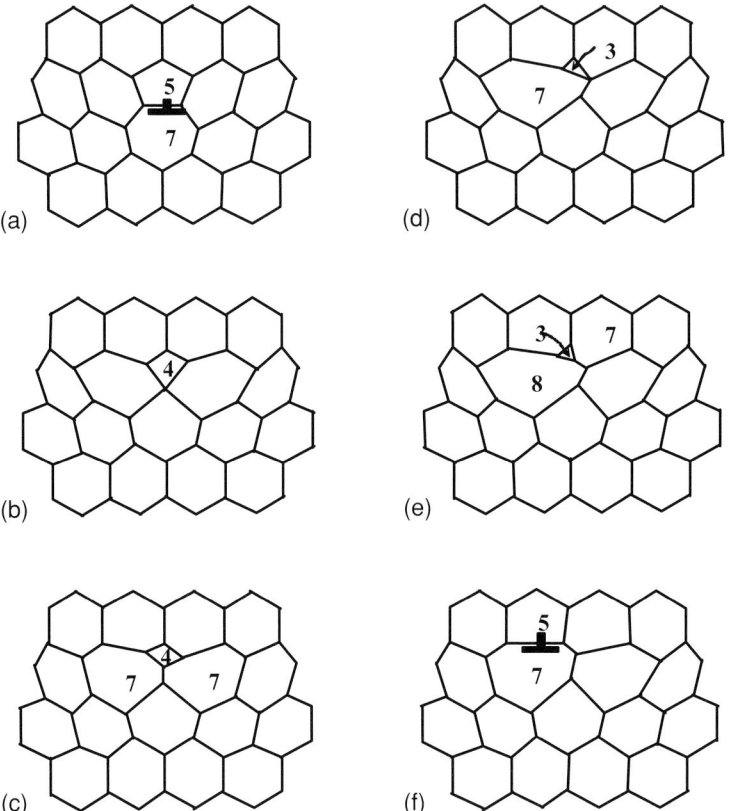

Figure 3.4. Topological transformations of grain array with a cellular dislocation, a pair of neighboring grains with 5 and 7 grain boundaries (in the spirit of the Hillert (1965) theory). These transformations result in elimination of one grain, climb of cellular dislocation, and increase in the mean grain size.

The second term on the left-hand side of formula (3.12) describes the grain-size change in time t during grain growth, specified by the velocity $v = dR/dt$ and driven by grain-boundary curvature. Feltham (1957) and Hillert (1965) derived the parabolic law $R = Kt^{1/2}$ for grain-growth kinetics.

Further development of grain-growth theories was given by Rhines and Craig (1974) who introduced the two important notions, grain-boundary sweeping and structural gradient. The first notion, after the Doherty (1975) modification, is specified by the "sweep constant," meaning the number of grains which vanish when grain boundaries sweep through a volume equal to that of the mean grain volume. The second notion, after the Doherty (1975) modification, is specified by the mean curvature of grain boundaries per grain as "structural gradient" representing the tendency to grain growth in a system. With these notions, Rhines and Craig (1974) obtained the exponent $m = 0.33$ in formula (3.9) for grain-growth kinetics.

Kurtz and Carpay (1980) developed a statistical theory of grain growth, involving the concept of topological classes of grains. Each topological class of grains in a polycrystalline aggregate includes all grains specified by the same number of grain boundaries. During grain

3.3 Grain-growth theories for conventional grain size materials

growth, grains undergo topological transformations resulting in their transfers between different topological classes. Sizes of grains belonging to one topological class are assumed to be log-normally distributed. Theory (see Kurtz and Carpay, 1980) effectively combines topological requirements, energy characteristics, and statistical character of grain growth.

However, neither the theory (Kurtz and Carpay, 1980) of topological classes, nor other statistical deterministic theories (Feltham, 1957; Hillert, 1965; Louat, 1974; Rhines and Craig, 1974) were successful in explaining all the grain-growth attributes (a)–(d) experimentally observed in a wide variety of materials. It was done by the theory (Chen, 1987; Pande, 1987; Pande and Dantsker, 1990; 1991; 1994; Pande et al., 2001; Pande and Rajagopal, 2001) treating grain growth as a stochastic process. The key idea of this approach is to involve a random ("noise") term, parallel with conventional deterministic term(s), in the evolution of a given grain in a polycrystalline aggregate. In doing so, it is assumed that the random term reflects the stochastic nature of internal processes and topological factors influencing growth of an individual grain sharing its boundaries with neighbors. With these processes and factors taken into account, the growing grain shows unpredictable deviations from the deterministic behavior caused by its effective average environment and described by mean-field theories. Following the stochastic theory (Pande et al., 2001; Pande and Rajagopal, 2001), grain-growth kinetics in its simplest form handling with two variables, the grain size R and time t, obeys the Langevin equation:

$$dR/dt = A(R, t) + [B(R, t)]^{1/2} T(t). \tag{3.13}$$

Here the grain-size change dR/dt in time is governed by both its average growth $A(R, t)$ and fluctuations $B(R, t)$ around the given grain, whereas $T(t)$ represents the random process associated with sharing of grain boundaries. With the appropriate choice of the terms on the right-hand side of Eq. (3.13), Pande et al. (2001) and Pande and Rajagopal (2001) effectively described the experimentally measured grain growth attributes (a)–(c) and derived the parabolic law $d = Kt^{1/2}$ for grain-growth kinetics.

The classical theories of grain growth in conventional coarse-grained polycrystals commonly assume both triple junctions to be characterized by infinite mobility and grain growth to occur through grain-boundary migration. With these assumptions, grain-boundary curvature serves as the main parameter of the driving force for grain growth keeping equilibrium angles (120°) made by grain boundaries at triple junctions. However, the assumptions under consideration are questionable especially in the case of nanocrystalline materials.

First of all, experiments (Galina et al., 1987; Czubayko et al., 1998) and computer simulations (Upmanyu et al., 2002) of grain growth in coarse-grained polycrystals are indicative of a finite mobility of triple junctions of grain boundaries. The finite mobility is related to the fact that triple-junction movement is accompanied by transformations of grain-boundary defect structures. For instance, the triple-junction movement shown in Figure 3.5 is accompanied by transformations of grain-boundary dislocations and re-arrangement of point defect distributions in the vicinity of the triple junction. These processes are slow and/or need energetic barriers to be overcome. The convergence of dislocations A and B (Figure 3.5a) resulting in the formation of a new grain-boundary dislocation C (Figure 3.5b) during triple-junction movement is characterized by an energetic barrier associated with re-arrangement of dislocation stress fields. Re-arrangement of point-defect distributions in

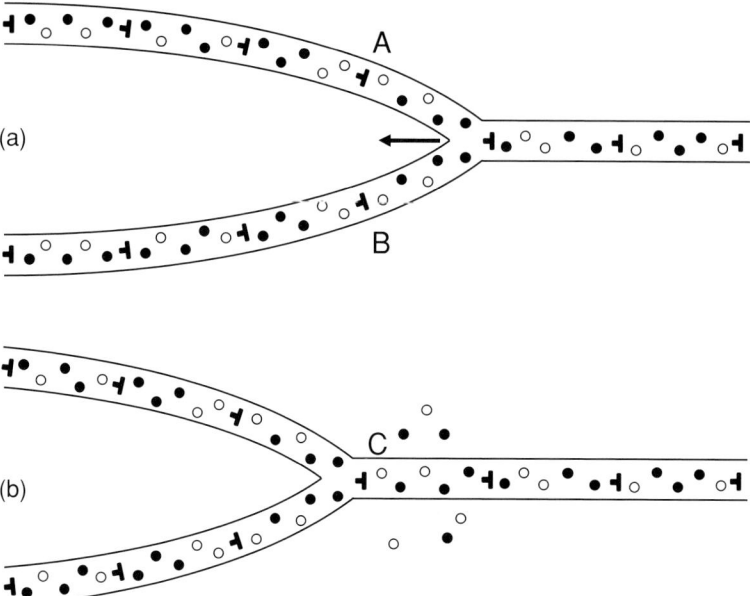

Figure 3.5. Movement of triple junction. (a) Initial and (b) new positions of a triple junction are shown. Its movement is accompanied by convergence of grain-boundary dislocations A and B, resulting in the formation of a new grain boundary C, and re-arrangement of vacancies (open circles) and impurities (full circles) in the vicinity of the triple junction.

the vicinity of the moving triple junction (Figure 3.5) is controlled by slow bulk and grain-boundary diffusion processes (Rabkin, 1999). As a corollary, triple-junction mobility is finite and, generally speaking, different from that of grain boundaries. In particular, the rate of triple-junction movement (accompanied by structural transformations shown in Figure 3.5) can be essentially lower than that of grain-boundary migration.

Following theoretical models (Rabkin, 1999; Gottstein et al., 2000), the drag effect of triple junctions drastically influences the grain-growth kinetics at certain conditions. As shown by these models and computer simulations (Upmanyu et al., 2002), the finite mobility of triple junctions causes deviations of triple-junction angles from equilibrium 120° angles and thus changes both grain-boundary curvature and the associated driving force for grain growth. When the finite mobility of triple junctions is lower than the grain-boundary mobility, triple junctions cause the drag effect contributing to hampering force for grain growth. In general, the effects of triple junctions result in a reduced rate of microstructure evolution during grain growth (Gottstein et al., 2000). More than that, according to analytical calculations made by Gottstein et al. (2000), for certain low values of the mobility of triple junctions, their drag effect completely retards shrinkage of grains in 2D model polycrystals.

Computer simulations (Upmanyu et al., 2002) show that the drag effect of triple junctions on 3D grain growth increases with decreasing the grain size and/or temperature. In particular, the drag effect of triple junctions is expected drastically to influence grain growth in nanocrystalline materials where the density of triple junctions is extremely high. Theoretical estimates (Pande and Masumura, 2003) indicate that the triple-junction drag effect

3.3 Grain-growth theories for conventional grain size materials

is capable of completely suppressing grain growth in 2D model nanocrystalline structures. These theoretical results are interesting in the interpretation of experimental data for high thermal stability of several nanocrystalline materials (see Okuda *et al.* (2001) and other experimental data discussed in the next section).

Grain boundaries in nanocrystalline materials are very short, nanoscale structural elements. This factor is also important for understanding the nature of structural stability of the nanocrystalline state. Grain boundaries are stress sources as they commonly contain defects (grain-boundary dislocations, etc.) inducing stress fields (Sutton and Balluffi, 1996). In nanocrystalline materials, there is the strong elastic interaction between neighboring nanoscale grain boundaries, because (extremely short) distances between them are close to the characteristic length scales of their stress fields. With the elastic interaction, stable grain-boundary defect configurations are formed that minimize the elastic energy of grain arrays in nanocrystalline materials (Ovid'ko *et al.*, 2006). Such grain arrays exhibit a certain structural stability, because there is an energy barrier needed to destroy or transform low-energy defect configurations at neighboring nanoscale grain boundaries. Thus, both very short grain boundaries and triple junctions distributed with very high density, hamper grain growth in nanocrystalline materials.

At the same time, the presence of triple-junction and grain-boundary phases characterized by large volume fractions essentially increases the mean energy density (per unit volume) of nanocrystalline materials and thereby the driving forces for grain growth, compared to the situation with conventional coarse-grained materials. However, with high diffusivity along grain boundaries and triple junctions even at low temperatures, it is possible effectively to decrease their energy and thus retard grain growth through segregation of impurities at grain boundaries (see discussion in the next section).

Classical theories of grain growth in coarse-grained polycrystals commonly assume all grain boundaries in a sample to have the same energy density (per unit area of grain boundary). However, the energy density of a grain boundary in a real polycrystal is highly sensitive to its misorientation and crystallography parameters (Sutton and Balluffi, 1996). The reduction in the energy of grain boundaries with the change of their misorientation parameters drives crystal lattice rotations in grains (Li, 1962; Harris *et al.*, 1998; Moldovan *et al.*, 2001; Haslam *et al.*, 2003). In doing so, the rate and geometric parameters of lattice rotations are controlled by diffusion along grain boundaries and dislocation-transformation processes in grain boundaries (Li, 1962; Harris *et al.*, 1998; Moldovan *et al.*, 2001; Haslam *et al.*, 2003; Gutkin *et al.*, 2003; Gutkin and Ovid'ko 2004). In particular, when grain size decreases, diffusion-assisted crystal-lattice rotations in grains are enhanced (Moldovan *et al.*, 2001). A crystal-lattice rotation in a grain, resulting in both disappearance of a grain boundary and coalescence of its adjacent grains, can conduct grain growth (Li, 1962; Nichols *et al.*, 1993; Haslam *et al.*, 2003). This mechanism for grain growth is hardly effective in coarse-grained polycrystals, because it demands a high-energy structural transformation of a large-scale grain as a whole. At the same time, this mechanism is capable of essentially contributing to grain growth in nanocrystalline materials where crystal-lattice rotations at the nanoscale level are enhanced (Moldovan *et al.*, 2001), compared to those in coarse grains. The role of crystal-lattice rotations in grain growth and deformation processes in nanocrystalline materials under thermal treatment and mechanical load represents the subject of intensively growing interest in nanomaterials science (see Mukherjee, 2002; Murayama et al., 2002; Ovid'ko, 2002; 2005; Haslam et al., 2003; Shan et al., 2004; Soer et al., 2004; Jin et al., 2004; Gutkin and Ovid'ko, 2004). We will discuss this subject with the focus placed on rotational plastic deformation in nanocrystalline materials in Chapter 5.

To summarize, classical grain-growth theories for conventional grain-size materials commonly operate with both topological requirements on grain arrays and grain-boundary curvature as the main parameter of the driving force for grain growth. Grain-boundary migration driven by grain-boundary curvature is viewed to be the dominant mechanism for grain growth. The dominant hampering force is related to a finite mobility of grain boundaries. Statistical deterministic and stochastic aspects are involved in mean-field and stochastic grain growth theories for conventional grain-size materials, respectively. In recent years, particular attention has been paid to the effects of triple junctions (whose mobility and energy parameters are different from those of grain boundaries) on grain growth. From theoretical analysis (Rabkin, 1999; Gottstein et al., 2000) and computer simulations (Upmanyu et al., 2002) it has been revealed that the effects of triple junctions result in a reduced rate of microstructure evolution during grain growth.

Nanocrystalline materials contain high-density ensembles of triple junctions and very short grain boundaries, both drastically affecting grain growth. The drag effect of triple junctions and the formation of stable defect configurations at neighboring nanoscale grain boundaries cause large hampering forces for grain growth and thereby enhance structural stability of the nanocrystalline matter. At the same time, the presence of triple-junction and grain-boundary phases characterized by large volume fractions essentially increases the mean energy density (per unit volume) of nanocrystalline materials and thereby the driving forces for grain growth, compared to the situation with conventional coarse-grained materials. Also, besides grain-boundary migration (being the dominant mechanism for grain growth in coarse-grained polycrystals), crystal-lattice rotations in nanoscale grains can effectively carry grain growth in nanocrystalline materials. The driving force for such lattice rotations is related to a decrease in the energy of a grain boundary with varying its misorientation parameters.

Thus, the set of the driving and hampering forces for grain growth is richer, and their average magnitudes are larger in nanocrystalline materials, compared to coarse-grained polycrystals. As a corollary, the balance between the driving and hampering forces for grain growth in nanocrystallne materials is significantly different from that in coarse-grained polycrystals. All these factors open wide perspectives in regulation and control of nanoscale grain growth and structural stability of the nanocrystalline matter. Several effective approaches for suppression of grain growth and stabilization of the structure in nanocrystalline materials will be discussed in the next sections.

3.4 Grain growth at ambient temperature in nanocrystalline materials

Significant grain growth, that is doubling the initial grain size in 24 h, has been observed at room temperature in a number of pure, relatively low-melting-temperature elements such as Sn, Pb, Al, and Mg (Birringer, 1989). Because of the large amount of grain-boundary enthalpy stored in the large grain-boundary area in nanocrystalline materials, a high driving force for grain growth is expected. Günther *et al.* (1992) studied grain growth in pure Cu, Ag, and Pd, and found grain growth occurring at much lower temperatures than those observed for recrystallization of the elements after heavy cold deformation. In fact, grain growth in Cu and Pd was observed even at room temperature. This is particularly dramatic for Pd which has a high melting temperature of 1552 °C such that room temperature is only 0.16 of the melting temperature – a very low homologous temperature. In all these cases the grain growth was observed to be abnormal. That is, a few grains grew to micron sizes while most grains remained nanoscale. Gertsman and Birringer (1994) studied abnormal grain growth in nanocrystalline Cu prepared by the inert-gas-condensation method with bulk densities of 93%, 96%, and 97%. After holding the samples for times of more than one month at room temperature, abnormal grain growth was observed in all samples. The abnormally coarse grains exhibited a range of sizes; most were <1 μm, but some were >2 μm. The nanocrystalline grains surrounding the coarser grains were about 10–50 nm in size. However, the coarser-grained regions represented only a few percent of the sample volume, with the nanocrystalline grains remaining stable in most of the samples. The growth rate for the average grain size, determined by X-ray diffraction line-broadening analysis, was larger for the samples with the higher average densities, implying that porosity was hindering grain-boundary migration. The general explanation given for abnormal grain growth is an inhomogeneous structure for the as-processed samples such that grain-growth inhibitors such as pores, impurities, or even grain-boundary structure, are not evenly distributed and rapid grain growth can occur where such inhibitors are absent owing to the large driving force and high grain-boundary mobility in these regions.

3.5 Inhibition of grain growth in nanocrystalline materials

In spite of the high driving force for grain growth and the observation of grain growth, at least abnormal grain growth, even at very low homologous temperatures, significant stabilization of nanocrystalline grain structures has been observed in many materials. The one common feature of such materials is that they are multicomponent, that is, either alloys or contain impurities. There are two basic ways in which grain growth can be reduced. The first is the *kinetic approach* in which the grain boundaries are pinned in various ways to decrease grain-boundary mobility. The second is the *thermodynamic approach* in which the driving force for grain growth is lowered by reducing the grain-boundary energy. These two approaches are discussed below in the context of nanocrystalline microstructures.

3.5.1 Kinetic approach

In the kinetic approach, the grain-boundary mobility is reduced by various possible mechanisms.

- Porosity drag. Porosity can reduce grain-boundary mobility. Hofler and Averback (1990) demonstrated that grain growth in ceramic TiO_2 prepared by the inert-gas-condensation method can be dramatically influenced by the level of porosity. Samples annealed for 20 h at 700 °C with a porosity of 25% had a grain size of 30 nm compared to a grain size of about 500 nm for samples with the lower porosity of 10%. From a practical point of view porosity is not a feasible method for enhancing grain-size stability since, in order to obtain good densities, higher sintering temperatures are needed which then result in grain growth. In addition, porosity can have major deleterious effects on mechanical behavior, as will be discussed more in Chapter 4.

- Second phase drag. There have been a number of examples of grain-boundary pinning in nanocrystalline materials by the presence of small second-phase particles – the Zener drag mechanism (Humphreys and Hatherly, 1996). The expression for the pinning pressure exerted on the grain boundary by small particles is:

$$P_Z = \frac{3F\gamma}{2r} \qquad (3.14)$$

where P_Z is the pinning pressure exerted by the particles on a unit area of the boundary, F is the volume fraction of randomly distributed spherical particles of radius r, and γ is the specific grain-boundary energy. It is clear that a high volume of small particles is desirable to increase the pinning pressure and thus to impede grain-boundary mobility. The stabilization in 5 nm grain size Ni–P alloys, which were produced by electrodeposition, did not take effect until after an increase of the grain size by a factor of 2–3 (Boylan et al., 1991). The solid solution of the Ni–1.2% P alloy did not measurably grow and was stable as a solid solution up to an annealing temperature of 473 K. For annealing temperatures of 523 K and 623 K, the grain size initially increased by a factor of 2–3, before the grain size stabilized, which coincided with the precipitation of a second phase, Ni_3P. Although other retarding effects in reducing the grain-boundary mobility are considered (solute drag), the pinning of the grain boundaries by Ni_3P as the second phase was believed to be the major factor of the grain-size stabilization at 573 K and 623 K. The same effect was found for electroplated 10 nm Ni with an addition of 0.12 wt.% S (El-Sherik et al., 1992), where it was attributed to the formation of nickel sulfide precipitates. Another example of nanocrystalline grain-size stabilization by second-phase particles is the work of Perez et al., 1998. The authors studied the thermal stability of cryomilled Fe–10 wt.% Al and found that a grain size of about 20 nm was maintained after annealing to about 600 °C and even after annealing at 1100 °C the grain size grew only to about 50 nm. This excellent grain-size stability was attributed to the presence of nanoscale Al_2O_3 and AlN dispersoids. Figure 3.6 illustrates their results, wherein the grain size of Fe is compared with that for the Fe–10 wt.% Al alloy as a function of heat treatment temperature. Another example of small-particle-induced grain-size stabilization is for an Al–base alloy by Shaw et al. (2003). An $Al_{93}Fe_3Cr_2Ti_2$ alloy was prepared by mechanical alloying. The as-milled alloy was a solid solution of Al. After annealing at various temperatures the internal strain was released, intermetallic compound particles precipitated, and there was some grain growth. The precipitation of a variety of intermetallics with Cr, Fe, or Ti was followed with several analytical techniques. There was essentially no grain growth at temperatures up to about 300 °C, with the grain-size distribution from TEM being

3.5 Inhibition of grain growth in nanocrystalline materials

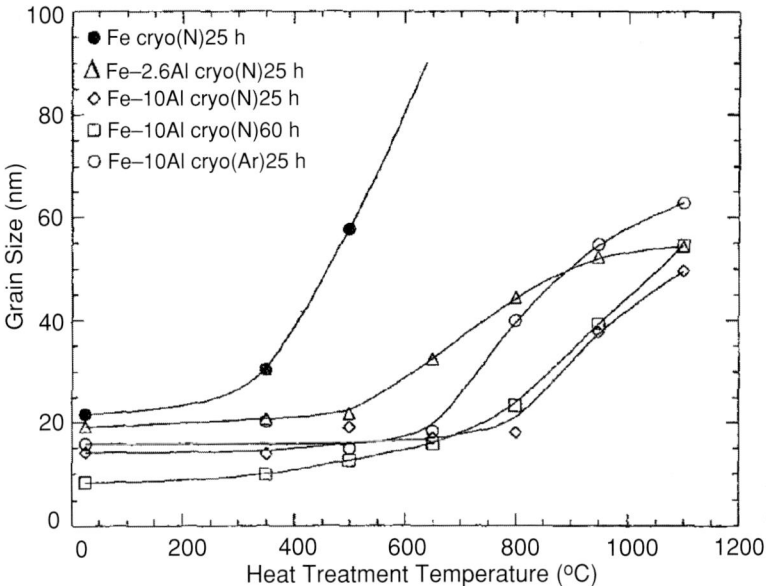

Figure 3.6. Grain size of cryomilled powders after 1 h heat treatment. Reprinted with permission. (Perez et al., 1998).

about 6–45 nm. Even after heating to 450 °C ($0.77T_m$), the grain-size distribution was still nanoscale, that is, from 20nm to 100 nm. The authors believe that the inhibition of grain growth at the lower annealing temperature was caused by solute drag on the grain boundaries, but at the higher temperatures was caused by pinning of the boundaries by the nanoscale intermetallic precipitates.

- Solute drag. In the case of grain growth in nanocrystalline materials containing solutes or impurities, the segregation of the solute atoms to the grain boundaries may depend upon the grain size. This problem was treated by Michels et al. (1999), with several experimental examples given. The equilibrium segregation to grain boundaries can be attained by diffusion of solute atoms through the lattice to the boundaries, or by the entrapment of solute at the boundaries as they move. Since in nanocrystalline materials grain growth usually occurs at temperatures too low for significant lattice diffusion, it is suggested that the latter mechanism is more likely. This mechanism of solute entrapment by moving boundaries was used by Knauth et al. (1993) to explain the differences in DSC results for grain growth in nanocrystalline Ni and in Ni–1at.% Si. While grain growth started at about the same temperature in both materials, the growth in the Ni–1at.%Si alloy slowed dramatically, presumably as the result of entrapment of Si atoms at the grain boundaries and their effect on the boundary mobility. Michels et al. (1999) provided additional evidence for solute entrapment in moving grain boundaries for the Pd–19 at.% Zr nanocrystalline alloy. The grain size increased with annealing time at 600 °C from about 5 nm to a saturation value of about 16 nm. At the same time, the solute composition at grain boundaries (determined by lattice-parameter measurements) increased from 19% Zr to about 27% Zr.

- Chemical ordering. Reduced grain-growth has been observed for ordered nanocrystalline intermetallic compounds. For example, the thermal stability of Fe_3Si with and without 5% Nb additions was studied by Gao and Fultz (1993; 1994). The mechanical alloying of the alloys resulted in disordered products with an initial grain size <10 nm for the $(Fe_3Si)_{95}Nb_5$ alloy. Three different processes were identified to occur simultaneously when the samples were annealed isothermally at temperatures between 350 °C and 500 °C. In an initial stage, the grain growth is thought to take place uninhibited. After that, the segregation of Nb to the grain boundaries, occurring at the same time as grain growth, is believed to provide the alloy with considerable stabilization of the grain size over the Fe_3Si alloy. The other process that is thought to inhibit the grain growth is the chemical ordering process to form the DO_3 structure of the ordered Fe_3Si, which did not happen homogeneously, but rather by nucleation and growth. It was also suggested that the segregation of the Nb atoms to the grain boundaries and the ordering process are interdependent. The ordering process is simplified by the diffusion of the Nb atoms out of the grains, because the ordering is easier in Fe_3Si compared to $(Fe_3Si)_{95}Nb_5$. Bansal *et al.* (1995) replaced the Nb additions with Mn, which is not expected to segregate to the grain boundaries since it replaces Fe atoms on specific sites in the ordered DO_3 structure. The ordering kinetics of the alloy with Mn were found to be accelerated compared to the binary alloy. At the same time, the grain growth was slowed down, which was linked to the enhanced ordering kinetics. The authors suspected decreasing grain-boundary mobility with an increase of the ordering parameter of the alloy upon annealing. They conclude that once ordering had taken place, the grain growth is inhibited, hypothesizing that the grain-boundary mobility is decreased in the ordered compound. This may be because diffusion is typically slower in ordered structures compared with their disordered counterparts.
- Grain-size stabilization. Grain-size dependent stabilization of a nanocrystalline microstructure has been experimentally observed (Lu, 1993; Krill *et al.*, 2001), and more recently predicted by theoretical models. Estrin *et al.* (2000) suggest that at small grain sizes the rate-controlling step for boundary migration is transport of excess volume, located at the grain boundaries, away from the moving boundaries. In computer simulations of Upmanyu *et al.* (1998), it was shown that the excess volume released from the boundaries during grain growth is incorporated into the crystalline lattice in the form of vacancies. This increase in the nonequilibrium vacancy concentration gives rise to an increase in free energy that counteracts the decrease in free energy associated with the reduction in grain boundary area during grain growth. These theories lead to a prediction of a linear dependence of the grain size on annealing time when the grain size is below some critical value. Linear growth kinetics are in agreement with experimental work of Krill *et al.* (2001), on nanocrystalline Fe.

3.5.2 Thermodynamic approach

Since the driving force for grain growth is directly proportional to the grain-boundary energy, reducing the grain-boundary energy should minimize grain growth. Addition of

3.5 Inhibition of grain growth in nanocrystalline materials

solute atoms that segregate to the grain boundaries will affect the grain-boundary energy. The energy varies with the overall solute concentration, c_s, according to the Gibbs (adsorption) equation

$$\partial \gamma_b / \partial \ln c_s = -RT\Gamma_s, \quad (3.15)$$

where T is the temperature, and Γ_s is the interfacial excess of the solute atoms. For grain boundary segregation of solute, $\Gamma_s > 0$ and therefore γ_b will decrease with increasing c_s. Hondros and Seah (1983) reported a decrease in γ_b for several binary alloys. The initial slope of the γ_b vs. c_s curves is more negative the greater the difference in atomic size is between the solute and solvent atoms. This suggests that elastic strain energy induces solute segregation to grain boundaries (Hondros and Seah, 1983). Based on this, solute additions to induce the lowering of the grain-boundary energy and stabilize the grain size against coarsening should be much larger, or smaller than the host atom. Weissmuller (1993; 1994) has applied these ideas to nanocrystalline materials and quantified the effects of segregation on grain-size stability. He predicts that, for alloy systems with a large heat of segregation, the nanocrystalline alloy is in a metastable state for a particular grain size which decreases with increasing concentration of the solute element. In the metastable state, the specific grain boundary energy is zero. Subsequently, Kirchheim (2002) and Liu and Kirchheim (2004) have extended these concepts and have compared their predictions with experimental data. The grain size at the metastable thermodynamic equilibrium was found in their analysis to be determined by the grain-boundary energy, the enthalpy change of grain-boundary segregation, and the solute excess of an equivalent grain-boundary monolayer at saturation. Good agreement for these predictions of the temperature dependence of the metastable grain size was obtained for $Pd_{100-x}Zr_x$ alloys (Krill and Birringer, 2001). This is illustrated in Figure 3.7. Other experimental studies of the stabilization of nanocrystalline grain size by the lowering of the grain-boundary energy due to solute segregation have been reported for Ni–P alloys (Farber et al., 2000), RuAl with Fe impurities (Liu and Mucklich, 2001), Nb–Cu alloys (Abe et al., 1992), Ti–Cu alloys (Abe and Johnson, 1992), Y–Fe alloys (Weissmuller et al., 1992), and TiO_2 with Ca (Terwilliger and Chiang, 1995). In those alloy systems that are metastable solid solutions due to nonequilibrium processing, the grain-size stabilization appears to be dominated by the segregation of solute to the grain boundaries and the lowering of the grain-boundary energy at lower annealing temperatures. At higher temperatures when precipitation of solute-rich phases can occur, dramatic increases in grain growth are typically observed.

The thermodynamic approach to grain-size stabilization by segregation of solute, impurity atoms to the grain boundaries appears to be an effective method for stabilization of nanocrystalline grain size. However, in terms of mechanical behavior it is not yet clear whether such grain boundary segregation will not always lead to an embrittlement of the alloy. More research is needed on this topic.

A special case of thermodynamic stabilization of nanocomposites is the formation of a nanostructure in a spinodally decomposed quasibinary system. Because such nanocomposites have so far been prepared only as thin films, this topic will be discussed in Section 3.7.

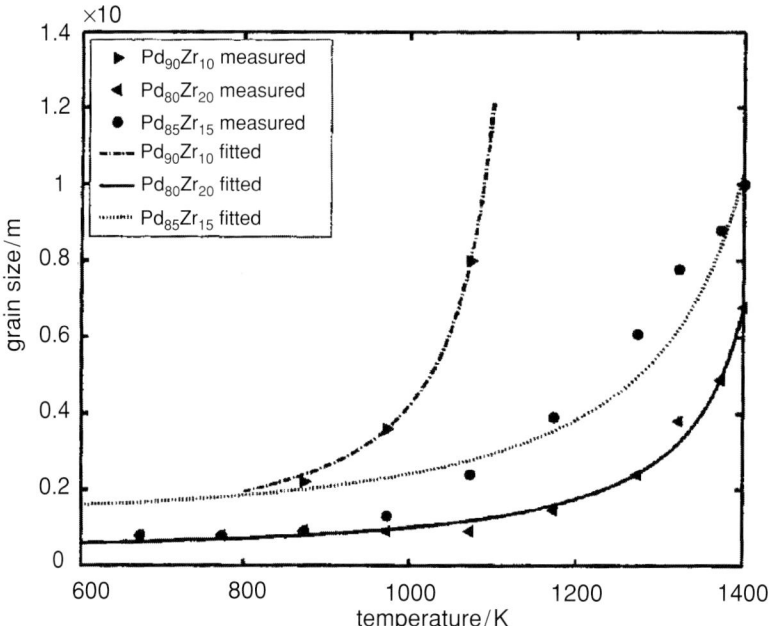

Figure 3.7. Theoretical fits of grain size vs. annealing for nanocrystalline $Pd_{100-x}Zr_x$ alloys. (From Liu and Kirchheim (2004), reproduced with permission of Elsevier.)

3.6 Experimental studies of isothermal grain-growth kinetics in nanocrystalline materials

There have now been a number of studies of the isothermal grain-growth kinetics in nanocrystalline materials. One approach to presenting the data is to assume Eq. (3.9) holds and defines the grain-growth exponent as $n = 1/m (\leq 0.5)$. Isothermal grain-growth data from the literature for nanocrystalline materials have been analyzed by fitting to Eq. (3.9) and are presented in Figure 3.8, where n is plotted against the reduced annealing temperature T/T_m. Here, T_m is the melting or liquidus temperature of the given material. The lines connecting data for a given material/reference are drawn only to aid the reader. While there is a good deal of variation from $n < 0.1$ to $n = 0.5$ the general trend is for n to increase toward the ideal 0.5 value with increasing reduced annealing temperature. For comparison, for conventional grain size (<5–10 µm) materials, a plot of n vs. T/T_m is given in Figure 3.9 from the paper by Hu and Rath (1970). These data extend to higher T/T_m values but exhibit similar trends to those for nanocrystalline materials, with n increasing toward 0.5 as T/T_m increases. In fact, the ideal value of $n = 0.5$ has been observed only for conventional grain-size materials in high-purity elements such as Cd, Fe, and Sn at high values of T/T_m (Vandermeer and Hu, 1994).

Often the activation energy of grain growth is compared to the activation energy of lattice and grain-boundary diffusion in order to help determine the underlying mechanisms involved. Malow and Koch (1996a; 1996b) summarized activation energies for grain growth

3.6 Experimental studies of isothermal grain-growth kinetics

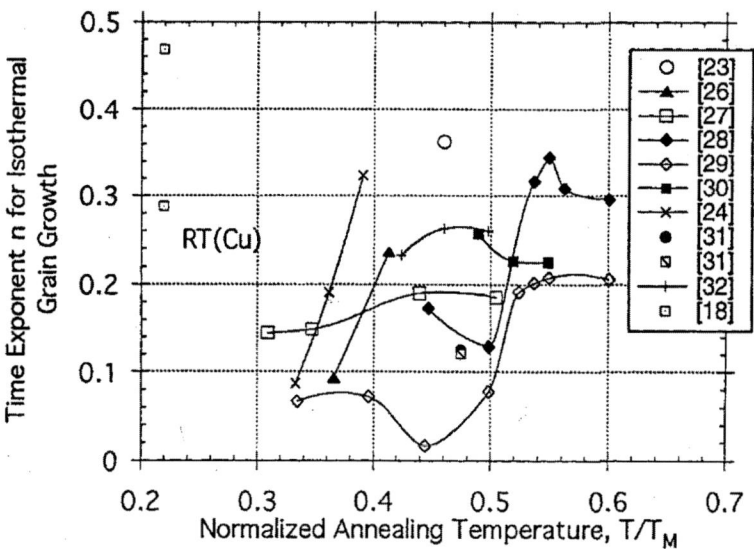

Figure 3.8. Time exponent for isothermal grain growth of various nanocrystalline materials as a function of the reduced annealing temperature. (From Malow and Koch (1996a), reproduced with permission of TMS.)

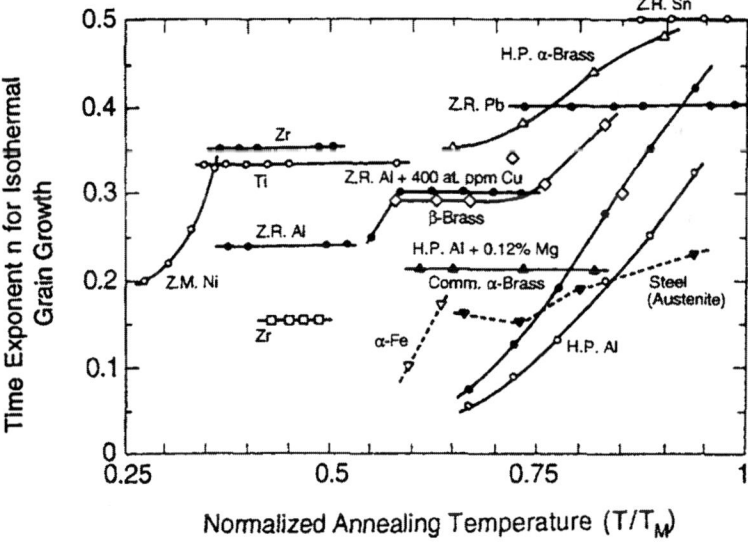

Figure 3.9. Time exponent for isothermal grain growth of polycrystalline metals and alloys as a function of the reduced annealing temperature. (From Hu and Rath (1970), reproduced with permission of TMS.)

of nanocrystalline materials and compared them to diffusion data. In some cases estimates for grain-boundary diffusion activation energies had to be made owing to lack of experimental data. In these cases the activation energy for grain-boundary diffusion was assumed to be 0.5 that for lattice diffusion. It was observed from this compilation of data that the majority of activation-energy values for grain growth in nanocrystalline materials compared more closely to those for grain-boundary diffusion as opposed to lattice diffusion. There are a few exceptions to this general trend. Grain growth in nanocrystalline Fe produced by ball milling was analyzed by using two different models. Each of the two models could be used only to deduce an activation energy for grain growth for one part of the range of annealing temperatures. The first model yielded grain-growth exponents n that exhibited the same general trend as zone-refined polycrystalline Fe with n increasing toward 0.5 as T/T_m increases and an activation energy for the lower four annealing temperatures of 125 kJ/mol. The second model, which takes pinning forces on the grain-boundary migration into account, gave an activation energy of 248 kJ/mol. The former value is lower than either the activation energy for grain-boundary or lattice diffusion in Fe, and the latter is close to the values found for grain growth in polycrystalline Fe and that for lattice self-diffusion in Fe. These two values suggest two different mechanisms for grain growth at the lower vs. the higher annealing temperatures. Eckert *et al.* (1992) found that the activation energy for grain growth in Pd compared closely to that of lattice self-diffusion in Pd, while for Ni they found that the activation energy was close to that of grain-boundary self-diffusion in Ni. The difference was attributed to a lower excess volume in Pd than in Ni which leads to a higher activation energy for grain growth in Pd. In contrast to these results on Ni, Knauth *et al.* (1993) also studied the grain growth in Ni and Ni–1 at.% Si, as described in Section 3.5.1. The activation energy for grain growth determined by the Kissinger analysis from DSC data of 220 kJ/mol is between the activation energy for grain-boundary diffusion and lattice self-diffusion for Ni. The authors suspect that a solute drag effect hinders the grain-boundary motion. Similar results were obtained for grain growth in Ni–1.2 wt.% P, where the Ni_3P precipitates were thought to have inhibited the grain-boundary motion.

Studies of grain growth in electrodeposited Ni by TEM and DSC revealed interesting behavior (Wang *et al.*, 1997; Klement *et al.*, 1995). The coarsening of the grain structure was found to have proceeded in three steps. At low temperatures (353–562 K), the first step was given as "nucleation" and abnormal grain growth. It was speculated that the "nucleation" of the sites (grains) of abnormal growth was connected to low-angle grain boundaries in the as-deposited material. Two or more grains with only little misorientation would have to rearrange their orientation only slightly through relaxation into lower-energy configurations, and could merge into significantly larger grains ("subgrain coalescence"). The second and third steps of the microstructural evolution in these Ni foils were normal grain growth between 562 K and 593 K, and finally growth towards equilibrium between 643 K and 773 K. Although the grain growth process is not a transformation process in the usual sense, the authors contend that the classical transformation theory can be used phenomenologically in order to analyze the DSC signals originating from the abnormal grain-growth process. The activation energy that was deduced in that manner had a value close to the grain-boundary self-diffusion for Ni.

Instead of comparing the activation energies of grain growth in nanocrystalline materials with diffusion data that had been obtained in coarse-grained materials, it would be more appropriate to compare the grain growth in nanocrystalline materials with diffusion behavior in nanocrystalline materials. Diffusion in nanocrystalline materials has been reviewed by Wurschum *et al.* (2002). While early studies of diffusion on nanocrystalline

3.6 Experimental studies of isothermal grain-growth kinetics

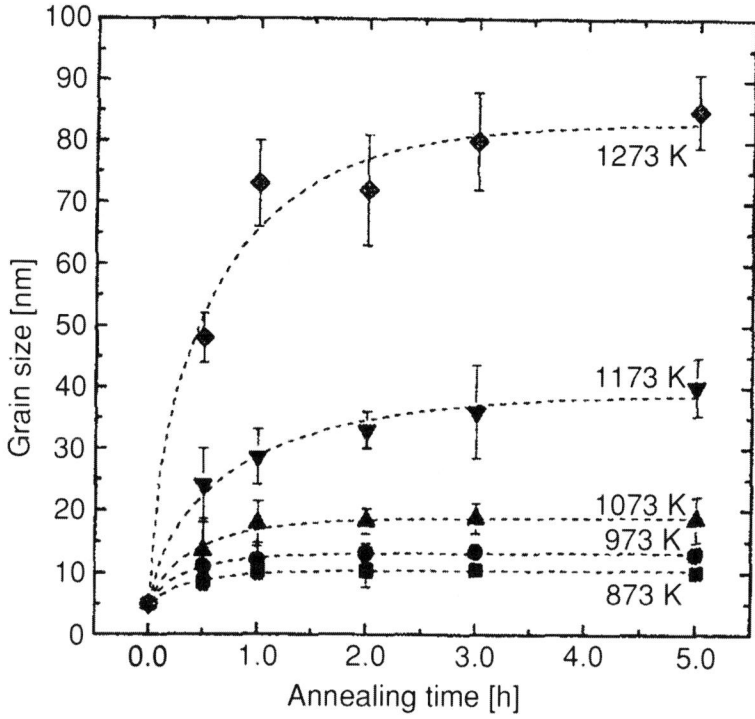

Figure 3.10. Grain size as a function of annealing time at different temperatures for RuAl. (From Liu and Mucklich (2001), reproduced with permission of Elsevier.)

materials prepared by the inert-gas-condensation method, and containing significant porosity, often indicated much higher diffusivities, low activation energies for diffusion than even grain-boundary diffusion in conventional grain-size materials, more recent studies on highly dense nanocrystalline materials in general give diffusion data that are comparable to grain-boundary diffusion data for conventional grain-size materials. This is consistent with the activation energies for grain growth in pure nanocrystalline materials which exhibit reasonable agreement with those for grain-boundary diffusion (Malow and Koch, 1996a; 1996b). The larger values for grain growth observed in some multicomponent nanocrystalline materials are probably the result of the inhibition of grain-boundary motion by various boundary-pinning mechanisms.

The typical behavior observed for isothermal grain-growth studies in nanocrystalline materials shows an initial increase in grain size with annealing time followed by a decreased grain-growth rate and apparent growth stagnation ("lock-up") for longer annealing times. This common behavior is illustrated for RuAl alloys in Figure 3.10 (Liu and Mucklich, 2001). The authors of this work attribute the grain-growth stagnation to boundary pinning by Fe impurities in their samples. The stagnation of grain growth is also evident in plots of grain size vs. annealing temperature for various annealing times as shown in Figure 3.11, also from the above study of RuAl. While the mechanisms responsible for the grain-growth stagnation in nanocrystalline materials will be different in different materials, the average

118   3 Stability of structural nanocrystalline materials – grain growth

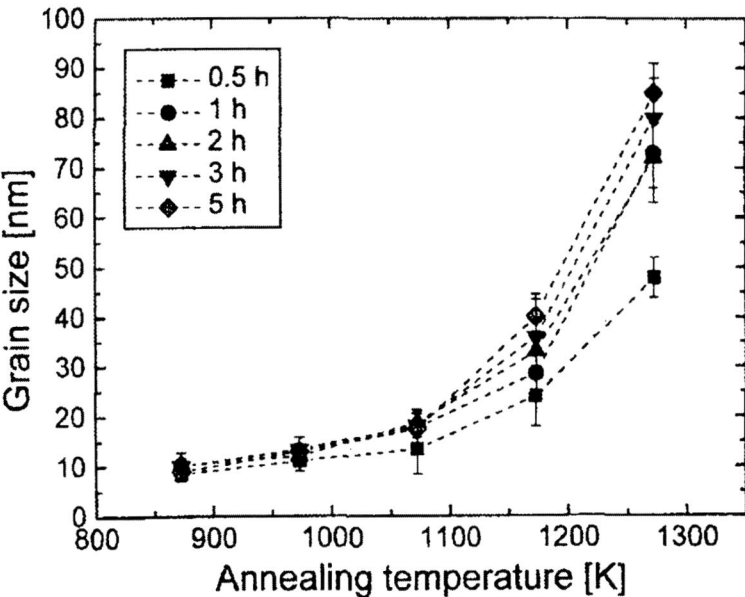

Figure 3.11. Grain size as a function of annealing temperature for different times for RuAl. (From Liu and Mucklich (2001), reproduced with permission of Elsevier.)

grain size appears to follow this behavior. Abnormal grain growth, as noted above, is also seen, especially in pure metals. In this case, the normal grain growth exhibits the stagnation behavior, but the abnormal grain growth can follow different kinetics. The abnormal grains are typically a small volume fraction of the material, however.

3.7 Thin films and coatings

3.7.1 Different behavior of nc-M(1)$_n$N/M(2) and nc-M$_n$N/a-Si$_3$N$_4$ and /a-BN nanocomposites

The two different approaches to superhard nanocomposite coatings that were discussed in Chapter 2, i.e. hardening by energetic ion bombardment and by the formation of a stable nanostructure, result in different thermal stability. Whereas the nc-M(1)$_n$N/M(2) coatings, in which the hardness enhancement was achieved by energetic ion bombardment during their deposition, lose the enhancement upon annealing to ≥ 450 °C, the nc-M$_n$N/a-Si$_3$N$_4$ (M = Ti, W, V, (Ti$_{1-x}$Al$_x$)N, nc-(Cr$_{1-x}$Al$_x$)N/a-Si$_3$N$_4$) and nc-TiN/a-BN nanocomposites prepared according to the generic design principle remain stable up to about 1100 °C. Figure 3.12a shows the dependence of the hardness of ZrN/Ni coatings with different Ni content on the temperature of isochronal annealing in forming gas (90 vol.% N$_2$ + 10 vol.% H$_2$) and Figure 3.12b shows the crystallite size. It is seen that, upon annealing above 400 °C, the hardness enhancement decreases to the bulk value whereas the crystallite size remains constant up to the highest temperature of 700 °C that was tested,

3.7 Thin films and coatings

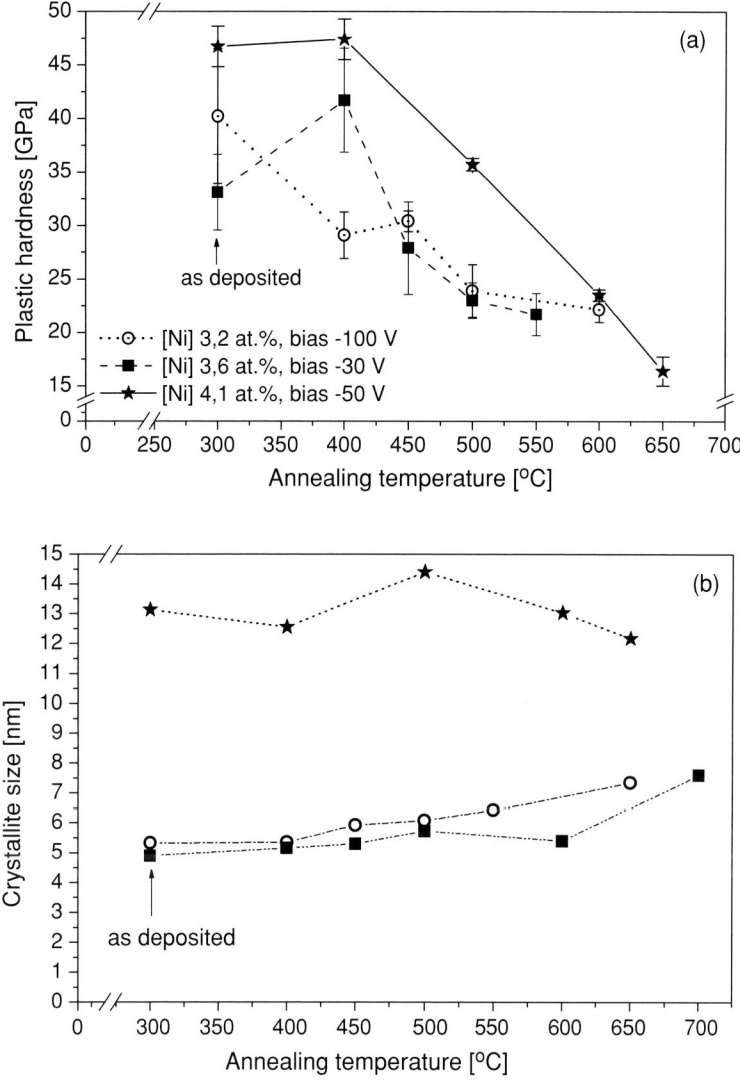

Figure 3.12. (a) Dependence of hardness and (b) of the crystallite size of the ZrN–Ni coatings on the temperature of isochronal annealing in forming gas for 30 min in each step (Karvankova et al., 2001).

because at higher temperatures diffusion of elements from the steel substrate into the coatings occurs. The decrease of the hardness enhancement results from the annealing of the complex effects of the ion bombardment, such as vacancies, interstitials, compressive stress and others which contribute to that enhancement. Similar behavior was observed for other nc-M(1)$_n$N/M(2) coatings (Karvankova et al., 2001) as well as for ordinary ceramic coatings, such as TiN, (TiAlV)N and TiB$_2$ (see references in Veprek, 1999; Veprek et al., 2005a).

120   3 Stability of structural nanocrystalline materials – grain growth

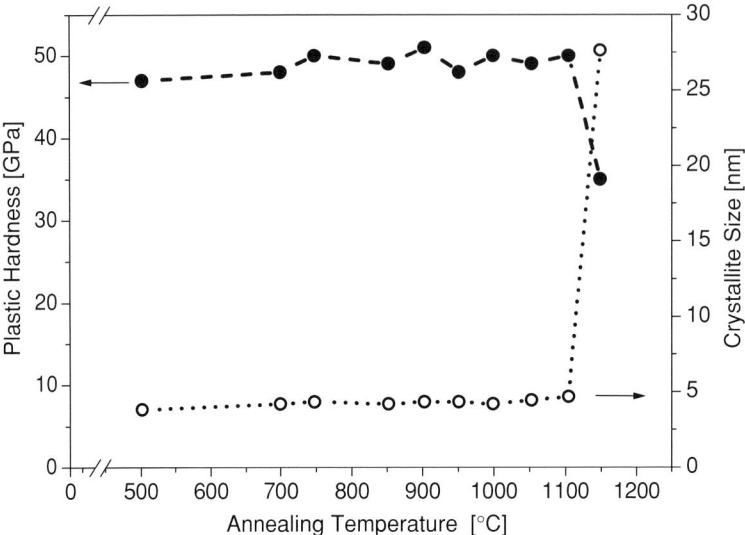

Figure 3.13. Dependence of plastic hardness and crystallite size of nc-TiN/a-Si$_3$N$_4$ superhard nanocomposites, deposited by plasma CVD, on the annealing temperature. The annealing was done in pure nitrogen for 30 min at each step (Veprek et al., 2005a).

Figure 3.13 shows an example of the high thermal stability of the superhard nanocomposites nc-TiN/a-Si$_3$N$_4$ deposited by plasma CVD under conditions where the formation of the nanostructure was completed during the deposition (Veprek et al., 1998; Niederhofer et al., 1999; Männling et al., 2001; Veprek et al., 2005a). A similar stability was found also for other nc-M$_n$N/a-Si$_3$N$_4$ nanocomposites that were deposited under the conditions outlined in the design concept (Veprek and Reiprich 1995) and in more detail later on (Veprek et al., 1998; 2005a; 2005b; Männling et al., 2001). It is seen that both the plastic hardness measured at room temperature after each annealing step as well as the crystallite size remain constant up to 1100 °C. Only above this temperature are coarsening and the concomitant decrease of the hardness found. This temperature corresponds to 66% of the decomposition temperature of Si$_3$N$_4$ of 2151 K where the saturation pressure of nitrogen reaches 1 atm (Chase et al., 1985). The free energy of formation of Si$_3$N$_4$ of -647 kJ/mol normalized to the number of atoms per mol amounts to -92 kJ. This is significantly smaller than that of TiN of $-309 : 2 = 154.5$ kJ. Therefore Si$_3$N$_4$ is less stable than TiN whose decomposition temperature lies above 3000 K (Chase et al., 1985).

As outlined in the earlier sections of this chapter, the thermal stability of nanocrystalline materials prepared via the consolidation route is limited owing to coarsening which occurs when the diffusion becomes significant, i.e. at the so-called Taman temperature of about 40% of the melting point or even less. The high stability of the nanocomposites discussed here is the result of the high immiscibility of the system up to the decomposition temperatures of the less-stable phase. This will be discussed in more detail in the following section. Such a high thermal stability was found only for coatings which were deposited either by plasma CVD or by reactive sputtering at a sufficiently high temperature of

3.7 Thin films and coatings

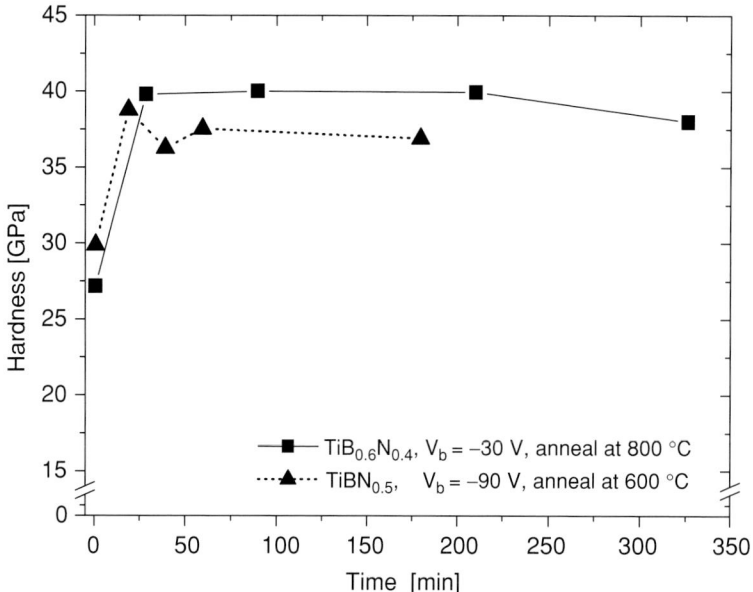

Figure 3.14. Self-hardening of Ti–B–N coatings deposited by means of magnetron sputtering at room temperature upon annealing. (From Hammer *et al.* (1994), with permission from the authors and with permission of Elsevier.)

550–600 °C, and sufficient nitrogen pressure of ≥ 0.002 mbar (Prochazka *et al.*, 2004; Veprek *et al.*, 2005a), which is necessary in order to achieve the strong, thermodynamically driven segregation of the phases during the deposition.

A special case represents the nc-TiN/a-BN and nc-TiN/a-BN/a-TiB$_2$ coatings which, from the thermodynamic point of view, should also show high thermal stability comparable with or even higher than nc-TiN/a-Si$_3$N$_4$. However, minor impurities of oxygen, which are present in the annealing furnace under flow of pure nitrogen owing to water desorption from the walls, cause oxidation of the boron-containing phases. Because boron oxide melts at about 460 °C, the BN and TiB$_2$ do not provide any efficient diffusion barrier to oxygen. It was shown that the maximum thermal stability upon annealing in nitrogen belongs to the nc-TiN/a-BN nanocomposites with about one monolayer of BN between the TiN nanocrystals (Veprek *et al.*, 2005a; Karvankova *et al.*, 2004; 2006). Thus, the practically achievable thermal stability of these nanocomposites is limited by their insufficient oxidation resistance.

When the deposition of "M–Si–N" coatings with the correct composition and sufficient purity is done at a lower temperature, the segregation of the phases will not be completed during the deposition and the hardness enhancement will be either low or absent. However, upon annealing after the deposition, such coatings will show a "self-hardening," i.e. an increase of the hardness upon post-annealing. A representative example is the annealing of Ti–B–N coatings deposited by means of magnetron sputtering at room temperature as reported by Hammer *et al.* (1994), as shown in Figure 3.14. The as-deposited coatings were amorphous and their hardness was relatively low. However, upon annealing at 600 °C and 800 °C they recrystalized forming nanocomposites, the hardness increased to about 40 GPa,

122 3 Stability of structural nanocrystalline materials – grain growth

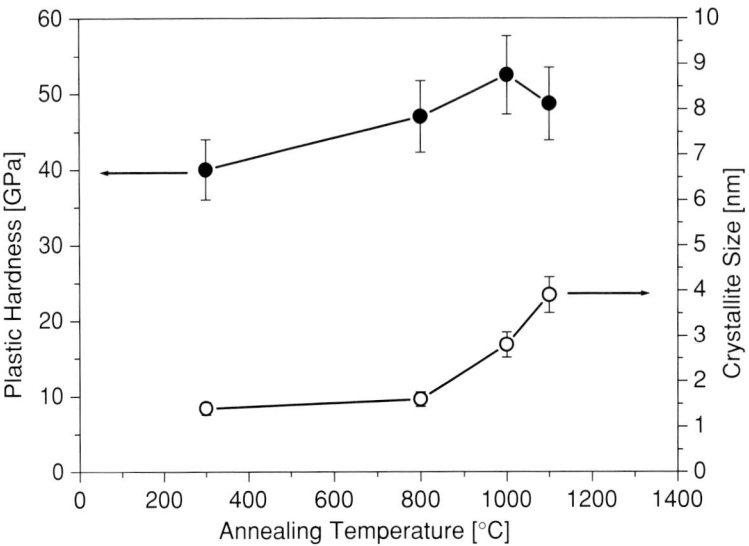

Figure 3.15. Self-hardening of industrial nc-$(Ti_{1-x}Al_x)$N/a-Si_3N_4 nanocomposites upon annealing in nitrogen at 1 atm for a period of 30 min (Männling et al., 2001).

and remained fairly stable upon a prolonged annealing (for further details see Hammer et al. (1994)).

During the plasma-assisted PVD of coatings on machining tools made of high-speed steel (HSS), the temperature has to be kept relatively low in order to avoid overheating of the edges, which always have a higher temperature than the bulk part of that tool. This problem arises because of the focusing effect of the space-charge sheath which results in a higher ion current to the sharp edge. Such coatings also show self-hardening upon post-annealing (Männling et al., 2001). Figure 3.15 shows an example of the behavior of nc-$(Ti_{1-x}Al_x)$N/a-Si_3N_4 nanocomposite coating deposited by means of vacuum arc evaporation in a large-scale industrial coating unit at a low temperature. The as-deposited coatings had a hardness of about 40 GPa which increased to more than 50 GPa upon annealing. Simultaneously, the crystallite size adjusted at a value of about 3–4 nm, which is found in this type of coatings deposited by plasma CVD (see Figure 3.13). In another example shown in (Veprek et al., 2004a), when the coating was deposited at somewhat higher temperature but without applied substrate bias so that the crystallite size of the as-deposited coatings was larger, the crystallite size decreased to 3–4 nm upon annealing when the self-hardening occurred.

It is also interesting to notice that the hardness of the nc-$(Ti_{1-x}Al_x)$N/a-Si_3N_4 nanocomposites remains very high even after annealing at 1200 °C. The metastable solid solution $(Ti_{1-x}Al_x)$N coatings decompose to fcc-TiN and h-AlN and soften upon annealing above 900 °C. The mechanism of this decomposition was recently studied in much detail by Mayrhofer and co-workers who have shown that, above about 800 °C, a surface-mediated, spinodal decomposition into cubic TiN and AlN occurs. Upon further annealing to $\geq$1000 °C, these transform into the softer h-AlN (Mayrhofer et al., 2003). However, as seen in Figure 3.15, the nc-$(Ti_{1-x}Al_x)$N/a-Si_3N_4 nanocomposite coatings do not show such

3.7 Thin films and coatings

behavior. In the case of another example of such coatings with a somewhat higher content of Si_3N_4, the hardness remained above 40 GPa and the fcc structure of the $(Ti_{1-x}Al_x)N$ remained preserved even after annealing to 1200 °C. The onset of softening and coarsening was observed only at 1250 °C, when a significant diffusion of cobalt from the substrate made of cemented carbide occurred (Veprek et al., 2004a).

These examples show that the formation of the stable nanostructure consisting of hard transition-metal nanocrystals "glued" together by about one monolayer of Si_3N_4 avoids not only coarsening and softening at elevated temperatures, but also the segregation of other elements towards the interface. Moreover, as will be discussed in Chapter 6, this dense and strong interface also provides the nanocomposites with a high oxidation resistance which is very important for dry and fast machining, for example, because at high temperatures of ≥ 800 °C which occur under such conditions, the chemical wear dominates.

The self-hardening is associated with the completion of the phase segregation and the formation of the stable nanostructure. More insight into this process was provided by the recent studies of internal friction in these nanocomposites upon annealing (Li et al., 2005). The internal friction is determined from the energy losses of a mechanical oscillator, such as a torsion pendulum or vibrating reed. For a given frequency, it appears as a loss peak at a certain temperature, which depends on the characteristic relaxation time

$$\tau = \tau_0 \exp -\left(\frac{E_{act}}{k_B T}\right). \tag{3.16}$$

Here τ_0 is the pre-exponential factor, E_{act} is the activation energy, k_B is the Boltzmann constant, and T is the temperature. This relaxation time is associated with reversible (anelastic) processes, such as atomic jumps (e.g. hydrogen in metals, carbon in α-Fe, etc.), rearrangement of small structural units and others within the material, which leads to a dissipation of the energy of the mechanical vibration of the oscillator into phonons. Figure 3.16a shows an example of the friction measurement with stable nc-TiN/a-Si_3N_4 nanocomposite deposited by means of plasma CVD which does not show any self-hardening (see Figure 3.13). One notices the absence of any friction peak. From the resonant frequency of the coatings shown also in that figure (which can be calculated from the measured resonant frequency of the substrate plus coatings (for details see Li et al., 2005)) one can also determine Young's modulus, which, in this case is in the range of about 450 GPa, in agreement with other independent measurements. However, in the case of the nc-$(Ti_{1-x}Al_x)N$/a-Si_3N_4 coatings, which show self-hardening (see Figure 3.16b), one can clearly see an internal friction peak that vanishes upon annealing to about 750 °C when the self-hardening is completed. The activation energy of about 0.7–0.9 eV and pre-exponential factor τ_0 of the order of 10^{-10} s, which were measured on a series of coatings, show that the internal friction measured at low temperature is associated with thermally activated, reversible, anelastic processes occurring within the grain boundaries and involving either clusters of atoms or a kind of reversible, anelastic short-range sliding of triple points of grain boundaries or their parts. The finding that upon annealing this peak decreases in intensity and vanishes when the self-hardening is completed shows that these relaxation processes are caused by an incomplete phase segregation in coatings deposited at a too-low temperature.

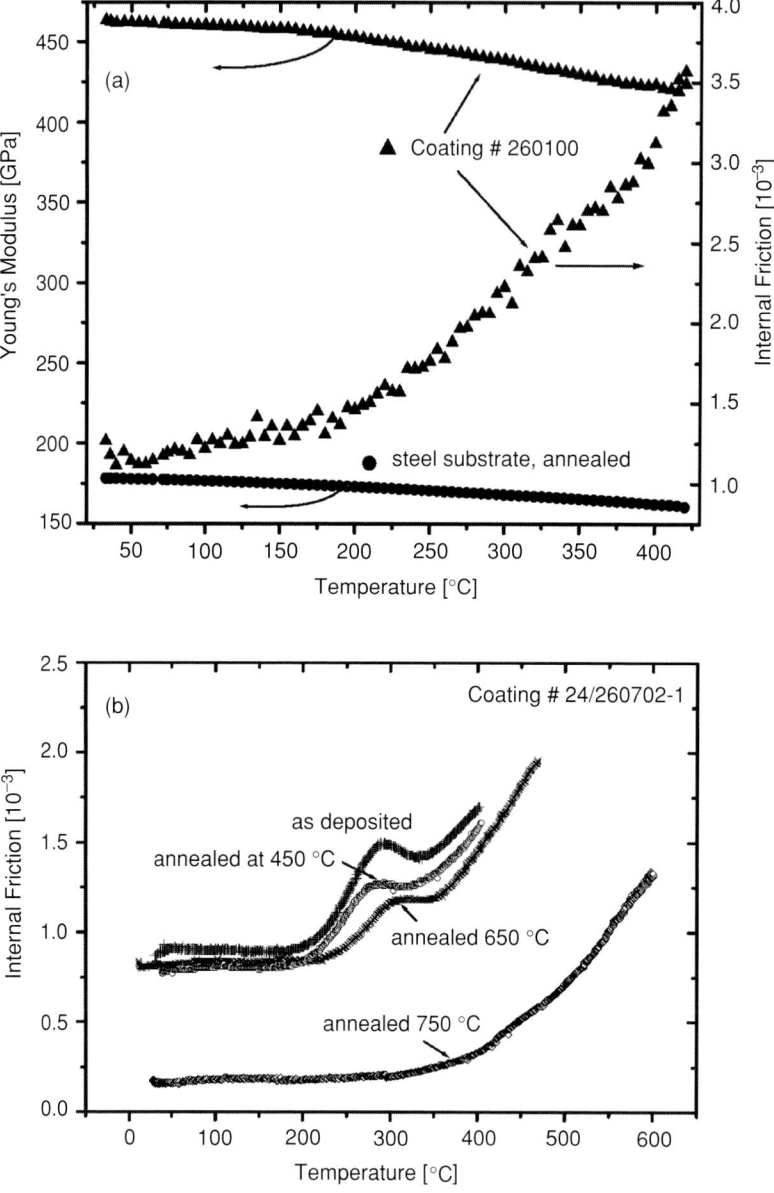

Figure 3.16. (a) Absence of any internal friction peak in the stable nc-TiN/a-Si$_3$N$_4$ superhard nanocomposite coating deposited by plasma CVD (see Figure 3.13). (b) In nc-(Ti$_{1-x}$Al$_x$)N/a-Si$_3$N$_4$ coatings that show self-hardening upon annealing (see Figure 3.15), a friction peak is seen at a temperature of about 280 °C, and it vanishes upon annealing to a temperature of about 750 °C when the self-hardening is completed. (c) Although the hardness of about 50 GPa of coatings deposited by reactive sputtering at 600 °C (see Figure 2.27) remains stable upon annealing up to 1100 °C (Prochazka et al., 2004; Veprek et al., 2005a), a weak internal friction peak is observed.

3.7 Thin films and coatings

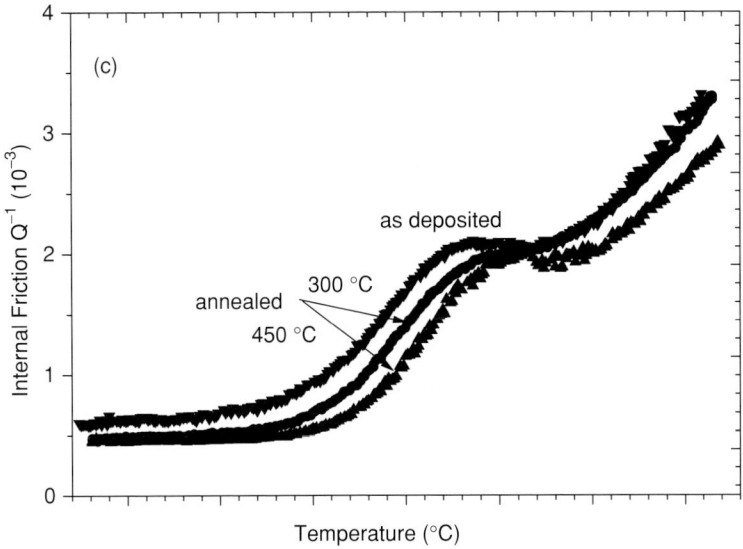

Figure 3.16. (cont.)

The internal friction measurements provide much higher sensitivity than the measurement of hardness and crystallite size. This is illustrated by Figure 3.16c, which was measured on the superhard nanocomposites deposited by magnetron sputtering at the high temperature of 600 °C and high purity having a hardness of ≥ 50 GPa (see Figure 2.27). Upon annealing, these coatings show no observable changes of the hardness and crystallite size (Prochazka et al., 2004; Veprek et al., 2005a). In spite of that one can see from Figure 3.16c that the formation of the nanostructure was not fully completed during the deposition by reactive magnetron sputtering, in contrast to the coatings deposited by plasma CVD (Figure 3.16a). This suggests that the complex chemistry taking place on the surface of the growing film and much higher nitrogen pressure used under plasma CVD conditions enhances the kinetics of the spinodal phase segregation.

A high thermal stability and a certain amount of self-hardening upon annealing to 750 °C and 1000 °C was reported also for the heterostructures (both epitaxial and polycrystalline) consisting of transition metal nitrides and borides and attributed to a recrystallization (Barnett et al. 2003).

3.7.2 Formation of strong nanostructures with high thermal stability via thermodynamically driven phase decomposition

The generic design concept as formulated by Veprek and Reiprich is based on a thermodynamically driven decomposition into the immiscible stoichiometric nitrides, such as TiN and Si_3N_4. The phase segregation is kinetically rate controlled by diffusion. This results in a compositionally modulated nanostructure (Veprek and Reiprich, 1995). In their paper these authors did not discuss if this process occurs by nucleation and growth or by spinodal mechanism.

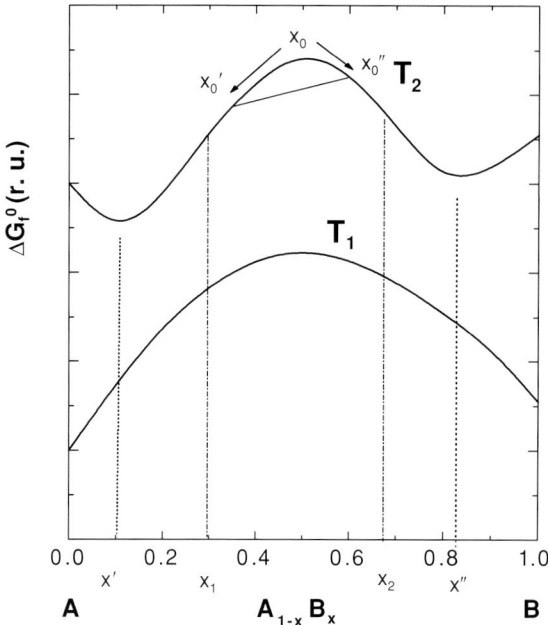

Figure 3.17. Schematics of the Gibbs free energy of the formation of mixed phases of the nitrides (e.g. TiN + Si_3N_4) vs. the composition for a high (T_2) and lower (T_1) temperature. A similar dependence is expected (and found, see text below) for a low and high nitrogen pressure at the same temperature.

Figure 3.17 shows examples of the dependence of the Gibbs free energy of the mixed system $A_{1-x}B_x$ on the composition for the case of partial miscibility (upper curve) and total immiscibility (lower curve). If the second derivative of the Gibbs free energy of the mixed phase ΔG_f^0 with the composition x is negative,

$$\frac{\delta^2 G_f^0(x)}{\delta x^2} < 0, \tag{3.17}$$

the system meets the conditions of being "chemically" spinodal, i.e. inherently instable to infinitesimal fluctuations of the composition in its initial stage of decomposition. Thus, the phase segregation occurs spontaneously without any activation. This is the case in the whole range of the composition for the lower curve, but only within the concentration range $x_1 < x < x_2$ for the upper one. Within the range $x' < x < x_1$ and $x_2 < x < x''$ of the upper curve, an infinitesimal fluctuation results in an increase of the free energy. The system there is metastable, and the segregation occurs by nucleation and growth with activation energy of nucleation $E_{act} > 0$.

The thermodynamic calculations of the chemical spinodal can be performed relatively simply for systems that form solid solutions. The heat of de-mixing, $\Delta H > 0$, can be measured by differential scanning calorimetry (DSC). For metallic alloys, the enthalpy of de-mixing is relatively low, of about 2–20 kJ/mol. In transition-metal alloy systems, the largest ("de-mixing") heat of formation of 65 kJ/mol was reported for Ag–W (Zhang and Liu, 2003). Relatively low values of the heat of de-mixing were reported also for

3.7 Thin films and coatings

the $Co_2TiO_4 - CoAl_2O_4$ system (Burkert et al., 1992). With a typical periodicity of the spinodally decomposed systems of the order of 5–10 nm (see Burkert et al., 1992) and corresponding specific interface area of the order of 10^4 to 10^3 m^2/mol (depending also on the morphology) it is clear that spinodal decomposition in such systems can occur only if the final phases are coherent. A much higher free energy of de-mixing is needed if the final phases are either semicoherent or incoherent, as in many oxide glasses, for example (Zarzycki, 1991).

Here we shall limit our discussion to the Ti–Si–N system under conditions which are relevant for the deposition and thermal stability of the nc-TiN/a-Si$_3$N$_4$ nanocomposites (Zhang and Veprek, 2006). In the case of a stoichiometric transition-metal nitride (TiN, W$_n$N, VN, etc.) in combination with a covalent nitride (Si$_3$N$_4$, BN, AlN) which are immiscible (Rogl and Schuster, 1992), the calculations of the chemical spinodal are difficult. Moreover, the majority of papers that dealt with spinodal decomposition in solid alloys did not need to consider the activity of a gaseous phase. However, a simple consideration of the thermodynamics of the system represented by reaction (3.18), as shown already by Veprek and Reiprich (1995) and Veprek et al. (1996) and more recently with more details (Veprek et al., 2006; see also Figure 2.23),

$$x\text{TiN} + (x \cdot y/3)\text{Si}_3\text{N}_4 = x\text{TiSi}_y + \{(x/2) + (2x \cdot y/3)\}\text{N}_2, \quad (3.18)$$

clearly shows that, at a given temperature, the chemical equilibrium shifts to the right-hand side of reaction (3.18) when the pressure of nitrogen decreases. Under vacuum and a high temperature, the nitrides are unstable. For example, as mentioned above, the saturation pressure of nitrogen over Si$_3$N$_4$ reaches 1 atm at a temperature of 2151 K (Chase et al., 1985). Consequently, the miscibility of such systems depends on both the temperature and the nitrogen pressure, which has to be accounted for when discussing the possible spinodal nature of such phase segregation. In the Ti–Si–N system, the mutual solubility increases with increasing temperature and decreasing nitrogen pressure as illustrated schematically by the upper curve in Figure 3.17. Therefore, the formation of the superhard nc-TiN/a-Si$_3$N$_4$ nanocomposite with a high thermal stability requires a sufficiently high nitrogen pressure during the deposition.

The calculation of the free energy of a mixed Ti–Si–N system with nitrogen content corresponding to the stoichiometric TiN and Si$_3$N$_4$ is difficult because of their immiscibility and resulting lack of the thermodynamic data on the mixed phases. Recently, Zhang and Veprek solved this problem by using the sublattice model (Zhang and Veprek, 2006). Accordingly, the free energy of the mixed state consisting of two sublattices, TiN and Si$_3$N$_4$, can be written as

$$G_{(A,B)_aC_c} = y_A^0 G_{A_aC_c} + y_B^0 G_{B_aC_c} + aRT(y_A \ln y_A + y_B \ln y_B) + ay_A y_B I_{AB}. \quad (3.19)$$

Here, y_X are the site fractions of the elements, ^0G_X are the free energies of formation of the pure stoichiometric phases, a and I_{AB} is a numerical constant and the interaction parameter, respectively. The first two terms on the right-hand side of Eq. (3.19) correspond to the contribution of the pure phases, the second one is the mixing entropy term and the last term describes the contribution of the enthalpy of mixing. The results of the calculation for several representative pressures of nitrogen and a temperature of 600 °C (873 K) are shown

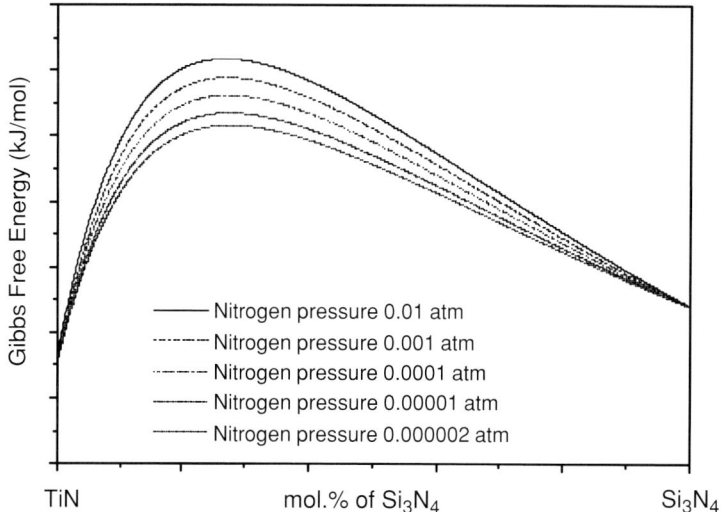

Figure 3.18. Calculated Gibbs free energy of the mixed Ti–Si–N system at a temperature of 873 K under different pressures of nitrogen as indicated (Zhang and Veprek, 2005).

in Figure 3.18 (Zhang and Veprek, 2005). It is clearly seen that the system is chemically spinodal within the whole range of composition and nitrogen pressures down to 0.002 mbar. For the composition of about 7–9 at.% of Si corresponding to the maximum hardness (see Figure 2.22), the de-mixing energy is larger than 200 kJ/mol.

The very large de-mixing energy and the spinodal nature of the curve in Figure 3.18 alone are not sufficient to prove that the phase segregation will indeed occur via spinodal mechanism, because in the solid also the concentration gradient and elastic strain energy terms have to be considered (Cahn and Hilliard, 1958; Cahn, 1967; 1999; Porter and Easterling, 2001; Binder, 1991; Wagner and Kampmann, 1991). In their paper, Zhang and Veprek (2005) discuss arguments in favor of the spinodal decomposition in the Ti–Si–N system. We refer the reader to that paper for further details.

References

Abe, Y. R., and Johnson, W. L. (1992). *Mater. Sci. Forum*, **88–90**, 513.
Abe, Y. R., Holzer, J. C., and Johnson, W. L. (1992). *Mater. Res. Soc. Symp. Proc.*, **238**, 721.
Atkinson, H. V. (1988). *Acta Metall.*, **36**, 469.
Bansal, C., Gao, Z., and Fultz, B. (1995). *NanoStructured Mater.*, **5**, 327.
Barnett, S. A., Madan, A., Kim, I., and Martin K. (2003). *MRS Bulletin*, **28**, 169.
Bertaut, E. F. (1950). *Acta Crystallogr.*, **3**, 14; *ibid.*, **5**, 117.
Binder, K. (1991). In *Materials Science and Technology, Vol. 5, Phase Transformations in Materials*, chapter 7, ed. Cahn, R. W., Haasen, P., and Kramer, E. J. p. 405.
Birringer, R. (1989). *Mater. Sci. Engr. A*, **A117**, 33.

Boylan, K., Ostrander, D., Erb, U., Palumbo, G., and Aust, K. T. (1991). *Scripta Metall. Mater.*, **25**, 2711.

Burke, J. E., and Turnbull, D. (1952). *Progr. Metal Phys.*, **3**, 220.

Burkert, N., Grüne, R., Schmalzried, H., and Rahman, S. (1992). *Ber. Bunsenges. Phys. Chem.*, **96**, 1603.

Cahn, J. W. (1967). *Trans. Metall. Soc. of AIME*, **242**, 166.

Cahn, J. W. (1999). *Scand. J. Metallurgy*, **20**, 9.

Cahn, J. W., and Hillard, J. E. (1958). *J. Chem. Phys.*, **28**, 258.

Campbell, J. H., and Fauchet, M. (1986). *Solid State Comm.*, **58**, 739.

Chang, H., Altstetter, C. J., and Averback, R. S. (1992). *J. Mater. Res.*, **7**, 2962.

Chase, M. W., Davies, C. A., Downey, J. R., Frurip, D. J., McDonald, R. A., and Syverud, A. N. (1985). *JANAF Thermochemical Tables*, 3rd edition, *J. Phys. Chem. Data*, **14** (Supplement No. 1).

Chen, I. W. (1987). *Acta Metall.*, **35**, 1723.

Chen, L. C., and Spaepen, F. (1988). *Nature*, **336**, 366.

Chen, L. C., and Spaepen, F. (1991). *J. Appl. Phys.*, **69**, 679.

Chen, L. C., and Spaepen, F. (1992). *NanoStructured Mater.*, **1**, 59.

Czubayko, U., Sursaeva, V. G., Gottstein, G., and Shvindlerman, L. S. (1998). *Acta Mater.*, **46**, 5863.

Delhez, R., de Keijser, Th. H., and Mittemeijer, E. J. (1980). In *Accuracy in Powder Diffraction*, NBS Special Publication No. 567, ed. Block, S., and Hubbard, C. R. Washington, D.C.: NBS, p. 213.

Doherty, R. D. (1975). *Metall. Trans.*, **A 6**, 588.

Eckert, J., Holzer, J. C., Krill, C. E. III., and Johnson, W. L. (1992). *J. Mater. Res.*, **7**, 1751.

El-Sherik, A. M., Boylan, D., Erb, U., Palumbo, G., and Aust, K. T. (1992). *Mater. Res. Soc. Symp. Proc.*, **238**, 727.

Enzo, S., Fagherazzi, G., Benedetti, A., and Polizzi, S. (1988). *J. Appl. Phys.*, **21**, 536.

Estrin, Y., Gottstein, G., Rabkin, E., and Shvindlerman, L. S. (2000). *Scripta Mater.*, **43**, 141.

Farber, B., Cadel, E., Menand, A., Schmitz, G., and Kirchheim, R. (2000). *Acta Mater.*, **48**, 789.

Feltham, P. (1957). *Acta Metall.*, **5**, 97.

Galina, A. V., Fradkov, V. Y., and Shvindlerman, L. S. (1987). *Phys. Met. Metall.*, **63**, 1220.

Gao, Z., and Fultz, B. (1993). *NanoStructured Mater.*, **2**, 231.

Gao, Z., and Fultz, B. (1994). *NanoStructured Mater.*, **4**, 939.

Gertsman, V. Y., and Birringer, R. (1994). *Scripta Metall. Mater.*, **30**, 577.

Gottstein, G., King, A. H., and Shvindlerman, L. S. (2000). *Acta Mater.*, **48**, 397.

Guinier, A. (1963). *X-ray Diffraction*, San Francisco: W. H. Freeman and Co., p. 121.

Günther, B., Kumpmann, A., and Kunze, H.-D. (1992). *Scripta Metall. Mater.*, **27**, 833.

Gutkin, M. Yu., and Ovid'ko, I. A. (2004). *Plastic Deformation in Nanocrystalline Materials*. Berlin, Heidelberg, New York: Springer.

Gutkin, M. Yu., Ovid'ko, I. A., and Skiba, N. V. (2003). *Acta Mater.*, **51**, 4059.

Hammer, P., Steiner, A., Villa, R., Baker, M., Gibson, P. N., and Haupt, J. (1994). *Surf. Coat. Technol.*, **68–69**, 194.

Haslam, A. J., Moldovan, D., Yamakov, V., Wolf, D., Phillpot, S. R., and Gleiter, H. (2003). *Acta Mater.*, **51**, 2097.
Harris, K. E., Singh, V. V., and King, A. H. (1998). *Acta Mater.*, **46**, 2623.
Hillert, M. (1965). *Acta Metall.*, **13**, 227.
Hofler, H. J., and Averback, R. S. (1990). *Scripta Metall. Mater.*, **24**, 2401.
Hondros, E. D., and Seah, M. P. (1983). In *Physical Metallurgy*, 3rd edition, ed. Cahn R. W., and Haasen, P. Netherlands: Elsevier Sci. Pub. BV, p. 856.
Hu, H., and Rath, B. B. (1970). *Metall.* Trans., **1**, 3181.
Humphreys, F. J., and Hatherly, M. (1996). *Recrystallization and Related Annealing Phenomena*, chapter 9. Tarrytown, NY: Elsevier Science Inc., p. 281.
Hunderi, O., and Ryum, N. (1980). *J. Mater. Sci.*, **5**, 1104.
Iqbal, Z., and Veprek, S. (1982). *J. Phys. C: Solid St. Phys.*, **15**, 377.
Iqbal, Z., Veprek, S., Webb, A. P., and Cappezuto, P. (1981). *Sol. State Commun.*, **37**, 993.
Jin, M., Minor, A. M., Stach, E. A., and Morris Jr, J. W. (2004). *Acta Mater.*, **52**, 5381.
Karvankova, P., Männling, H.-D., Eggs, C., and Veprek, S. (2001). *Surf. Coat. Technol.*, **146–147**, 280.
Karvankova, P., Veprek-Heijman, M. G. J., Zawrah, M. F., and Veprek, S. (2004). *Surf. Coat. Technol.*, **467**, 133.
Karvankova, P., Veprek-Heijman, M. G. J., Azinovic, D., and Veprek, S. (2006). *Surf. Coat. Technol.*, **200**, 2978.
Kirchheim, R. (2002). *Acta Mater.*, **50**, 413.
Kissinger, H. E. (1957). *Anal. Chem.*, **29**, 1702.
Klement, U., Erb, U., El-Sherik, A. M., and Aust, K. T. (1995). *Mater. Sci. Eng. A*, **A203**, 177.
Klug, H. P., and Alexander, L. E. (1974). *X-ray Diffraction Procedures for Polycrystalline and Amorphous Materials*, 2nd edition, chapter 9. New York: Wiley.
Knauth, P., Charai, A., and Gas, P. (1993). *Scripta Metall. Mater.*, **28**, 325.
Krill, C. E., and Birringer, R. (1998). *Phil. Mag. A*, **77**, 621.
Krill, C. E., and Birringer, R. (2001). In *Recrystallization and Grain Growth*, ed. Gottstein, G., and Molodov, A. D. Berlin: Springer-Verlag, p. 205.
Krill, C. E., Helfen, L., Michels, D., *et al.* (2001). *Phys. Rev. Lett.*, **86**, 842.
Kurtz, S. K., and Carpay, F. M. A. (1980). *J. Appl. Phys.*, **51**, 5725.
Li, J. C. M. (1962). *J. Appl. Phys.*, **33**, 2958.
Li, S. Z., Fang, Q. F., Liu, Q., Li, Z. S., Gao, J., Nesladek, P., Prochazka, J., Veprek-Heijman, M. G. J., and Veprek, S. (2005). *Composites Science and Technology*, **65**, 735.
Liu, F., and Kirchheim, R. (2004). *Scripta Mater.*, **51**, 521.
Liu, K. W., and Mucklich, F. (2001). *Acta Mater.*, **49**, 395.
Louat, N. P. (1974). *Acta Metall.*, **22**, 721.
Lu, K. (1993). *NanoStructured Mater.*, **2**, 643.
Malow, T. R., and Koch, C. C. (1996a). In *Synthesis and Processing of Nanocrystalline Powder*, ed. Bourell, D. L. Warrendale, PA: TMS, pp. 33–44.
Malow, T. R., and Koch, C. C. (1996b). *Mater. Sci. Forum*, **225–227**, 595.

Männling, H.-D., Patil, D. S., Moto, K., Jilek, M., and Veprek, S. (2001). *Surf. Coat. Technol.*, **146–147**, 263.

Mayrhofer, P. H., Hörling, A., Karlsson, L., *et al.* (2003). *Appl. Phys. Lett.*, **83**, 2049.

Melendres, C. A., Narayanasamy, A., Maroni, V. A., and Siegel R. W. (1989). *J. Mater. Res.*, **4**, 1246.

Michels, A., Krill, C. E., Ehrhardt, H., Birringer, R., and Wu, D. T. (1999). *Acta Mater.*, **47**, 2143.

Moldovan, D., Wolf, D., and Phillpot, S. R. (2001). *Acta Mater.*, **49**, 3521.

Moral J. E., and Ashby, M. F. (1974). *Acta Metall.*, **22**, 567.

Mukherjee A. K. (2002). *Mater. Sci. Eng.*, **A 322**, 1.

Mullins W. W. (1956). *J. Appl. Phys.*, **27**, 900.

Mullins W. W. (1998). *Acta Mater.*, **46**, 6219.

Murayama, M., Howe, J. M., Hidaka, H., and Takaki, S. (2002). *Science*, **295**, 2433.

Nichols, C. S., Mansuri, C. M., Townsend, S. J., and Smith, D. A. (1993). *Acta Metall. Mater.*, **41**, 1861.

Niederhofer, A., Nesladek, P., Männling, H.-D., Moto, K., Veprek, S., and Jilek, M. (1999). *Surf. Coat. Technol.*, **120–121**, 173.

Obraztsova, E. D. (1994). In *Nanophase Materials*, ed. Hadjipanayis, G. C., and Siegel, R. W. Dordrecht, Netherlands: Kluwer Academic Publishing, p. 438.

Okuda, S., Kobiyama, M., Inami, T., and Takamura, S. (2001). *Scr. Mater.*, **44**, 2009.

Ossadnik, Ch., Veprek, S., and Gregora, I. (1999). *Thin Solid Films*, **337**, 148.

Ovid'ko, I. A. (2002). *Science*, **295**, 2386.

Ovid'ko, I. A. (2005). *Int. Mater. Rev.*, **50**, 65.

Ovid'ko, I. A., Pande, C. S., and Masumura, R. A. (2006). In *Handbook on Nanomaterials*, ed. Gogotsi, Y. G. Florida: CRC, p. 531.

Pande, C. S. (1987). *Acta Metall.*, **35**, 2671.

Pande, C. S., and Dantsker, E. (1990). *Acta Metall.*, **38**, 945.

Pande, C. S., and Dantsker, E. (1991). *Acta Metall.*, **39**, 1359.

Pande, C. S., and Dantsker, E. (1994). *Acta Metall. Mater.*, **42**, 2899.

Pande, C. S., and Masumura, R. A. (2003). In *Nanostructures: Synthesis, Functional Properties and Applications*, ed. Tsakalakos, T., Ovid'ko, I. A., and Vasudevan A. K., Dordrecht: Kluwer, p. 169.

Pande, C. S., and Rajagopal, A. K. (2001). *Acta Mater.*, **49**, 1805.

Pande, C. S., Masumura, R. A., and Marsh, S. P. (2001). *Phil. Mag.*, **A 81**, 1229.

Perez, R. J., Jiang, H. G., Dogan, C. P., and Lavernia, E. J. (1998). *Metall. Mater*. Trans. A, **29A**, 2469.

Porter, D. A., and Easterling, K. E. (2001). *Phase Transformations in Metals and Alloys*. Cheltenham, UK: Nelson Thomas Ltd, p. 291.

Prochazka, J., Karvankova, P., Veprek-Heijman, M. G. J., and Veprek, S. (2004). *Mater. Sci. Eng.*, **A 384**, 102.

Rabkin, E. (1999). *Interface Sci.*, **7**, 297.

Rhines, F. N., and Craig, K. R. (1974). *Metall.*Trans., **A 5**, 413.

Richter, H., Wang, Z. P., and Ley, L. (1981). *Solid State Commun.*, **39**, 625.

Rogl, P., and Schuster, J. C. (1992). *Phase Diagrams of Ternary Boron Nitride and Silicon Nitride Systems*. Materials Park, Ohio: ASM The Materials Society.
Sarott, F.-A., Iqbal, Z., and Veprek, S. (1982). *Solid State Commun.*, **42**, 465.
Sattler, K., Raina, G., Ge, M., Venkatiswaran, N., Xhie, J., Liao, Y. X., and Siegel, R. W. (1994). *J. Appl. Phys.*, **76**, 546.
Shan, Zh., Stach, E. A., Wiezorek, J. M. K., Knapp, J. A., Follstaedt, D. M., and Mao, S. X. (2004). *Science*, **305**, 654.
Shaw, L., Luo, H., Villegas, J., and Miracle, D. (2003). *Acta Mater.*, **51**, 2647.
Shen, T. D., Koch, C. C., McCormick, T. L., Nemanich, R. J., Huang, J. Y., and Huang, J. G. (1995). *J. Mater. Res.*, **10**, 139.
Smith, C. S. (1948). *Transactions AIME*, **175**, 15.
Smith, C. S. (1953). *Acta Metall.*, **1**, 295.
Smith, C. S. (1964). *Metall. Rev.*, **9**, 1.
Soer, W. A., De Hosson, J. T. M., Minor, A. M., Morris Jr, J. W., and Stach, E. A. (2004). *Acta Mater.*, **52**, 5783.
Suryanarayana, C. (1995). *Intl. Mater. Rev.*, **40**, 41.
Sutton, A. P., and Balluffi, R. W. (1996). *Grain Boundaries in Crystalline Materials*. Oxford: Oxford Science Publications.
Terwilliger, C. D., and Chiang, Y. M. (1995). *Acta Mater.*, **43**, 319.
Ungar, T., and Borbely, A. (1996). *Appl. Phys. Lett.*, **69**, 3173.
Ungar, T., Ott, S., Sanders, P. G., Borbely, A., and Weertman, J. R. (1998). *Acta Mater.*, **46**, 3693.
Upmanyu, M., Srolovitz, D. J., Shvindlerman, L. S., and Gottstein, G. (1998). *Interface Sci.*, **6**, 287.
Upmanyu, M., Srolovitz, D. J., Shvindlerman, L. S., and Gottstein, G. (2002). *Acta Mater.*, **50**, 1405.
Vandermeer, R. A., and Hu, H. (1994). *Acta Metall. Mater.*, **42**, 3071.
Veprek, S. (1999). *J. Vac. Sci. Technol.*, **A 17**, 2401.
Veprek, S., and Reiprich, S. (1995). *Thin Solid Films*, **268**, 64.
Veprek, S., Sarott, F.-A., and Iqbal, Z. (1987). *Phys. Rev.*, **B 36**, 3344.
Veprek, S., Reiprich, S., and Li, S. H. (1995). *Appl. Phys. Lett.*, **66**, 2640.
Veprek, S., Haussmann, M., and Reiprich, S. (1996). *J. Vac. Sci. Technol.*, **A 14**, 46.
Veprek, S., Nesladek, P., Niederhofer, A., Glatz, F., Jilek, M., and Sima, M. (1998). *Surf. Coat. Technol.*, **108–109**, 138.
Veprek, S., Männling, H.-D., Jilek, M., and Holubar, P. (2004a). *Mater. Sci. Eng.*, **A 366**, 202.
Veprek, S., Männling, H.-D., Niederhofer, A., Ma, D., and Mukherjee, S. (2004b). *J. Vac. Sci. Technol.*, **B 22**, L5.
Veprek, S., Veprek-Heijman, G. M. J., Karvankova, P., and Prochazka, J. (2005a). *Thin Solid Films*, **476**, 1.
Veprek, S., Männling, H.-D., Karvankova, P., and Prochazka, J. (2006). *Surf. Coat. Technol.*, **200**, 3876.
Verhoeven, J. D. (1986). *Scanning electron microscopy. In Metals Handbook*, 9th edition, Vol. 10. Metals Park, OH: ASM, p. 490.

References

Wagner, R., and Kampmann, R. (1991). In *Materials Science and Technology, Vol. 5, Phase Transformations in Materials*, chapter 4, ed. Cahn, R. W., Haasen, P., and Kramer, E. J., p. 213.

Wang, N., Wang, Z., Aust, K. T., and Erb, U. (1997). *Acta Mater.*, **45**, 1655.

Warren, B. E. (1990). *X-ray Diffraction*. New York: Dover Publication Inc., p. 262.

Weissmuller, J. (1993). *NanoStructured Mater.*, **3**, 261.

Weissmuller, J. (1994). *J. Mater. Res.*, **9**, 4.

Weissmuller, J. (1996). In *Synthesis and Processing of Nanocrystalline Powder*, ed. Bourell, D. L. Warrendale, PA: TMS, p. 3.

Weissmuller, J., Krauss, W., Haubold, T., Birringer, R., and Gleiter, H. (1992). *NanoStructured Mater.*, **1**, 439.

Williamson, G. K., and Hall, W. H. (1953). *Acta Metall.*, **1**, 22.

Wurschum, R., Brossmann, U., and Schaefer, H.-E. (2002). In *Nanostructured Materials: Processing, Properties, and Applications*, ed. Koch, C. C. Norwich, NY: William Andrew Pub., p. 267.

Youssef, K. M. (2003). *Synthesis, structure, and properties of nanocrystalline zinc by pulsed-current electrodeposition*. Ph.D. Thesis, North Carolina State University, p. 166.

Zarzycki, J. (1991). *Glasses and the Vitreous State*. Cambridge: Cambridge University Press, p. 161.

Zhang, R. F., and Liu, B. X. (2003). *J. Mater. Res.*, **18**, 1499.

Zhang, R. F., and Veprek, S. (2006). *Mater. Sci. Eng.*, **A424**, 128.

4 Mechanical properties of structural nanocrystalline materials – experimental observations

In this chapter we will describe and discuss the experimental evidence for the mechanical behavior of nanocrystalline materials. This will include pure metals, alloys, intermetallic compounds, ceramics, and multiphase materials. The range of mechanical properties for which measurements have been made will be covered. While the models and theoretical explanations for the various phenomena believed responsible for mechanical properties of nanocrystalline materials will be emphasized in Chapter 5, some discussion of deformation mechanisms must necessarily accompany the description of experimental results.

4.1 Elastic properties of nanostructured materials

The early measurements of the elastic constants on nanocrystalline materials prepared by the inert-gas-condensation method gave values, for example for Young's modulus, E, which were significantly lower than values for conventional grain size materials (Suryanarayana, 1995). While various reasons were given for the lower values of E, it was suggested by Krstic and co-workers (1993) that the presence of extrinsic defects, e.g. pores and cracks, was responsible for the low values of E in nanocrystalline materials compacted from powders. This conclusion was based upon the observation that nanocrystalline NiP produced by electroplating with negligible porosity levels had an E value comparable to fully dense conventional grain size Ni (Wong *et al.*, 1993). Krstic *et al.* (1993) and Boccaccini *et al.* (1993) developed theories to account for the decrease in E with porosity which agree with E vs.% porosity data on nanocrystalline Fe produced by inert-gas condensation and warm consolidation (Fougere *et al.*, 1995). Young's moduli of nanocrystalline Fe, Cu, Ni, and Cu–Ni alloys prepared by mechanical milling/alloying of powders were measured by a nanoindentation technique (Shen *et al.*, 1995). The E values for nanocrystalline Cu, Ni, and Cu–Ni alloys with grain sizes ranging from 17 nm to 26 nm were essentially identical to those of the corresponding conventional grain size materials. The dependence of E for nanocrystalline Fe on grain size corresponded well to a model (Shen *et al.*, 1995) which suggests that the change in

4.1 Elastic properties of nanostructured materials

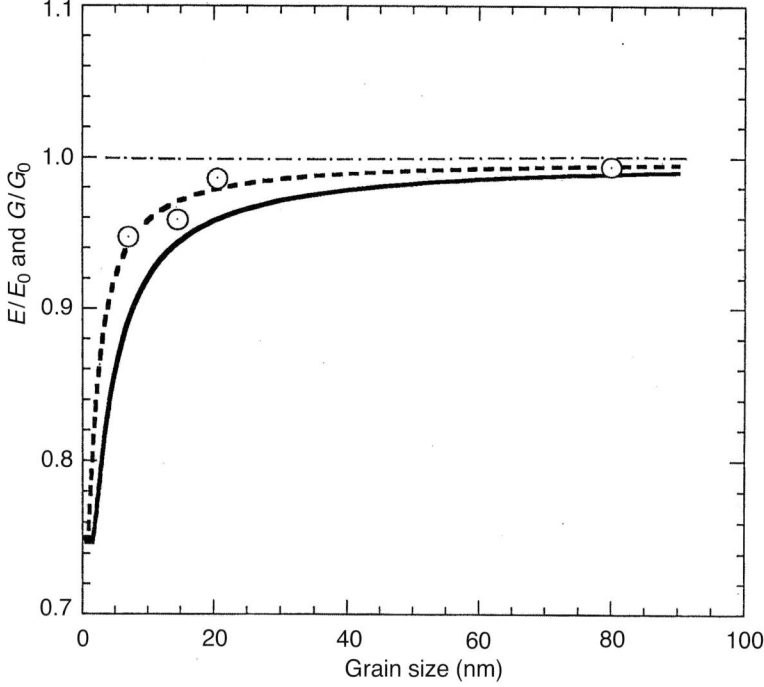

Figure 4.1. Calculated ratios of Young's (E) and shear (G) moduli of nanocrystalline material to those of polycrystals as a function of grain size. The dashed and solid curves correspond to grain-boundary thicknesses of 0.5 nm and 1.0 nm respectively. The open circles show E/E_0 values of nanocrystalline Fe vs. grain size. (From Shen et al. (1995), reproduced with permission of MRS.)

E and G (the shear modulus) of nanocrystalline materials, which are porosity free, with a grain size larger than about 5 nm, should be very limited. This simple model involved assuming that the moduli consist of contributions from the grain interior, grain boundary, and grain triple junctions. The prediction of the change of E and G with grain size based on this model is given in Figure 4.1 for assumed grain-boundary thicknesses of 0.5 nm and 1.0 nm. The data for nanocrystalline Fe closely fit the curve for 0.5 nm grain-boundary thickness. It is concluded that the intrinsic moduli of nanocrystalline materials are essentially the same as those for conventional grain size materials until the grain size becomes very small, e.g. <5 nm, such that the number of atoms associated with the grain boundaries and triple junctions becomes very large.

Weertman (2002) points out that besides the possibility of porosity, measuring techniques had contributed to the low values of elastic moduli reported in the earlier work. In the case of the modulus results deduced from stress–strain curves, a poor method for measuring the strain contributed to the apparent low values (Nieman et al., 1991). The subsequent use of a miniature strain gage glued to the small tensile specimens led to reasonable results (Sanders et al., 1997).

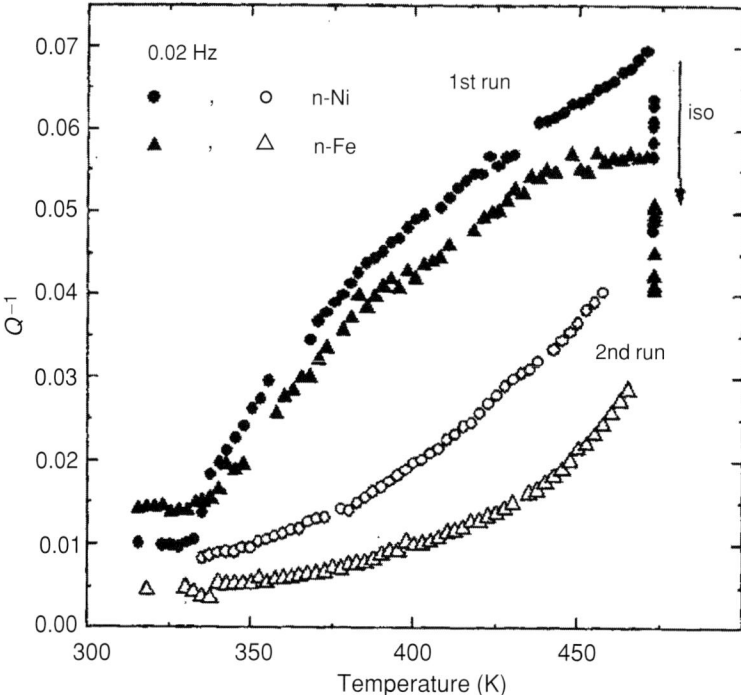

Figure 4.2. Internal friction Q^{-1} at 0.02 Hz in nanocrystalline Ni (circles) and Fe (triangles). Solid symbols refer to the first run. Open symbols refer to the second run after annealing at 200 °C for 30 min. (From Bonetti et al. (1999), reproduced with permission of Elsevier.)

4.2 Anelastic properties

Anelasticity is defined as the dependence of elastic strain on both stress and time. This can result in a lag of strain behind stress. In materials subjected to cyclic stress, the anelastic effect causes internal damping. Internal friction or mechanical damping has been a valuable research tool for the study of the properties of lattice imperfections in crystalline solids. There are a number of mechanisms by which vibrational energy can be dissipated internally by a material. These include the thermoelastic effect, stress-induced ordering of interstitial and substitutional solute atoms, grain-boundary relaxation, motion of dislocations, and certain phase transformations. A large literature on relaxation methods exists and has provided useful information regarding a variety of mechanisms in solids. These results have been reviewed in sources such as Nowack and Berry (1972). Anelastic measurements have now been made on a number of nanocrystalline metals. Bonetti et al. (1999) studied the anelastic behavior of nanocrystalline Fe and Ni in the quasistatic and low-frequency anelastic regime. They measured the elastic energy-dissipation coefficient (Q^{-1}) and the stress relaxation as a function of frequency and temperature in a range of temperatures where grain growth was not expected. A significant change in the magnitude of Q^{-1} was observed after annealing for 30 min. at 200 °C even though no grain growth occurred. Figure 4.2 shows these results

4.2 Anelastic properties

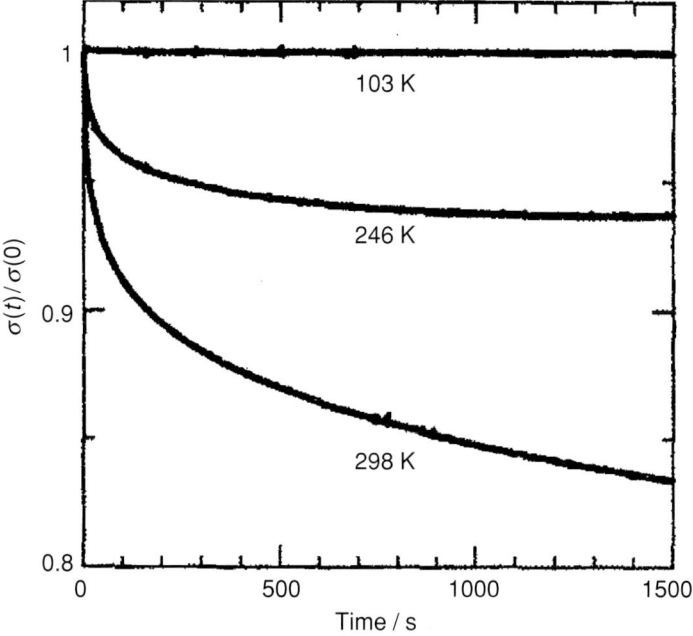

Figure 4.3. Stress relaxation measurements on a nanocrystalline Au sample with about 26 nm grain size. The value of $\sigma(0)$ was about 0.14 GPa. (From Sakai et al. (1999), reproduced with permission of Elsevier.)

for nanocrystalline Ni and Fe. The activation energies determined for these processes were comparable to those of grain-boundary diffusion for the respective metals. The authors suggest that the mechanical response may originate from a viscous behavior at the grain boundaries mediated by atomic diffusion in the grain-boundary disordered regions. In the nanocrystalline Fe the samples with initial grain size of 40 nm showed lower damping values than samples with smaller, 15 nm grain sizes. However, no difference in damping was observed between 40 nm and 20 nm grain size Ni. Sakai et al. (1999) studied the elastic and anelastic properties of nanocrystalline Au prepared by the inert-gas-condensation method with grain sizes ranging from 26 nm to 60 nm. They observed stress relaxation for test temperatures greater than about 220 K. This is illustrated in Figure 4.3. Below this temperature, values for Young's modulus, E, from a vibrating reed technique are similar to results from stress–strain curves, and the temperature dependence is similar to that for conventional-grain-size Au. Above 220 K the vibrating reed results continue to follow the coarse-grain values, but modulus values from the stress–strain curves, obtained at a very low strain rate, decrease with increasing temperature at a faster rate. A stress–strain curve at 77 K at a strain rate of 10^{-4}/s is linear over a wide strain range, but at 293 K the plot continuously curves away from the initial slope, indicating the presence of creep. The authors suggest that the deformation process near room temperature in their nanocrystalline Au is governed by the strain rate as well as the stress level. More recently, Tanimoto et al. (2004) have reviewed anelasticity measurements in nanocrystalline materials. They give references for anelastic measurements on nanocrystalline Pd, Cu, Al, Ni, Fe, and Ag. In most of these experiments the anelastic behavior at temperatures above 300 K revealed a steep increase in the internal friction, Q^{-1}, at slightly elevated temperature followed by its decrease once grain growth occurred. As in the work of Bonetti *et al.* referred to above, the activation energies estimated for the steep increase in Q^{-1} were comparable with those reported for the grain-boundary diffusion in the conventional-grain-size metals. This might suggest that nanocrystalline metals behave similar to conventional-grain-size metals with the exception of the increased volume fraction of grain boundaries. However, Tanimoto *et al.* (2004) found a large anelastic strain in quasistatic tensile tests (creep tests) at room temperature for nanocrystalline Au. After unloading from an applied stress a strain is composed of an anelastic strain and a plastic creep strain. The anelastic strain component is believed to be associated with a certain anelastic process in the grain boundaries. In an earlier study by the authors (Sakai *et al.*, 2001) the activation energy for the early stage of the grain boundary anelasticity was found to be 0.2 eV, which is much lower than the activation energy of 0.88 eV reported for grain-boundary diffusion in conventional-grain-size Au. The authors quote molecular dynamics simulations for grain-boundary diffusion and migration that report the cooperative motions of many atoms which can result in a low activation energy. They suggest that such cooperative atomic motions of many atoms may be responsible for the large anelastic effect observed in their nanocrystalline Au samples. The unusual anelastic behavior of nanocrystalline Au reported by the authors suggests that anelasticity studies can provide useful information on the defect structure of nanocrystalline materials and that more detailed studies are needed.

4.3 Hardness and strength

This section will describe and discuss the experimental evidence for the hardness and strength of nanocrystalline metals and alloys, ceramics and intermetallics, and multiphase materials. The behavior of ultrahard thin nanocrystalline films will be covered in Section 4.9.

4.3.1 Hardness and strength of nanocrystalline metals and alloys

Because of the difficulties in preparing artifact-free nanocrystalline materials of sufficient size for the standard mechanical testing methods, the most measured mechanical property of nanocrystalline materials as a function of grain size is hardness. The large number of experimental studies of the hardness of nanocrystalline materials have been reviewed by a number of authors. These include Morris, 1998; Siegel and Fougere, 1994; Siegel, 1997; Weertman and Averback, 1996; Weertman, 2002a,b.; Milligan, 2003; and Kumar *et al.*, 2003. It is clear from the data on hardness and strength that the hardness (strength) of nanocrystalline metals can be much higher than the conventional-grain-size counterpart. The hardness increases by reducing grain size to the smallest nanoscale values, about 10 nm, can be factors of about 2 to over 10. As an extreme example, the yield strength of nanocrystalline Cu (23 nm) has been found to be 22 times that of conventional-grain-size (20 μm) Cu (Youssef *et al.*, 2004). While such significant increases in hardness and, where

4.3 Hardness and strength

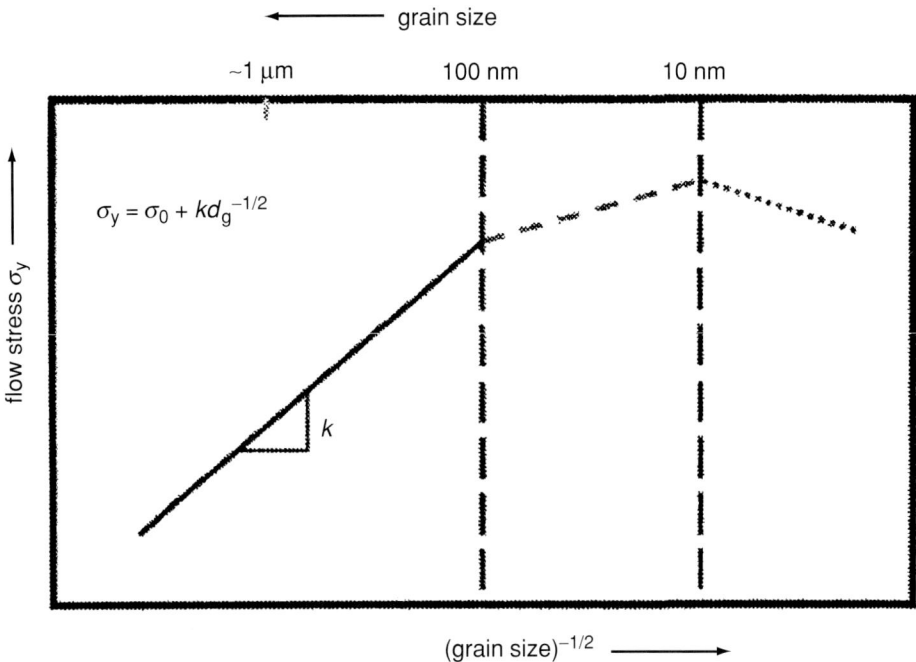

Figure 4.4. Schematic of the variation of yield stress as a function of grain size. (After Kumar et al. (2003), reproduced with permission of Elsevier.)

measured, strength, have been documented in a number of nanocrystalline metals and alloys, there is no general agreement on the mechanism(s) for this hardening. For conventional-grain-size materials (grain sizes >1 μm) the empirical Hall–Petch equation (Hall, 1951; Petch, 1953) predicts that

$$\sigma_y = \sigma_0 + k\,d^{-1/2},$$

where σ_y is the yield strength, σ_0 is a friction stress below which dislocations will not move in a single crystal, k is the Hall–Petch slope, and d is the grain size. A similar expression is given for hardness. Hall–Petch behavior is well documented for conventional-grain-size materials as the grain size is reduced. However, and this is dependent on the material, as the grain size is reduced below about 100 nm, the slope of the Hall–Petch curve decreases and at the finest grain sizes can level out to zero, or in some cases exhibit a negative slope. A schematic drawing of this behavior is given in Figure 4.4. The mechanism for the Hall–Petch behavior in conventional-grain-size materials has been generally explained by a pile-up of dislocations in one grain producing a stress concentration that activates a dislocation source in the adjacent grain (Cottrell, 1958). Alternative mechanisms have involved the activation of grain-boundary dislocation sources (Li, 1963) or elastic and plastic incompatibility stress between neighboring grains (Meyers and Ashworth, 1982). All these models involve dislocation motion or generation. At the smallest grain sizes (∼10 nm) it is believed dislocation activity ceases. At these small grain sizes the dislocation image forces are sufficient to

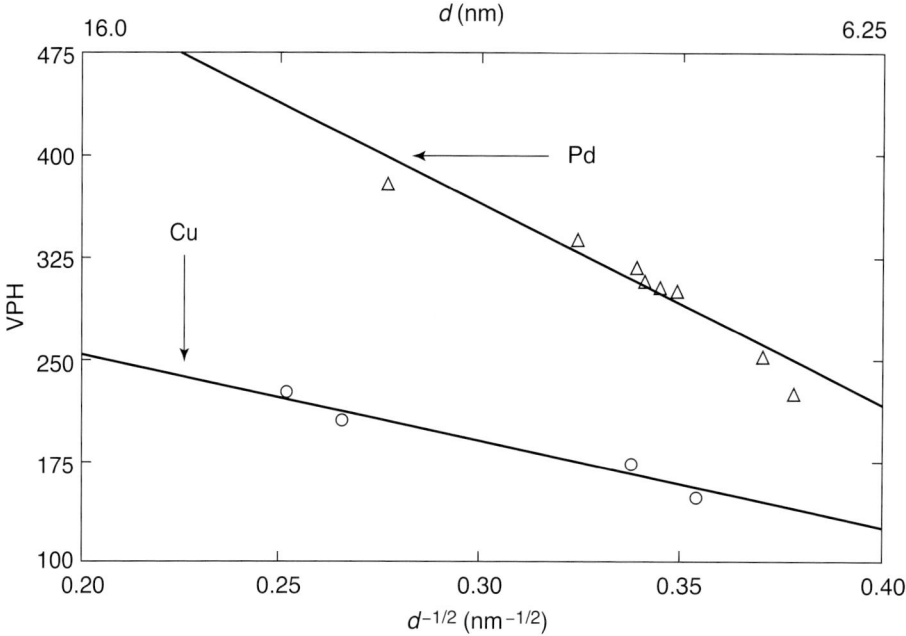

Figure 4.5. Hardness vs. grain size, $d^{-1/2}$, for nanocrystalline Cu and Pd. (From Chokshi *et al.* (1989), reproduced with permission of Elsevier.)

eliminate dislocations by moving them into the grain boundaries. In addition, dislocation multiplication mechanisms such as the Frank–Read source would require stresses of the order of the theoretical strength. It is predicted therefore that dislocations are absent in the smallest nanocrystals and deformation must involve other than conventional dislocation creation and motion.

The first report of an inverse Hall–Petch effect in nanostructured materials was given by Chokshi *et al.* (1989) on nanocrystalline Cu and Pd prepared by the inert-gas-condensation method. A clear inverse Hall–Petch behavior was observed for both nanocrystalline Cu and Pd for grain sizes less than 16 nm or 14 nm, respectively, as shown in Figure 4.5. The authors rationalized these results as being caused by the occurrence of diffusional creep at room temperature. It was suggested that an "equicohesive" grain size demarcates low-temperature behavior (positive Hall–Petch slope) from high-temperature behavior (inverse Hall–Petch), analogous to the well-known equicohesive temperature at a given grain size. While other instances of an inverse Hall–Petch effect were subsequently observed by other investigators it was also noted that in most of these reports nanocrystalline grain size was varied by annealing the smallest grain size samples to obtain grain growth and therefore a range of grain sizes. It is suggested (Morris, 1998) that thermally treating nanophase samples in the as-produced condition may result in such changes in structure as densification, stress relief, phase transformations, or grain-boundary structure, all of which may lead to the observed inverse Hall–Petch behavior. Only a small number of reports of the inverse Hall–Petch effect have been for as-produced nanocrystalline samples with a range of grain sizes.

4.3 Hardness and strength

Other problems with experimental verification of the inverse Hall–Petch effects include the accurate measurement of grain sizes and grain size distribution at the nanoscale. Koch and Narayan (2001) reviewed the experimental evidence for the inverse Hall–Petch effect. They concluded that only a few of the experiments in which an inverse Hall–Petch effect has been reported are free from obvious or possible artifacts. However, these few, along with the predictions of computer simulations, suggest that the effect is real. In analogy to the elastic behavior in nanostructured materials only deviating from that of conventional-grain-size materials when the grain size is less than about 10 nm, the clear experimental observations for the inverse Hall–Petch effect only occur at such grain sizes. That is the observed softening with decreasing grain size is restricted to sizes of the order of 10 nm.

Solid solution hardening is a fundamental metallurgical phenomenon that has been used to strengthen metals since ancient times. It is useful to consider the influence of alloying additions on the hardness of nanocrystalline metals. Shen and Koch (1996) prepared a series of nanocrystalline alloys by mechanical attrition. These were Ti(Cu), Nb(Cu), Fe(Cu), Ni(Cu), Cr(Cu), Cu(Fe), Cu(Co), and Cu(Ni) solid solutions, where the element in parentheses is the solute. The nearest-neighbor distance of the various nanocrystalline solid solutions changed continuously with the addition of alloying elements, indicating the continuous solid solution of solute into the matrix metals. The quantitative agreement of heat release with respect to the decomposition of nanocrystalline solid solutions with the thermodynamically calculated formation enthalpy of the corresponding solid solutions further verified the formation of extended nanocrystalline solid solutions. The grain size of the various nanocrystalline alloys ranged from about 10 nm to 22 nm, as revealed by X-ray diffraction line-broadening analysis. The addition of Cu into Ti and Nb tends to decrease the grain size of the solid solutions formed while the addition of Cu into Cr, Fe, and Ni does not decrease the grain size but exhibits more complex behavior. The r.m.s. strain increases with the addition of alloying elements for all of the solid solutions studied. The microhardness of nanocrystalline Ti(Cu), Nb(Cu), Cu(Ni), Cu(Fe), and Cu(Co) solid solutions increased with increasing solute concentration while the microhardness of nanocrystalline Cr(Cu), Fe(Cu), and Ni(Cu) solid solutions decreased with increasing Cu content. These results are illustrated for Cu(Fe) and Fe(Cu) alloys in Figure 4.6 and Figure 4.7 respectively. The influence of alloying elements on the hardness of the nanocrystalline solid solutions is different from that in conventional polycrystalline solid solutions. The strength of nanocrystalline solid solutions depends upon both solid-solution hardening and grain-boundary hardening, while the latter makes the major contribution to the total strength. The hardness increase resulting from the solid-solution hardness effect in nanocrystalline alloys is of the same magnitude as that in conventional polycrystalline alloys. The addition of alloying elements into the matrix for nanocrystalline alloys, however, may have a strong influence on the grain-boundary energy and grain-boundary diffusion coefficient and thus on the deformation and recovery mechanisms of the nanocrystalline alloys formed by mechanical attrition. The different deformation and recovery behaviors are responsible for the various grain sizes obtained by milling and thus for the decreased or increased grain-boundary-hardening effects for various nanocrystalline solid solutions. The total hardness of the nanocrystalline alloys formed by milling may be increased or decreased with the addition of alloying elements. The increase or decrease in hardness is dependent on the combined effects of the

4 Mechanical properties of structural nanocrystalline materials

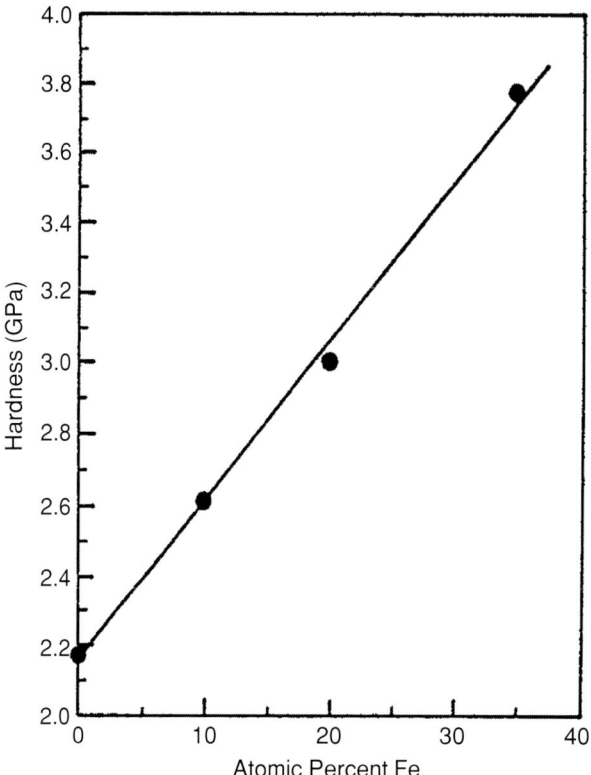

Figure 4.6. Hardness vs. atomic percent Fe for nanocrystalline Cu(Fe) alloys. (After Shen and Koch, 1996.)

hardness increase resulting from solid-solution hardening and the hardness increase or decrease resulting from the changes in grain size and therefore grain-boundary hardening. That is, if solute additions result in a smaller grain size the total effect is hardening, while if they increase the grain size, the result is softening.

While hardness was the main mechanical property measured in many of the earlier studies of nanostructured metals, tensile and compression tests have also been made. In studies on well-compacted nanocrystalline metals made by the inert-gas-condensation method it was found that good agreement was observed between yield stresses in compression and hardness (divided by 3 following the Tabor approximation). An example of this behavior for nanocrystalline Cu is given in Figure 4.8 after Weertman (2002). Tensile test data, however, resulted in severe underestimates of strength (Youngdahl et al., 1997; Sanders et al., 1997) which may be explained by the presence of defects such as voids or incomplete particulate bonding. In these relatively hard but, because of defects, low-toughness materials, the critical flaw size can be as small as about 1 μm (Sanders et al., 1997), which may be common to imperfectly compacted samples. Since compression tests are less sensitive to such defects as pores and cracks they are less likely to exhibit underestimated yield stresses as tensile tests can. Besides this artifact-related asymmetry in tension and compression yield stress values, there is evidence in some materials of another asymmetry due to lack of

4.3 Hardness and strength

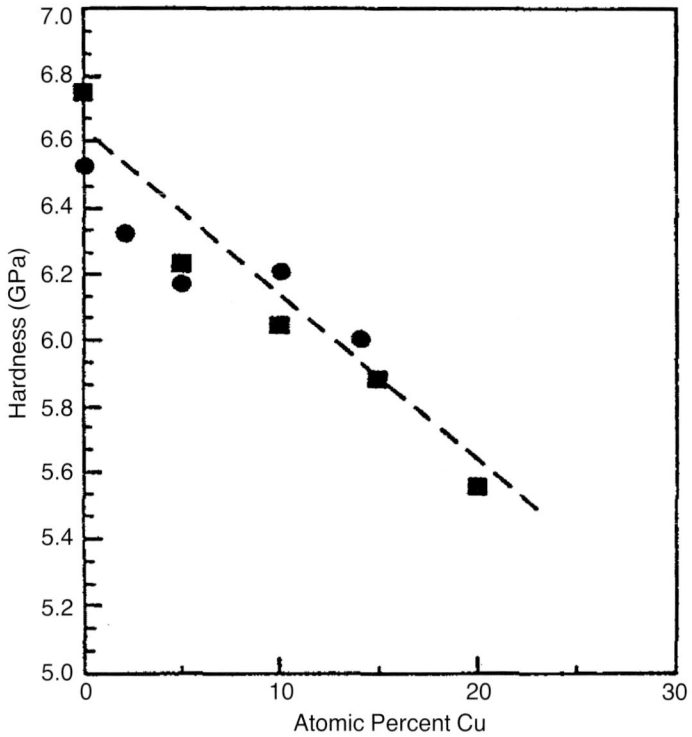

Figure 4.7. Hardness for nanocrystalline Fe(Cu) alloys vs. atomic percent Cu. (After Shen and Koch, 1996.)

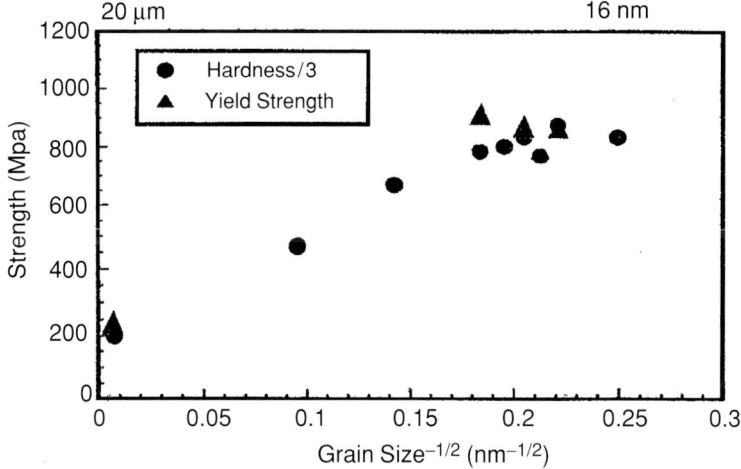

Figure 4.8. Hall–Petch plot for nanocrystalline Cu samples. Circles, hardness/3; triangles, compressive yield stress. (From Weertman (2002), reproduced with permission of William Andrew Publishing.)

strain hardening. The work of Carsley *et al.* (1998) on Fe–10% Cu alloys with grain sizes between 45 nm and 1.3 μm showed deformation by shear bands, similar to that observed in metallic glasses, and an asymmetry in compression and tension yield strengths. The yield strength in compression was about 30% greater than in tension. Wei *et al.* (2002) reported similar results for pure Fe for grain sizes less than 1 μm. In the Carsley *et al.* study the shear bands were not seen on the planes of maximum shear but at about 49° to the tensile axis in compression and between 52° and 54° in tension. The authors explained these results as probably caused by a pressure sensitive von Mises criterion for yielding in conjunction with a zero-extension shear localization criterion. These deviations of the shear-band orientations from the maximum shear angle of 45° had been observed previously for amorphous polymers, but not at the same angles seen in the nanocrystalline and micron grain-size Fe–10% Cu alloys. The study of Wei *et al.* (2002) on nanocrystalline Fe reported shear-band angles with the deformation axis of about 45°, that is along the planes containing the maximum shear stresses. The occurrence of shear bands in nanocrystalline metals has not been observed or reported in most cases. The examples for Fe and Fe–10% Cu tested in compression (Carsley *et al.*, 1995; Carsley *et al.*, 1998; Malow *et al.*, 1998; Wei *et al.*, 2002) or biaxial tension, miniaturized disk bend test (Malow and Koch, 1998) occur where no strain hardening is evident. Other evidence for shear bands in nanocrystalline metals is scant. Witney *et al.* (1995) observed protrusions in fatigued nanocrystalline Cu that suggested shear bands in the direction of maximum shear stress in tensile-tested nanocrystalline Cu (Nieman *et al.*, 1991). While it might be expected to see shear bands in the materials that exhibit little or no strain hardening, the data on nanocrystalline metals is sparse and conflicting. For example, localized deformation – presumably shear bands – was observed in nanocrystalline Pd after 40% compression (Sanders *et al.*, 1997), a material that exhibits strain hardening.

Asymmetry in tension and compression yield strength values have also been observed in "nanocrystalline" (grain size range 40 nm to 200 nm) Al alloys (Cheng *et al.*, 2003), albeit with a much smaller effect than in the nanocrystalline Fe and Fe–10% Cu studies described above. This alloy did not exhibit the shear instability.

Shear bands are also observed in conventional-grain-size and single-crystal materials, as well as in metallic glasses. It has been stated (Hutchinson, 1984) that shear bands may be an inherent feature of plastic deformation in polycrystalline materials beyond certain levels of strain. Shear bands are also enhanced at high strain rates. The degree of strain hardening is critical in suppression of shear bands. It should be pointed out that shear bands were first reported in a nanocrystalline Pd–20% Si alloy recrystallized from the amorphous state (Donovan and Stobbs, 1981). The crystal structure of this Pd–20% Si alloy was not reported.

In summary, the strength and hardness values of nanocrystalline metals can be much larger than those of conventional-grain-size metals, factors of over 20 times have been reported. Asymmetries in yield strength measured by compression vs. tensile tests can sometimes be attributed to processing artifacts, but in some cases this is apparently because of mechanical instabilities and shear bands which are the result of low or non-existent strain hardening. These effects may in turn be influenced by defects from processing and this area requires further clarification.

4.3 Hardness and strength

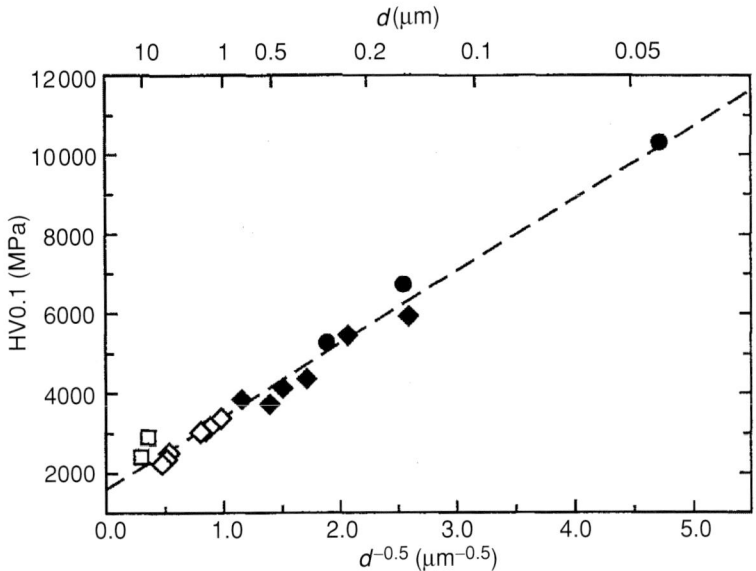

Figure 4.9. Microhardness of Ti–Al-based alloys: ●, mechanically alloyed Ti–46Al–5Si and Ti–20Al–17Si–9.5Si; ◆, mechanically alloyed binary Ti–Al; ◇, conventional powder-processed material; □, ingot material of similar composition. (From Bohn et al. (1995), reproduced with permission of TMS.)

4.3.2 Hardness and strength of nanocrystalline intermetallic compounds and ceramics

Intermetallic compounds exhibit Hall–Petch behavior in that they show increased strength with decreasing grain size (Sauthoff, 1994). Since most intermetallic compounds of interest exhibit at least limited plasticity, presumably dislocation-related mechanisms are responsible for the increased strength owing to the obstacles grain boundaries present to their propagation. However, a large part of the strength and hardness of intermetallics resides in their strong atomic bonding so the effects of grain size on strengthening are often less than for metals. The grain-size effects are larger in the so-called structural intermetallic compounds. TiAl intermetallics have been studied extensively owing to their promising properties for elevated temperature applications in aerospace or power-generation components. Limited studies have been conducted on nanocrystalline TiAl. A Hall–Petch plot for the hardness down to less than about 50 nm grain sizes in given in Figure 4.9 (Bohn et al., 1995). This nanocrystalline TiAl was prepared by mechanical attrition followed by powder consolidation by hot isostatic pressing. Chang et al. (1992) had previously prepared nanocrystalline TiAl by a modified inert-gas-condensation method that used magnetron sputtering. They observed Hall–Petch behavior similar to that of Bohn et al. down to grain sizes of about 40 nm. At smaller grain sizes inverse Hall–Petch behavior was observed. The initial compacted powders had grain sizes of 10–20 nm and were a mixture of amorphous and crystalline phases. Compaction at 250 °C maintained the grain size and still had an unknown volume of amorphous phase and about 12% porosity. Grain growth occurred on

146   4 Mechanical properties of structural nanocrystalline materials

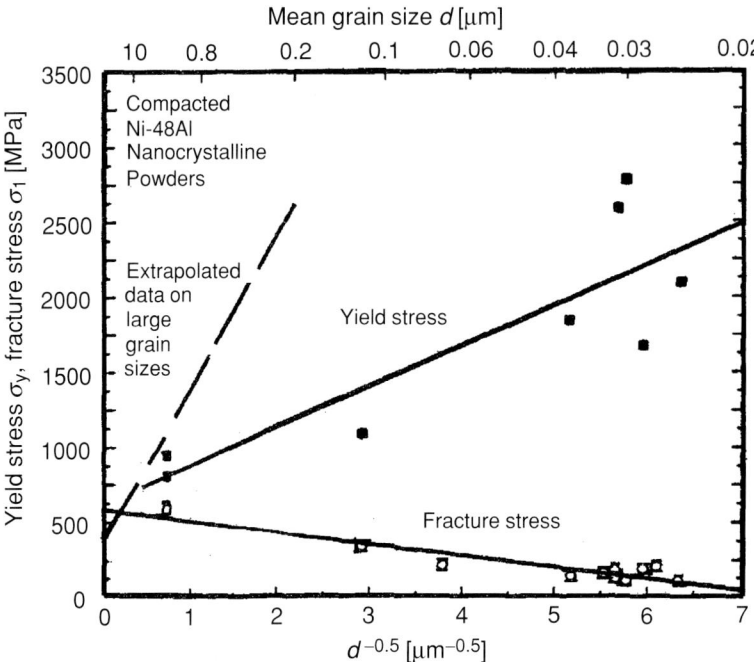

Figure 4.10. Yield stress (hardness/3) and fracture stress in NiAl as a function of grain size. (From Hoffmann and Birringer (1996), reproduced with permission of Elsevier.)

subsequent annealing. After sintering at temperatures above 450 °C the porosity decreased to 4% along with grain growth. While this study has often been quoted by those modeling the inverse Hall–Petch effect the softening at small grain sizes can not be unambiguously attributed to grain-size changes alone. It should be pointed out that TiAl has been studied at conventional grain sizes (about 100 nm to 1 μm) by several investigators whereby large variations in Hall–Petch slopes have been reported (Chu and Thompson, 1991; Vasudevan et al., 1989; Huang and Chesnutt, 1994) which suggests that the hardness/strength of this material is very sensitive to the details of the structure and microstructures.

Another "structural" intermetallic compound which has been studied in nanocrystalline form is NiAl. Hoffmann and Birringer (1996) prepared nanocrystalline NiAl by mechanical attrition of powder followed by consolidation by hot forging. The average composition of their NiAl compound was 48.4 at.% Al. The samples were very brittle so hardness/3 was used to estimate the yield stress as a function of grain size. The miniaturized disk bend test (MDBT) was used to measure the fracture strength. The yield stress, as measured by hardness, increased with decreasing grain size, while the fracture stress decreased as shown in Figure 4.10. This apparent lack of plasticity was presumably the result of the porosity as well as the inherent brittleness of NiAl. Hall–Petch data have been reported for conventional-grain-size NiAl and the slope of the Hall–Petch plot is very sensitive to stoichiometry, with a minimum slope, that is, almost grain-size independence, for the stoichiometric Ni–50 at.% Al composition and rises as the atomic percentage of Al decreases (Baker et al., 1991). The Hall–Petch plot from hardness data of Hoffmann and Birringer on nanocrystalline

4.3 Hardness and strength

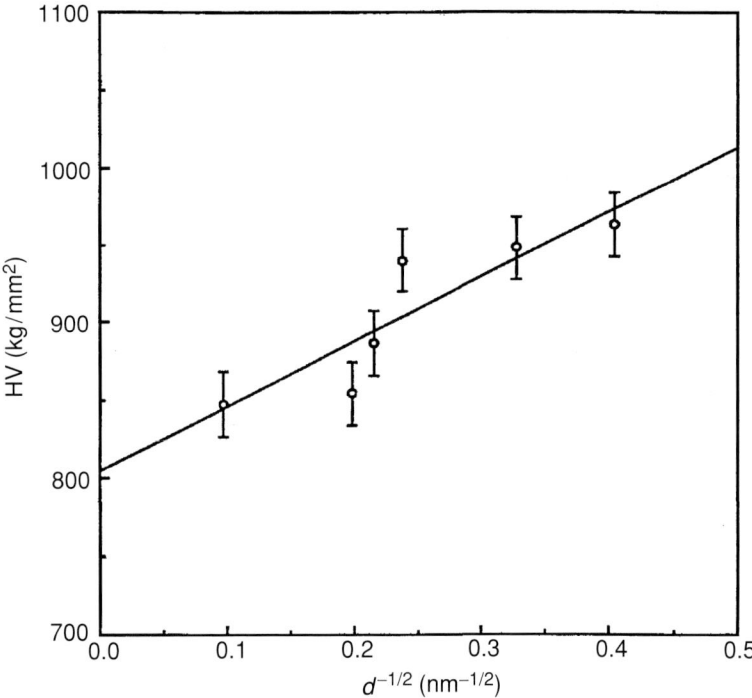

Figure 4.11. Hardness vs. $d^{-1/2}$ for Nb_3Sn, where d is the average grain size. (From Koch and Cho (1992), reproduced with permission of Elsevier.)

NiAl agrees with this composition dependence of the hardness/grain-size behavior. In this case the hardness of the nanocrystalline NiAl is about three times that of coarse grained NiAl.

Conrad et al. (2005) have recently presented hardness vs. grain-size data for several intermetallic compounds which all exhibit an inverse Hall–Petch effect at small grain sizes. Most of these materials show this effect only at the smallest grain sizes, less than 20 nm, with the exception of $NbAl_3$ and Nb_3Al. Kim and Okazaki (1992) prepared nanocrystalline $NbAl_3$ by mechanical alloying of Nb and Al powder followed by consolidation by electric discharge compaction. However, the grain size was varied by annealing, which presumably also changed the structure/chemistry. It is likely that incomplete compaction occurred at the finest grain sizes. Therefore, the inverse Hall–Petch effect beginning at about 60 nm may not be caused by inherent properties of the material.

An intermetallic compound that exhibited only about a 20% increase in hardness when nanoscale over conventional-grain-size samples is Nb_3Sn which was made nanocrystalline by mechanical attrition (Koch and Cho, 1992). The Hall–Petch plot for this intermetallic compound is given in Figure 4.11. The modest increase in hardness is at variance with the larger increases observed in most metals at the nanoscale, and is more consistent with the behavior of brittle ceramic materials. Nb_3Sn is a very brittle intermetallic compound with the A15 crystal structure. It typically fails in the elastic regime until very high test temperatures, about 1400 °C, are reached. Therefore, it appears to behave more like a ceramic.

Ceramic materials also exhibit an increase in hardness/strength with decreasing grain size but the mechanism is presumably much different than in materials such as metals that deform plastically. The conventional explanation for this behavior is that the intrinsic flaw size scales with the grain size (Barsoum, 1997). If cracks are inter-granular or intra-granular they would be limited by the grain size. Thus assuming the Griffith criterion for fracture the strength is expected to follow a $d^{-1/2}$ dependence as observed. However, as grain size decreases to the nanoscale the strength does not keep increasing as grain size decreases. For the fine-grained ceramics fracture occurs from preexistent flaws in the material and strength/hardness becomes relatively independent of grain size. Hofler and Averback (1990) carried out a study of the grain growth and hardness in TiO_2. They observed that the Vickers hardness of the nanocrystalline material was about the same as for single-crystal TiO_2 and the fracture toughness was independent of grain size. Conrad et al. (2005) have analyzed the apparent inverse Hall–Petch effect seen in TiN at grain sizes below about 50 nm. The grain-size softening observed could not be correlated to any one of three models suggested for the inverse Hall–Petch effect, namely, Coble creep, grain-boundary shear, or dislocation-line tension modification. Factors which the authors suggest might be applicable were an increase in the Taylor orientation factor M owing to corresponding decreases in texture with decreasing grain size, and/or processing defects.

4.3.3 *Hardness and strength of multiphase nanocrystalline materials*

Most commercial structural materials are multiphase and it is expected that multiphase nanocrystalline materials will also eventually be the materials of choice for applications. Much less work has been carried out on multiphase materials with a nanocrystalline matrix although from precipitation-hardened aluminum alloys to tempered alloy steels, conventional structural materials often contain nanoscale second phases. There have been a few studies of second-phase nanoscale particles in a nanocrystalline matrix, but more work has been carried out on nanoscale second phases in an amorphous matrix. This work has been reviewed by Eckert (2002). Most of the research has focused on multiphase nanoscale microstructures in Al-, Mg-, and Zr-based multicomponent systems. Most of these were prepared by crystallization of the amorphous phase obtained on cooling from the liquid phase. A smaller subset of studies has used mechanical attrition and annealing to develop a two-phase amorphous/nanocrystalline microstructure.

Inoue and co-workers have been responsible for much of the research in this area. They have synthesized a number of amorphous Al-based alloys by rapid solidification. They have reported on the formation of homogeneous dispersions of nanoscale Al particles on crystallization in an amorphous matrix (Inoue et al., 1994). In Al–Ni–Ln (Ln = lanthanide metal) dispersion of nanoscale Al particles results in increases of 1.3–1.4 times the already high strength values of the amorphous alloys. The authors attribute these high strength values to a combination of the nanoscale effect, perfect crystalline Al particles without internal defects (dislocations), and good interfaces between the amorphous matrix and the nanoscale Al particles. However, Greer (2001) has shown that the hardening on precipitation of Al from metallic glass matrices is likely to be the result of solute enrichment in the glassy matrix rather than dispersion hardening per se. Inoue and co-workers have

4.3 Hardness and strength

developed a variety of nonequilibrium microstructures in Al-based alloys which include: an amorphous single phase; an amorphous matrix with fine nanoscale (3–5 nm) Al particles; a nano-quasicrystalline structure consisting of nanoscale (15–30 nm) icosohedral particles surrounded by a nanoscale fcc Al phase without grain boundaries; a two-phase mixture of nanogranular amorphous ($d = 10$ nm) and Al ($d = 7$ nm) phases; and a nanogranular fcc Al phase ($d = 20$–30 nm) surrounded by an amorphous network. The strength levels of the multiphase nanoscale alloys are typically greater than the single phase amorphous alloy. The fracture strengths of the amorphous alloys are about 1200 MPa and the nanostructured Al particles raise this value to 1560 MPa (Eckert, 2002). Many of these alloys possess good ductility as well as high strength and will be discussed again under the sections on ductility and optimization of strength and ductility. Similar results on high-strength two-phase amorphous/nanocrystalline alloys have been reported for Mg-based and Zr-based alloys (Eckert, 2002).

Morris and Morris (1991) studied the mechanical properties of nanoscale dispersions in Cu-based alloys. Atomized powders of Cu–5 at.% Zr were ball milled and consolidated at 700–800 °C. The second-phase particles were zirconium oxides and carbides formed during processing. The microstructures as determined by TEM consisted of Cu grains less than 100 nm in size, ranging down to 40 nm with fine 5–10 nm oxide or carbide particles. The grain sizes obtained could be explained by the particle pinning of grain boundaries according to the Zener–Gladman models. For these materials, with grain sizes below about 70 nm the strength was considerably lower than would be predicted from dislocation theory based on dislocation–particle (Orowan) interactions.

He and Ma (1996) prepared a nanocomposite of Fe and Cu by ball milling powders of Cu and Fe together and then consolidating them to full density by a sinter forging technique. Since Fe and Cu are immiscible, after the sinter forging at temperatures from 450 °C to 500 °C (increasing with the percentage of Fe) grains of Cu ranged from about 45 nm to 30 nm while Fe grains were about 25–30 nm in size. An enhanced hardness was observed with a significant positive deviation from the rule-of-mixtures predictions by using nanocrystalline Cu and Fe as constituent phases. The authors attribute this effect to additional microstructural strengthening from the interface boundaries between the dissimilar fcc and bcc metals. That is, they suggest the bcc/fcc interfaces are stronger obstacles to dislocation motion than the grain boundaries in the individual metals.

Nanocomposites of Fe–85 vol.% Fe_3C were prepared by mechanical alloying by Goodwin *et al.* (1997). These nanocomposites were prepared in an attempt to make nanocrystalline white-iron-type material. After consolidation of the powders by field-assisted sintering the grain size retained was 45 nm. It is believed that this small grain size was responsible for the high hardness value of 10.8 GPa that was measured.

A nanostructured M50 steel was prepared by three different chemical synthesis methods (Gonsalves *et al.*, 1997). TEM of the compacted samples indicated a grain-size range of 5–70 nm with an average of about 30 nm. The hardness was 8536 MPa compared to 7840 MPa for conventional M50 steel. The yield strength was also higher at 3142 MPa compared to 2646 MPa for conventional material.

Zimmermann *et al.* (2002) have reported the results of second-phase SiC particles incorporated in electrodeposited nanocrystalline Ni. The SiC particles were not nanoscale, but

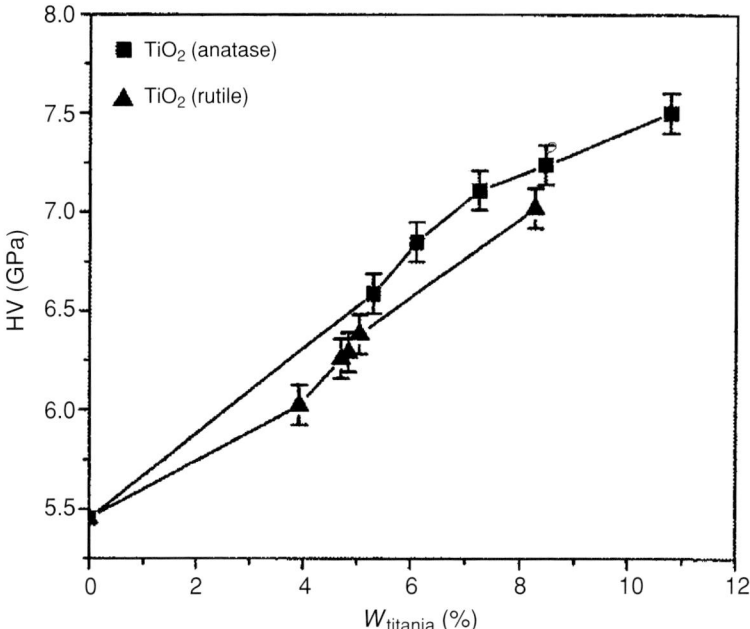

Figure 4.12. Dependence of microhardness of Ni–TiO$_2$ nanocomposites on TiO$_2$ content. (From Li et al. (2005), reproduced with permission of Elsevier.)

about 400 nm in size. The tensile and yield strengths of the nanocomposite with SiC content <2 vol.% were higher than that of pure nanocrystalline Ni of the same grain size. For example, a Ni sample with 10 nm grain size had yield strength and tensile strength values of 830 MPa and 1056 MPa respectively compared to 920 MPa and 1431 MPa for a 10 nm sample containing 1.8 vol.% SiC. Li et al. (2005) made titania–nickel nanocomposites by electrodeposition. The grain sizes of the nanocrystalline Ni and TiO$_2$ were determined by X-ray diffraction line-broadening analysis to be 10 nm and 12 nm respectively. No TEM results were reported. The hardness of the nanocomposite increased significantly, as shown in Figure 4.12. This shows the behavior of micron-size rutile dispersions as well as the nanocrystalline anatase dispersions. The hardness for the composites containing the smaller anatase particles show higher hardness than those with the larger rutile particles.

A very thermally stable nanocrystalline composite of Ni and TaC was obtained by the crystallization of sputtered amorphous Ni$_{63}$Ta$_{23}$C$_{14}$ (at.%) films (Wilde and Greer, 2001). The carbide composition approached stoichiometry with increasing annealing temperature and reached TaC$_{0.8}$ at 700 °C. The grain sizes for both the Ni and TaC grains remained small, growing only to about 17 nm for Ni and 13 nm for TaC at 700 °C. The high stability against coarsening was attributed to the mechanism of grain growth in a duplex structure. The hardness values of the amorphous films before devitrification and the hardness of the duplex Ni–TaC structure after maximum grain growth at 700 °C remained almost constant at about 12 GPa. With only about 35% of the volume of the composite comprised of the carbide phase, this matches the behavior of conventional WC–Co cemented carbides and may be a useful material for coating applications.

4.4 Ductility of nanocrystalline materials: optimization of strength and ductility

Ductility is defined as the ability of a material to undergo plastic deformation without fracture. It is of critical importance for engineering materials for both manufacturability and performance. Measures of ductility include percentage elongation (uniform plastic flow prior to mechanical instability – "necking" – or fracture) and percentage reduction in area. Fracture toughness is also some measure of potential ductility. Engineering materials exhibit wide variations in ductility which can often limit their application.

Ductility is a property of nanocrystalline materials which might be predicted to be enhanced by extrapolation of its grain-size dependence in conventional polycrystalline materials. It has been predicted that extrapolation of the grain size, or the scale of the microstructure, to the nanoscale will lead to an increase in ductility (Bohn *et al.*, 1995). As far as failure and ductility are concerned, this idea is based upon experience with conventional materials, where the yield and fracture stresses show different dependencies on grain size (Dieter, 1976). The fracture stress typically increases faster than the yield stress with decreasing grain size such that ductile/brittle transitions can occur. For example, the ductile/brittle transition temperature in mild steel can be lowered by about 40 °C by reducing the grain size by a factor of five. How ductility may be affected by the extreme reduction of grain size to the nanoscale can be considered from the point of view of the limitations to ductility in nanocrystalline materials. Three major limitations to ductility for nanocrystalline materials can be identified. These are: (1) artifacts from processing; (2) force instability in tension; and (3) crack nucleation or propagation instability. These will be discussed in turn.

The various methods for producing bulk nanocrystalline materials have been described in Chapter 2. Many of these methods involve the synthesis of nanoscale particulates, or powders with a nanocrystalline internal grain structure, followed by consolidation to bulk samples. These can be called "two-step" methods. The goal of particulate/powder compaction is to obtain theoretical density and complete particle bonding without significant coarsening of the nanocrystalline structure. Porosity was a major artifact, especially for earlier studies of metals made by the inert-gas-condensation method. Even when theoretical density is attained, complete particle bonding may be lacking. Nanocrystalline materials made by ball milling of powders can also suffer from lack of complete bonding after powder consolidation. The surfactant that is sometimes used to prevent excessive cold-welding during milling can prevent complete bonding during consolidation and therefore limit ductility. Another popular method for producing nanocrystalline materials is electrodeposition. While this method is "one-step" in that it does not require consolidation of particulates and the problems associated with this, most of the nanocrystalline materials made by electrodeposition have also exhibited very poor ductility. In order to attain a nanoscale grain structure additives are often needed in the electrodeposition bath, which control the growth of the deposited grains. The additives are typically organic materials, which can leave a residue of impurity atoms such as sulfur trapped at the grain boundaries. Methods to eliminate such possible impurities but still provide nanocrystalline microstructures include pulse-electrodeposition and the simultaneous plating of solute additions which do not segregate to grain boundaries.

The force instability in tension begins at maximum load as evidenced by "necking" – local deformation, reduction in cross-sectional area. This occurs when the load the sample can support owing to strain hardening becomes less than the applied load determined by the strength and the reduction in cross-sectional area. This leads to a criterion for instability during tensile loading known as the Considere criterion. The amount of uniform elongation depends upon the strain hardening such that true strain [$\varepsilon_u = n$ in a cylindrical specimen or $\varepsilon_u = 2n$ for a sheet), where n is the strain hardening coefficient. For an ideally plastic material (such as amorphous alloys) where $n = 0$, the necking instability would begin just as soon as yielding occurred. This criterion implies that the sample is mechanically stable until the rate of strain hardening falls to a level determined by the flow stress (and prior strain) at that time. Materials with a high capacity for strain hardening are therefore stable, while those with little capacity for strain hardening are potentially unstable.

The crack nucleation or propagation instability comes about if the imposed stress concentration at an existing flaw exceeds the critical toughness value of the material. This approach considers the stress intensity (K) at a pre-existing flaw (or J, the energy or work input required to reach that state) as the sample is increasingly loaded (Chan, 1990). Work must be supplied from the external source to produce the elastic stress concentration at the notch tip, to produce the local plastic strain as the notch starts to deform, and to achieve global plasticity in the case where the material shows some overall ductility. These terms vary with the initial flaw size, considered to be the same as the grain (powder) size, and assuming the flow stress and fracture toughness to be constants, it should be possible at sufficiently fine grain sizes to achieve global plasticity before brittle, local crack propagation sets in. Given the strong increase in flow stress with decreasing grain size at the nanoscale, the competition between plastic flow and fracture is difficult to predict.

Elongation to failure in tension is plotted vs. grain size in Figure 4.13 for a variety of metals and alloys. It is clear that for most metals with grain sizes below about 30 nm the elongation to failure values are very low, typically less than 2%–3%. These are for metals that in conventional-grain-size condition have elongations to failure of the order of 50% or more. Since this graph was originally published (Koch *et al.*, 1999) several new datum points have been added that show significant ductility for grain sizes of ≤30 nm. These more recent results, which provide optimized strength and ductility behavior in metallic materials, will be described below.

4.4.1 Optimization of strength and ductility in metallic materials

Minimization or elimination of processing artifacts is the key to revealing the inherent mechanical properties of nanocrystalline materials and potentially optimizing high strength and good ductility. As described in Chapter 2, one-step processing methods are more likely to provide nanocrystalline samples free from artifacts. The non-trivial problem is to produce nanocrystalline materials with fine grain sizes, that is, less than about 50 nm, without

4.4 Ductility of nanocrystalline materials

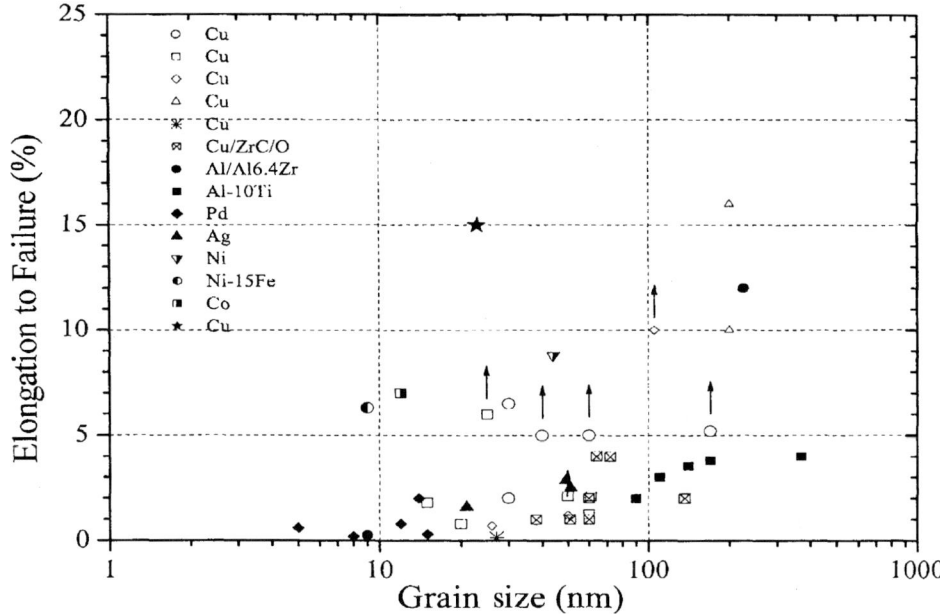

Figure 4.13. Elongation to failure in tension versus grain size for a variety of metals and alloys (Koch *et al.*, 2005), reprinted with permission.

introduction of artifacts. Many strategies needed to obtain nanocrystalline microstructures lead invariably to artifact creation. In the case of two-step processing, which requires a subsequent consolidation step, perfect artifact-free bulk samples are difficult to obtain. Even if theoretical densities are reached, particulate bonding may be weak and a favored location for fracture under applied stress. Unfortunately, one-step processes are not without problems. Vapor-deposition processes such as sputtering, electron beam evaporation, and pulse laser ablation can provide essentially artifact-free nanocrystalline materials to the finest grain sizes (<10 nm) but are typically limited to thin films. Therefore, they are not of interest for production of bulk samples for mechanical testing or applications.

Electrodeposition, as described above, often requires additives to give nanocrystalline grain sizes in terms of biasing nucleation over growth of the depositing grains. These additives can lead to embrittlement and are presumably responsible for the very low ductilities typically observed in nanocrystalline electrodeposits. However, if additives can be avoided and other deposition parameters optimized to give nanocrystalline structures, an artifact-free material may be provided. Such materials represent some of the breakthrough materials to be discussed that combine high strengths and good ductilities.

Production of nanocrystalline materials by severe plastic deformation methods have been described in Chapter 2. With the exception of ball milling of powders, these techniques can be classified as one-step processes since bulk materials are obtained and therefore do not need particulate consolidation. Of the variety of severe plastic deformation methods, it appears that only ball milling, high-pressure torsion, and accumulative roll-bonding can regularly produce average grain sizes below 100 nm – that is, nanocrystalline materials.

Figure 4.14. Cu spheres with nanocrystalline microstructures prepared by combined liquid-nitrogen temperature and room-temperature milling.

The problems with ball milling as a two-step process that requires consolidation have been discussed above. The samples produced by high-pressure torsion are typically limited in size, which complicates the mechanical testing that might be carried out on them. In order to obtain nanocrystalline structures with accumulative roll-bonding, many cycles of rolling, cleaning, and stacking of foils is needed which suggests a problem for practical scale-up for bulk materials.

While ball milling of powders is usually a two-step process, a recent ball-milling strategy has been developed that provides for *in situ* powder consolidation during milling (Zhu *et al.*, 2003). This is accomplished by a judicious selection of milling temperatures (liquid-nitrogen and room-temperature combinations), milling times at each temperature, and ball/powder ratios. In certain ductile metals it is found that solid spheres up to 10–15 mm in diameter can be formed; these can then be pressed into disks for a variety of mechanical tests. Metals and alloys that have been processed in this way include Zn, Cu, Al, and several Cu and Al alloys. An example of spherical Cu balls that were produced in this manner is given in Figure 4.14. The interior of these balls has a nanocrystalline microstructure with an average grain size of about 23 nm.

Crystallization of amorphous precursors can be used to obtain nanocrystalline materials as another one-step process. Potential difficulties with this method include the possible residual amorphous material that may be present and affect the inherent mechanical properties of the sample. Another possible problem is that many crystallized amorphous alloys contain brittle intermetallic phases. However, some examples of successful utilization of this method will be given below.

Even if processing artifacts can be eliminated, strain hardening is needed in order to minimize mechanical instabilities that lead to premature local deformation (necking) and failure. The ability to strain harden therefore becomes an important criterion for ductility

4.4 Ductility of nanocrystalline materials

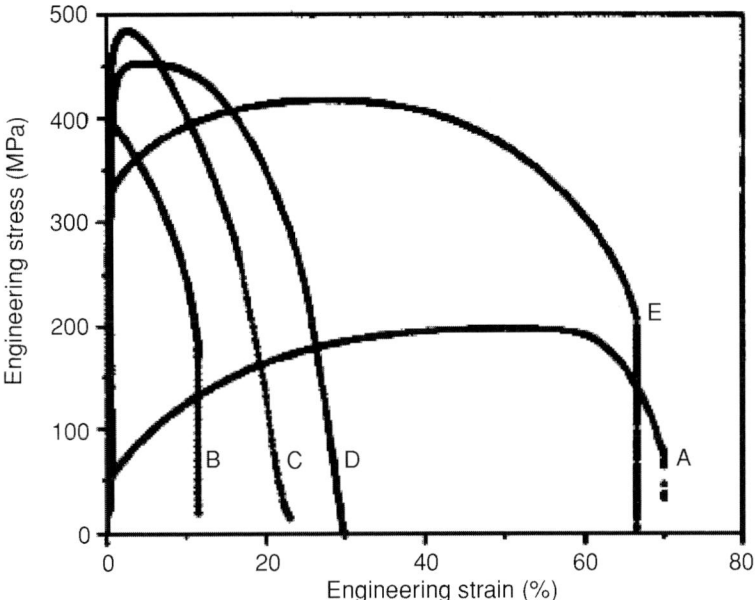

Figure 4.15. Stress–strain curves for Cu with different microstructures. Curve E represents bi-modal nanostructured Cu. (From Wang *et al.* (2002), reproduced with permission of *Nature*.)

in nanostructured materials. It is not clear whether nanocrystalline materials at the smaller grain sizes can strain harden by the usual dislocation mechanisms. An approach that has been used to provide strain hardening in nanostructured materials is to introduce a bi-modal grain-size distribution by appropriate processing methods. The supposition was that the larger grains should deform by the usual dislocation mechanisms and provide strain hardening, while the smaller nanoscale grains would provide the strength and hardness.

Wang *et al.* (2002) have demonstrated the above strategy of a bi-modal grain-size distribution to provide a combination of high strength and high-tensile ductility in Cu. The procedure used is described in Chapter 2 and involves rolling at liquid-nitrogen temperature to obtain a high dislocation density, followed by low-temperature annealing to give a microstructure consisting of about 25% volume fraction of grains 1–3 μm in diameter in a matrix of nanocrystalline and sub-micron-size grains. This work on Cu provided stress–strain curves for annealed coarse-grained Cu, Cu rolled to 95% at room temperature, Cu rolled to 93% at liquid nitrogen temperature, and these samples annealed for 3 min at either 180 °C or 200 °C. These stress–strain curves are presented in Figure 4.15. The optimum properties, curve E in the figure, were obtained for the mixed-grain-size material with 1–3 μm grains imbedded in the matrix of nanoscale and sub-micron-size grains. This material had a high yield stress of about 340 MPa, a total elongation to failure of 65% and uniform elongation of about 30%. The ductility was thus comparable to that of annealed conventional-grain-size Cu, but the yield strength was almost seven times higher.

Another example of using bi-modal grain-size distributions to optimize strength and ductility was reported for Al–Mg alloys. A commercial Al–Mg alloy, 5083 (Tellkamp *et al.*, 2001) and an Al–7.5% Mg alloy (Witkin *et al.*, 2003) were prepared by cryomilling followed by powder compaction by hot isostatic pressing and extrusion. The cryomilling of Al alloy

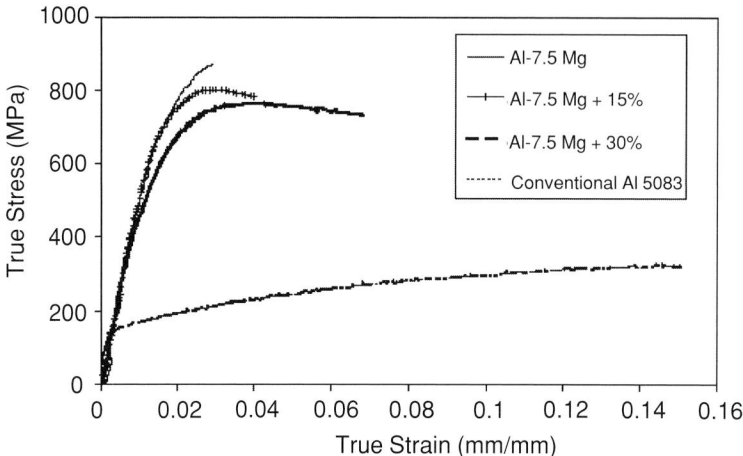

Figure 4.16. Tensile curves of extruded Al–7.5 % Mg samples. (From Witkin *et al.* (2003), reproduced with permission of Elsevier.)

5083 resulted in a nanoscale microstructure with average grain size about 30 nm (Tellkamp *et al.*, 2001). After hipping and extrusion the grain size remained mostly nanoscale, at an average value of about 35 nm. However, some larger grains were also observed in the TEM analysis. The stability of the nanoscale grain size during the elevated temperature compaction steps was attributed to the large number of various precipitates including several intermetallic compounds such as Mg_2Si and Al_3Mg_2 as well as compounds formed from interstitial impurity atoms, namely AlN and Al_2O_3 which presumably retard grain growth by Zener pinning of the grain boundaries. A few larger micron-size grains were formed by secondary recrystallization. These large grains were believed to be responsible for the good ductility observed in these materials along with large increases in strength. Guided by these results, an Al–7.5% Mg alloy was cryomilled to nanostructured grain sizes (Witkin *et al.*, 2003). The cryomilled powder was then combined with either 15% or 30% by volume of unmilled alloy powder, which was made by powder atomization and had micron-scale grain sizes. The powders were then consolidated by hipping and extrusion to bulk samples for tensile testing. The tensile stress–strain curves for the Al–7.5% Mg alloys are plotted in Figure 4.16 and compared to that for conventional Al 5083 alloy. It is clear that the additions of larger grains to the nanocrystalline matrix have increased the tensile ductility from about 1.4% to 5.4% elongation, with some decrease in strength values but still about four times the yield strength of conventional Al 5083.

Bi-modal grain-size distributions were also obtained by crystallization of amorphous alloys at different annealing temperatures and times by Sergueeva *et al.* (2004). The grain-size distribution effect was then studied on NiTi and Vitroperm ($Fe_{73.5}Cu_1Nb_3Si_{15.5}B_7$). The tensile stress–strain behavior was measured at $0.4\,T_m$. The presence of larger grains embedded in the nanocrystalline matrices in these materials reduced the strength but increased the strain hardening and increased the ductility compared to the materials with a homogeneous grain-size distribution.

While the above strategies for optimization of strength and ductility using a bi-modal grain-size distribution compromise some strength for ductility, there are several recent

4.4 Ductility of nanocrystalline materials

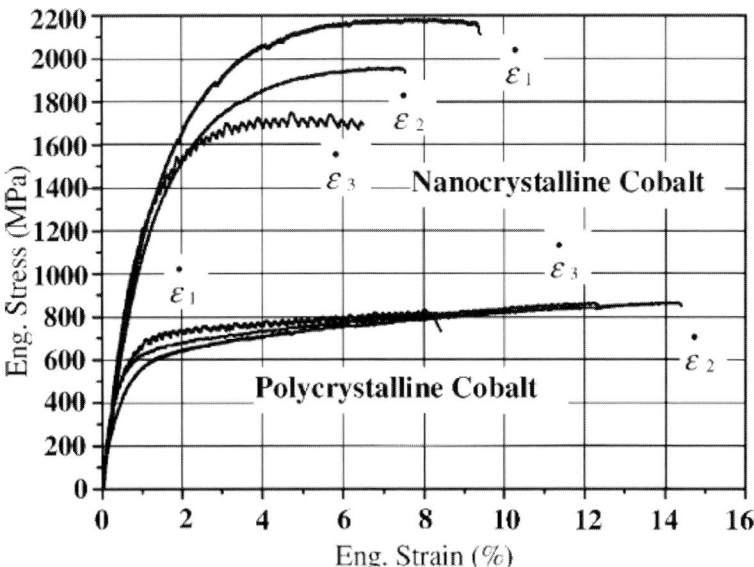

Figure 4.17. Representative stress–strain curves for nanocrystalline Co and polycrystalline Co (annealed at 800 °C) at strain rates $\varepsilon_1 = 1 \times 10^{-4}$, $\varepsilon_2 = 5 \times 10^{-4}$, and $\varepsilon_3 = 2.5 \times 10^{-3}$ (s^{-1}). (From Karimpoor et al. (2003), reproduced with permission of Elsevier.)

results on nanostructured materials where strength levels are high and good ductility can still be achieved. These results are for nanocrystalline materials with small grain sizes (<30 nm) and with relatively narrow grain size distributions such that no grains >50 nm are present.

The first example of combined high strength and good ductility for a nanostructured metal was for electrodeposited Co. It was prepared with a small average grain size of about 12 nm and with a fairly narrow grain-size distribution of ±7 nm (Karimpoor et al., 2003). This material had the hcp structure with no trace of the fcc phase, that is, it had the equilibrium structure for room temperature. The hardness, yield strength, and ultimate tensile strength for this nanostructured Co were 2–3 times higher than for conventional-grain-size Co. The nanocrystalline Co exhibited elongation to fracture values of 6%–9%, which are comparable to those for the conventional-grain-size Co. The stress–strain curves for this nanostructured Co and conventional-grain-size Co are given in Figure 4.17. Of great interest is the dependence of the mechanical behavior of the nanostructured Co on the strain rate of the tensile tests. Applying lower strain rates resulted in higher flow stress and tensile strength at relatively constant yield strength. This behavior is in contrast to the usual response of a material in which dislocation slip is the dominant deformation mechanism. In such materials higher strain rates result in higher tensile strength. The authors suggest that this response of nanocrystalline Co to changes in strain rate is typical of materials which deform predominantly by deformation twinning. That is, higher strain rates result in lower flow stress and tensile strength. They therefore suggest that the dominant deformation mechanism in their nanocrystalline Co is twinning. More studies, in particular high-resolution TEM or *in situ* TEM under stress, are needed to confirm these ideas about twinning deformation.

Li and Ebrahimi (2004) have prepared nanocrystalline Ni and Ni–Fe alloys by electrodeposition without the use of any additives that might induce embrittlement. Their samples

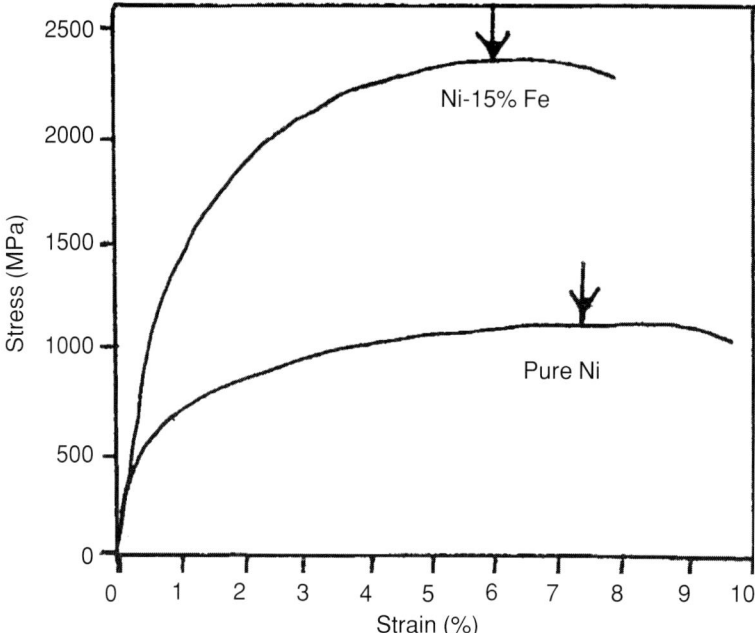

Figure 4.18. Engineering stress–strain curves for nanocrystalline Ni and Ni–15 % Fe alloy. Arrows indicate the ultimate tensile stresses. (From Li and Ebrahimi (2004), reproduced with permission of the American Institute of Physics.)

exhibited excellent strength values along with good ductility. Stress–strain curves for the pure nanocrystalline Ni and a Ni–15% Fe alloy are given in Figure 4.18. The Ni sample had a grain size of 44 nm and the Ni–15% Fe sample had a 9 nm grain size. The fracture behavior of the alloys was very different even though both exhibited good values of total percentage elongation. The Ni sample had an elongation of about 9% and also showed significant reduction in area and ductile fracture behavior consistent with that for other ductile fcc metals and deformation by dislocation motion. While the Ni–15% Fe sample also showed reasonable elongation of about 6%, the reduction in area was negligible and fracture appeared to be brittle. TEM revealed grain-boundary cracking. The authors suggest that the Ni–15% Fe alloy with the average grain size of 9 nm was below the "cross-over" grain size from dislocation dominated deformation processes to grain-boundary deformation processes such as grain-boundary sliding. In spite of this apparent brittle fracture behavior, good ductility along with high strength were observed, suggesting processing artifacts did not affect the mechanical properties.

The *in situ* consolidation of ball-milled powders in several metals has allowed for the production of artifact free samples for tensile testing. Bulk nanocrystalline Cu spheres were synthesized using a combination of liquid-nitrogen-temperature and room-temperature milling (Youssef *et al.*, 2004). Spheres with diameters up to about 8 mm were obtained that could be pressed into disks from which samples for mechanical testing could be machined. TEM results shown in Figure 4.19a indicate that the consolidated Cu consists of equiaxed nanograins oriented randomly, as can be seen from the corresponding selected

4.4 Ductility of nanocrystalline materials

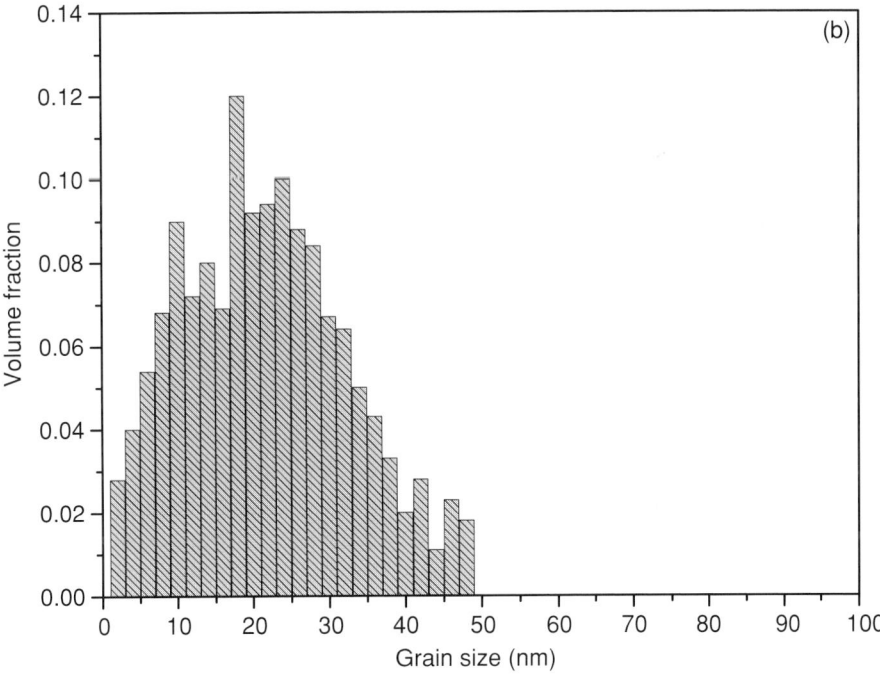

Figure 4.19. TEM observations of the typical microstructure in the *in situ* consolidated nanocrystalline Cu. The bright-field TEM micrograph (a) and the SADP (the upper left inset in (a)) show roughly equiaxed grains with random orientations. The statistical distribution of grain size (b) was obtained from multiple dark-field images of the same sample (Koch *et al.*, 2005).

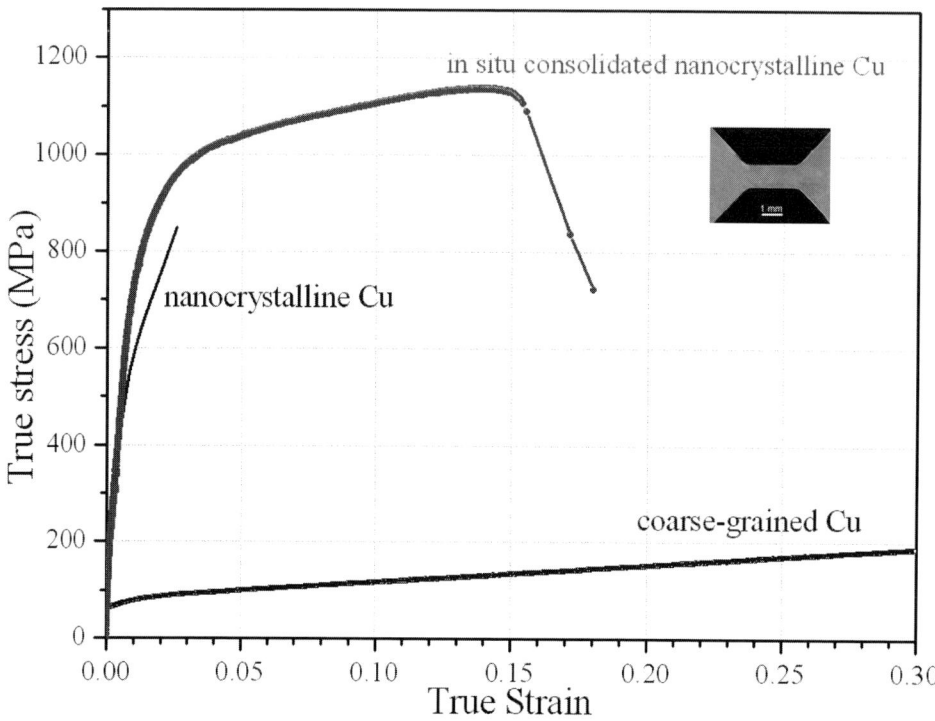

Figure 4.20. A typical tensile stress–strain curve for the bulk *in situ* consolidated nanocrystalline Cu sample in comparison with that of a coarse-grained polycrystalline Cu sample (about 80 μm average grain size) and a nanocrystalline Cu sample prepared by the inert-gas-condensation and compaction technique (mean grain size 26 nm) (Koch *et al.*, 2005).

area diffraction pattern, the left inset in Figure 4.19a. Statistical analysis of multiple dark field images reveals a monotonic lognormal grain size distribution with an average grain size of 23 nm (Figure 4.19b). Density measurements, scanning electron microscopy of the sample surfaces, and TEM analysis show that no porosity is introduced during the *in situ* consolidation of nanocrystalline Cu. The chemical analysis of the consolidated nanocrystalline Cu indicated that the oxygen content increased from 0.10 at.% in the starting powder to 0.29 at.% in the final product. The measured Fe contamination was less than 0.1 at.%. Therefore, it may be concluded that the nanocrystalline Cu made by the above procedure is free of artifacts in that there is no porosity, no debonding, and minimal impurity contamination. Tensile test data for the *in situ* consolidated nanocrystalline Cu is compared with the stress–strain curve for conventional Cu and nanocrystalline Cu made by the inert-gas-condensation method in Figure 4.20. In the case of the nanocrystalline Cu, the 0.2% offset yield strength (σ_y) and the ultimate tensile strength (σ_u) reach values of 791 ± 12 MPa and 1120 ± 29 MPa respectively. This σ_y value is at least one order of magnitude higher than that of coarse-grained pure Cu samples, and σ_u of the nanocrystalline Cu is about five times higher than that of the coarse-grained Cu sample. The hardness value of this nanocrystalline Cu is 2.3 GPa, which is consistent with the Hall–Petch behavior of Cu. Therefore, it is concluded that the high values of hardness and yield strength are caused by the small grain size (23 nm). These strength values are comparable to the highest values

4.4 Ductility of nanocrystalline materials

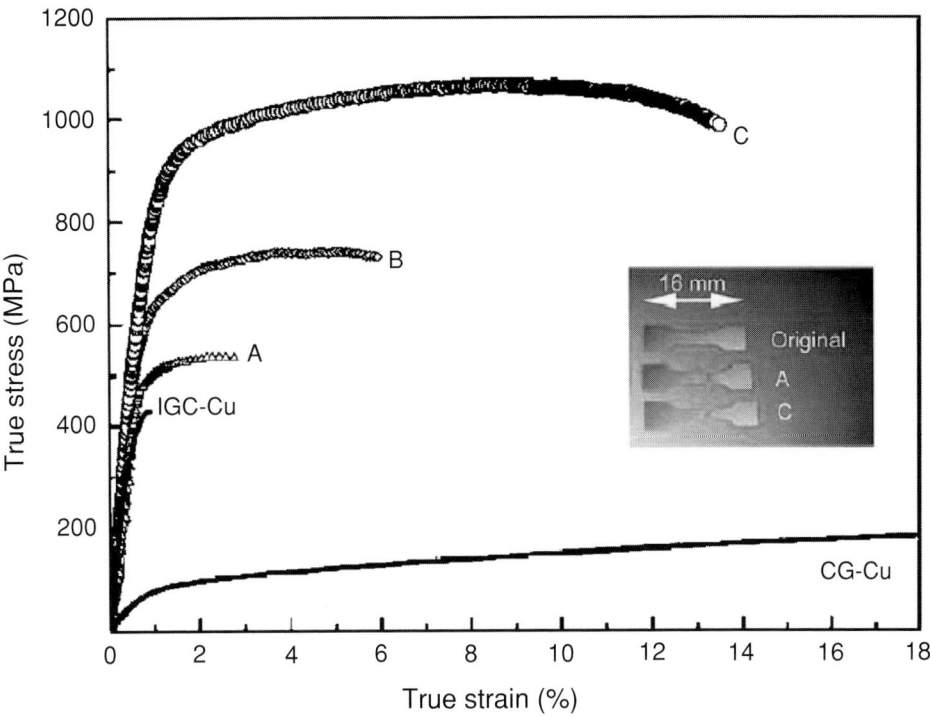

Figure 4.21. Tensile true stress–strain curves for three different as-deposited Cu samples (A, B, and C). Inset shows the specimen geometries before and after tensile tests. For comparison, tensile stress–strain curves for a coarse-grained (CG) Cu and a 3D nanocrystalline Cu made by using inert-gas-condensation (IGC) method are also included. (From Shen et al. (2005), reproduced with permission of Elsevier.)

observed for nanocrystalline Cu but more significant is the good ductility observed with 14% uniform elongation and 15% elongation to failure. This ductility is much greater than previously reported for nanocrystalline materials of this grain size. Another important feature of the stress–strain curve (Figure 4.20) is the large strain hardening observed in the plastic region which suggests a high lattice dislocation accumulation during the plastic deformation up to failure. Strain hardening is often limited in nanocrystalline materials at grain sizes where dislocation activity is believed to be difficult. Therefore, the optimization of both strength and ductility in the above nanocrystalline Cu is believed to be caused by the lack of any artifacts from processing and the ability to exhibit significant strain hardening.

While nanoscale grain boundaries have been the focus for increasing strength in studies of nanocrystalline materials, twin boundaries can also be an effective obstacle to dislocation motion and a potent stengthener. Lu and co-workers (Shen et al., 2005) have recently reported the synthesis of nanoscale growth twins in electrodeposited Cu. The Cu grain sizes were of the order of 400–500 nm and the twin lamellae thicknesses ranged from averages of about 100 nm down to <20 nm. The yield strength of the Cu followed Hall–Petch behavior with increased strength as twin lamellae spacing decreased. Increased ductility also was observed with decreasing twin lamellae spacing. Stress–strain curves for these materials are illustrated in Figure 4.21 and compared to nanocrystalline Cu made by the inert condensation and compaction process and conventional grain size Cu. The Cu with the finest twin lamellae spacing shows both high strength and ductility. Higher strain hardening than conventional-grain-size Cu is also noted.

4.4.2 In situ *TEM and X-ray diffraction studies of nanocrystalline materials*

While transmission electron microscopy (TEM) is a powerful tool for studying the microstructure of deformation in materials, there are special problems associated with its application to nanocrystalline materials (Milligan, 2003). These problems include overlapping of the nanoscale grains which makes high-resolution TEM difficult. Most postmortem TEM studies on deformed nanocrystalline materials in the early studies did not observe dislocations and theoretical speculations suggested that they might not exist because of large image forces that would pull them into boundaries along with the difficulty in activation of dislocation sources, such as the Frank–Read source. To overcome these problems, Milligan *et al.* (1993) and Ke *et al.* (1995) carried out *in situ* TEM studies on thin foils of Au and Ag which were sputtered. The thin foil dimensions solved the problem of grain overlap, and the examination of the thin foils deformed on a stage in the microscope allowed for direct examination of the deformation process as it occurred. Grain sizes were varied between 8 nm and 150 nm. At grain sizes of 25 nm or smaller no dislocations were observed. However, significant plasticity was observed which the authors attribute to diffusion assisted grain boundary deformation processes. Evidence for grain rotation was given. The larger-grained materials clearly showed dislocation activity *in situ* as well as postmortem. The authors concluded that there is a transition in deformation mechanisms with decreasing grain size from dislocation plasticity to grain-boundary plasticity, presumably grain-boundary sliding and rotation.

Youngdahl *et al.* (2001) carried out *in situ* TEM studies on nanocrystalline Cu foils made by the inert-gas-condensation method. They recorded their observations in the microscope on video tape. They observed what was believed to be dislocation activity as the deformation mechanism down to grain sizes of 30 nm. For grain sizes smaller than 30 nm overlap problems made the viewing of individual grains difficult. They observed no evidence for grain-boundary sliding or rotation in the range of grain sizes that could be studied.

Kumar *et al.* (2003) carried out a comprehensive study of the deformation structure of electrodeposited nanocrystalline Ni with a grain size of about 30 nm. In this case, dislocation debris was observed in postmortem TEM studies, but the dislocation density left in the specimens could not account for the high levels of plastic strain that had been given. Still images and video tape images were taken for *in situ* tensile tests in the TEM. A large amount of dislocation activity was observed. It appeared that dislocations were emitted from grain boundaries. Voids and wedge cracks were observed to form along grain boundaries and triple junctions owing to transgranular slip and unaccommodated grain-boundary sliding. It was concluded that the extensive plasticity was a combination of dislocation plasticity and nucleation and growth of voids. Possible evidence for some twinning was also reported.

Hugo *et al.* (2003) also used *in situ* TEM to study the deformation structure in Ni thin films produced by either d.c. magnetron sputtering or pulse laser deposition. The films

4.4 Ductility of nanocrystalline materials

produced by sputtering exhibited brittle behavior which was attributed to high porosity found at the grain boundaries. The films made by pulse laser ablation were ductile. In both kinds of Ni films, intense dislocation activity was observed in terms of both nucleation and propagation of dislocations. This was the case for even grains as small as about 10 nm. The dislocation motion was more easily identified in larger grains (~30 nm) but abrupt changes in contrast in the entire small grain were interpreted as dislocation motion. However, multiple dislocations were also observed in grains as small as about 10 nm. No grains were conclusively observed to rotate. These observations at least indirectly indicated that grain-boundary sliding contributed only a small amount of strain to the overall ductility. This is contrary to some thinking about the grain sizes at which dislocation deformation can still occur (see Milligan, 2003) and may be related to the possibility that dislocation activity may be enhanced in thin films compared to bulk material as has been shown in molecular dynamics simulations (Derlet and Van Swygenhoven, 2002).

The above study was extended with more detailed *in situ* tensile straining in the TEM on Ni films 1.5–10 μm in thickness by Mitra *et al.*, 2004. The straining was stopped for TEM observations, with the load still applied. Contrast changes which were believed to be clear evidence for dislocation activity were seen in grains as small as 20 nm. Parallel arrays of roughly equally spaced dislocations were observed that were spaced about 5–10 nm apart.

In order to determine the origin of the optimized mechanical properties of the nanocrystalline Cu prepared by *in situ* consolidation of ball-milled powder described above, Cu samples were studied by *in situ* TEM tensile straining. The sub-size dog-bone-shaped tensile specimens were incrementally strained in discrete steps until a crack was formed. The crack typically advanced normal to the loading direction and the region ahead of the crack tip was monitored. Upon straining, bright-field TEM observations showed rapid changes in contrast that took place continuously in almost all the grains. Similar observations have been identified previously as evidence of dislocation activity (Youngdahl *et al.*, 2001; Kumar *et al.*, 2003; Hugo *et al.*, 2003). Figure 4.22 shows bright-field TEM micrographs of a single nanoscale grain (~23 nm wide) that incrementally strained at the crack tip (the loading and crack growth directions are indicated by black arrows). At a given strain the nanograin appears to have a significant number of dislocations (Figure 4.22a). With increasing strain (Figure 4.22b) more dislocations are generated from the grain boundary, glide through the grain, and pile up to the center of the grain (white arrows). Fewer dislocations are seen in Figure 4.22c than in Figure 4.22b which may be caused by contrast variations during straining and/or by annihilation of some dislocations due to thinning of the nanograin at its center. The dislocation pile-up is observed in the grain interior even after the nanograin failed upon further straining, that is after the applied stress was removed (Figure 4.22d). The above experimental results illustrate the existence of lattice dislocation pile-ups and glide activity during straining of nanograins as well as after the external stress has been released. These observations may explain why strain hardening is observed in these nanocrystalline Cu samples. Figure 4.23 shows the bright-field TEM image of another nanograin (~8 nm wide and 15 nm long) away from the crack tip. This grain contains trapped dislocations, which also verifies the important role of dislocations on the deformation process of this nanocrystalline Cu even in grains with these small dimensions. As noted above, other reports (see Mitra *et al.*, 2004) show evidence for dislocation activity in grain sizes in this

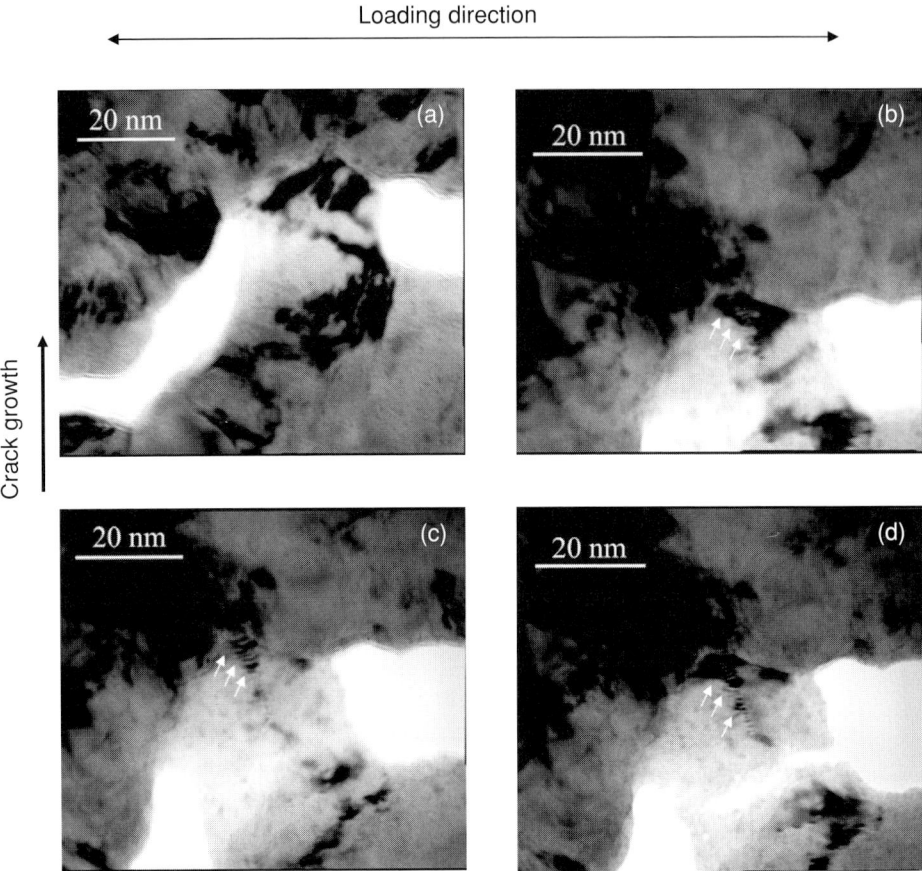

Figure 4.22. BFTEM micrographs show the dislocation activity of a nanocrystalline Cu grain during *in situ* TEM deformation. The loading and crack growth directions are shown at the upper and left sides of the figure, respectively. (a) The nanograin at certain straining, which shows dislocations inside the nano grain. (b) Dislocation generation and pile-up on the grain boundary upon further straining (white arrows). (c) More straining of the nanograin showing also the dislocations pile-up. (d) Further straining initiates a crack in the nanograin (black arrow), while the piled-up dislocations still exist (Koch *et al.*, 2005). Reprinted with permission.

range under applied stress. For smaller grain sizes, less than about 10–15 nm, it has been predicted by many (see Ovidko, 2003) that deformation occurs by grain-boundary mediated plasticity such as grain-boundary sliding or grain rotation. The above *in situ* TEM studies, with the exception of those on nanocrystalline Au foils (Milligan *et al.*, 1993; Ke *et al.*, 1995) have not seen evidence for grain-boundary deformation. However, Shan *et al.* (2004) have reported recently the observation of grain-boundary-mediated deformation in nanocrystalline Ni (average grain size of 10 nm) by examining successive video frames in dark-field TEM mode.

Additional evidence for possible grain rotation and/or sliding comes from the high-resolution TEM study of nanocrystalline Pd prepared by the inert-gas-condensation method

4.4 Ductility of nanocrystalline materials

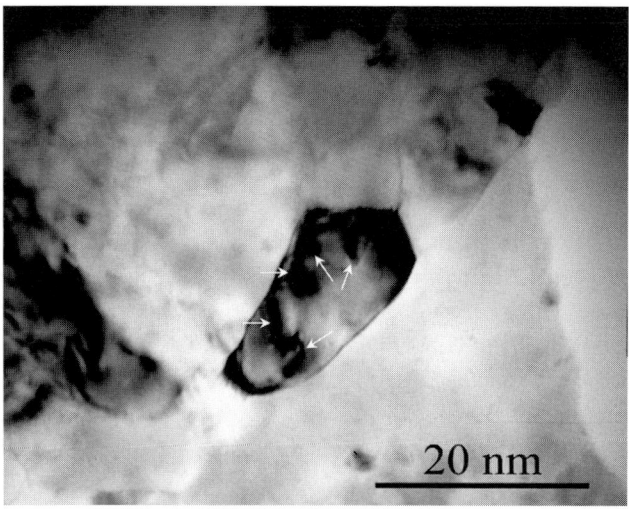

Figure 4.23. A typical bright-field TEM micrograph of trapped dislocations in a Cu nanograin formed during *in situ* TEM deformation (Koch *et al.*, 2005). Reprinted with permission.

(Rosner *et al.*, 2004). The nanocrystalline Pd with about 15 nm grain size was cold rolled up to a true strain of 0.32. Significant amounts of twinning on {111} planes were found and Shockley partial dislocation observed. In each grain twinning was observed only on a single set of parallel planes. This must mean that only one of the five independent slip systems needed for general deformation is operable which suggests that grain rotation and grain-boundary sliding must be occurring together with twinning.

Markmann *et al.* (2003) had previously studied the structure of nanocrystalline Pd with grain sizes in the range of 10–30 nm at a low strain rate, essentially creep, and a high strain rate, rolling. The creep results will be discussed later in this chapter. X-ray analysis of the material that showed an increase in the stacking fault density provided conclusive evidence for the activity of lattice dislocations in small nanocrystalline grains in a bulk sample. However, no rolling texture was observed and the grain shapes remained equiaxed after large deformations which suggested that grain-boundary sliding must also be contributing to the deformation.

An *in situ* X-ray diffraction peak profile analysis was carried out by Budrovic *et al.* (2004) on nanocrystalline Ni. They observed an increase in line broadening which was attributed to inhomogeneous strain in the sample, since the other major cause of line broadening, decrease in grain size, was not evident. This increased line broadening was essentially completely reversible, that is, when the applied stress was removed the increased broadening decreased back to the unstressed level. The recoverable peak broadening observed during deformation of the nanocrystalline Ni (grain size about 26 nm) suggests a deformation mechanism that increases total inhomogeneous strain during loading, but this inhomogeneous strain is recovered on unloading. This is consistent with no residual dislocation network being accumulated as is consistent with some TEM results on nanocrystalline Ni.

4.4.3 Experimental observation of twinning

Twinning is an alternative deformation mechanism to dislocation slip and may be important in the deformation of nanocrystalline materials. There have been a number of reports of twinning in nanocrystalline metals which are very different from the behavior in conventional grain size metals. This was first shown for the occurrence of twinning in several fcc structure metals such as Al and Pd which do not exhibit deformation twinning in conventional grain size or single-crystal form. Deformation twinning was observed in nanocrystalline Al films prepared by magnetron sputtering with grain sizes of about 10–30 nm (Chen *et al.*, 2003). Since Al is a metal with a very high stacking fault energy, no evidence of twinning had been reported – at subambient temperatures or high strain rates as can be the case for lower stacking fault energy metals. The films were deformed by either nanoindentation or mechanical grinding. In both cases twins were observed by TEM and HRTEM. A dislocation model was proposed for the occurrence of twinning based on classical dislocation theory that predicts that the critical shear stress needed to initiate a Shockley partial twinning dislocation becomes smaller, and therefore favored, over that needed to nucleate a perfect dislocation when the grain size is reduced to the 10–20 nm level. Deformation twinning was also reported in cryogenically ball milled Al by Liao *et al.* (2003). Two types of deformation twins were observed. One was the heterogeneous nucleation and growth of twins from grain boundaries, and the other was twin lamella formation via the dissociation and migration of grain boundary segments. As described earlier, Rosner *et al.* (2004) have also observed twinning in Pd, a metal with a high stacking fault energy.

Twinning has also been observed in several other nanocrystalline metals under conditions where it is not seen for the same conventional-grain-size metal. For example, twinning has been observed in nanocrystalline Cu at room temperature and slow strain rate where, in conventional-grain-size Cu, twinning is seen only at subambient temperatures and high strain rates or strains (Liao *et al.*, 2004). Kumar *et al.* (2003) observed twins during the deformation of nanocrystalline Ni in their *in situ* deformation experiments in TEM. They caution that twinning in nanocrystalline materials is seen under unusual conditions of high strains or in thin foils. The separation of observations of deformation twins from growth twins is also a concern.

While most of the observations of twinning in nanocrystalline metals have been for fcc structure metals, there is at least one example of twinning in nanocrystalline bcc Ta metal (Wang *et al.*, 2005). Ta films with grain sizes about 10–30 nm were prepared by magnetron sputtering. Deformation was carried out by nanoindentation. High-resolution TEM was used to study the deformed region under the indenter. The observation of multiple twin intersections suggested that the mechanism of deformation twinning in this bcc material is different from that in fcc nanocrystalline metals. Conventional coarse-grained Ta does not deform by twinning except at very high strain rates and/or low temperatures.

4.4.4 Strain rate and test temperature effects

Important variables which help our understanding of deformation mechanisms in conventional-grain-size materials are test temperature and strain rate. These variables should

4.4 Ductility of nanocrystalline materials

also help understanding of nanocrystalline materials. To date there have been only limited studies on the influence of test temperature dependence of deformation behavior in nanocrystalline materials. One study of nanocrystalline Cu (Huang et al., 1997) involved hardness measurements from −180 °C to 60 °C. A continuous increase in hardness was noted with decreasing temperature from about 1.5 GPa at 60 °C to 4.2 GPa at −180 °C. This more than two-fold increase in hardness over this temperature range can be compared with the more modest increase for conventional-grain-size Cu of about 30%. The magnitude of the hardness increase for nanocrystalline Cu is comparable to that for conventional-grain-size Fe, but the temperature dependence is different. Nanocrystalline Cu increases hardness steadily with decreasing temperature but with a somewhat lower rate of decrease at the lower temperatures. Conventional-grain-size Fe, however, exhibits its greatest rate of hardness increase at the lower temperature regime. The temperature sensitivity of hardness/strength for conventional-grain-size metals is generally attributed to the nature of the Peierls lattice stress and depends upon the dislocation structure. It is typically large for bcc metals and small for fcc metals. The very different temperature sensitivity for nanocrystalline Cu suggests again a different deformation mechanism from conventional dislocation dominated deformation. Tensile tests have been carried out on nanocrystalline Au samples (36 nm grain size) at 77 K and 293 K (Tanimoto et al., 1999). They found a marked increase of more than two fold in yield stress at 77 K compared to 293 K. They also noted a strain-rate dependence of the yield stress at room temperature. A linear dependence of strain rate and stress in creep tests in a range of applied stress suggested Coble-type creep at room temperature. Wang and Ma (2004) also observed a much larger ratio of the yield stress at 77 K to that at room temperature for nanocrystalline fcc Ni and fcc Co than their conventional-grain-size counterparts. That is, the ratio was 1.47 for nanocrystalline Ni compared to 1.20 for conventional-grain-size Ni, and 1.80 for nanocrystalline Co compared to 1.41 for conventional-grain-size Co. The authors attribute the strong temperature dependence to the unusually small activation volume, $\Delta V^* \sim 10\,b^3$ (b is the Burgers vector), measured in strain rate change tests. Grain-boundary dislocation nucleation was proposed as the thermally activated deformation mechanism in these nanocrystalline metal grains.

Strain-rate sensitivity is another important variable from which to deduce deformation behavior. Jia et al. (2000) have carried out compression tests on nanocrystalline Fe (80 nm grain size) over a range of strain rates from 10^{-4} s^{-1} to 3×10^3 s^{-1}. Essentially no change in flow stress was noted for nanocrystalline Fe in contrast to about a 100% flow stress increase over the same strain-rate range observed for conventional grain size Fe (Ostwaldt et al., 1997). These results are illustrated in Figure 4.24. Yoo et al. (1999) carried out high strain rate compression (Kolsky bar) tests on conventional and nanocrystalline-grain-size (63 nm) Ta. Increased flow stress was not observed with increasing strain rate as is common in bcc metals with a high Peierls stress, which is consistent with the above results on nanocrystalline Fe. These limited results on nanocrystalline bcc Fe and Ta are very different from the usual strain rate sensitivity in conventional-grain-size or single-crystal bcc metals. Wei et al. (2004) have reported similar results for nanocrystalline bcc V with about 100 nm grain size. Wei et al. (2004) have summarized the strain-rate sensitivity, m, as a function of grain size for several bcc metals. They show a decrease in m from values of about 0.05 for conventional grain sizes to less than 0.01 for grain sizes below 100 nm.

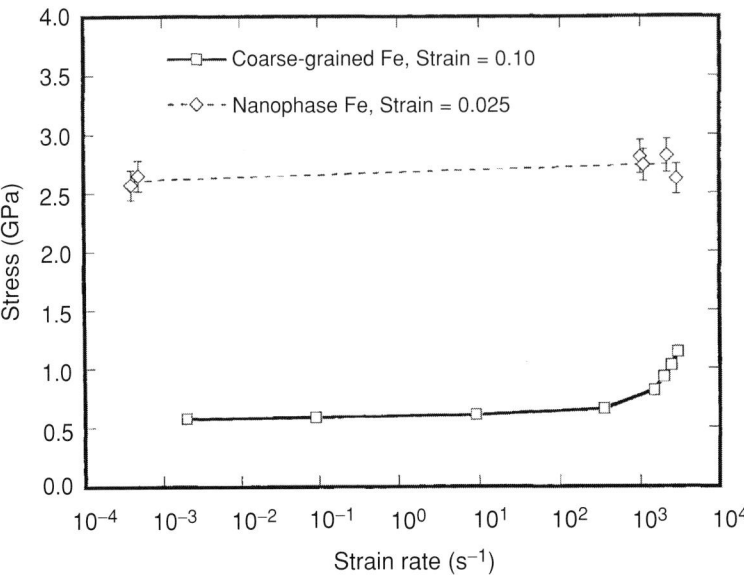

Figure 4.24. Comparison of the strain-rate dependence of the flow stress for conventional-grain-size and nanocrystalline Fe. (From Jia et al. (2000), reproduced with permission of Elsevier.)

In conventional bcc metals the non-planar dislocation core structure results in a very large Peierls stress at low homologous temperatures which makes the yield and flow stress strain rate and temperature sensitive. The authors suggest reasons for this behavior which will be discussed in Chapter 5.

More studies have been carried out on the strain rate dependence of yield and flow stresses in fcc metals. Dalla Torre et al. (2002) studied the tensile properties of electrodeposited nanocrystalline Ni (average grain size about 20 nm) at strain rates from 10^{-5} s^{-1} to 10^3 s^{-1}. The nanocrystalline Ni showed a much larger increase in ultimate tensile strength with increasing strain rates (for strain rates >10) than conventional-grain-size Ni. These results are illustrated in Figure 4.25. These results are in marked contrast to those shown in Figure 4.24 for bcc Fe. The nanocrystalline Ni also exhibited a higher strain-rate sensitivity for all applied strain rates and a lower work-hardening rate than conventional-grain-size Ni. At the highest strain rates, shear banding with local grain growth was observed in the nanocrystalline Ni. Schwaiger et al. (2003) also studied electrodeposited nanocrystalline Ni (about 40 nm grain size) along with 200 nm Ni and conventional-grain-size Ni. Depth-sensing indentation and tensile tests were used. The hardness of the nanocrystalline Ni increased with load rate in the indentation tests while that for the 200 nm Ni remained constant. The flow stress for the nanocrystalline Ni also increased with increasing strain rates for rates from 3×10^{-4} to 3×10^{-1}. At these strain rates Dalla Torre et al. (2002) did not see any increases in strength with increasing strain rate in their nanocrystalline Ni. However, the strain-rate sensitivity, m, in both studies was similar, rising from about 0.005 for coarse-grained Ni to about 0.02 for the nanocrystalline samples. Cu is another fcc metal that shows increased strain-rate sensitivity as the grain size is decreased

4.5 Superplasticity of nanocrystalline materials

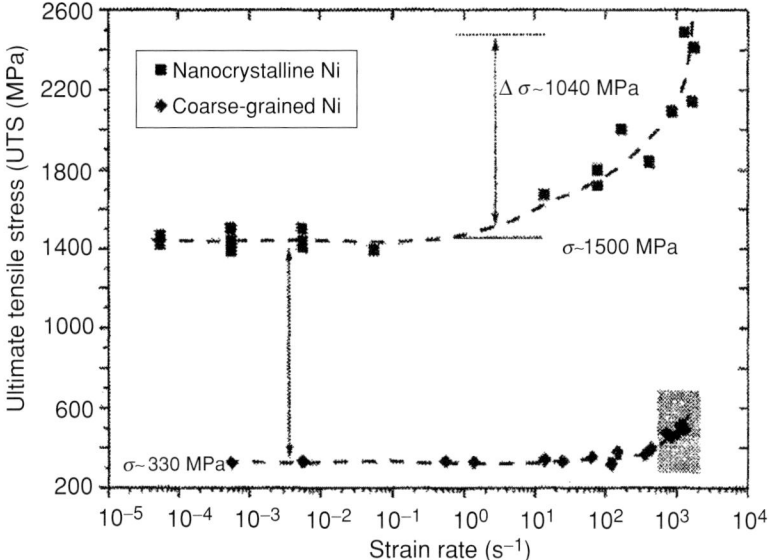

Figure 4.25. Ultimate tensile strength vs. strain rate for nanocrystalline and coarse-grained Ni. (From Dalla Torre *et al.* (2002), reproduced with permission of Elsevier.)

(Wei *et al.*, 2004). Electrodeposited nanocrystalline Cu with a grain size of about 28 nm was studied by Lu *et al.* (2001). A larger strain-rate sensitivity, *m*, value of 0.036 compared to coarse-grained Cu of about 0.006 was seen. An unusual strain-rate dependence of the tensile ductility was observed in that the fracture strain increased with increasing strain rate instead of decreasing as in coarse-grained Cu. Asaro and Suresh (2005) have described mechanistic models to explain the low values for activation volume associated with the high strain-rate sensitivity of fcc nanocrystalline metals. These models will be discussed in Chapter 5.

The strain-rate effects have been less studied for hcp nanocrystalline materials. Nanocrystalline Co made by electrodeposition showed little effect of strain rate on the yield strength but the ultimate tensile strength decreased with increasing strain rate over the limited range of 1×10^{-4} s^{-1} to 2.5×10^{-3} s^{-1} (Karimpoor *et al.*, 2003). In contrast to these results, Zhang *et al.* (2002) found that the ultimate tensile strength of Zn with an average grain size of 240 nm, so not really nanocrystalline, increased with increasing strain rate over the small range of 1×10^{-4} to 4×10^{-3}. More studies are needed on the strain dependence of strength and strain sensitivity in nanocrystalline hcp metals.

4.5 Superplasticity of nanocrystalline materials

Superplasticity is the phenomenon observed in some metallic materials and ceramics wherein very large uniform elongations in tension are seen with values typically greater than about 200% and sometimes in excess of 2000%. The criteria for superplasticity in conventional-grain-size materials are: a deformation temperature usually at least 0.5 of the melting temperature; a strain rate which optimizes the strain-rate sensitivity, m, to values of at least 0.33; fine, equiaxed grains of <10 μm diameter; little grain growth at the test temperature. Typical superplastic strain rates for metals are 10^{-5} s^{-1} to 10^{-2} s^{-1} and even lower by about a factor of 10 for ceramics. These slow strain rates limit production and limit superplastic forming to low-volume part production such as in the aerospace industry. In the micron-grain-size materials it has been shown that reducing the grain size serves to increase the strain rate range over which superplastic behavior can be observed, as well as lowering the temperature where it can occur (Edington et al., 1976). A general equation for elevated temperature crystalline plasticity including superplastic deformation can be written as

$$\dot{\varepsilon} = A \frac{DGb}{kT} \left(\frac{b}{d}\right)^p \left(\frac{\sigma}{G}\right)^n,$$

where $\dot{\varepsilon}$ is the steady-state strain rate, D is the appropriate diffusivity (lattice or grain boundary), G is the shear modulus, b is the Burgers vector, k is the Boltzmann constant, T is the test temperature, d is the grain size, p is the grain-size exponent, σ is the applied stress, n is the stress exponent, and A is a constant (Mukherjee, 2002). This equation shows that as stated above the slow strain rates associated with micron-size grains can be accelerated by several orders of magnitude if the grain size is reduced to the nanoscale. Similarly, nanoscale grain sizes could achieve the same strain rates at lower temperatures. These effects have been demonstrated in nanocrystalline materials. Reviews of the superplastic behavior of nanocrystalline materials have been given by Mayo (1997) and Mohamed and Li (2001).

Taketani et al. (1994) studied the superplastic properties of an Al–Ni–Mm–Zr alloy (Mm = misch metal). The grain size for this material ranged from 70 nm to 200 nm. The strain rates for superplastic behavior (> ~300% elongation) were about 10^{-1} s^{-1} to 10^1 s^{-1} at a test temperature of 873 K. These are much faster than for the conventional (micron)-grain-size Al alloys. Mishra et al. (1997) studied the deformation of nanocrystalline Pb–62% Sn and Zn–22% Al that were prepared by high-pressure torsion straining. Tensile superplasticity was observed at room temperature in the nanocrystalline Pb–Sn alloy and at 393 K in the nanocrystalline Zn–Al alloy. The strain-rate sensitivity value, m, was 0.45 for Pb–Sn alloy at room temperature indicating a grain-boundary sliding mechanism for the superplastic behavior. No superplasticity was observed in a conventional-grain-size Zn–Al sample under the test conditions used. This supports the idea that superplasticity can be observed at faster strain rates in nanocrystalline materials compared to micron-grain-size materials. However, the test homologous temperatures were 0.64 and 0.54 for the Pb–Sn and Zn–Al alloys respectively. Superplasticity is observed at these temperatures in these materials for micron-grain-size alloys. In fact superplasticity is seen for the Zn–22 wt.% Al alloy even at room temperature (Edington et al., 1976).

Mishra et al. (1998) investigated the superplastic behavior of nanocrystalline Ni$_3$Al prepared by severe plastic deformation using high-pressure torsion straining. This material had an average grain size of about 50 nm and exhibited large tensile elongations at 650 °C and 725 °C at a strain rate of 10^{-3} s^{-1}. This behavior, not seen in conventional-grain-size

4.5 Superplasticity of nanocrystalline materials

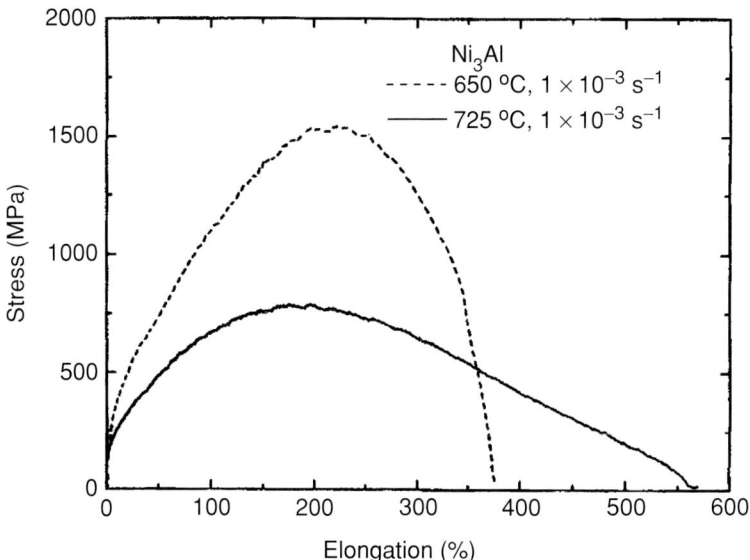

Figure 4.26. The variation of true flow stress with elongation at 650 °C and 725 °C and a strain rate of 1×10^{-3} s^{-1} for nanocrystalline Ni$_3$Al. (From Mishra et al. (1998), reproduced with permission of Elsevier.)

Ni$_3$Al, is illustrated in Figure 4.26. Mukhopadhyay et al. (1990) had previously studied the superplastic behavior of Ni$_3$Al that had a grain size of 6 μm. They observed superplasticity at a temperature of 1100 °C and a strain rate of 8.9×10^{-4} s^{-1}. Compared with these results the work of Mishra et al. (1998) on the nanocrystalline material showed an extensive strain-hardening stage and very high flow stresses. Mishra et al. (1998) proposed an "exhaustion plasticity" model to explain these results. Their samples made by high-pressure torsion contained a high dislocation density. They assume that the applied stress moves the pre-existing dislocations and grain-boundary sliding is accommodated by the movement of these dislocations. As the easy paths of grain-boundary sliding and dislocation motion become exhausted, the flow stress increases which results in the observed strain-hardening behavior. The authors believe that the peak flow stress represents the critical stress required to nucleate new dislocations during grain-boundary sliding.

McFadden et al. (1999) reported on superplasticity in three materials. These were nanocrystalline Ni made by electrodeposition, with a 20 nm grain size, nanocrystalline Ni$_3$Al with a 50 nm grain size, and an Al alloy, 1420-Al with a 100 nm grain size. The Ni$_3$Al sample results were essentially the same as reported in Mishra et al. (1998). The Ni sample showed a marked increase in ductility, into the superplastic range of >200%, for test temperatures greater than 280 °C. This temperature represents a very low homologous temperature of only 0.36 for the onset of superplasticity which usually only occurs at homologous temperatures of >0.5. However, unlike conventional superplasticity behavior, substantial grain growth was observed to occur during the tests at 350 °C from the starting 20 nm grain size to 1.3 μm along the tensile axis and 0.64 μm transverse to the tensile axis. The stress–strain curves for the nanocrystalline Ni at several test temperatures are given

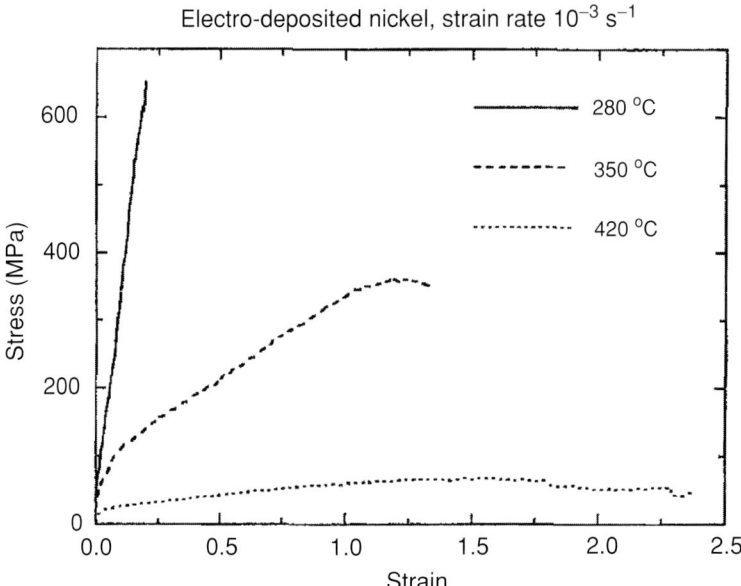

Figure 4.27. Stress–strain data for electrodeposited nanocrystalline Ni at a strain rate of 10^{-3} s^{-1} and the test temperatures indicated. (From McFadden *et al.* (1999), reproduced with permission of *Nature*.)

in Figure 4.27. The grain growth during elevated temperature testing was more modest for the Ni$_3$Al and the Al alloy since the grain growth was presumably inhibited by the ordered structure and grain-boundary precipitates respectively.

Han and Lavernia (2005) have recently reported on the high-strain-rate superplasticity in a "nanostructured" Al–Mg alloy. This alloy was prepared by cryomilling followed by extrusion. While the mean grain size was 90 nm the grain-size distribution extended up to about 250 nm so it is questionable whether to consider this a nanocrystalline material. The superplastic behavior at 573 K showed almost 300% elongation at a strain rate of about 10^{-1} s^{-1}. The microstructure after testing at 573 K was not significantly changed in keeping with conventional superplastic behavior. While the strain-rate sensitivity exponent, m, was only 0.2, less than usually observed for conventional superplasticity (typically, $m > 0.3$) the other aspects of the behavior such as no strain hardening and no grain growth are more like conventional superplasticity than the behavior observed in those few tests discussed above on nanocrystalline metals and alloys.

While the data on superplastic behavior of nanocrystalline materials are limited, the following features appear to be common. Nanocrystalline materials exhibit higher strain-rate superplasticity than micron-grain-size samples and at lower temperatures, with Ni showing the lowest homologous temperature of just 0.36. Extensive strain hardening, high flow stresses, and grain growth are also observed during tensile testing. These phenomena are not common to superplastic behavior in micron-grain-size materials. Suggested mechanisms for the apparent differences in superplasticity for nanocrystalline materials and their micron-grain-size counterparts have been given by Mukherjee (2002).

4.6 Creep of nanocrystalline materials

Time dependent deformation of a material is known as *creep* (see Meyers and Chawla, 1999). Creep is most important in conventional-grain-size structural materials in applications at elevated temperatures, typically at homologous temperatures of 0.5 or higher. A variety of microscopic mechanisms have been identified for creep deformation that are a function of the applied stress, test temperature, and microstructure. Deformation-mechanism (Weertman–Ashby) maps are used to describe the regimes in which various mechanisms occur in different materials. The important regimes are: (1) dislocation glide, which occurs at the highest stress levels and is almost independent of temperature; (2) dislocation creep, including glide and climb (both controlled by diffusion), which is found at high stress levels and relatively high homologous temperatures; (3) Nabarro–Herring creep (vacancy movement in the grains) at lower stresses and high temperatures; and (4) Coble creep (vacancy movement within grain boundaries) at lower stresses and lower temperatures. The fields in these deformation maps are also dependent upon the strain rate and the grain size. For example, Coble creep would be favored at small grain sizes. The expression for Coble creep is given by the following

$$\frac{d\varepsilon}{dt} = \frac{B\sigma \Omega \Delta D_b}{d^3 kT}$$

where B is a constant, σ is the applied stress, Ω is the atomic volume, d is the grain size, Δ is the grain boundary width, D_b is the boundary diffusion coefficient, and k is Boltzmann's constant. With increased values for D_b predicted for nanocrystalline materials and the nanoscale d in the denominator cubed, it was expected that the creep rate in nanocrystalline materials would be many orders of magnitude greater than that in conventional-grain-size materials (Nieman *et al.*, 1990). Early investigations of creep in nanocrystalline materials focused on this prediction. Nanocrystalline Cu and Pd samples prepared by the inert-gas-condensation method to densities of about 97% were subjected to creep deformation at room temperature (Nieman *et al.*, 1990; Nieman *et al.*, 1991). The creep rates observed in these studies were much lower than would be predicted for Coble creep. In fact, there were no significant differences between the creep rates of the nanocrystalline metals and their coarse-grained counterparts. While the creep rates and stresses were not exactly comparable in the nanocrystalline and coarse-grained samples it was concluded that creep was not enhanced in the nanocrystalline materials at room temperature. Creep curves for the nanocrystalline and coarse-grained Cu samples are given in Figure 4.28. The predicted diffusional creep was not observed and the shape of the creep curve was well described by the equation for logarithmic (exhaustion) creep. Subsequently, Sanders *et al.* (1997) also carried out creep tests on nanocrystalline Cu, Pd, and Al–Zr made by an improved inert-gas-condensation and compaction method that gave somewhat higher densities than in the earlier work. Creep tests were made for homologous temperatures ranging from 0.24 to 0.64. In this case the measured creep rates were two to four orders of magnitude smaller than the values predicted by the expression for Coble creep. At moderate temperatures the creep rates were comparable to or even lower than the corresponding rates for the coarse-grain counterparts. With the exception of tests performed at the highest temperatures, all creep curves could be fit by the expression for logarithmic (exhaustion) creep. The as-prepared samples were highly

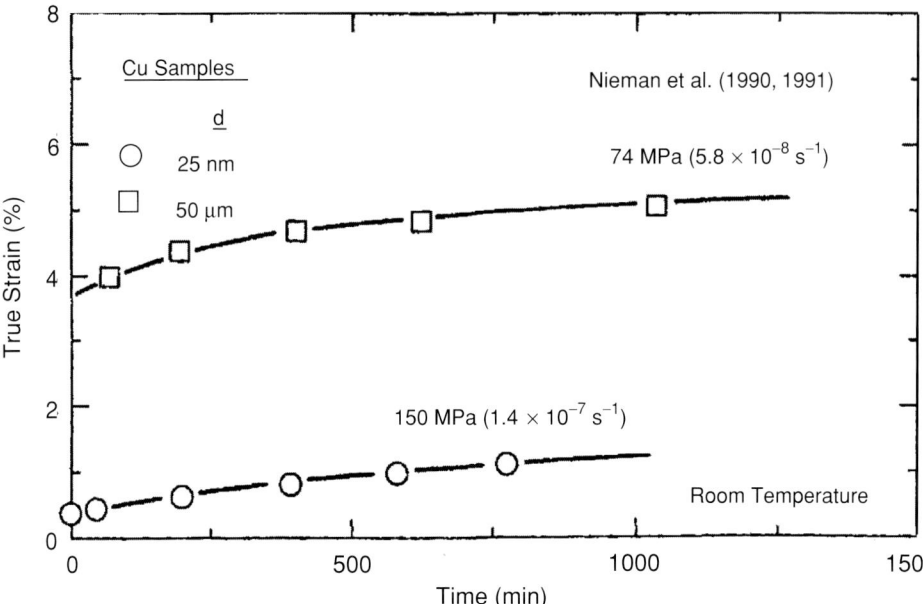

Figure 4.28. True strain vs. time, creep, for Cu samples at room temperature (Mohamed and Li, 2001). (Data from Nieman *et al.* (1990); Nieman *et al.* (1991) reproduced with permission of Elsevier.)

twinned and contained many low-energy grain boundaries so the low creep rates were attributed to the prevalence of these low-energy boundaries and low dislocation activity due to the small grain sizes. Significant grain growth was observed at the higher creep temperatures.

Creep of "nanocrystalline" Cu prepared by electrodeposition was reported by Cai *et al.* (2000). The quotation marks are put around the term nanocrystalline in the above sentence because while the authors claim a grain size of about 30 nm, in fact these were low-angle subgrain boundaries and the high-angle grains were of the order of microns. Tensile creep measurements were made in the temperature range 20–50 °C. Unlike the work of Nieman *et al.* (1990; 1991), the authors found a steady-state creep rate that was proportional to an effective stress $\sigma_e = (\sigma - \sigma_0)$, where σ is the applied stress, and σ_0 is the threshold stress. The activation energy measured (0.72 eV) was comparable to that measured for grain-boundary diffusion in nanocrystalline Cu. The creep rates were found to be of the same order of magnitude as those calculated from the equation for Coble creep. However, Li *et al.* (2004) studied creep in compression on the same "nanocrystalline" Cu prepared by electrodeposition and supplied by Professor K. Lu as well as on several sub-micron-grain-size and coarse-grain Cu samples. They found that the minimum creep rates in tension were much larger than those in compression. This was explained as the likely increase in tension due to the occurrence of tensile fracture during the test. They also determined that the high-angle grain boundaries were about 10 μm as opposed to the 30 nm subgrain boundaries. This work suggests that Coble creep is unlikely in this material at temperatures slightly above room temperature.

4.6 Creep of nanocrystalline materials

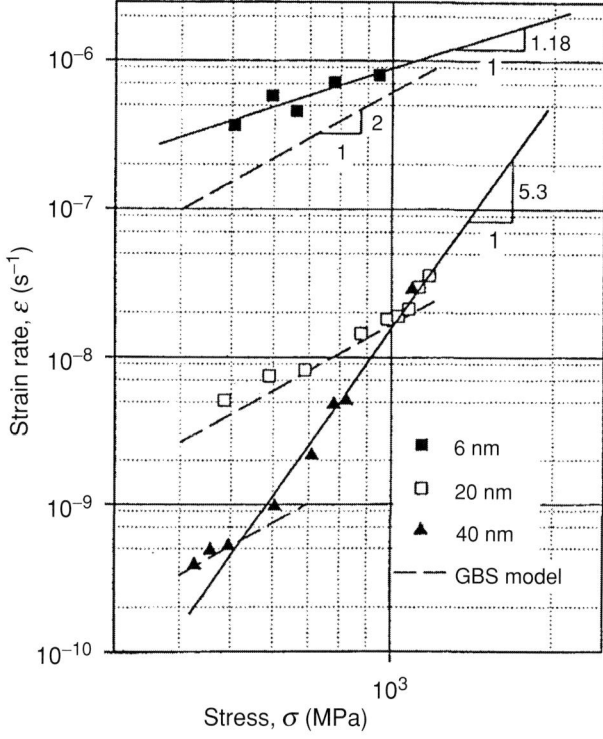

Figure 4.29. Strain rate as a function of applied stress for electrodeposited nanocrystalline Ni. (From Wang *et al.* (1997), reproduced with permission of Elsevier.)

Markmann *et al.* (2003) reported on the creep behavior of nanocrystalline Pd made by the inert-gas-condensation and compaction method which had a grain size of about 8 nm as prepared. Tensile creep measurements were made at 313 K. The strain rate in the creep experiments agreed fairly well with the predictions of Coble creep considering the uncertainty in the grain-boundary-diffusion data for Pd.

Tanimoto *et al.* (1999) also reported an influence of Coble-type creep for low strain rate deformation at room temperature for nanocrystalline Au samples (20–60 nm grain size) prepared by the inert-gas-condensation and compaction method.

Nickel is another nanocrystalline metal for which creep studies have been carried out. Wang *et al.* (1997) studied the effect of grain size on the steady-state creep rate tested at room temperature. Their specimens were prepared by electrodeposition with grain sizes of 6, 20, and 40 nm. The samples did not contain any porosity so at least in this sense were artifact-free. The creep data in the form of strain rate vs. applied stress for the three grain size samples are presented in Figure 4.29. The authors compared various constitutive equations to their data and found that the equation for grain-boundary sliding controlled by grain-boundary diffusion fits their data the best. The grain-boundary-sliding model, however, underestimated the creep rate for the 6 nm grain-size sample and the authors suggest that some contribution to creep from intercrystalline components is also taking place. For the larger grain-size samples and at higher applied stress levels a higher slope of about 5.3 was found which suggests power law or dislocation creep. Criticisms of this work (Mohamed and Li, 2001) include the fact that the total creep strains measured were not large enough to determine the presence of steady-state creep and that no measurement of activation energies for creep were made. Yin *et al.* (2001) have also studied the creep behavior of nanocrystalline Ni prepared by electrodeposition at room temperature and 373 K. The as-received grain size of these specimens was about 30 nm and no grain growth was observed after creep testing at room temperature. "Slight" grain growth was observed after the tests at 373 K. The calculated creep rate at room temperature based on the Coble creep mechanism was consistent with the experimental data. That is, a plot of minimum strain rate as a function of applied stress gave a slope of 1.1 similar to the results of Wang *et al.* (1997) on their smallest grain-size samples. However, the results at 373 K gave a much larger slope of about 6.5 which suggests other creep mechanisms such as dislocation creep. Subsequently, Yin *et al.* (2005) have continued a study of electrodeposited Ni with varying interstitial element (S, B) content. The creep-test results showed that both the minimum creep rate and creep strain significantly decreased with increasing sulfur content or by boron doping. The creep rate exhibited much stronger dependence on applied stress at 373 K and 473 K compared to the creep rate at room temperature. The authors explained the creep behavior using a grain-boundary sliding model. The calculated back stresses suggested that the interstitial elements, S and B, at the grain boundaries retarded the sliding of grain-boundary dislocations thereby enhancing the creep resistance.

Creep measurements were made on nanocrystalline alloys prepared by crystallization of amorphous precursors by Wang *et al.* (1994), Deng *et al.* (1995), and Xiao and Kong (1997). The work of Wang *et al.* (1994) was on a Ni–20 at.% P alloy. This alloy composition is at a deep eutectic in the phase diagram and can be easily made amorphous. Owing to its low melting temperature the creep-testing temperatures of 543 K and 593 K represent homologous temperatures of 0.47 and 0.51 respectively. The recrystallized microstructure consisted of the majority Ni_3P phase in a Ni matrix. The grain size of the Ni_3P phase was estimated to be about 28 nm from X-ray line-broadening analysis and the Ni phase "about one-half" of this after annealing at 603 K. Samples annealed at 773 K had Ni_3P grain size of 257 nm and were compared with the nanocrystalline samples. The creep curves for both sets of samples exhibited an instantaneous strain on loading, a region of decreasing strain with time, and an apparent steady-state region. The creep rate of the nanocrystalline material was higher than the sub-micron-grain-size samples. The stress exponent for the nanocrystalline samples was 1.2, which is close to the value of 1.0 predicted for Coble creep. A low activation energy of 69 kJ/mol was noted which is less than the value for grain-boundary diffusion in pure Ni. The corresponding values for the sub-micron-grain-size samples were a stress exponent of 2.5 and activation energy of 106 kJ/mol. Based upon these results the authors conclude that the creep in the nanocrystalline material is controlled by grain-boundary diffusion (Coble creep) but creep in the sub-micron-grain-size samples is more complex and may be the result of dislocation creep combined with other mechanisms such as grain-boundary sliding and grain-boundary diffusion. Criticisms of these conclusions (Mohamed and Li, 2001) include the observations of instantaneous strain and primary creep which are not characteristic of diffusional creep and the limited range of creep strains to define steady-state creep. Xiao and Kong (1997) carried out similar experiments to the above on recrystallized $Fe_{78}B_{13}Si_9$ (at.%) wherein the microstructure consisted of bcc α-Fe(Si) with Fe_2B. The grain size of the Fe(Si) was about 27 nm and the Fe_2B was smaller. Sub-micron-grain-size samples were obtained by annealing. The results of this study were very similar to those for the Ni–P alloys described above.

An early study of creep in a nanocrystalline ceramic material is that of Hahn and Averback (1991) on nanocrystalline TiO_2. This material was prepared by the inert-gas-condensation and compaction method with an original grain size of about 40 nm. The specimens were creep tested at temperatures from 600 °C to 800 °C in compression. No steady-state creep was observed. Grain growth occurred during the creep tests such that, depending on the temperature, the grain size measured after the tests ranged from 120 nm to 1000 nm. While the authors suggested that their results were caused by an interface reaction mechanism, the significant grain growth during the creep tests means that nanocrystalline materials were not really being tested.

A more recent study of creep in nanocrystalline materials that is strongly influenced by grain growth during the test is for nanocrystalline Cu made by the inert-gas-condensation and compaction method using microhardness measurements to determine indentation creep as a function of time, temperature, and grain size (Zhang et al., 2004). For nanocrystalline Cu (45 nm and 80 nm samples) and sub-micron-grain-size Cu (200 nm) the Vickers hardness decreased approximately linearly with the logarithm of the dwell-time. At short dwell-times the hardness increased significantly with decreasing grain size and with decreasing temperature. At longer dwell-times the influence of these variables became much less important. TEM examination of the regions under the indenter showed that rapid grain growth occurred. For example, the sample with an initial grain size of 45 nm showed grain growth under the indenter for a dwell-time of 30 min up to grain sizes of several hundred nanometers while the unindented microstructure remained constant. The remarkable observation was that such grain growth under the high stresses under the indenter took place even at -190 °C. The possibility of grain growth during high stress creep tests must therefore be a consideration for interpretation of such data.

The above results for creep studies on nanocrystalline materials suggest that much more work is needed on well-characterized materials and at temperatures where grain growth may be minimized in order to assess the mechanisms responsible for creep deformation.

4.7 Fatigue of nanocrystalline materials

Fatigue is typically defined as a degradation of mechanical properties leading to failure of a material or a component under cyclic loading. Fatigue is a problem that can affect any structural material which is exposed to a cyclic load. Therefore, knowledge of fatigue behavior is critical to most potential structural applications of nanocrystalline materials.

The fatigue behavior of "nanocrystalline" materials has been reviewed by Vinogradov and Agnew (2004). However, this review is mainly concerned with sub-micron-grain-size materials that were produced by severe plastic-deformation methods. Therefore the results may not be strictly applicable to nanocrystalline materials as we have defined them with grain sizes <100 nm. Also the heavily cold-worked structures common to materials made by severe plastic-deformation methods may not be completely representative of the behavior of nanocrystalline materials made by other techniques. A summary of the experiments on the sub-micron-grain-size materials made by severe plastic-deformation methods reveals a general enhancement of high cyclic fatigue life. However, the low cyclic fatigue life appears shorter than that of conventional-grain-size counterparts, perhaps because of some loss of ductility. The sub-micron-grain-size materials had significantly faster fatigue crack growth rates than their conventional-grain-size counterparts.

The number of fatigue experiments on true nanocrystalline materials is very limited. Witney et al. (1995) studied the fatigue behavior of nanocrystalline Cu made by the inert gas condensation and compaction method. Samples with as-prepared grain sizes of about 19 nm and 23 nm were tested by cycling the tensile load between a tensile maximum that ranged from 50% to 80% of the yield stress. The minimum stress was 10 MPa. Fatigue failure occurred with the cycling to 80% of the yield stress after about 6×10^5 cycles at a rate of 1 Hz. A moderate increase in grain size was observed after cycling of about 30% (e.g. 19–25 nm). Slight elongation was observed after the prolonged fatigue tests. The amount of strain was comparable to that seen in room-temperature creep at stresses of the same value as the maximum stresses in fatigue (Nieman et al., 1991). Examination of sample gauge lengths after fatigue testing showed that a number of intense shear bands had formed which led to protrusions, presumably also to intrusions, and eventually to crack formation similar to the fatigue behavior of single crystal or coarse grain size materials.

Palumbo et al. (1997) examined the fatigue behavior of electrodeposited nanocrystalline Ni with average grain sizes in the range 50–100 nm. The fatigue behavior was evaluated in air at room temperature and at 300 °C with high cycle fatigue tests performed in fully reversed bending ($R = 1$) at frequencies in the range 0.5–25 Hz. The fatigue performance at room temperature was compared to those of several conventional-grain-size Ni samples and found to be within the scatter band of the data. The authors say that the fatigue performance at 300 °C was not degraded.

A more recent report on the fatigue life and fatigue crack-growth characteristics has been given for nanocrystalline electrodeposited Ni with 20–40 nm grain size (Hanlon et al., 2003). The authors compared the stress–life (S–N) fatigue response of fully dense electrodeposited nanocrystalline Ni with that of electrodeposited sub-micron (about 300 nm) grain-size Ni and conventional-grain-size Ni. They performed fatigue experiments at zero tension–zero loading (R = ratio of the minimum load to the maximum load of the fatigue cycle = 0) at a cycle frequency of 1 Hz. The results are shown in Figure 4.30 which indicate that grain refinement has a beneficial effect on the S–N fatigue response of Ni. Nanocrystalline Ni has a slightly higher increase in the tensile stress range needed to achieve a fixed life and in the endurance limit (defined at 2×10^6 fatigue cycles) compared to the sub-micron-grain-size Ni, which in turn is higher than conventional-grain-size Ni. The authors also conducted fatigue crack-growth experiments on nanocrystalline Ni using edge-notched specimens that were subjected to cyclic tension at different R ratios. Figure 4.31 illustrates the variation of fatigue crack growth rate da/dN as a function of the stress intensity factor range ΔK ($\Delta K = K_{max} - K_{min}$, where K_{max} and K_{min} are the maximum and minimum stress intensity factors corresponding to the maximum and minimum loads, respectively).

4.7 Fatigue of nanocrystalline materials

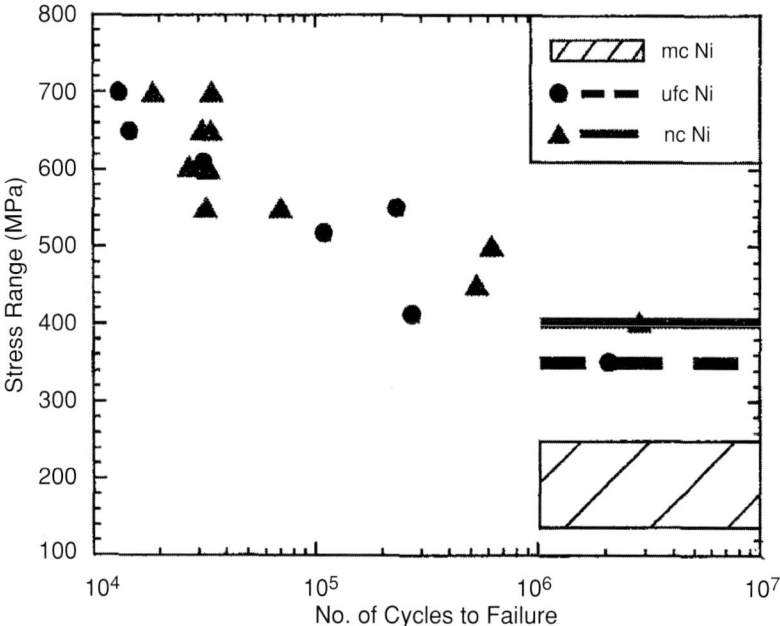

Figure 4.30. A comparison of the *S–N* fatigue response showing the stress range versus number of cycles to failure for the nanocrystalline (nc), sub-micron-grain-size (ufc), and conventional-grain-size (mc) Ni. (From Hanlon *et al.* (2003), reproduced with permission of Elsevier.)

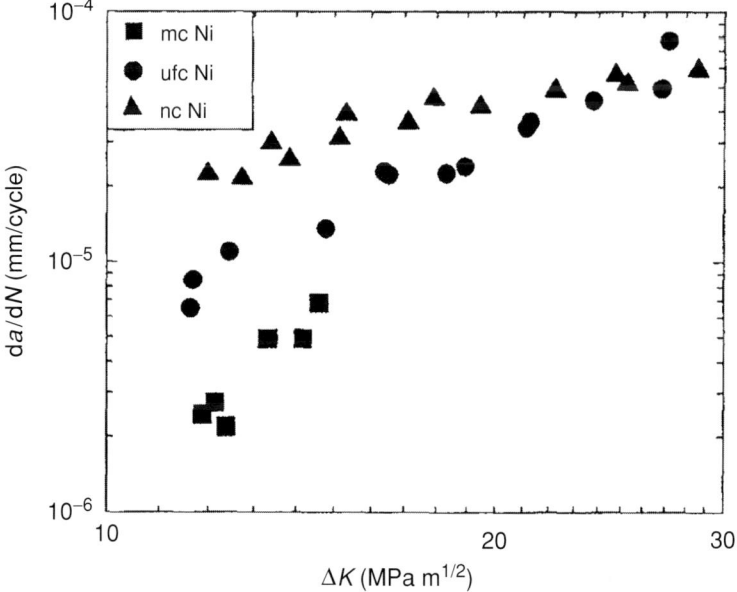

Figure 4.31. Fatigue crack growth rate, da/dN, as a function of the stress intensity factor range, ΔK, for nanocrystalline (nm), micron-grain-size (ufc), and conventional-grain-size (mc) Ni at $R = 0.3$ at a fatigue frequency of 10 Hz at room temperature. (From Hanlon *et al.* (2003), reproduced with permission of Elsevier.)

The fatigue crack growth in the nanocrystalline Ni is clearly faster than in the larger grain samples in the intermediate regime of fatigue fracture. So while the nanocrystalline structure appears to increase the resistance to failure under stress-controlled fatigue, the resistance to fatigue crack growth was decreased. These same trends were observed for the micron-grain-size materials formed by severe plastic deformation as reviewed by Vinogradov and Agnew (2004). It should be pointed out that this nanocrystalline Ni, while well characterized and with a relatively narrow grain-size distribution, is relatively brittle, with elongation in tension of about 3%. It would be of great interest to investigate the fatigue response of nanocrystalline materials which have the combination of high strength and good ductility.

4.8 Fracture and fracture toughness of nanocrystalline materials

The fracture behavior of nanocrystalline materials is not well defined because of all the questions regarding intrinsic versus extrinsic factors that may affect the ductility and therefore fracture of these materials. Milligan (2003, p. 539) has stated that in most of the literature on the mechanical behavior of nanocrystalline materials (as of mid 2002) the results reported may not be representative of intrinsic material response because of possible artifacts from processing and incomplete microstructural characterization including lack of measurements of grain-size distribution. In those examples of nanocrystalline materials that apparently do not contain processing artifacts and have optimized values of strength and ductility, fracture behavior has been briefly considered. Li and Ebrahimi (2004) studied the fracture behavior in their nanocrystalline Ni and Ni–15% Fe alloy. The Ni sample with 44 nm grain size and about 9% elongation in tension showed deep microvoids on the fracture surface, which strongly suggested fracture by the microvoid coalescence mechanism. A large reduction in area was also observed that is common for ductile fcc metals. In contrast to this behavior, the Ni–15% Fe sample, although having reasonable elongation values in tension of about 6%, showed no reduction in area and the fracture surface showed shallow microvoids. TEM studies revealed intergranular cracks within the microvoid regions. It was found that the main crack was formed by the formation of multiple microcracks and their coalescence. This Ni–15% Fe sample had an average grain size of about 9 nm and it was suggested that its deformation behavior was controlled by grain boundary sliding. The results of Youssef *et al.* (2004) on artifact-free nanocrystalline (23 nm) Cu with optimized strength and ductility revealed a fracture surface with complete dimpled morphology. The dimple size ranged from about 100 nm to 400 nm. This fracture behavior is consistent with ductile fcc metals in spite of the high strength observed. These very limited results suggest that when dislocation controlled deformation is dominant, even in very strong fcc nanocrystalline materials, ductile fracture behavior can be obtained.

There is very little published data on the fracture toughness of nanocrystalline materials. The chief problem which has prevented "valid" measurements of fracture toughness is the dimensions of available samples of nanocrystalline materials for fracture toughness tests. While some processing methods can provide large specimens in two dimensions, the thickness of, for example, electrodeposited samples is still generally limited. One example of fracture toughness measurements on a sample with sufficient dimensions (5 mm thick,

4.8 Fracture and fracture toughness of nanocrystalline materials

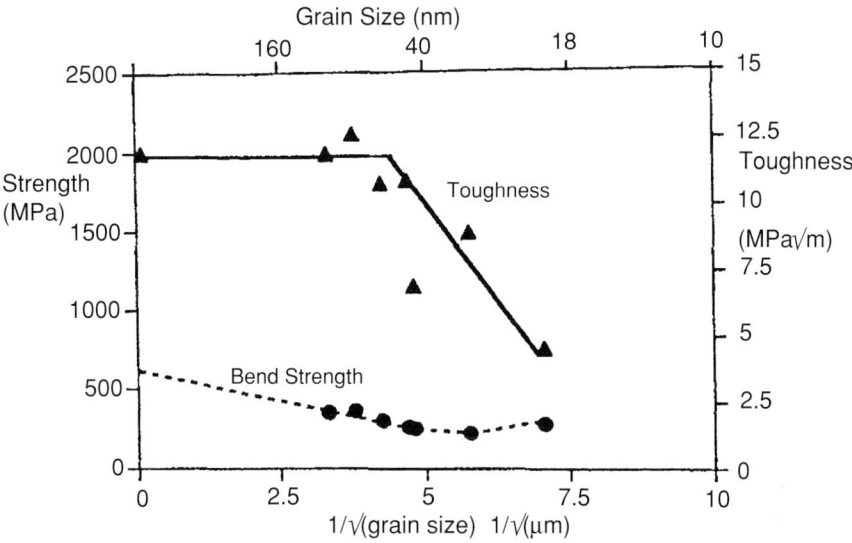

Figure 4.32. Fracture toughness of compacted FeAl samples as a function of grain size. (From Morris-Munoz et al. (1999), reproduced with permission of Elsevier.)

3 mm wide, 20 mm gauge length) for three-point bending fracture toughness was reported for nanocrystalline Fe–Al alloys by Morris-Munoz et al. (1999). The fracture toughness of this material stayed fairly constant at about 12 MPa$\sqrt{m}$ as the grain size was decreased from conventional down to about 40 nm. The fracture toughness then decreased steadily from this value to a minimum value of about 5 MPa$\sqrt{m}$ at a grain size of 18 nm. This is illustrated in Figure 4.32. It was not clear whether this loss of toughness with decreasing grain size was the result of flaws from processing such as incomplete particulate bonding at the smallest grain sizes, or of inherent properties of the material. Mirshams et al. (2001) made fracture toughness measurements on electrodeposited nanocrystalline Ni samples with about 20–50 nm grain sizes and about 220–350 μm in thickness. The R-curve response was studied by plotting the stress intensity factor or strain energy release rate against the increment in crack length with the objective of assessing the resistance of the material to quasistatic crack growth in a subcritical manner for specimens in plane stress. The crack growth resistance decreased with increasing annealing temperature for the nanocrystalline Ni. Carbon doping also reduced the crack growth resistance. The authors report plane stress steady-state quasistatic high fracture toughness values for K_R of 120 MPa$\sqrt{m}$ for the nanocrystalline Ni which is higher than their measurements of conventional polycrystalline Ni of 62 MPa$\sqrt{m}$. This high value of fracture toughness was observed in spite of very little ductility in the nanocrystalline samples of about 2% elongation in tension compared with 35% elongation in tension in the conventional polycrystalline Ni. These results are intriguing but raise a number of questions regarding the structure of the materials tested and the validity of the use of stress intensity factor to characterize fracture response (Kumar et al., 2003). It is clear that further experimental studies of fracture toughness are needed in order to develop an understanding of the mechanisms controlling fracture toughness in nanocrystalline materials.

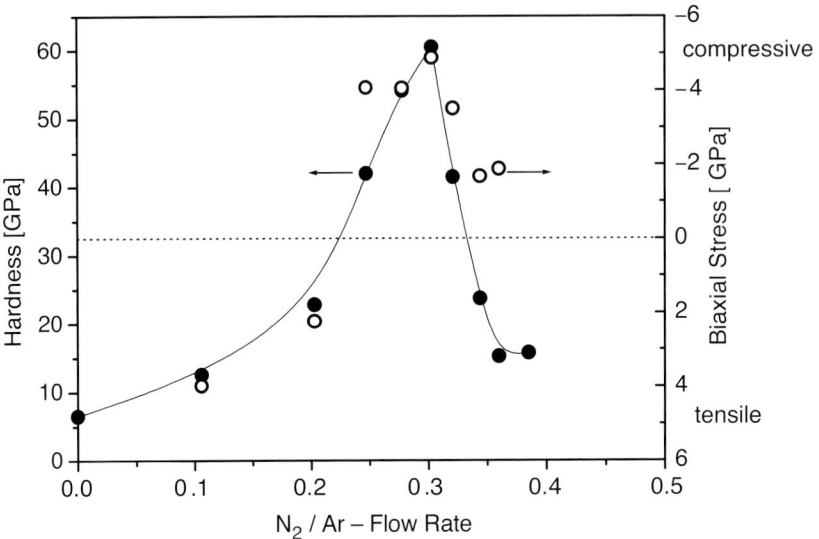

Figure 4.33. Hardness and compressive stress of TiN$_x$ coatings deposited by magnetron sputtering vs. the ratio of the N$_2$:Ar flow rate (Veprek et al., 2005a, after Musil et al., 1988).

4.9 Mechanical properties of superhard nanostructured coatings

There are different mechanisms of hardness enhancement in the three different classes of superhard coatings that were introduced in Chapters 1 and 2. We shall discuss them in the following sections separately.

4.9.1 Hardness enhancement due to energetic ion bombardment during the deposition

In the case of coatings which were hardened by energetic ion bombardment during their deposition, the hardness enhancement is a complex, not yet fully understood, synergetic effect of densification, formation of point and extended defects, and built-up of a compressive stress (see Veprek et al., 1987). Often in the literature, the hardness enhancement is attributed to the compressive stress alone, because one finds a correlation between the hardness and the compressive stress, as shown for example in Figure 4.33 (Veprek et al., 2005a; Valvoda et al., 1988; Musil et al., 1988). However, this correlation is only incidental, owing to the fact that both the hardness enhancement and compressive stress are caused by the energetic ion bombardment. A compressive (or tensile) stress can enhance (reduce) the hardness only by the amount equal to its value, because it acts against the shear stress induced under the indenter during the indentation. Therefore, the flow stress σ_{fl} in coatings with biaxial stress, σ_{biax}, is enhanced as compared to the yield stress σ_Y of the stress-free coatings by $\sigma_{fl} = \sigma_Y - \sigma_{biax}$ (notice that $\sigma_{biax} < 0$ for compressive and $\sigma_{biax} > 0$ for tensile stress). Obviously, a compressive stress of about 5 GPa cannot cause a hardness enhancement from 20–22 GPa for bulk TiN to 60 GPa as found in the coatings of Musil et al.

4.9 Mechanical properties of superhard

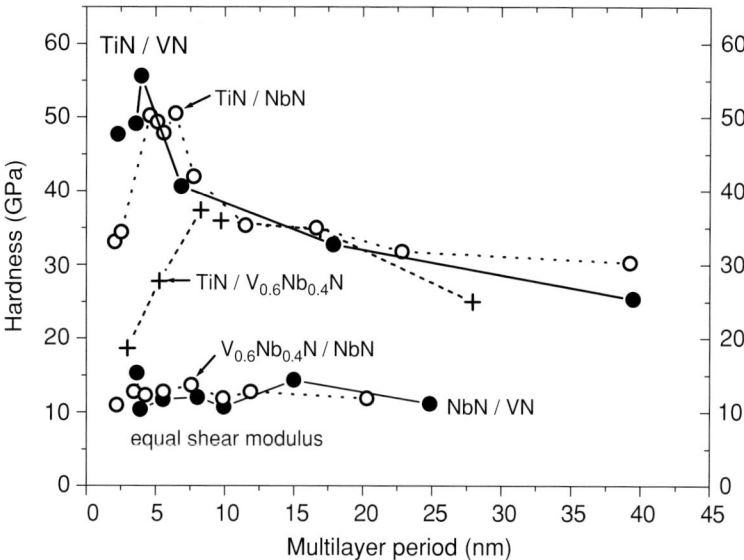

Figure 4.34. Dependence of the hardness of a variety of heterostructures on the superlattice period. VN and NbN have different elastic moduli and lattice periods as compared with TiN whereas the lattice mismatch of $(V_{0.6}Nb_{0.4})N$ to TiN is small. NbN/VN and NbN/$(V_{0.6}Nb_{0.4})N$ have nearly the same values of the shear modulus (Helmersson *et al.*, 1987; Shinin *et al.*, 1992; Shinin and Barnett, 1995; Chu and Barnett, 1995). Compiled from the original data, with the permission of the authors.

(see Figure 4.33). More details about this problem and references to the original literature have been reported (Veprek and Argon, 2002; Ljungcrantz *et al.*, 1995; Tsui *et al.*, 1996; Bolshakov *et al.*, 1996; Swadener *et al.*, 2001). As mentioned in Chapter 3, the hardness enhancement of these coatings is lost upon annealing to 450–600 °C. Studies on a possible enhancement of tensile strength of these coatings were not reported so far.

4.9.2 The mechanism of hardness enhancement in heterostructures

The strengthening of the heterostructures is caused by a reduction of the dislocation activity, because for a small period of ≤20 nm, there is no dislocation multiplication source operating within the layers, and the elastic mirror force induced within the material with a higher elastic modulus makes it more difficult for a dislocation moving within the weaker layer to cross the interface into the stronger layer (Koehler, 1970). Lehocky (1978a; 1978b) studied metallic heterostructures which were not superhard, but he was able to prepare specimens for tensile measurements and to show that their tensile strength was indeed enhanced as predicted by the theory of Koehler.

In the case of the ceramic superhard heterostructures that were prepared later as thin films, and which are difficult to remove from the substrate for the tensile measurements without breaking, because of the internal stress, only the hardness could be measured by indentation (Helmersson *et al.*, 1987; Shinin *et al.*, 1992; Barnett, 1993; Shinin and Barnett, 1995; Chu and Barnett, 1995; Barnett and Madan, 1999; Barnett *et al.*, 2003). Figure 4.34

illustrates the importance of the selected combination of the materials on the properties of the heterostructures. VN and NbN have different elastic moduli and slight lattice mismatch as compared with TiN. Therefore, the highest hardness enhancement is found. $(V_{0.6}Nb_{0.4})N$ and TiN have also different values of shear moduli but smaller lattice mismatch which results in a somewhat smaller, yet significant hardness enhancement. NbN/VN and NbN/$(V_{0.6}Nb_{0.4})$N have nearly the same values of the shear modulus which causes a decrease of the resistance of the heterostructure against dislocation movement across the interface, and concomitant absence of any hardness enhancement.

The decrease of the hardness observed for the TiN/VN and TiN/NbN heterostructures for period smaller than about 5 nm is an artifact due to the roughness of the interface (Chu and Barnett, 1995; Barnett and Madan, 1999). When either materials of a larger immiscibility are used, or the heterostructures made of refractory materials are annealed, this decrease of hardness becomes smaller or vanishes (Barnett and Madan, 1999; Barnett et al., 2003).

4.9.3 Hardness and other mechanical properties of superhard nanocomposites nc-M_nN/a-Si_3N_4 and nc-TiN/a-BN

4.9.3.1 The role of the interfacial Si_3N_4 layer

As shown in Chapter 2, the hardness of the nc-M_nN/a-Si_3N_4 nanocomposites deposited either by plasma CVD or by reactive sputtering at a sufficiently high temperature of 550–600 °C reaches a maximum at a silicon content of about 7–8 at.%. By means of XRD and XPS, the phases of the quasibinary system were identified as stoichiometric TiN and Si_3N_4 (Veprek and Reiprich, 1995; Veprek et al., 2000b) which are immiscible. Therefore, Veprek and Reiprich estimated the thickness of the Si_3N_4 layer covering the TiN nanocrystals to be about 0.3–0.5 nm, i.e. about one monolayer or slightly more. More detailed investigation confirmed this prediction. One example is shown in Figure 4.35 for the nc-TiN/a-Si_3N_4 system, and similar results were found also for all other nc-Me_nN/a-Si_3N_4 systems with Si_3N_4 as the covalent nitride phase which were studied in detail (Niederhofer et al., 2001). Furthermore, the maximum hardness is found in the superhard nanocomposites nc-TiN/a-BN and nc-TiN/a-BN/a-TiB_2 deposited by plasma CVD also for coverage of the TiN nanocrystals by about one monolayer of BN (Karvankova et al., 2003; 2006). As an explanation of this finding it was suggested that the structurally flexible, covalently bonded Si_3N_4 (or BN) decreases the interfacial energy of the ionic TiN. The one monolayer of Si_3N_4 is further stabilized by a partial electron transfer from TiN which has a lower group electronegativity than the thin slab of Si_3N_4 with a widened band gap (Veprek and Reiprich, 1995; Veprek et al., 1995a; 1995b). The reason for the appearance of the hardness maximum at a coverage of about one monolayer of Si_3N_4 was suggested as follows. At a lower Si_3N_4 content, the interfaces between the TiN nanocrystals are not fully filled and grain-boundary sliding is not efficiently prevented. At one monolayer, the Si_3N_4 forms a sharp, semi-coherent interface (tensile misfit in Si_3N_4 of several percent) (see Veprek et al., 2005a) to TiN which, together with the band gap widening and concomitant partial electron transfer stabilizes the nanostructure (Veprek et al., 1995b). With Si_3N_4 thickness further increasing, the strain energy of the semi-coherent interface increases leading to a destabilization of the system.

4.9 Mechanical properties of superhard

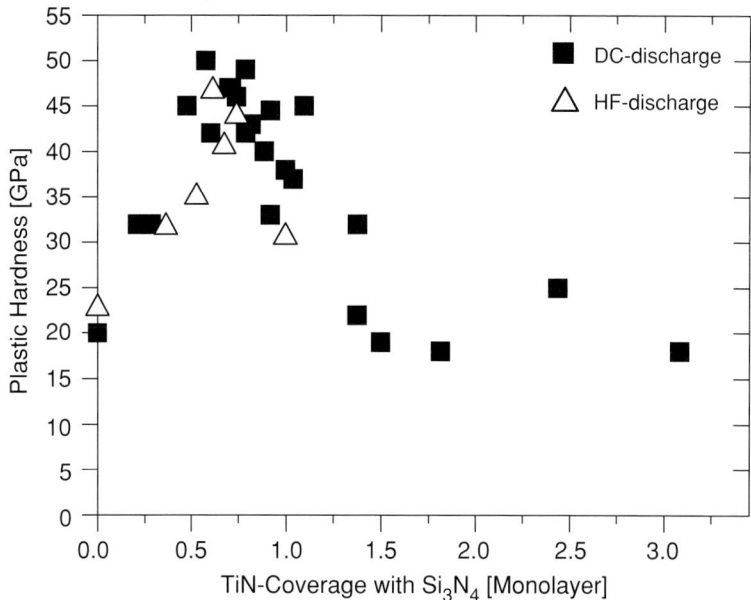

Figure 4.35. Dependence of the hardness of binary nc-TiN/a-Si$_3$N$_4$ nanocomposites on the coverage of the TiN nanocrystals with Si$_3$N$_4$ for coatings deposited in an HF discharge under low energetic ion bombardment (Veprek and Reiprich, 1995) and in a d.c. discharge with energetic ion bombardment (Niederhofer et al., 2001).

The concept of one monolayer of Si$_3$N$_4$ yielding the highest hardness was recently supported by both experimental work on epitaxial heterostructures and first principle calculations which will be briefly summarized here, because they contribute to the understanding of the origin of the high hardness enhancement.

Odén (2004), Söderberg et al. (2005) and Hu et al. (2005) prepared epitaxial heterostructures consisting of TiN layers with a thin SiN$_x$ interfacial layer of a different thickness in between. Figure 4.36, from the work of Odén and co-workers (Odén, 2004; Söderberg et al., 2005), shows a sharp maximum of the hardness for a SiN$_x$ thickness of 0.3 nm corresponding to one monolayer. The thickness of TiN of 4.5 nm, similar to the crystallite size in the nanocomposites (Veprek and Reiprich, 1995), was kept constant in this series. In the work of Hu et al. (2005), the curve is broader with maximum hardness and elastic modulus at an SiN$_x$ thickness of about 7 and 4 nm, respectively. A comparison of the cross-sectional TEM micrographs from the papers of the two groups reveals that those of Odén et al. are of significantly higher quality. This explains in a natural way the sharper maximum of hardness at 0.3 nm, i.e. about one monolayer of Si$_3$N$_4$, in their work. The mechanism of the hardening of the heterostructures was discussed by the authors in terms of the Koehler's model.

Deeper understanding of the mechanism of the hardness enhancement comes from the recent first-principle density functional theory calculations of Hao et al. (2006). These authors considered 55 different TiN–Si$_3$N$_4$–TiN and TiN–TiSi$_x$–TiN sandwiches with different stoichiometry and atomic configurations. The α-, β-, and γ-Si$_3$N$_4$ were the

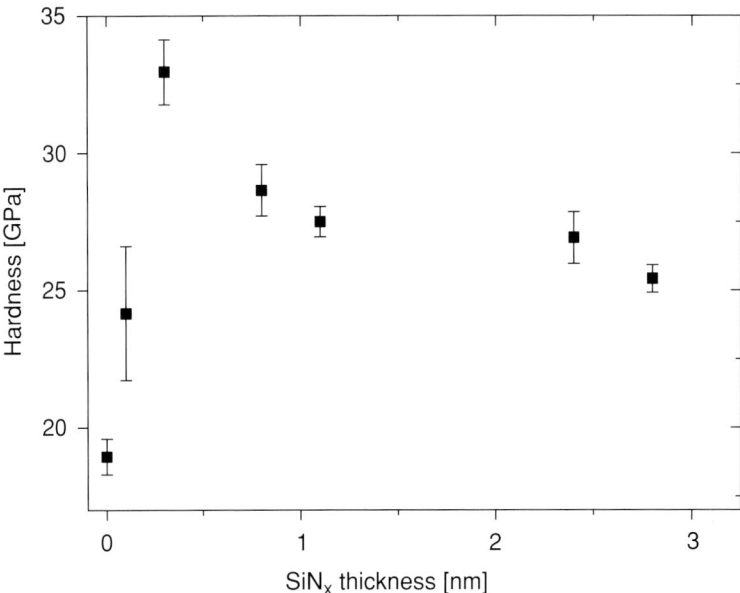

Figure 4.36. Dependence of the hardness of TiN/SiN$_x$/TiN epitaxial heterostructures on the thickness of the SiN$_x$. The thickness of the TiN layers of 4.5 nm, similar to the crystallite size in the nanocomposites, was kept constant. (From Odén (2004) and Södeberg et al. (2005), reproduced with the permission of the authors.)

thermodynamically stable compounds for a sufficiently high, temperature-dependent, activity of nitrogen, whereas C49–TiSi$_2$ was most stable at low nitrogen activity. The calculations fully confirmed the earlier published results, that under the conditions used for the deposition of the nanocomposites by means of plasma CVD (Veprek and Reiprich, 1995) and plasma PVD (Zhang and Veprek, 2006), the Si$_3$N$_4$ is the thermodynamically stable phase, whereas TiSi$_2$ forms only at much lower nitrogen activity. Therefore, Hao et al. (2006) concluded, in agreement with these papers, that kinetic constraints, and possibly impurities, are most probably responsible for lack of the reproducibility of the results of Veprek et al. as claimed by some authors recently.

The calculations also showed that the most stable sandwich is that with one monolayer of β-like Si$_3$N$_4$ followed by the γ-like one. The decohesion energy of that interfacial layer lies between the ideal decohesion energy of TiN (111) and (110), i.e. it is higher than the decohesion energy of bulk β-Si$_3$N$_4$. This is caused by a partial negative charge transfer to the Si-atoms within that interface, in agreement with the simple earlier reasoning based on the increase of the energy band splitting and concomitant electron negativity in nanosized Si$_3$N$_4$ interfacial phase (Veprek et al., 1995b). The high quality of the DFT calculations of Hao et al., as documented by the high precision of their calculations for bulk TiN, Si$_3$N$_4$, and TiSi$_2$, makes their conclusions highly reliable. Recently, Patscheider et al. have used XPS to study the partial negative charge transfer to the Si$_3$N$_4$ interfacial layer in the TiN/Si$_3$N$_4$ heterostructures (Patscheider, 2004). They could show that there is indeed an indication of a negative charging of that interfacial layer.

4.9 Mechanical properties of superhard

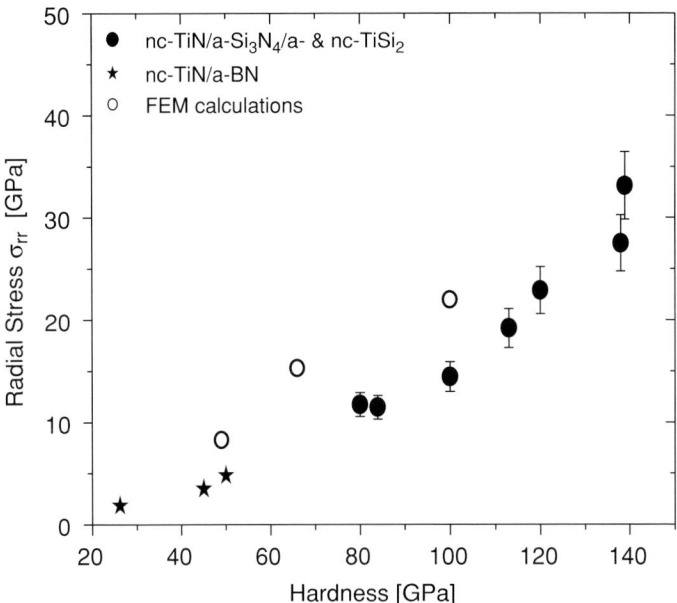

Figure 4.37. Radial stress calculated from the measured indentation curves according to the Hertzian theory, which the superhard coatings sustain without any cracks formation, vs. hardness (Veprek et al., 2003a).

According to the calculations of Hao et al., the nc-M_nN/a-Si_3N_4 nanocomposites with well-relaxed nanostructure during their formation by the spinodal decomposition, i.e. prepared according to the generic design principle (Veprek and Reiprich, 1995), and free of impurities (Veprek et al., 2004; 2005d), should approach the ideal strength of strong ceramic materials, as already estimated by Argon and Veprek (2002) and later on confirmed experimentally by Veprek et al. (2003a). Figure 4.37 shows the tensile stress at the periphery of the contact between the diamond Vickers indenter and the coatings, which was determined from the measured indentation curves using the Hertzian theory, for crack-free indentations into a variety of nanocomposites (Veprek et al., 2003a). The calculations by means of the finite element method (FEM) support these experimental results. These results are in agreement with the recent theoretical calculations of Hao et al. and they confirm that these nanocomposites are operating close to the ideal strength of flaw-free materials.

This can be illustrated by a simple consideration of the Griffith thermodynamic criterion for propagation of a crack (Hertzberg, 1989; Anderson, 1996). The critical stress σ_c at which a planar crack of length $2a$ in a brittle material with Young's modulus E and surface energy γ starts to grow is

$$\sigma_c = \sqrt{\frac{2\gamma \cdot E_Y}{\pi a}}.$$

When the crystallite size decreases to few nanometers, the maximum size of a planar crack is smaller than that, and the Griffith formula approaches that of ideal strength with a approaching the interatomic distance a_0 (Hertzberg, 1989; Anderson, 1996). Consequently, the strength of such material should approach the ideal decohesion strength of flaw-free material (Veprek and Argon, 2002; Veprek et al., 2003a). This is of course only a rough estimate which does not say anything about the mechanism of plastic deformation in the nanocomposites.

Within the context of the discussion of the role of one monolayer interfacial layer that acts as a strong "glue" between the nanocrystals we should briefly mention also the question of critical activation volume needed for inducing plastic deformations. From the discussion of the Hall–Petch effect in the foregoing sections it is obvious that there must be a critical minimum volume within which plastic deformation by crystal plasticity can be induced. Recently, Van Vliet et al. (2003) have shown by nanoindentation into bulk, single crystals of ductile, soft metals, that when loaded at a scale of about 10 nm, they behave ideally and deform reversibly. Only when a certain critical deformation volume is exceeded, discrete plastic events occur. A similar result was reported by Schuh et al. (2003) for metallic glasses. It has to be emphasized, however, that the origin and mechanisms of the discrete plastic events in single crystals and in metallic glasses, which are observed at the nanoscale, are physically different phenomena (for details see Van Vliet et al., 2003; Schuh et al., 2003). From the molecular dynamic simulations of plastic deformation in silicon glass by Demkowicz and Argon it follows, that the plastic events, which occur within discrete, small volume elements of a size of about 3 nm, similar to metallic glasses, are triggered by a simultaneous movement of several (typically 5–8) atoms (Demkowicz and Argon, 2004). Because this is not possible to achieve within a one-monolayer thin Si_3N_4 slab, we have herewith another reason why such a slab has to approach the ideal strength.

4.9.3.2 Possible mechanism of plastic deformation in superhard nanocomposites

Plastic deformation in ductile crystalline materials occurs by dislocation activity which decreases with decreasing crystallite size leading to strengthening according to the Hall–Petch mechanism, as discussed in Section 4.3. In small nanocrystals, the dislocation activity is absent and their strength approaches the ideal one. Thus, the strength and hardness of the nanocrystalline materials are limited by the strength of the grain boundaries. Plastic deformation of such materials usually occurs by grain-boundary sliding, as discussed in Section 4.3. As discussed above, the strength of the interface in the nanocomposites under consideration can also approach the ideal one. Thus, the question arises as to which of the phases, the nanocrystals or the X-ray amorphous Si_3N_4, will carry plastic (i.e. permanent) deformation which is observed, e.g. upon indentation, into the hardest nanocomposites which approach the ideal strength.

Ceramics and glasses are considered to be brittle materials which, upon application of a critical load, fracture by catastrophic crack growth. However, when a glass is subjected to a sufficiently large but sub-critical load for a sufficiently long time, it can deform plastically, i.e. irreversibly. The mechanism of plastic deformation of metallic glasses occurs because of localized shear transformations within small volume elements of a few nanometers in size. Under certain conditions, these localized events can form shear bands. The macroscopic plastic deformation in metallic glasses occurs only in constrained compression or shear (not in tension) and the details of the mechanism depend on the strain rate

4.9 Mechanical properties of superhard

(Schuh et al. (2003), and references therein). In covalent glasses, such as silicon, the plastic deformation occurs by a viscous-like flow within the highly coordinated (average coordination number 5–6), liquid-like component (Demkowicz and Argon, 2004). By analogy, it was suggested, that plastic deformation in the superhard nanocomposites should occur also by shear transformation within the "disordered," liquid-like interfacial component (Argon and Veprek, 2002; Veprek and Argon, 2002). This appears to apply to nanocomposites with a not-fully-relaxed structure, whereas in the well-relaxed, ultrahard nanocomposites whose tensile strength approaches the ideal one, both components, i.e. the nanocrystals and the one-monolayer interfacial layer probably operate close to ideal shear resistance (Prilliman et al., 2006). More future work is needed in order to resolve this question.

When the ultrahard nanocomposites are tested by indentation, the pressure under the indenter reaches values of that hardness, i.e. several tens of giga-pascals. Under such conditions, large enhancement of elastic moduli and, therefore, of the resistance against elastic deformation occurs (Argon and Veprek, 2002; Veprek and Argon, 2002). For example, the bulk modulus, $B(P)$, increases linearly with pressure P as described by the universal binding energy relation (UBER) (Rose et al., 1984; Smith et al., 1987) in agreement with many experimental data, the proportionality factor, dB/dP, being between 4 and 6 for the majority of solids:

$$B(P) \cong B(0) + (4-6) \cdot P.$$

For the nc-TiN/a-Si$_3$N$_4$ nanocomposites with zero-pressure bulk modulus $B(0) \approx 285$ GPa, $dB/dP = 5$ and indentation hardness of about 50 GPa, the value of bulk modulus under the indenter increases by a factor of ≤ 1.8. Moreover, also the flow stress, i.e. resistance of the material against plastic deformation, is enhanced under the high pressure in a similar way (Ratko Veprek et al., 2006a). For example, for a material with indentation static hardness of about 40 GPa, the stiffening reaches a factor of about 1.6. In order to correctly describe the mechanical behaviour of such materials, all these effects were accounted for in order to formulate a new constitutive material's model which was used in advanced FEM modeling (Veprek et al., 2006a). Figure 4.38 shows as one example the contour plots of the volumetric strain calculated by the UBER model in comparison with the linear one. It is obvious that the mechanical properties of the superhard nanocomposites cannot be correctly described by linear elasticity. This has, of course, many consequences including the question of the applicability of the theory of the hardness measurements by the automated load-depth-indentation technique. Instead, it is recommended to use the classical indentation hardness measurement technique based on the contact area of remaining plastic deformation developed many years ago (Tabor, 1951).

Because this modeling is in an early stage of its development, we refer the reader to the original work for further details. Here we only emphasize that this stiffening explains in a simple way the experimental observation according to which the elastic recovery, i.e. the fraction of elastic deformation to the total one of these ceramic materials increases with increasing hardness. For example, the elastic recovery of nanocomposites with hardenss of about 40–50 GPa amounts to about 80% (Veprek et al., 1995b), whereas that of ultrahard nanocomposites with hardness of 80–100 GPa reaches 94% (Veprek et al., 2000a; 2000b). In contrast, the elastic recovery of conventional ceramics usually decreases with increasing hardness.

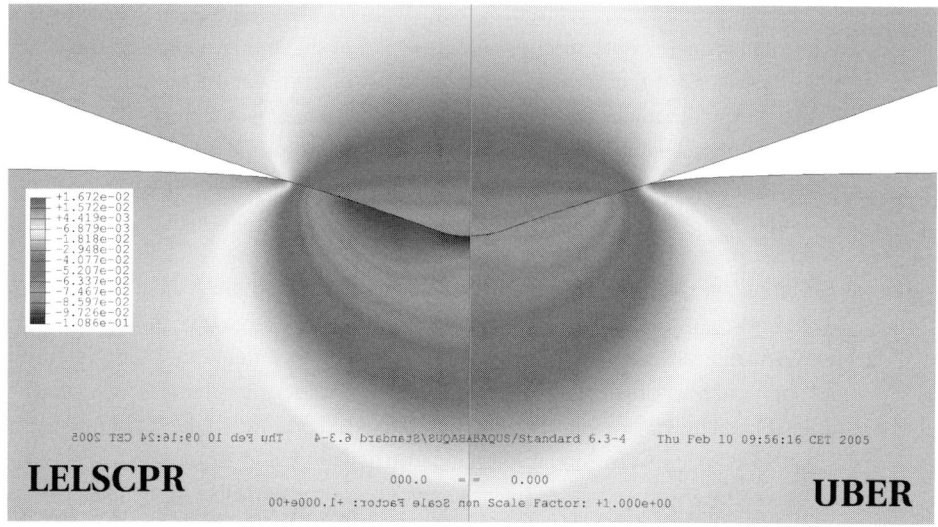

Figure 4.38. Contour plot of the volumetric strain calculated by FEM using the LELSCPR model (left), which can account for large strains but uses linear elasticity (constant values of elastic moduli and of yield stress), and the UBER model (right), which, besides handling large strains accounts also for the pressure enhanced elastic moduli and flow stress (Veprek *et al.*, 2006a).

4.9.3.3 Hardness of the nc-M_nN/a-Si_3N_4 nanocomposites and the role of impurities

The maximum hardness achieved in the quasibinary nc-M_nN/a-Si_3N_4, nc-TiN/a-BN and similar nanocomposites is about 50–60 GPa. Only in the quasiternary nc-TiN/a-Si_3N_4/a- and nc-$TiSi_2$ nanocomposites a hardness of 80 to ≥100 GPa was reported (Veprek *et al.*, 2000a; 2000b). Because the measurements of such high hardness on thin coatings may be subject to a number of artifacts (Veprek *et al.*, 2003b), these very high values of hardness of the coatings were compared with that of diamond, as shown in Figure 4.39 (Veprek *et al.*, 2005a). Using a more advanced indentometer, Fischer-Cripps *et al.* (2006) recently measured the hardness of that diamond and on several superhard coatings prepared by Veprek *et al.* with hardness between 40 and 50 GPa and obtained the same results as reported by these researchers earlier.

The hardness of these coatings strongly depends on the impurity content, as shown in Figure 4.40 for oxygen in a series of coatings deposited by means of plasma CVD in two different apparatuses and by reactive sputtering in three different apparatuses (Veprek *et al.*, 2004; Prochazka *et al.*, 2004; Veprek *et al.*, 2005a; 2005b; Veprek *et al.*, 2006b). In spite of some scattering of the data, particularly at a low-oxygen-impurity content, it is clear that whenever the oxygen impurity content is ≥0.4 at.% the hardness is below 40 GPa. The dotted line in Figure 4.40 shows the reciprocal coverage of the nanocrystals with oxygen impurities. One notices, that there are on average 20 and 4 oxygen atoms per nanocrystal for oxygen content of 0.5 and 0.1 at.%, respectively (Veprek *et al.*, 2005b). Because an O^{2-} ion within the interfacial layer (impurities are unstable within few nanometers small nanocrystals) represents a large defect owing to its large size and possibly also electronic interaction, it is understandable that the strength of the interface is reduced by only an oxygen content

4.9 Mechanical properties of superhard

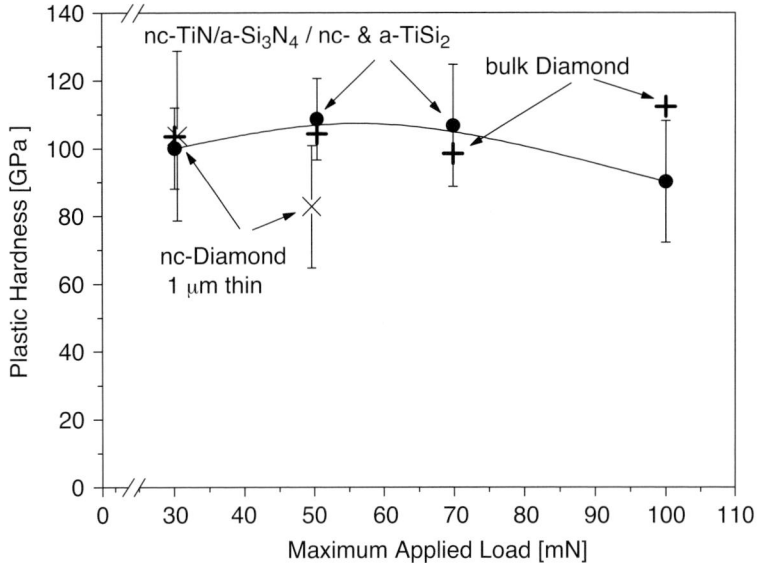

Figure 4.39. Hardness of ultrahard nc-TiN/a-Si$_3$N$_4$/a- and nc-TiSi$_2$ coatings in comparison with bulk industrial diamond and with single-phase, nanocrystalline diamond film (Veprek *et al.*, 2005a).

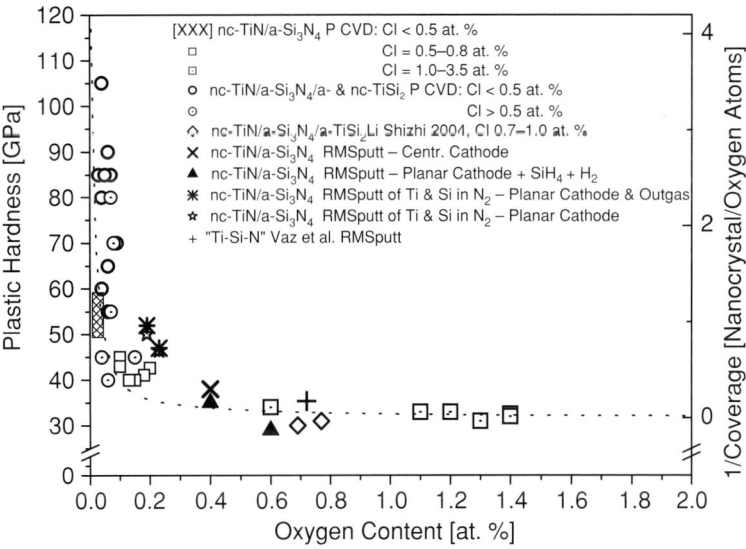

Figure 4.40. Effect of oxygen on the degradation of hardness in nc-TiN/a-Si$_3$N$_4$ nanocomposites deposited by plasma CVD and by reactive magnetron sputtering in five different apparatuses from three different countries: "P CVD" – nanocomposites deposited by means of plasma chemical vapor deposition as described by Veprek and Reiprich (1995) and Veprek *et al.* (2004b) with chlorine content as indicated; "Li Shizhi" – P CVD as described by Li *et al.* (1992); "RMSputt – Planar Cathode" – reactive magnetron sputtering of either Ti and Si target or Ti target combined with PCVD of Si from SiH$_4$ + H$_2$ (see Prochazka *et al.*, 2004); "RMSputt – Centr. Cathode" – a prototype of an industrial-like coating equipment with a central cathode (Männling, 2003); "Ti-Si-N" Vaz *et al.*" (Vaz *et al.*, 2001). The dotted line corresponds to the reciprocal coverage of the nanocrystals with oxygen atoms (see text) (Veprek *et al.*, 2005b).

of 100 p.p.m., and, for more than 0.5 at.% (5000 p.p.m.), the strength of the interface is dominated by the oxygen-related bonds.

This is not surprising because there are many examples known and documented in the literature where a few hundred parts per million (p.p.m.) of impurities in a metal caused embrittlement and catastrophic failure of parts. In the majority of such cases, the understanding of the degradation of the properties of the material is not known at the atomistic level. For example, it has been known for more than 100 years that less than 100 p.p.m. of bismuth causes catastrophic embrittlement of otherwise very ductile copper, but the mechanism is still not unambiguously understood (for several other examples, see the references quoted in Veprek *et al.* (2005d)). Recently, Schweinfest *et al.* argued on the basis of first principle calculations that this is a simple size effect of the larger Bi atoms as compared to Cu atoms (Schweinfest *et al.*, 2004). However, Duscher *et al.* (2004), who used a combination of high-resolution TEM, Cu-L$_3$ near-edge electron energy loss spectroscopy and first principle calculations, provided strong arguments for electronic effects. Therefore, much work is needed to fully understand the strong effect of minor oxygen impurities on the degradation of the hardness in the nc-TiN/a-Si$_3$N$_4$ nanocomposites. The simple size effect of O^{2-} can surely explain a significant part of it, but electronic effects are very likely to contribute to the observed softening as well (Veprek *et al.*, 2005d).

It should be emphasized that here we are discussing the hardness enhancement due to the formation of the nanostructure by spinodal decomposition, which has a high thermal stability. Reports about a similar hardness enhancement obtained in coatings which were deposited at a low temperature (Diserens *et al.*, 1998; 1999) or even room temperature (Hu *et al.*, 2002) will most probably not show such a high sensitivity to impurities because that hardness enhancement is, for the deposition temperature of $\leq 350\,^\circ$C used by Diserens *et al.* at least to a part, and for room temperature used by Hu *et al.* predominantly, caused by energetic ion bombardment during the deposition. Unfortunately, in the majority of papers, neither the thermal stability nor the impurity content were reported. Therefore, great care has to be exercised in order to distinguish between these two different mechanisms of hardness enhancement.

4.9.4 *Toughness of hard and superhard nanocomposites*

Fracture toughness is the resistance of a material against the propagation of a pre-existing, well-defined crack. It is expressed either in terms of energy release rate K_C or stress intensity factor K_I. For the latter, $K_I = \sigma (\pi \cdot a)^{1/2}$, where σ is the stress needed to propagate a pre-existing crack of length a (Hertzberg, 1989; Anderson, 1996). Fracture toughness of brittle materials, such as glasses and ceramics, is conveniently measured by indentation. However, for thin films, this technique is less suitable because, for a correct measurement, the thickness of the film should be larger than the length of the crack in order to eliminate the disturbing effects of the substrate from which cracks can be initiated.

There are various mechanisms of toughening of materials in general, and of thin films in particular. For example, hard nanocrystals of a transition metal carbide can be imbedded into amorphous, diamond-like carbon (Voevodin *et al.*, 1997; Voevodin *et al.*, 2002; Voevodin and Zabinski, 2005), or hard transition-metal nitrides can be combined with a ductile,

4.9 Mechanical properties of superhard

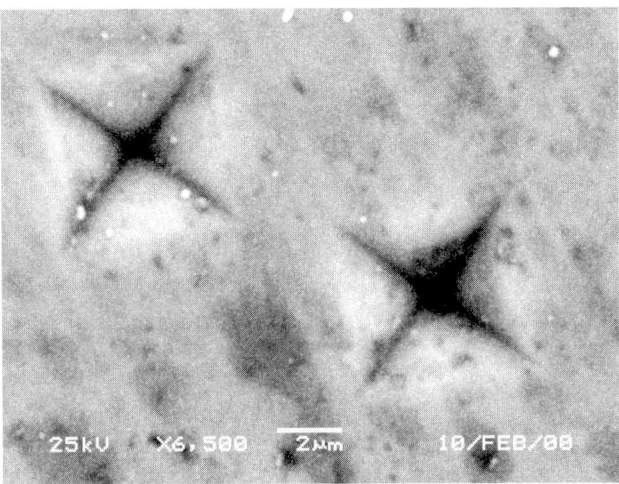

Figure 4.41. Indentation into a 6.1 μm thin nc-TiN/a-Si$_3$N$_4$,a- and nc-TiSi$_2$ coatings deposited on a soft steel (c. 2 GPa) substrate (Veprek et al., 2003a).

soft metal (Baker et al., 2005). These coatings show an excellent resistance against brittle fracture, particularly under impact loading. This makes them of high interest for applications as functional materials in a variety of structural machine parts. A frequently used technique of toughening of thin coatings for industrial applications is the deposition of multilayers of materials with different elastic moduli (Holleck et al., 1990). We refer to a recent review for further details (Zhang et al., 2005).

The superhard nanocomposites prepared according to a generic design principle by spinodal phase segregation show also very high resistance against crack formation upon indentation. This is illustrated in Figure 4.41, which shows two indentations into a 6.1 μm thin nc-TiN/a-Si$_3$N$_4$/a- and nc-TiSi$_2$ ultrahard nanocomposite coatings deposited on a soft steel substrate. The indentations are aligned almost along their diagonal where, upon the indentation, the maximum stress is concentrated and, therefore, the radial cracks usually start to form and propagate. There are no cracks seen on the SEM micrograph in Figure 4.41. The absence of cracking was checked also on the loading curves as well as by Hertzian analysis of the unloading ones (Veprek et al., 2003b). The indentation depth of almost 2 μm corresponds to a large strain of $\geq 10\%$. A large fraction of the observed deformation is elastic because of the excellent adherence of the coating to the soft substrate, which deformed plastically and thus prevented the coating from elastically recovering. When the coating was cut into a thin strip by sputtering with energetic gallium ions (see Figure 4.42) the elastic strain partially relaxed and cracks with a complex structure formed (Strunk, 2003). This example shows that these nanocomposites indeed have a very high elastic limit approaching the ideal elastic limit of flaw-free materials and very high resistance against brittle fracture (Veprek et al., 2003a).

However, this is not evidence of a high fracture toughness as defined by the energy-release rate of the stress intensity factor with a pre-fabricated crack. The high resistance against brittle fracture observed in these nanocomposites may simply be caused by a high

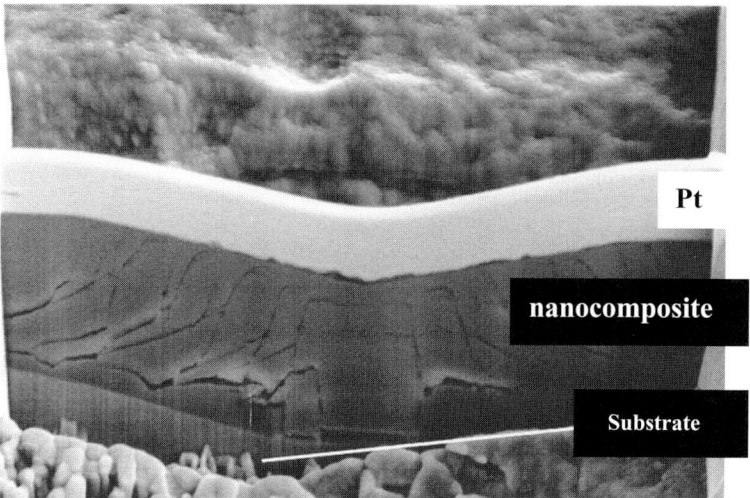

Figure 4.42. Cross-sectional view of an indentation into the coating shown previously. The coating was protected by a narrow, thick platinum strip against the energetic gallium ions that were used to remove the nanocomposite layer and part of the substrate in the front and in the back (Strunk (2003), with permission.)

threshold for crack initiation. Because the maximum size of a planar crack (a flaw within the interfacial layer) is less than the size of the nanocrystals of 3–4 nm, the stress concentration factor

$$\frac{\sigma_{tip}}{\sigma_{appl}} = 1 + 2\sqrt{\frac{a}{\rho}} \approx 3$$

(σ_{appl} is the applied uniaxial applied stress and σ_{tip} is the stress concentrated on the tip of the crack with a tip radius ρ) is very small for a nanocrack of the size a 1 nm with atomically sharp tip radius ρ 0.3 nm. For comparison, in a fine-grade, sub-micron ceramic with a 100–1000 nm the stress concentration factor will be 20–60, i.e. it will be much easier to initiate cracks from a flaw.

4.10 Summary

The mechanical behavior of nanocrystalline materials has been a research topic of intense and ever growing interest in the past decade with a world-wide research community devoted to solving its many intriguing problems. This chapter has attempted to review the experimental situation as of mid 2005. New results are continually being reported which may make some of these conclusions obsolete by the time of publication. The theoretical situation will be described in depth in Chapter 5.

It seems clear that the elastic properties of nanocrystalline materials do not differ from those of conventional-grain-size materials until the nanoscale grain size is reduced to about 10 nm or less so that a substantial fraction of atoms is associated with the grain boundaries

4.10 Summary

and can lower the average elastic moduli. The limited studies of anelastic properties may help in understanding of the mechanical response of nanocrystalline materials. Hardness and strength of nanocrystalline materials are well documented to be higher than those of conventional-grain-size materials. In the case of pure metals they may be 10–20 times greater. The "inverse Hall–Petch" effect is probably real but exists only at the smallest grain sizes (<10 nm) where grain-boundary deformation mechanisms may be controlling the behavior. Solid solution hardening in substitutional nanocrystalline solid solutions is typically overwhelmed by the grain-size-hardening effect. Superhard nanocrystalline composite films have been produced which are already finding applications in the cutting tool industry.

Asymmetries in yield strength measured by compression vs. tensile tests can sometimes be attributed to processing artifacts, but in some cases this is apparently caused by mechanical instabilities and shear bands which are the result of low or non-existent strain hardening. These effects may be influenced by processing variables. Multiphase or nanocomposite materials offer the promise of superior properties. To date most of the research in this area has been devoted to crystallization of bulk metallic glasses and the multiphase structures that can result.

Until a few years ago, the consensus opinion regarding the ductility of nanocrystalline metals and alloys was that while high strength could be obtained, ductility was always nil at the smallest grain sizes (<30 nm). However, recently there have been several examples of nanocrystalline metals and alloys that exhibit both high strength and good ductility. These observations may be strongly linked to improved processing methods, although the precise reasons for the excellent behavior of the materials in question are not completely understood. Strain hardening is a necessary criterion for ductile behavior and this implies that dislocation-controlled deformation may be dominant. *In situ* deformation studies in TEM or in X-ray synchrotron sources have shown that dislocation activity can exist in nanocrystalline materials down to grain sizes of at least about 20 nm.

Twinning is another deformation mechanism that may be very important in nanocrystalline materials. Twinning has been observed in several nanocrystalline metals with high stacking fault energies (Al, Pd) which do not usually exhibit twinning in conventional-grain-size form.

Strain rate and test temperature effects point to differences in the behavior of nanocrystalline fcc and bcc metals and their differences from conventional-grain-size metals. For example, the strain-rate sensitivity, m, is found to increase with decreasing grain size for fcc metals while it apparently decreases with decreasing grain size for bcc metals.

Nanocrystalline materials exhibit higher strain-rate superplasticity than micron-grain-size samples and at lower temperatures. Extensive strain hardening, high flow stresses, and grain growth are observed during tensile testing, making the superplastic behavior very different from that in micron-grain-size materials.

Limited studies of the creep of nanocrystalline materials suggest that much more work is needed on well-characterized materials and at temperatures where grain growth may be minimized in order to assess the mechanisms responsible for creep deformation. Very few studies have been reported for fatigue of nanocrystalline materials. These results show improved resistance to failure under stress-controlled fatigue but decreased resistance to fatigue crack growth. Because of the need for larger samples for valid measurements, there are only a few measurements of fracture toughness of nanocrystalline materials. High values of fracture toughness have been reported in one case. More work is needed on the fracture toughness of nanocrystalline materials.

Processing of large, artifact-free bulk samples that are truly nanocrystalline (grain sizes less than 100 nm, ideally less than 50 nm) is a key challenge to allow for the eventual testing of the inherent mechanical properties of these fascinating materials. So far, almost flaw-free nanocomposite materials whose strength approaches the ideal one have been prepared only as thin films by a self-organization that occurs upon spinodal phase segregation.

References

Anderson, T. L. (1996). *Fracture Mechanics*. Boca Raton: CRC Press.
Argon, A. S., and Veprek, S. (2002). *Mater. Res. Soc. Symp. Proceedings*, **697**, 3.
Asaro, R. J., and Suresh, S. (2005). *Actual Mater.*, **53**, 3369–3382.
Baker, I., Nagpal, P., Liu, F., and Munroe, P. R. (1991). *Acta Metall. Mater.*, **39**, 1637–1644.
Baker, M., Kench, P. J., Tsotsos, C., Gibson, P. N., Leyland, A., and Matthews (2005). *J. Vac. Sci. Technol.*, **A 23**, 1–11.
Barnett, S. A. (1993). In *Physics of Thin Films Vol. 17: Mechanic and Dielectric Properties*, ed. Francombe, M. H., and Vossen, J. L. Boston: Academic Press, p. 2.
Barnet, S. and Madan, A. (1999). *Phys. World*, **11**, 45.
Barnett, S. A., Madan, A., Kim, I., and Martin, K. (2003). *MRS Bulletin*, **28**, 169.
Barsoum, M. W. (1997). *Fundamentals of Ceramics*. New York: McGraw-Hill, pp. 414–417.
Boccaccini, A. R., Ondracek, G., Mazilu, P., and Windelburg, D. (1993). *J. Mech. Behav. Mater.*, **4**, 119.
Bohn, R., Haubold, T., Birringer, R., and Gleiter, H. (1991). *Scripta Metall. Mater.*, **25**, 811–816.
Bohn, R., Oehring, M., Pfullmann, Th., Appel, F., and Bormann, R. (1995). In *Processing and Properties of Nanocrystalline Materials*, ed. Suryanarayana, C., Singh, J., and Froes, F. H. Warrendale, PA: TMS, pp. 355–366.
Bolshakov, A., Oliver, W. C., and Pharr, G. M. (1996). *J. Mater. Res.*, **11**, 760.
Bonetti, E., Campari, E. G., Del Bianco, L., Pasquini, L., and Sampaolesi, E. (1999). *NanoStructured Mater.*, **11**, 709–720.
Budrovic, Z., Van Swygenhoven, H., Derlet, P. M., Van Petegem, S., and Schmitt, B. (2004). *Science*, **304**, 273–276.
Cai, B., Kong, Q. P., Lu, L., and Lu, K. (2000). *Mater. Sci. Engr. A*, **286**, 188–192.
Carsley, J. E., Milligan, W. W., Hackney, S. A., and Aifantis, E. C. (1995). *Metall. Mater. Trans. A*, **26A**, 2479.
Carsley, J. E., Fisher, A., Milligan, W. W., and Aifantis, E. C. (1998). *Metall. Mater. Trans. A*, **29**, 2261–2271.
Chan, K. S. (1990). *Scripta Metall. Mater.*, **24**, 1725–1730.
Chang, H., Altstetter, C. J., and Averback, R. S. (1992). *J. Mater. Res.*, **7**, 2962–2970.

References

Chen, M., Ma, E., Hemker, K. J., Sheng, H., Wang, Y., and Cheng, X. (2003). *Science*, **300**, 1275–1277.
Cheng, S., Spencer, J. A., and Milligan, W. W. (2003). *Acta Mater.*, **51**, 4505–4518.
Chokshi, A. H., Rosen, A., Karch, J., and Gleiter, H. (1989). *Scripta Mater.*, **23**, 1679.
Chu, W.-Y., and Thompson, A. W. (1991). *Scripta Metall. Mater.*, **25**, 641–644.
Chu, X., and Barnett, S. A. (1995). *J. Appl. Phys.*, **77**, 4403.
Conrad, H., Narayan, J., and Jung, K. (2005). *Inter. J. Refractory & Hard Materials* (in press).
Cottrell, A. H. (1958). *Trans. TMS-AIME*, **212**, 192.
Dalla Torre, F., Van Swygenhoven, H., and Victoria, M. (2002). *Acta Mater.*, **50**, 3957–3970.
Demkowicz, M. J., and Argon, A. S. (2004). *Phys. Rev. Lett.*, **93**, 025505.
Deng, J., Wang, D. L., Kong, Q. P., and Shui, J. P. (1995). *Scripta Metall. Mater.*, **32**, 349–352.
Derlet, P. M., and Van Swygenhoven, H. (2002). *Phil. Mag. A*, **82**, 1–15.
Dieter, G. E. (1976). *Mechanical Metallurgy*, 2nd edition. McGraw-Hill, p. 270.
Diserens, M., Patscheider, J., and Lévy, F. (1998). *Surf. Coat. Technol.*, **108–109**, 241.
Diserens, M., Patscheider, J., and Lévy, F. (1999). *Surf. Coat. Technol.*, **120–121**, 158.
Duscher, G., Chisholm, M. F., Alber, U., and Ruhle, M. (2004). *Nature Mater.*, **3**, 621–626.
Donovan, P. E., and Stobbs, W. M. (1981). *Acta Metall.*, **29**, 1419.
Eckert, J. (2002). Structure formation and mechanical behavior of two-phase nanostructured materials. In *Nanostructured Materials: Processing, Properties and Applications*, ed. Koch, C. C. Norwich, NY; William Andrew Publ., pp. 423–525.
Edington, J. W., Melton, K. N., and Cutler, C. P. (1976). *Prog. Mater. Sci.*, **21**, 61–170.
Fischer-Cripps, A. C., Karvankova, P., and Veprek, S. (2006). *Surf. Coat. Technol.*, **200**, 5645.
Fougere, G. E., Riester, L., Ferber, M., Weertman, J. R., and Siegel, R. W. (1995). *Mater. Sci. Eng.*, **A204**, 1.
Gonsalves, K. E., Rangarajan, S. P., Law, C. C., Garcia-Ruiz, A., and Chow, G. M. (1997). *NanoStructured Mater.*, **9**, 169–172.
Goodwin, T. J., Yoo, S. H., Matteazzi, P., and Groza, J. R. (1997). *NanoStructured Mater.*, **8**, 559–566.
Greer, A. L. (2001). *Mater. Sci. Engr. A*, **304–306**, 68–72.
Hahn, H., and Averback, R. S. (1991). *J. Am. Ceram. Soc.*, **74**, 2918–2921.
Hall, E. O. (1951). *Proc. R. Soc. Lond.*, **364**, 474.
Han, B. Q., and Lavernia, E. J. (2005). *Adv. Eng. Mater.*, **7**, 247–250.
Hanlon, T., Kwon, Y.-N., and Suresh, S. (2003). *Scripta Mater.*, **49**, 675–680.
Hao, S., Delley, B., Veprek, S., and Stampfl, C. (2006). *Phys. Rev. Lett.*, **97**, 086102.
He, L., and Ma, E. (1996). *NanoStructured Mater.*, **7**, 327–339.
Helersson, U., Todorova, S., Barnett, S. A., Sundgren, J., Markert, L. C., and Greene, J. E. (1987). *J. Appl. Phys.*, **62**, 481.
Hertzberg, R. W. (1989). Deformation and Fracture Mechanics of Engineering Materials, 3rd edition. New York: Wiley.

Hoffmann, M., and Birringer, R. (1996). *Acta Mater.*, **44**, 2729–2736.
Hofler, H. J., and Averback, R. S. (1990). *Scripta Metall. Mater.*, **24**, 2401–2406.
Holleck, H., Lahres, M., Woll, P. (1990). *Surf. Coat. Technol.*, **41**, 179.
Hu, X., Han, Z., Li, G., and Gu, M. (2002). *J. Vac. Sci. Technol.*, **A20**, 1921.
Hu, X., Zhang, H., J. Dai, J., Li, G., and Gu, M. (2005). *J. Vac. Sci. Technol.*, **A23**, 114.
Huang, S. C., and Chesnutt, J. C. (1994). In *Intermetallic Compounds, Vol. 2, Practice*, ed. Westbrook, J. H., and Fleischer, R. L. Chichester, UK: John Wiley & Sons, Ltd., pp. 73–90.
Huang, Z., Gu, L. Y., and Weertman, J. R. (1997). *Scripta Mater.*, **42**, 1071–1075.
Hugo, R. C., Kung, H., Weertman, J. R., Mitra, R., Knapp, J. A., and Follstaedt, D. M. (2003). *Acta Mater.*, **51**, 1937–1943.
Hutchinson, J. W. (1984). *Scripta Metall.*, **18**, 421–422.
Inoue, A., Nakazato, K., Kawamura, Y., and Masumoto, T. (1994). *Mater. Sci. Engr. A*, **179/180**, 654–658.
Jia, D., Ramesh, K. T., and Ma, E. (2000). *Scripta Mater.*, **42**, 73–78.
Karimpoor, A. A., Erb, U., Aust, K. T., and Palumbo, G. (2003). *Scripta Mater.*, **49**, 651–656.
Karvankova, P., Veprek-Heijman, M. G. J., Zindulka, O., Bergmaier, A., and Veprek, S. (2003). *Surf. Coat. Technol.*, **163–164**, 149.
Karvankova, P., Veprek-Heijman, M. G. J., Azinovic, D., and Veprek, S. (2006). *Surf. Coat. Technol.*, **200**, 2978.
Ke, M., Hachney, S. A., Milligan, W. W., and Aifantis, E. C. (1995). *NanoStructured Mater.*, **5**, 689–698.
Kim, D. K., and Okazaki, K. (1992). *Mater. Sci. Forum*, **88–90**, 553–560.
Koch, C. C., and Cho, Y. S. (1992). *NanoStructured Mater.*, **1**, 207–212.
Koch, C. C., Morris, D. G., Lu, K., and Inoue, A. (1999). *MRS Bulletin*, **24**, 54–58.
Koch, C. C., and Narayan, J. (2001). *The Inverse Hall–Petch Effect – Fact or Artifact?* In *MRS Symp. Proc. Vol. 634*, ed. Farkas, D., Kung, H., Mayo, M., Van Swygenhoven, H., and Weertman, J. R. Warrendale, PA: MRS, pp. B5.1.1–B5.1.11.
Koch, C. C., Youssef, K. M., Scattergood, R. O., and Murty, K. L. (2005). *Adv. Engr. Mater.*, **7**, 787–794.
Koehler, J. S. (1970). *Phys. Rev.*, **B2**, 547.
Krstic, V., Erb, U., and Palumbo, G. (1993). *Scripta Metall. Mater.*, **29**, 1501.
Kumar, K. S., Suresh, S., Chisholm, M. F., Horton, J. A., and Wang, P. (2003). *Acta Mater.*, **51**, 387–405.
Kumar, K. S., Van Swygenhoven, H., and Suresh, S. (2003). *Acta Mater.*, **51**, 5743–5774.
Lehocky, S. L. (1978a). *J. Appl. Phys.*, **49**, 5479.
Lehocky, S. L. (1978b). *Phys. Rev. Lett.*, **41**, 1814.
Li, H., and Ebrahimi, F. (2004). *Appl. Phys. Lett.*, **84**, 4307–4309.
Li, J., Sun, Y., Sun, X., and Qiao, J. (2005). *Surface & Coatings Technology*, **192**, 331–335.
Li, J. C. M. (1963). *Trans. TMS-AIME*, **227**, 239.
Li, S. H., Shi, Y. L., and Peng, H. R. (1992). *Plasma Chem. Plasma Process.*, **12**, 287.
Li, Y. J., Blum, W., and Breutinger, F. (2004). *Mater. Sci. Engr. A*, **387–389**, 585–589.

References

Liao, X. Z., Zhao, Y. H., Srinivason, S. G., and Zhu, Y. T. (2004). *Appl. Phys. Lett.*, **84**, 592–594.

Liao, X. Z., Zhou, F., Lavernia, E. J., He, D. W., and Zhu, Y. T. (2003). *Appl. Phys. Lett.*, **83**, 5062–5064.

Ljungcrantz, H., Hultman, L., and Sundgren, J.-E. (1995). *J. Appl. Phys.*, **78**, 832.

Lu, L., Li, S. X., and Lu, K. (2001). *Scripta Mater.*, **45**, 1163–1169.

Malow, T. R., and Koch, C. C. (1998). *Metall. Mater. Trans. A*, **29A**, 2285–2295.

Malow, T. R., Koch, C. C., Miraglia, P. Q., and Murty, K. L. (1998). *Mater. Sci. Engr. A*, **252**, 36–43.

Männling, H.-D. (2003). Ph.D. Thesis, Technical University Munich.

Markmann, J., Bunzel, P., Rosner, H., Liu, K. W., Padmanabhan, K. A., Birringer, R., Gleiter, H., and Weissmuller, J. (2003). *Scripta Mater.*, **49**, 637–644.

Mayo, M. J. (1997). *NanoStructured Mater.*, **9**, 717–726.

McFadden, S. X., Mishra, R. S., Valiev, R. Z., Zhilyaev, A. P., and Mukherjee, A. K. (1999). *Nature*, **398**, 684–686.

Meyers, M. A., and Ashworth, E. (1982). *Phil. Mag.*, 737.

Meyers, M. A., and Chawla, K. K. (1999). *Mechanical Behavior of Materials*. New Jersey: Prentice-Hall, p. 541.

Milligan, W. W. (2003). Mechanical behavior of bulk nanocrystalline and ultrafine-grain metals. In *Comprehensive Structural Integrity*, ed. Milne, I., Ritchie, R. O., and Karihaloo, B. Amsterdam: Elsevier, pp. 529–550.

Milligan, W. W., Hackney, S. A., Ke, M., and Aifantis, E. C. (1993). *NanoStructured Mater.*, **2**, 267–276.

Mirshams, R. A., Xiao, C. H., Whang, S. H., and Yin, W. M. (2001). *Mater. Sci. Engr. A*, **315**, 21–27.

Mishra, R. S., Valiev, R. Z., and Mukherjee, A. K. (1997). *NanoStructured Mater.*, **9**, 473–476.

Mishra, R. S., Valiev, R. Z., McFadden, S. X., and Mukherjee, A. K. (1998). *Mater. Sci. Engr. A*, **252**, 174–178.

Mitra, R., Chiou, W.-A., and Weertman, J. R. (2004). *J. Mater. Res.*, **19**, 1029–1037.

Mohamed, F. A., and Li, Y. (2001). *Mater. Sci. Engr. A*, **298**, 1–15.

Morris, D. G. (1998). Mechanical behavior of nanostructured materials, *Materials Science Foundations 2*. Enfield NH: Trans. Tech. Publ., pp. 42–74.

Morris, D. G., and Morris, M. A. (1991). *Acta Metall. Mater.*, **39**, 1763.

Morris-Munoz, M. A., Dodge, A., and Morris, D. G. (1999). *NanoStructured Mater.*, **11**, 873–885.

Mukherjee, A. K. (2002). *Mater. Sci. Engr. A*, **322**, 1–22.

Mukhopadhyay, J., Kaschner, G., and Mukherjee, A. K. (1990). *Scripta Metall. Mater.*, **24**, 857–862.

Musil, J., Kadlec, S., Vyskocil, J., and Valvoda, V. (1988). *Thin Solid Films*, **167**, 107.

Niederhofer, A., Bolom, T., Nesladek, P., Moto, K., Eggs, C., Patil, D. S., and Veprek, S. (2001). *Surf. Coat. Technol.*, **146–147**, 183.

Nieman, G. W., Weertman, J. R., and Siegel, R. W. (1990). *Scripta Metall. Mater.*, **24**, 145–150.

Nieman, G. W., Weertman, J. R., and Siegel, R. W. (1991). *J. Mater. Res.*, **6**, 1012–1027.
Nowack, A. S., and Berry, B. S. (1972). *Anelastic Relaxation in Crystalline Solids*. New York: Academic Press.
Odén, M. (2004). Invited paper at the *51st Int. Symp. of the American Vacuum Society*, Anaheim, November 14–19, 2004.
Ostwaldt, D., Klepaczko, J. R., and Klimanik, P. J. (1997). *PHYS IV FRANCE*, **7**, C3–385.
Ovidko, I. A. (2003). *Phil. Mag. Lett.*, **83**, 611–620.
Palumbo, G., Gonzalez, F., Brennenstuhl, A. M., Erb, U., Shmayda, W., and Lichtenberger, P. C. (1997). *NanoStructured Mater.*, **9**, 737–746.
Patscheider (2004). Private communication, unpublished.
Petch, N. J. (1953). *J. Iron Steel Inst.*, **174**, 25.
Prilliman, S. G., Erdonmez, C. K., Clark, S. M., Alivisatos, A. P., Karvankova, P., and Veprek, S. (2006). *Mater. Sci. Eng. A*, **437**, 379.
Prochazka, J., Karvankova, P., Veprek-Heijman, M. G. J., and Veprek, S. (2004). *Mater. Sci. Eng. A*, **384**, 102.
Rose, J. H., Smith, J. R., Guinea, F., and Ferrante, J. (1984). *Phys. Rev. B*, **29**, 2963.
Rosner, H., Markmann, J., and Weissmuller, J. (2004). *Phil. Mag. Lett.*, **84**, 321–334.
Sakai, S., Tanimoto, H., and Mizubayashi, H. (1999). *Acta Mater.*, **47**, 211–217.
Sakai, S., Tanimoto, H., Otsuka, K., Yamada, T., Koda, Y., Kita, E., and Mizubayashi, H. (2001). *Scripta Mater.*, **45**, 1313.
Sanders, P. G., Rittner, M., Kiedaisch, E., Weertman, J. R., Kung, H., and Lu, Y. C. (1997). *NanoStructured Mater.*, **9**, 433–440.
Sanders, P. G., Eastman, J. A., and Weertman, J. R. (1997). *Acta Mater.*, **45**, 4019.
Sauthoff, G. (1994). Plastic deformation. In *Intermetallic Compounds, Principles and Practice*, ed. Westbrook, J. H., and Fleischer, R. L. Chichester: John Wiley and Sons, pp. 924–925.
Schuh, C. A., Argon, S. A., Nieh, T. G., and Wadsworth, J. (2003). *Phil. Mag.*, **83**, 2585.
Schwaiger, R., Moser, B., Dao, M., Chollacoop, N., and Suresh, S. (2003). *Acta Mater.*, **51**, 5159–5172.
Schweinfest, R., Paxton, A. T., and Finnis, M. W. (2004). *Nature*, **432**, 1008.
Sergueeva, A. V., Mara, N. A., and Mukherjee, A. K. (2004). *Mat. Res. Symp., Proc.*, **821**, P9.8.1–P9.8.7.
Shan, Z., Stach, E. A., Wiezorek, J. M. K., Knapp, J. A., Follstaedt, D. M., and Mao, S. X. (2004). *Science*, **305**, 654–657.
Shen, T. D., and Koch, C. C. (1996). *Acta Mater.*, **44**, 753–761.
Shen, T. D., Koch, C. C., Tsui, T. Y., and Pharr, G. M. (1995). *J. Mater. Res.*, **10**, 2892.
Shen, Y. F., Lu, L., Lu, Q. H., Jin, Z. H., and Lu, K. (2005). *Scripta Mater.*, **52**, 989–994.
Shinin, M., Hultman, L., and Barnett, S. A. (1992). *J. Mater. Res.*, **7**, 902.
Shinin, M., and Barnett, S. A. (1995). *Appl. Phys. Lett.*, **64**, 61.
Siegel, R. W. (1997). *Mater. Sci. Forum*, **235–238**, 851–860.
Siegel, R. W., and Fougere, G. E. (1994). In *Nanophase Materials: Synthesis Properties, Applications*, ed. Hadjipanayis, G. C., and Seigel, R. W. Dordrecht, the Netherlands: Kluwer Acad. Publ., p. 233.

Smith, J. R., Ferrante, J., Vinet, P., Gray, J. G., Richter, R., and Rose, J. (1987). In *Chemistry and Physics of Fracture*, ed. Latanisos, R. M., and Jones, R. H. Dordrecht: Martinus Nijhoff, p. 329.

Söderberg, H., Molina, J., Hultman, L., and Odén, M. (2005). *J. Appl. Phys.*, **97**, 114–327.

Strunk, H. P. (2003). Institute of Materials Science, University Erlangen-Nürnberg, Germany, unpublished results.

Suryanarayana, C. (1995). *Int. Mater. Rev.*, **40**, 41–64.

Swadener, J. G., Taljat, B., and Pharr, G. M. (2001). *J. Mater. Res.*, **16**, 2901.

Tabor, D. (1951). *The Hardness of Metals*. Oxford: Clarendon Press.

Taketani, K., Uoya, A., Ohtera, K., Uehara, T., Higashi, K., Inoue, A., and Masumoto, T. (1994). *J. Mater. Sci.*, **29**, 6513–6517.

Tanimoto, H., Sakai, S., and Mizubayashi, H. (2004). *Mater. Sci. Engr. A.*, **370**, 135–141.

Tanimoto, H., Sakai, S., and Mizubayashi, H. (1999). *NanoStructured Mater.*, **12**, 751–756.

Tellkamp, V. L., Melmed, A., and Lavernia, E. J. (2001). *Metall. Mater. Trans. A*, **32A**, 2335.

Tsui T. Y., Oliver W. C., and Pharr, G. M. (1996). *J. Mater. Res.*, **11**, 752.

Valvoda, V., Kuzel, R., Cerny, R., and Musil, J. (1988). *Thin Solid Films*, **156**, 63.

Van Vliet, K. J., Li, J., Zhu, T., Yip, S., and Suresh, S. (2003). *Phys. Rev. B*, **67**, 104105.

Vasudevan, V. K., Court, S. A., Kurath, P., and Fraser, H. L. (1989). *Scripta Metall.*, **23**, 467–469.

Vaz, P., Rebouta, L., Godeau, Ph., Girardeau, T., Pacaud, J., Riviere, J. P., and Traverse, A. (2001). *Surf. Coat. Technol.*, **146–147**, 274.

Veprek, S., and Argon, A. S. (2002). *J. Vac. Sci. Technol.*, **B20**, 650.

Veprek, S., and Reiprich, S. (1995). *Thin Solid Films*, **268**, 64.

Veprek, S. Sarott, F.-A., and Iqbal, Z. (1987). *Phys. Rev.*, **B36**, 3344.

Veprek, S., Reiprich, S., and Li, S. H. (1995a). *Appl. Phys. Lett.*, **66**, 2640.

Veprek, S., Haussmann, M., and Reiprich, S. (1995b). *J. Vac. Sci. Technol.*, **A14**, 46.

Veprek, S., Niederhofer, A., Moto, K., Nesladek, P., Männling, H.-D., and Bolom, T. (2000a). *Mater. Res. Soc. Symp. Proc.*, **581**, 321.

Veprek, S., Niederhofer, A., Moto, K., Bolom, T., Männling, H.-D., Nesladek, P., Dollinger, G., and Bergmaier, A. (2000b). *Surf. Coat. Technol.*, **133–134**, 152.

Veprek, S., Mukherjee, S., Karvankova, P., Männling, H.-D., He, J. L., Moto, K., Prochazka, J., and Argon, A. S. (2003a). *J. Vac. Sci. Technol.*, **A21**, 532.

Veprek, S., Mukherjee, S., Karvankova, P., Männling, H.-D., He, J. L., Moto, K., Prochazka, J., and Argon, A. S. (2003b). *Thin Solid Films*, **436**, 220.

Veprek, S., Männling, H.-D., Niederhofer, A., Ma, D., and Mukherjee, S. (2004). *J. Vac. Sci. Technol*, **B22**, L5.

Veprek, S. Veprek-Heijman, G. M. J., Karvankova, P., and Prochazka, J. (2005a). *Thin Solid Films*, **476**, 1.

Veprek, S., Männling, H.-D., Karvankova, P., and Prochazka, J. (2005b). *Surf. Coat. Technol.*, in press.

Veprek, R. G., Parks, D. M., Argon, A. S., and Veprek, S. (2005c). *Mater. Sci. Eng. A*, submitted.

Veprek, S., Karvankova, P., and Veprek-Heijman, M. G. J. (2005d). *J. Vac. Sci. Technol.*, **B23**, L17.

Vinogradov, A. Y., and Agnew, S. R. (2004). Nanocrystalline materials: fatigue. In *Dekker Encyclopedia of Nanoscience and Nanotechnology*, Marcel Dekker, Inc., pp. 2269–2288.

Voevodin, A. A., Prasad, S. V., and Zabinski, J. S. (1997). *J. Appl. Phys.*, **82**, 855.

Voevodin, A. A., Fitz, T. A., Hu, J. J., and Zabinski, J. S. (2002). *J. Vac. Sci. Technol.*, **A20**, 1434.

Voevodin, A. A., and Zabinski, J. S. (2005). *Surf. Coat. Technol.*, **65**, 741.

Wang, D. L., Kong, Q. P., and Shui, J. P. (1994). *Scripta Metall. Mater.*, **31**, 47–54.

Wang, N., Wang, Z., Aust, K. T., and Erb, U. (1997). *Mater. Sci. Engr. A*, **237**, 150–158.

Wang, Y. M., and Ma, E. (2004). *Appl. Phys. Lett.*, **85**, 2750–2752.

Wang, Y., Chen, M., Zhou, F., and Ma, E. (2002). *Nature*, **419**, 912–915.

Wang, Y. M., Hodge, A. M., Biener, J., Hamza, A. V., Barnes, D. E., Liu, K., and Nieh, T. G. (2005). *Appl. Phys. Lett.*, **86**, 101915-1-3.

Weertman, J. R. (2002). In *Nanostructured Materials: Processing, Properties, and Applications*, ed. Koch, C. C. Norwich, NY: William Andrew Pub., pp. 397–421.

Weertman, J. R., and Averback, R. S. (1996). In *Nanomaterials: Synthesis, Properties, and Applications*, ed. Edelstein, A. S., and Cammarata, R. C. Bristol: Institute of Physics Publ., p. 323.

Wei, Q., Cheng, S., Ramesh, K. T., and Ma, E. (2004). *Mater. Sci. Engr. A.*, **381**, 71–79

Wei, Q., Jia, D., Ramesh, K. T., and Ma, E. (2002). *Appl. Phys. Lett.*, **81**, 1240–1242.

Wilde, J. R., and Greer, A. L. (2001). *Mater. Sci. Engr. A.*, **304–306**, 932–936.

Witkin, D., Lee, Z., Rodreguez, R., Nutt, S., and Lavernia, E. J. (2003). *Scripta Mater.*, **49**, 297–302.

Witney, A. B., Sanders, P. G., Weertman, J. R., and Eastman, J. A. (1995). *Scripta Metall. Mater.*, **33**, 2025–2030.

Wong., L., Ostrander, D., Erb, U., Palumbo, G., and Aust, K. T. (1993). In *Nanophases and Nanocrystalline Structures*, ed. Shull, R. D., and Sanchez, J. M. Warrendale, PA: TMS, p. 85.

Xiao, M., and Kong, Q. P. (1997). *Scripta Mater.*, **36**, 299–303.

Yin, W. M., Whang, S. H., Mirshams, R., and Xiao, C. H. (2001). *Mater. Sci. Engr. A*, **301**, 18–22.

Yin, W. M., Whang, S. H., and Mirshams, R. A. (2005). *Acta Mater.*, **53**, 383–392.

Yoo, S. H., Sudarshan, T. S., Sethuram, K., Subhash, G., and Aifantis, E. C. (1999). *NanoStructured Mater.*, **12**, 23.

Youngdahl, C. J., Sanders, P. G., Eastman, J. A., and Weertman, J. R. (1997). *Scripta Mater.*, **37**, 809.

Youngdahl, C. J., Weertman, J. R., Hugo, R. C., and Kung, H. H. (2001). *Scripta Mater.*, **44**, 1475–1478.

Youssef, K. M., Scattergood, R. O., Murty, K. L., and Koch, C. C. (2004). *Appl. Phys. Lett.*, **85**, 929–931.

Youssef, K. M., Scattergood, R. O., Murty, K., Horton, J. A., and Koch, C. C. (2005). *Appl. Phys. Lett.*, **87**, 091904-1–091904-13.

Zhang, K., Weertman, J. R., and Eastman, J. A. (2004). **85**, 5197–5199.

References

Zhang, X., Wang, H., Scattergood, R. O., Narayan, J., Koch, C. C., Sergueeva, A. V., and Mukherjee, A. K. (2002). *Appl. Phys. Lett.*, **81**, 823–825.

Zhang, S., Sun, D., Fu, Y., and Du. H. (2005). *Surf. Coat. Technol.*, **198**, 2.

Zhang, R. F., and Veprek, S. (2006). *Mater. Sci. Eng. A.*, **424**, 128.

Zhu, X. K., Zhang, X., Wang, H., Sergueeva, A. V., Mukherjee, A. K., Scattergood, R. O., Narayan, J., and Koch, C. C. (2003). *Scripta Mater.*, **49**, 429–433.

Zimmermann, A. F., Palumbo, G., Aust, K. T., and Erb, U. (2002). *Mater. Sci. Engr. A*, **328**, 137–146.

5 Mechanical properties of structural nanocrystalline materials – theory and simulations

5.1 Introduction

The rapidly growing scientific and technological interest in structural nanocrystalline bulk materials and coatings arises from their outstanding mechanical properties opening a range of new applications; see, for example, reviews (Koch *et al.*, 1999; Gleiter, 2000; Gutkin *et al.*, 2001; Mohamed and Li, 2001; Padmanabhan, 2001; Veprek and Argon, 2002; Kumar *et al.*, 2003a; Milligan, 2003; Ovid'ko, 2004; Valiev, 2004; Chokshi and Kottada, 2006; Han *et al.*, 2005; Ovid'ko, 2005a,b; Wolf *et al.*, 2005) and books (Roco *et al.*, 2000; Chow *et al.*, 2000; Farkas *et al.*, 2001; Berndt *et al.*, 2003; Komarneni *et al.*, 2003; Gutkin and Ovid'ko, 2004a). These outstanding mechanical properties are caused by the interface and nanoscale effects associated with structural peculiarities of nanocrystalline materials where the volume fraction of the interfacial phase is extremely high, and grain size d does not exceed 100 nm. For instance, nanocrystalline bulk materials and coatings often exhibit extremely high strength, superhardness and good fatigue resistance desired for numerous applications; see Chapter 4 and the literature (Siegel and Fougere, 1995; Hahn and Padmanabhan, 1995; Koch *et al.*, 1999; Gleiter, 2000; Gutkin *et al.*, 2001; Mohamed and Li, 2001; Niederhofer *et al.*, 2001; Padmanabhan, 2001; Veprek and Argon, 2002; Kumar *et al.*, 2003a; Milligan, 2003; Patscheider, 2003; Valiev, 2004; Chokshi and Kottada, 2006; Han *et al.*, 2005; Ovid'ko, 2005a,b; Wolf *et al.*, 2005). At the same time, in most cases, nanocrystalline materials show low tensile ductility at room temperature, which essentially limits their practical utility. However, recently, several examples of substantial tensile ductility and even superplasticity of nanocrystalline materials have been reported (Mayo 1997, 1998; Mishra *et al.*, 1998; McFadden *et al.*, 1999; Islamgaliev *et al.*, 2001; Mishra *et al.*, 2001; Valiev *et al.*, 2001; Mukherjee, 2002a,b; Valiev *et al.*, 2002; Wang *et al.*, 2002; Champion *et al.*, 2003; He *et al.*, 2003; Karimpoor *et al.*, 2003; Kumar *et al.*, 2003b; Ma, 2003; Zhan *et al.*, 2003; He *et al.*, 2004; Wang and Ma, 2004a; Youssef *et al.*, 2004, 2005, 2006). These experimental data form a basis for the prognosis that nanocrystalline materials with unique combination of high strength and good ductility will be "ideal" materials for a wide range of structural and functional applications in the twentyfirst century.

5.1 Introduction

The mechanical properties of nanocrystalline materials are strongly influenced by their definitive (generic) structural peculiarities which are nanoscopic size of grains and extremely high volume fraction of the interfacial phase. In the first approximation, these peculiarities are characterized by mean grain size d whose range $d < 100$ nm specifies the nanocrystalline matter. Besides the definitive structural peculiarities, there are structural peculiarities that can be different in different nanocrystalline materials with the same mean grain size and are sensitive to both material characteristics and fabrication technology parameters. Examples of such non-generic structural peculiarities of nanocrystalline bulk materials and coatings are typical grain shape, grain-size distribution, non-composite/composite structure, shape characteristics of nanoparticles in nanocomposites, the presence/absence of fabrication-induced flaws, dominant type and structure of grain and interphase boundaries. The structural peculiarities, their sensitivity to material characteristics and fabrication technology parameters as well as their effects on the deformation behavior of nanocrystalline materials cause the "fabrication–structure–properties" relationships which are of crucial importance for performance of structural nanocrystalline materials. Identification of these relationships, especially those providing optimization of strength and ductility, is a formidable challenge because it requires detailed knowledge of many structural parameters as well as how this information is coupled to the fabrication technology, phase content and mechanical properties of the nanocrystalline matter. It is the subject of intensive multidisciplinary research efforts including experimental characterization, computer simulations, and theoretical analysis.

In recent years, significant progress has been reached in the experimental study of the structure and its evolution in deformed nanocrystalline materials (see a detailed discussion in Chapter 4). However, in many cases, the nanocrystalline structures and deformation mechanisms operating in them cannot be unambiguously identified with the help of contemporary experimental methods, because of high precision demands on experiments at the atomic and nanoscale levels. In these circumstances, analytical theoretical models and computer simulations of the structure, plastic deformation and fracture processes are very important for understanding the fundamental nature of the outstanding mechanical properties of nanocrystalline materials. Besides, the approach involving theoretical models and computer simulations can serve as a basis for development of high technologies exploiting the unique mechanical characteristics of nanocrystalline materials.

Analytical theoretical models of structure and mechanical properties of nanocrystalline bulk materials and coatings are based on both standard analytical methods (used in the theory of structure and mechanical properties of conventional coarse-grained polycrystals and composite solids; see Hirth and Lothe, 1982; Sutton and Balluffi, 1996) and new theoretical concepts (taking into account the structural peculiarities and their effects on the mechanical properties of nanocrystalline materials). Theoretical models of nanocrystalline structures and their mechanical properties include, in particular, geometric and continuum models. Geometric models address structural geometry of grains (crystallites), solid/solid interfaces (grain and interphase boundaries) and their ensembles in nanocrystalline materials, involving in consideration nanoscale effects. Defects in grains and interfaces are treated as local violations of (translational and/or rotational) symmetries of crystal lattice and interfaces, respectively. The continuum models use results of geometric models and/or atomistic simulations of nanocrystalline structures in a description of defects, grain and interphase boundaries as sources of internal stress fields and carriers of plastic flow, strongly influencing structure evolution, plastic deformation and fracture processes in nanocrystalline materials under mechanical load.

The continuum models commonly exploit the standard methods of the elasticity theory of defects in solids. Besides, diffusion equations and methods of the gradient theory of defects in nanocrystalline solids are used in analysis of diffusion processes and processes in which defect cores are essential, respectively. Generally speaking, continuum models provide a link between the atomic and nanoscale structures of nanocrystalline materials as well as their macroscopic mechanical characteristics. Special attention is paid to defect structure transformations, plastic flow and fracture processes at the nanoscale level whose crucial role is inherent to nanocrystalline materials. Combined description of plastic deformation and fracture mechanisms operating in nanocrystalline materials at the first step involves geometric-model-based definition of defect configurations that serve as carriers of plastic flow and induce fracture at the nanoscale level. Then, continuum-model-based analysis is given of transformations/evolution of the defect configurations under mechanical load. With results of the combined theoretical geometric and continuum study, the macroscopic mechanical characteristics of nanocrystalline materials are estimated.

The nanocrystalline material is specified by the presence of high-density ensembles of interfaces and defects in its "vacuum" state, the state with the lowest level of the free energy or another relevant thermodynamic potential. In this case, both the notion of excited state and the difference between the excited and "vacuum" states should be understood and described in terms of transformations of interfaces, defects and their ensembles. Limitations of analytical theoretical models are inherently related to their focus on typical structures, defect configurations and processes. Non-typical structures, defect configurations and processes are commonly ignored, despite the fact that they sometimes can play a significant role. Besides, theoretical analytical models are not always effective in a simultaneous description of structures and processes at different length scales. In these circumstances, results of analytical models should be approved by their comparison with experimental data and results of atomistic simulations. In fact, the combined approach involving experimental, theoretical and computer studies is most effective in the understanding of very complicated multiscale processes responsible for the unique mechanical properties of nanocrystalline materials.

Let us briefly discuss general methodology, validity and limitations of computer simulations of nanocrystalline structures and their mechanical properties. In recent years, impressive results in atomistic simulations of deformation and fracture mechanisms operating in nanocrystalline materials have been obtained. These results are based on dramatic enhancement of computer processor speed and development of parallel computing architectures during past decades. The most effective approach to atomistic simulations of nanocrystalline structures during plastic deformation is recognized to be molecular dynamics (Van Swygenhoven *et al.*, 2003; Wolf *et al.*, 2005). With this approach, the real-time behaviors of diverse nanocrystalline structures under mechanical load have been simulated. The molecular dynamics operates with Newton's equations for systems consisting of millions of atoms

5.1 Introduction

whose interactions are calculated using prescribed interatomic interaction potential functions. This approach accounts for crystal lattice anharmonicity and highly inhomogeneous internal stresses in deformed atomic structures (for details, see Van Swygenhoven et al., 2003; Wolf et al., 2005). Besides these attractive advantages, there are fundamental limitations of the approach, including the extremely short time window inherent to molecular dynamics simulations of real-time behavior and validity of prescribed interatomic interaction potentials. For instance, typical values of deformation time scale and strain rate handled by molecular dynamics are around 10 ns (10^{-8} s) and 10^7 s^{-1}, respectively. These values are unrealistic for most experiments on plastic deformation of nanocrystalline materials with typical strain rates ranging from 10^{-5} s^{-1} to 10^3 s^{-1}; see Chapter 4. As a corollary, results of computer simulations of deformation processes in nanocrystalline structures are discussive. In particular, unrealistically high strain-rate deformation simulated by molecular dynamics hardly involves thermally activated processes that often play a very important role in deformation and fracture of diverse materials especially at intermediate and high temperatures. As to interatomic interactions, they are described by interaction potential functions of empirical or semi-empirical origin. Their relevance is commonly tested by simulations of defect-free solids, in which case the many-body character of electronic bonding between atoms and its complex variations in the vicinities of crystal lattice defects are not fully captured. Also, most atomistic simulations handle monatomic nanostructured solids. Their application to analysis of evolution of polyatomic and composite nanostructures is limited to difficulties in a correct description of interatomic interaction potentials in the case of two or more sorts of atoms. Nevertheless, molecular-dynamics-based atomistic simulations of plastic deformation of nanocrystalline materials provide important information on *possible* structural transformations occurring in mechanically loaded nanocrystalline materials. This approach is treated to be one of effective methods of nanomaterials science and expected to grow rapidly in the future in parallel with progress in hardware and software of computer simulations.

The main aim of this chapter is to review results of computer simulations and theoretical models that describe nanocrystalline structures and their evolution during plastic deformation. Focus will be placed on deformation and fracture mechanisms operating at the atomic and nanoscale levels, and their effects on the macroscopic mechanical characteristics of structural nanocrystalline materials, with experimental data in this area (see Chapter 4) taken into account. For brevity, we will concentrate our consideration on final results of computer simulations and theoretical models, while their technical and mathematical details will be just briefly discussed.

The chapter is organized as follows. Section 5.2 addresses specific structural peculiarities of single-phase and composite nanocrystalline materials. In fact, Section 5.2 specifies nanocrystalline materials as the subject of theoretical approach and computer simulations. Section 5.3 briefly presents the list of key experimental data on mechanical behavior of nanocrystalline materials and discusses the general theoretical concepts on plastic deformation and fracture in these materials, based on experimental facts and atomistic simulations. Nanoscale and interface effects on lattice dislocation slip are analysed in Section 5.4, using results of theoretical models and atomistic simulations. Theoretical models describing competition between lattice dislocation slip and diffusional deformation mechanisms in nanocrystalline materials are considered in Section 5.5. Section 5.6 deals with theoretical models and computer simulations of grain-boundary sliding in nanocrystalline materials and its role in plastic flow localization. Section 5.7 addresses the interaction between different deformation mechanisms in mechanically loaded nanocrystalline materials. The focuses are placed on the interaction manifesting itself in both the emission of lattice dislocations from grain boundaries and twin deformation mode (associated with activity of partial dislocations emitted from grain boundaries). Section 5.8 considers the interaction between different deformation mechanisms with particular attention being paid to the rotational deformation mode effectively operating because of the interaction in mechanically loaded nanocrystalline materials. Theoretical models and computer simulations of fracture processes in deformed nanocrystalline materials are considered in Section 5.9. Key theoretical representations on strain rate sensitivity, ductility and superplasticity of nanocrystalline materials are discussed in Section 5.10. Theoretical models of diffusion processes in deformed nanocrystalline materials are discussed in Section 5.11. Section 5.12 contains concluding remarks and outlines key topics for future computer and theoretical studies of deformation and fracture processes in nanocrystalline materials.

5.2 Specific structural features of nanocrystalline materials

In general, structural nanocrystalline materials can be divided into single-phase and composite nanocrystalline materials, depending on their phase content. Single-phase and composite nanocrystalline materials are compositionally homogeneous and inhomogeneous solids with nanoscale grains (nanocrystallites), respectively. Below we consider the specific structural features of single-phase and composite nanocrystalline materials (see subsections 5.2.1 and 5.2.2, respectively), with particular attention being paid to those features that affect the outstanding mechanical properties of nanocrystalline materials.

5.2.1 *Specific structural features of single-phase nanocrystalline materials*

Single-phase nanocrystalline materials are compositionally homogeneous solids consisting of nanoscale grains (nanocrystallites) divided by grain boundaries (Figure 5.1). Grains are characterized by the grain size $d < 100$ nm and have the crystalline structure. Crystal lattices of grains are misoriented relative to each other. Therefore, neighboring grains are divided by grain boundaries, plane or faceted layers (with thickness being around 1 nm) that carry a geometric mismatch between adjacent misoriented crystalline grains (Figure 5.1). Grain boundaries have atomic structure and properties being different from those of grains. Grain boundaries join at triple junctions that are tube-like regions with diameter being about 1–2 nm (Figure 5.1). Triple junctions are recognized as line defects with structure and properties commonly being different from those of grain boundaries that they adjoin (Palumbo and Aust, 1989; King, 1999; Gottstein *et al.*, 2000; Caro and Van Swygenhoven, 2001). With the nanoscale range of grain sizes, the volume fractions occupied by the grain-boundary and triple-junction phases in nanocrystalline materials are very high. It is one of the definitive structural features of nanocrystalline materials.

5.2 Specific structural features of nanocrystalline materials

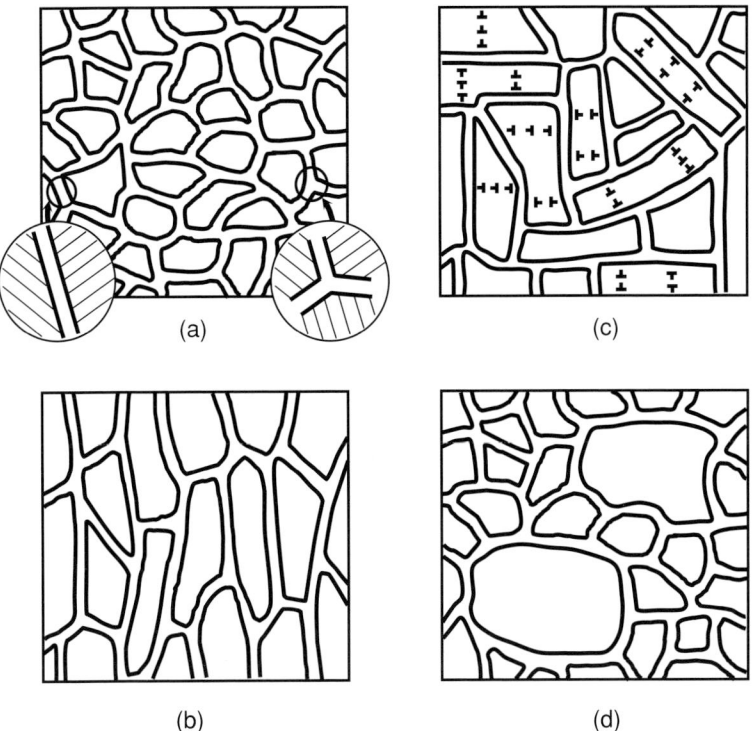

Figure 5.1. Typical single-phase nanocrystalline structures. (a) Nanocrystalline materials consisting of tentatively equiaxed grains (nanocrystallites) of the same phase. Large-scale view of typical structural elements of nanocrystalline materials – short grain boundaries and their triple junctions – are shown in the bottom part of the figure. (b) Nanocrystalline materials consisting of columnar grains of the same phase. (c) Single-phase nanocrystalline materials having elongated grains with high-angle grain boundaries, which contain equiaxed subgrains divided by low-angle boundaries. (d) Single-phase nanocrystalline materials with bimodal structure consisting of tentatively equiaxed grains with nanoscopic and microscopic sizes.

The fact that grains are characterized by nanoscopic sizes serves as another definitive structural feature of nanocrystalline structures. In most cases, nanocrystalline materials consist of approximately equiaxed grains (Figure 5.1a). That is, all the three sizes of grains are almost the same. However, there are examples of different geometry of grains (see Chapter 2). For instance, electrodeposited nanocrystalline materials commonly consist of columnar grains (Figure 5.1b) (Kumar et al., 2003a,b), while nanocrystalline materials fabricated by the cryomilling method typically have elongated grains with high-angle grain boundaries, which contain equiaxed subgrains divided by low-angle boundaries (Figure 5.1c) (Zghal et al., 2002). Besides, in recent years, nanocrystalline materials with a bimodal structure consisting of approximately equiaxed grains with nanoscopic and microscopic sizes have been fabricated (Figure 5.1d) (Tellkamp et al., 2001; Wang et al., 2002; Sergueeva et al., 2004; Wang and Ma 2004a,b; Han et al., 2005).

As to grain boundaries, they are also typical structural elements of conventional coarse-grained polycrystals where the volume fraction of the grain boundary phase is low ($<1\%$).

210 5 Mechanical properties of structural nanocrystalline materials

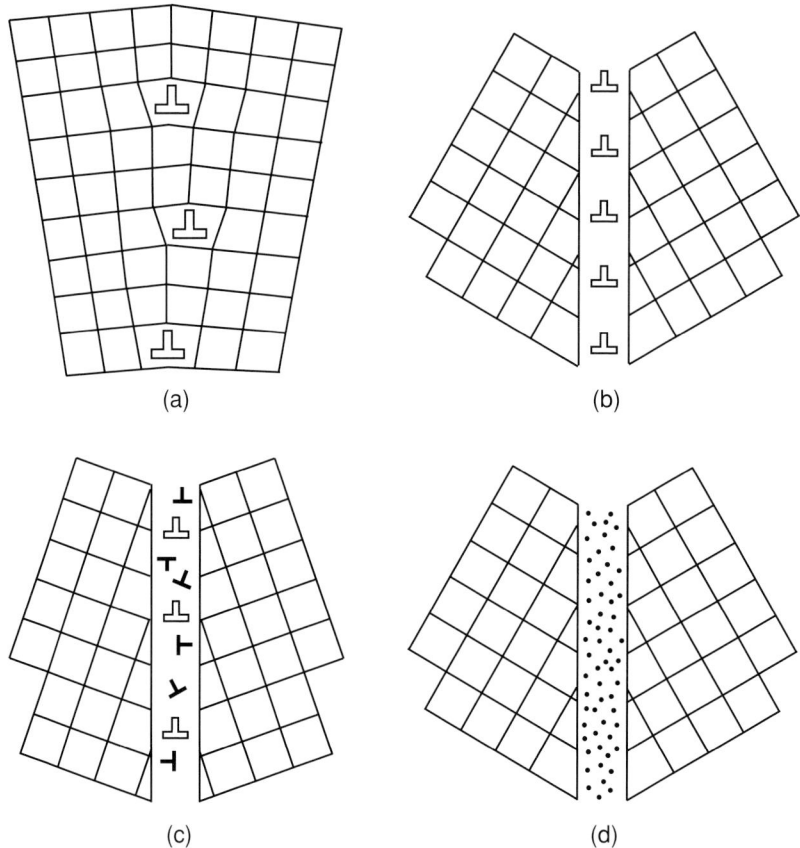

Figure 5.2. Types of grain boundaries: (a) low-angle periodic boundaries; (b) high-angle periodic boundaries; (c) nonequilibrium grain boundaries containing high-density ensembles of defects; and (d) amorphous grain boundaries (schematically).

Grain boundaries in single-phase polycrystalline materials represent the traditional subject of experimental and theoretical studies in materials science and solid state physics; see, e.g., Sutton and Balluffi (1996). In these studies, it is well recognized that both the structure and properties of grain boundaries in polycrystalline materials are essentially different from those of the bulk phase (crystalline grain interiors). Atoms in the bulk phase form equilibrium crystal lattices corresponding to minimum of the free energy or another relevant thermodynamic potential of the infinite crystal. Atoms in grain boundaries are arranged in a more complicated and disordered manner, because there are restrictions on their arrangement, imposed by the adjacent crystal lattices of misoriented grains.

Grain boundary structures in coarse-grained polycrystals can be divided into the following key categories: low-angle periodic boundaries (Figure 5.2a) (Sutton and Balluffi, 1996), high-angle periodic boundaries (Figure 5.2b) (Sutton and Balluffi, 1996), non-equilibrium grain boundaries containing high-density ensembles of defects (Figure 5.2c) (Nazarov *et al.*, 1993; Valiev *et al.*, 2000; Ovid'ko, 2004; Valiev, 2004), and amorphous grain boundaries

5.2 Specific structural features of nanocrystalline materials

(Figure 5.2d) (see Clarke, 1987; Sutton and Balluffi, 1996, and references therein). Low-angle periodic grain boundaries are characterized by low misorientation angles (tentatively <15°) between the adjacent grains and represent periodic arrays of lattice dislocations (Sutton and Balluffi, 1996). For illustration, a low-angle tilt boundary is shown as a periodic wall of edge lattice dislocations in Figure 5.2a. High-angle periodic boundaries represent narrow regions with thickness being around 1 nm and the atomic structure being very different from that of the adjacent crystalline grains (Sutton and Balluffi, 1996; Valiev et al., 2000; Ovid'ko, 2004; Valiev, 2004) (Figure 5.2b). High-angle periodic boundaries commonly contain periodic arrays of grain boundary dislocations (with small Burgers vectors being displacement-shift-complete lattice vectors which characterize grain boundary translation symmetries) (Figure 5.2b). These grain-boundary dislocations are called "equilibrium" or, in other terms, intrinsic grain-boundary dislocations and associated with the misorientation mismatch between the adjacent grains. Nonequilibrium grain boundaries are often formed in coarse-grained polycrystalline materials under high-strain deformation, in which case grain boundaries intensively absorb lattice dislocations (Valiev et al., 2000; Valiev, 2004). Besides intrinsic grain-boundary dislocations, high-density ensembles of extrinsic grain-boundary dislocations (resulting from splitting of absorbed lattice dislocations) and other defects are present in non-equilibrium grain boundaries (Figure 5.2c). Extrinsic grain-boundary dislocations, after some relaxation time interval, can disappear via annihilation of dislocations with opposite Burgers vectors and other transformations. As a result of the relaxation process whose intensity is controlled by temperature, non-equilibrium grain boundaries (Figure 5.2c) transform into their equilibrium states (Figure 5.2b). Amorphous grain boundaries (Figure 5.2d) frequently exist in covalently bonded ceramic materials where their chemical composition is different from that of the adjacent crystalline grains (see Clarke, 1987; Sutton and Balluffi, 1996, and references therein).

The fact that the structure and behavior of grain boundaries are very different from those of the bulk phase has been used in the pioneering paper (Birringer et al., 1984) reporting about fabrication of nanocrystalline materials and defining them as a new state of matter. Birringer et al. (1984) assumed that grain boundaries with their very specific structural and behavioral features cause very specific structural and behavioral features of nanocrystalline materials defined as polycrystals with finest grains and extremely high volume fraction of the grain boundary phase. In early studies, the characteristic grain size range of the nanocrystalline matter was defined to be from 3 nm to 20 nm (Birringer et al., 1984; Gleiter, 1989). Later, the notion of nanocrystalline materials has been extended to that of nanostructured materials as diverse solids – nanocrystalline materials, nanocomposites, semi-conductor quantum dots and wires, nanoparticles, carbon and non-carbon nanotubes, fullerenes, nanoscale multilayers, and others – with structural elements (commonly crystallites) having at least one dimension smaller than 100 nm (see, e.g., Gleiter, 2000; Roco et al., 2000). In this book, we consider single-phase and composite nanocrystalline materials for structural applications.

In general, nanocrystalline materials have specific structural features causing their specific behavior and properties. We distinguish the following key specific features of single-phase nanocrystalline materials, differentiating them from conventional coarse-grained polycrystals.

(1) Grains have nanoscopic sizes. Grain size d does not exceed 100 nm.
(2) The volume fraction of the grain boundary phase is high in nanocrystalline materials compared to coarse-grained polycrystals. In particular, the volume fraction occupied by grain boundaries reaches a value of 10% or more in nanocrystalline materials with grain size lower than 10 nm.
(3) Grain boundaries are very short. Values of grain boundary length do not exceed 100 nm.
(4) The volume fraction of triple junctions of grain boundaries is high in nanocrystalline solids compared to coarse-grained polycrystals.
(5) The structure of grain boundaries in nanocrystalline materials fabricated by severe plastic deformation and other highly nonequilibrium methods is nonequilibrium with high densities of disorderedly arranged grain boundary defects.

Let us give a brief general discussion of these structural features and their effects on the properties of nanocrystalline materials. Since nanocrystalline materials contain nanoscale grains (see the structural feature (1)), processes sensitive to the length scale occur in these materials in a way different from that in conventional coarse-grained polycrystals. For instance, the generation and movement of lattice dislocations are hampered in nanoscale grains (Gryaznov *et al.*, 1989, 1991; Evans and Hirth, 1992; Siegel, 1994; Romanov, 1995; Chokshi and Kottada, 2006). Since nanocrystalline materials contain high-density ensembles of grain boundaries (see the structural feature (2)), grain boundaries crucially influence the properties of nanocrystalline materials. The fact that grain boundaries are very short in nanocrystalline materials (see the structural feature (3)) has several consequences for their structural geometry and behavior. We will discuss them below.

First of all, nanosized grain boundaries in nanocrystalline materials with the finest grains are characterized by length values close to those of periods that characterize periodic grain-boundary structures in coarse-grained polycrystals. In these circumstances, nanosized grain boundaries in their equilibrium (low-energy) state in nanocrystalline materials often can not be defined as periodic; they are either quasiperiodic (Ovid'ko, 2000) or have the structure with inhomogeneities of the grain-boundary dislocation distribution in the vicinities of triple junctions (Nazarov *et al.*, 2003). Also, nanosized grain boundaries easily undergo structural transformations related to changes in their length and form. In particular, the enhanced local migration of grain boundaries facilitates grain growth and grain-boundary sliding processes in nanocrystalline materials. On the other hand, however, there is a strong elastic interaction between neighboring nanosized grain boundaries, because (extremely short) distances between them are close to the characteristic scales of their stress fields. With this interaction, grain-boundary structures are formed that minimize the elastic energy of grain-boundary ensembles in nanocrystalline materials. The low-energy ensembles exhibit a certain structural stability, because there is an energy barrier needed to destroy low-energy grain-boundary structures. This factor hampers grain growth and other processes associated with structural transformations of grain boundaries in nanocrystalline materials (see a discussion in Chapter 3).

Now let us discuss the structural feature (4) of nanocrystalline materials. Grain boundaries form triple junctions distributed with high densities in nanocrystalline materials

5.2 Specific structural features of nanocrystalline materials

(Figure 5.1). At the same time, triple junctions have both the structure and properties different from those of grain boundaries that they adjoin (King, 1999). In particular, triple junctions often contain defects that influence the structure of adjacent grain boundaries (Gutkin et al., 2004a,b; Gutkin and Ovid'ko, 2004a) and, as shown in experiments (Palumbo and Aust, 1989; Yin et al., 1997; Kumar et al., 2003b), theoretical models (Osipov and Ovid'ko, 1992; Gutkin and Ovid'ko, 1994; Ovid'ko and Sheinerman, 2004a) and computer simulations (Farkas et al., 2002; Samaras et al., 2002) are capable of enhancing the structural and chemical transformations in the vicinities of triple junctions. Also, following experimental data (Bokstein et al., 2001), the diffusion along triple junction tubes is much faster than the diffusion along grain-boundary planes and the bulk diffusion. In the case of plastic flow occurring through grain-boundary sliding, triple junctions serve as its obstacles and thereby control this deformation mode. In the context discussed, the effects of triple junctions are essential in nanocrystalline materials where their volume fraction is very high.

Also, conditions of fabrications of single-phase nanocrystalline materials strongly affect their structure and behavior. Nanocrystalline materials are often fabricated by highly nonequilibrium methods causing the formation of so-called extrinsic grain boundary defects, that is, defects which are not geometrically necessary at grain boundaries. In these circumstances, the density of nonequilibrium grain boundaries (Figure 5.2c) in nanocrystalline materials is much higher than that in conventional coarse-grained polycrystals (see structural feature (5) of nanocrystalline materials). Nonequilibrium grain boundaries with high densities of extrinsic defects (Figure 5.2c) exhibit specific properties that are different from those of equilibrium grain boundaries containing intrinsic defects only (Valiev et al., 2000; Ovid'ko, 2004; Valiev, 2004). It is believed that the nonequilibrium grain boundaries are responsible for the unusual diffusion and deformation behaviors of nanocrystalline materials produced by severe plastic deformation (Valiev et al., 2000; Kolobov et al., 1999, 2000; Kolobov, 2002; Nazarov, 2003).

To summarize, nanocrystalline materials are characterized by the specific structural features (1)–(5) that strongly affect their mechanical and other properties. In particular, these structural features essentially influence deformation and fracture processes in the nanocrystalline matter. Theoretical models of the relationships between the structure and mechanical properties of the nanocrystalline matter will be considered in the next sections of this chapter. Now let us turn to a discussion of the specific structural features of composite nanocrystalline materials.

5.2.2 Specific structural features of composite nanocrystalline materials

Composite nanocrystalline materials are compositionally inhomogeneous solids containing nanoscale grains (nanocrystallites) of at least one component phase. Typical composite nanocrystalline bulk structures are shown in Figure 5.3. They are a nanocrystalline composite consisting of approximately equiaxed nanoscale grains of different phases (Figure 5.3a); a nanocrystalline composite consisting of grains of one phase, divided by grain boundaries with different chemical composition (second phase) (Figure 5.3b); a nanocrystalline composite consisting of large grains of one phase, embedded into a nanocrystalline matrix of

214   5 Mechanical properties of structural nanocrystalline materials

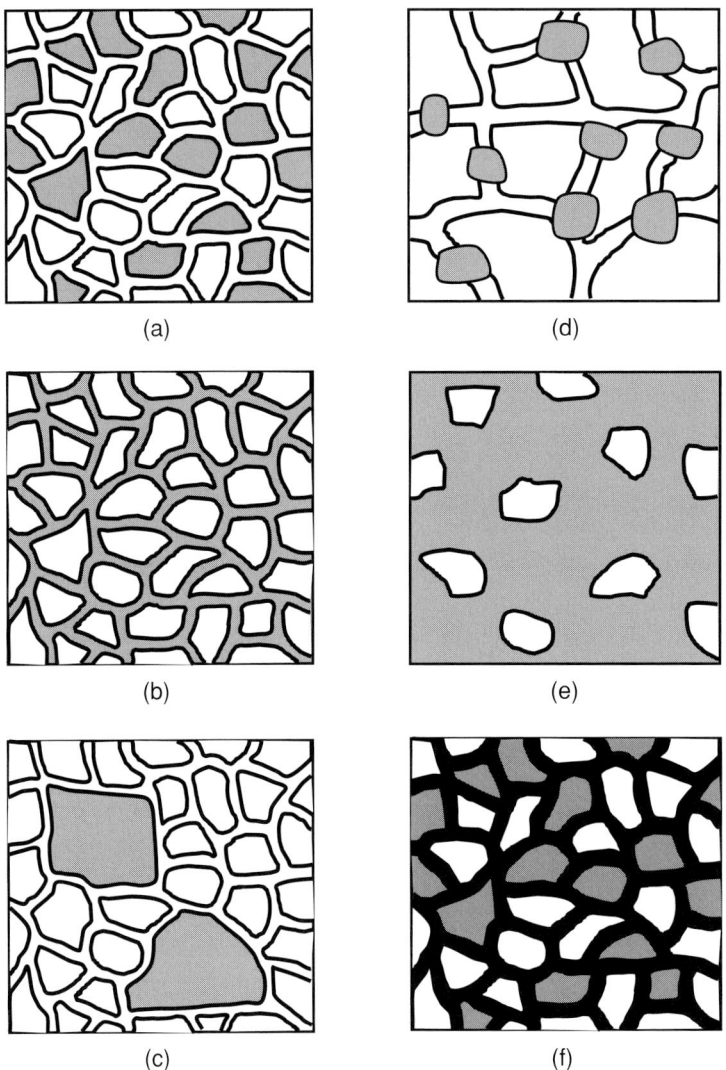

Figure 5.3. Typical composite nanocrystalline bulk structures. (a) Nanocrystalline composite materials consisting of tentatively equiaxed nanoscale grains of different phases. (b) Nanocrystalline materials consisting of grains of one phase, divided by grain boundaries with different chemical composition (second phase). (c) Nanocrystalline composite materials consisting of large grains of one phase, embedded into a nanocrystalline matrix of the second phase. (d) Nanocrystalline composite materials consisting of large grains of one phase with nanoscale particles of the second phase located at grain boundaries between large grains. (e) Nanocrystalline composite materials consisting of nanocrystallites of one phase, embedded into the amorphous matrix of the second phase. (f) Nanocrystalline materials consisting of grains of two phases, divided by grain boundaries with different chemical composition (third phase).

5.2 Specific structural features of nanocrystalline materials

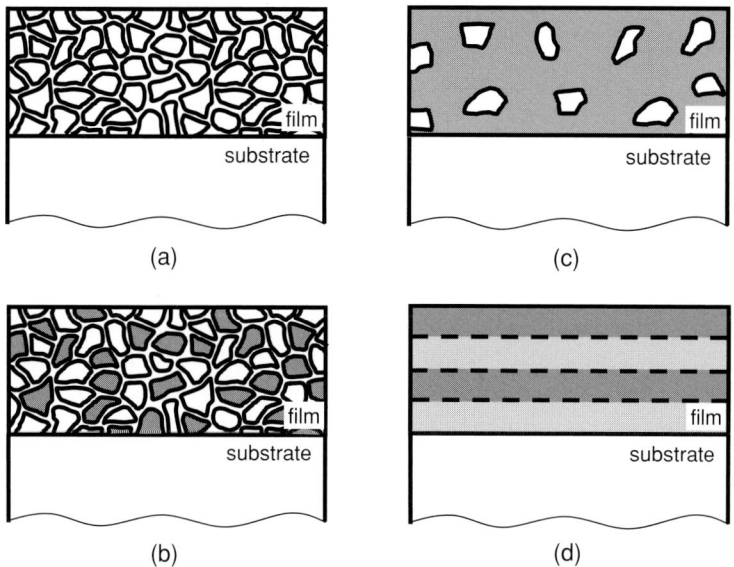

Figure 5.4. Typical nanocrystalline films and coatings. (a) Single-phase nanocrystalline film (coating) with tentatively equiaxed nanoscale grains on substrate. (b) Composite nanocrystalline film (coating) consisting of tentatively equiaxed nanoscale grains of two phases on substrate. (c) Composite nanocrystalline film (coating) consisting of second-phase nanoparticles embedded into a film matrix on substrate. (d) Multilayered coating consisting of alternate nanocrystalline layers with different chemical compositions on substrate.

the second phase (Figure 5.3c); a nanocrystalline composite consisting of large grains of one phase with nanoscale particles of the second phase located at grain boundaries between large grains (Figure 5.3d); a nanocrystalline composite consisting of nanocrystallites of one phase, embedded into the amorphous matrix of the second phase (Figure 5.3e); a nanocrystalline composite consisting of grains of two phases, divided by grain boundaries with a different chemical composition (third phase) (Figure 5.3f). These and other typical classes of nanocomposite solids are discussed in detail by Niihara *et al.* (1993) and Kuntz *et al.* (2004).

Besides composite nanocrystalline bulk materials (Figure 5.3), there are layered composite nanocrystalline structures including films and coatings deposited on substrates. A substrate material is different from material(s) of either thin film or thick coating deposited onto the substrate. As a corollary, layered nanocrystalline structures are composite, if even a thin film or thick coating is made of single-phase material. Typical layered composite nanocrystalline structures are shown in Figure 5.4. They are a single-phase nanocrystalline film/coating with approximately equiaxed nanoscale grains on substrate (second phase) (Figure 5.4a); a composite nanocrystalline film/coating consisting of tentatively equiaxed nanoscale grains of two phases on substrate (third phase) (Figure 5.4b); a composite nanocrystalline film/coating consisting of second-phase nanoparticles embedded into a film matrix on substrate (third phase) (Figure 5.4c); a multilayered coating consisting of alternate nanocrystalline layers with different chemical compositions on substrate (Figure 5.4d).

In composite nanocrystalline materials (Figures 5.3 and 5.4), interphase and grain boundaries represent interfaces that cause the specific effects on mechanical and other properties of these materials. Interphase boundaries are always characterized by geometric or, in other words, phase mismatch between different crystalline lattices of different phases matched at these boundaries. In most cases, interphase boundaries carry both misorientation and phase mismatches between adjacent crystallites matched at these boundaries. Crystalline structure and phase content of adjacent solids cause the existence of geometrically necessary or, in other terms, intrinsic defects at solid/solid interfaces. In particular, these intrinsic defects are associated with misorientation inherent to grain boundaries, and provide a partial relaxation of misfit stresses occurring owing to phase mismatch between different crystalline lattices of different phases matched at interphase boundaries. The intrinsic defects exist also at solid/solid interfaces in conventional materials (with coarse-grained structure), where interfaces are approximately plane and have large dimensions. Interphase boundaries in conventional materials are commonly modeled as plane interfaces of infinite extent. In doing so, ensembles of intrinsic interfacial defects are modeled as infinite, periodically arranged structures. However, these theoretical representations on interphase boundaries and interfacial defects are oversimplified in the case of nanocrystalline composite materials in which most interphase boundaries are curved and short (see Figures 5.3 and 5.4). Owing to curved geometry of interphase boundaries in nanocrystalline composite materials, high-density ensembles of their triple junctions exist in such materials. They are treated as defects of the special type that are capable of strongly influencing the properties of nanocrystalline composite materials.

Parallel with structural geometry and phase content, conditions of fabrications of nanocrystalline composite materials strongly affect interfacial defect structure of interphase boundaries. As with single-phase nanocrystalline materials, nanocrystalline composite materials are often fabricated by highly nonequilibrium methods causing the formation of extrinsic interfacial defects, that is, defects which are not geometrically necessary at interphase boundaries. By analogy with the notion of nonequilibrium grain boundaries (Valiev et al., 2000; Valiev, 2004), it is natural to define nonequilibrium interphase boundaries as those containing high-density ensembles of extrinsic defects. Such nonequilibrium interphase boundaries are expected to exist in nanocomposites fabricated at highly nonequilibrium conditions.

In general, theoretical representations on the structural features of interphase boundaries and their role in both plastic deformation and fracture processes in nanocomposite solids are not well developed, compared to those in the case of grain boundaries in single-phase nanocrystalline materials. Also, molecular dynamics simulations of the deformation behavior of nanocomposite solids are in their infancy, because of serious difficulties in correct selection of prescribed interatomic interaction potentials for heterophase structures. In these circumstances, our following consideration will be focused on mostly the theory and computer simulations of mechanical properties of single-phase nanocrystalline materials. The theory of mechanical properties of nanocomposite materials will be discussed briefly.

5.3 Basic concepts on plastic deformation processes in nanocrystalline materials

Nanocrystalline materials – solids composed of nanoscale grains (nanocrystallites with characteristic sizes not exceeding 100 nm) divided by grain and interphase boundaries (Figures 5.1, 5.3 and 5.4) – can exhibit outstanding mechanical properties. Experimental research in this area has been overviewed in Chapter 4 where a lot of experimental facts were presented and discussed. Because of both the high complexity of plastic deformation and fracture processes in nanocrystalline materials and limitations in experimental characterization at the nanoscale level, there are controversial experimental data as well as experimental facts whose interpretations are vague. Nevertheless, the aims of this chapter focus on the theory of the outstanding mechanical properties of nanocrystalline materials let us list the key facts and tendencies experimentally observed in mechanically loaded nanocrystalline materials (for details, see Chapter 4) and recognized as the experimental basis for theoretical concepts in this area. Such experimentally revealed facts and tendencies can be divided into the two categories associated with the macroscopic and nanoscale levels. The experimental facts related to the macroscopic deformation behavior are as follows.

(i) The yield stress and microhardness of nanocrystalline solids are extremely high. The yield stress and microhardness of nanocrystalline materials are commonly 2–10 times larger than the corresponding mechanical characteristics of coarse-grained polycrystalline materials with the same chemical composition. Record values of microhardness ($H_V \approx 70-100$ GPa) of nanocrystalline composite coatings are of the same order as microhardness ($H_V \approx 70-90$ GPa) of diamond (for details, see Chapter 4).

(ii) In the range of small grain sizes d being of the order of 10 nm, the dependence of the yield stress τ on d deviates from the classical Hall–Petch relationship given by the formula

$$\tau = \tau_0 + kd^{-1/2}, \tag{5.1}$$

with τ_0 and k being constant parameters. In doing so, the $\tau(d)$ dependence in nanocrystalline materials often shows two different behaviors, depending on their mode of processing. In the range of small grain sizes, heat-treated materials exhibit "inverse" Hall–Petch behavior (softening with reduction of grain size), while the yield stress or hardness of as-prepared materials slightly increases or saturates at grain size $d \leq 10$ nm showing little or no "inverse" Hall–Petch behavior (for details, see Weertman and Sanders, 1994; Volpp et al., 1997; and Chapter 4). (Notice that the aforesaid is not always true. For instance, laser-ablated nanocrystalline films show an inverse Hall–Petch effect (Koch and Narayan, 2001).)

(iii) Nanocrystalline materials commonly exhibit low tensile ductility at room temperature. Recently, however, nanocrystalline materials with very low porosity (Champion et al., 2003; Kumar et al., 2003b; Youssef et al., 2004, 2005, 2006), nanocrystalline materials with dendrite-like inclusions (He et al., 2003; Ma, 2003; He et al., 2004), and nanomaterials with bimodal structure (consisting of nano- and microsized grains)

(Tellkamp *et al.*, 2001; Wang *et al.*, 2002; Sergueeva *et al.*, 2004; Wang and Ma, 2004a,b; Han *et al.*, 2005) showing substantial ductility have been fabricated.

(iv) Some nanocrystalline materials show (tensile) superplasticity commonly at lower temperatures and higher strain rates compared to their coarse-grained counterparts (Mayo 1997, 1998; Mishra *et al.*, 1998; McFadden *et al.*, 1999; Islamgaliev *et al.*, 2001; Mishra *et al.*, 2001; Valiev *et al.*, 2001; Mukherjee 2002a,b; Valiev *et al.*, 2002; Zhan *et al.*, 2003). Superplastic deformation in nanocrystalline materials is characterized by very high values of the flow stress, strengthening at the first extended stage of deformation and plastic strain-to-failure ε_f reaching values of hundreds per cent.

The above experimental facts are indicative of the unique macroscopic deformation behavior of nanocrystalline materials. The effects of the structural features of nanocrystalline materials on their outstanding macroscopic mechanical characteristics can be extracted from the experimentally detected phenomena occurring in mechanically loaded nanocrystalline materials at the nano-scale level. The key experimental facts and tendencies observed at the nanoscale level are listed in the following points.

(v) Lattice dislocations – basic carriers of plastic flow in conventional coarse-grained polycrystals – exist in nanocrystalline materials with intermediate grains having the mean grain size d in the range from some critical value d_c ($= 10$–30 nm) to 100 nm, the upper limit of grains inherent to nanocrystalline materials. In nanocrystalline materials with the finest grains having the mean grain size $d < d_c$ ($= 10$–30 nm), lattice dislocations cease to be experimentally observed (for details, see Chapter 4).

(vi) Crystal lattice rotations occur in nano-scale grains in deformed nanocrystalline materials (Ke *et al.*, 1995; Mukherjee, 2002a,b; Murayama *et al.*, 2002; Shan *et al.*, 2004).

(vii) Plastic flow is often localized in shear bands (see Nieman *et al.*, 1991; Carsley *et al.*, 1995; Witney *et al.*, 1995; Andrievskii *et al.*, 1997; Andrievskii *et al.*, 1998; Wei *et al.*, 2002; Jia *et al.*, 2003; and other papers cited in Chapter 4). Structural characterization of shear bands in nanocrystalline Fe shows its brick-like grain structure with grain boundaries being approximately parallel and perpendicular to the shear direction (Wei *et al.*, 2002; Jia *et al.*, 2003).

(viii) Grain-boundary sliding essentially contributes to superplastic deformation of nanocrystalline materials (see Mukherjee, 2002a,b).

(ix) Grain boundaries emit perfect and partial lattice dislocations into grain interiors of deformed nanocrystalline materials (He and Lavernia, 2001; Mukherjee 2002a,b; Chen *et al.*, 2003; Kumar *et al.*, 2003b; Liao *et al.*, 2003a,b, 2004b). Also, deformation twins (associated with partial dislocation activity) have been experimentally observed in nanocrystalline materials (Liao *et al.*, 2003a,b, 2004a; Zhu *et al.*, 2005a; Wang *et al.*, 2005).

(x) With reducing grain size, there is a shift in fracture mode, from ductile (for nanocrystalline Ni with mean grain size around 44 nm) to brittle fracture (for nanocrystalline Ni–15%Fe with mean grain size around 9 nm) (Li and Ebrahimi, 2004).

The experimental data on the macroscopic deformation behavior of nanocrystalline materials, listed in points (i) and (ii), have become known since the late 1980s; see the articles

5.3 Basic concepts on plastic deformation processes

(Siegel and Fougere, 1995; Hahn and Padmanabhan, 1995) reviewing early studies in this area. Very high values of the yield stress (see point (i)) are explained to be related to the effect of grain boundaries as obstacles for movement of lattice dislocations being carriers of plastic flow in grain interiors. Besides, the dislocation generation by Frank–Read and other conventional dislocation sources is retarded. That is, grain refinement causes lattice dislocation movement and generation to be hampered, in which case the yield stress increases. This statement is in agreement with classical representations (Li and Chou, 1970; Armstrong and Head, 1965; Evans and Hirth, 1992) on the strengthening effect of grain boundaries in conventional coarse-grained polycrystals, described by the standard Hall–Petch relationship (1). However, mechanically loaded nanocrystalline materials exhibit deviations from the conventional grain-size strength relationship (1) at very small grains with $d \leq 30$ nm; see point (ii). This abnormal Hall–Petch relationship is a remarkable manifestation of specific behavioral features of nanocrystalline solids.

Much effort has been spent on theoretically describing the abnormal Hall–Petch relationship in nanocrystalline materials. In early theoretical studies, models of nanocrystalline materials as two-phase composites with nanograin interiors and grain boundaries being component phases were dominant (see Gryaznov *et al.*, 1993; Gutkin and Ovid'ko, 1993; Carsley *et al.*, 1995; Ovid'ko, 1997; Wang *et al.*, 1995; Kim, 1998; Konstantinidis and Aifantis, 1998; Kim *et al.*, 2000). They describe the yield stress τ and account for the macroscale phenomena (i) and (ii) in nanocrystalline materials, using the so-called rule of mixture. In this approach, the yield-stress τ of a nanocrystalline material is some weighted sum (mixture) of the yield stresses characterizing the grain-interior and grain-boundary phases, which strongly depends on the volume fraction of the grain-boundary phase and, therefore, the grain size d. The yield stress of the grain-boundary phase is assumed to be lower than that of the grain-interior phase, in which case the rule of mixture describes the deviations from the conventional Hall–Petch relationship in accordance with experimental data. Such models deal with macroscopically averaged characteristics and use the volume fraction of the grain-boundary phase as the control parameter. They do not account for the nanoscale phenomena (v)–(x) related to evolution of defects and grain-boundary structures, and can not be effective in explanation of the new macroscale phenomena (iii) and (iv) that definitely need to be described in terms of defects and transformations of grain boundary structures. In this context, the models based on the "rule-of-mixture" approach are too approximate. These models have been recently reviewed (Gutkin *et al.*, 2003a; Gutkin and Ovid'ko, 2004a,b).

In recent years, the problem of deviations from Hall–Petch behavior and, in general, unusual deformation characteristics of nanocrystalline materials in the range of finest grains was attacked by both molecular dynamics approach and theoretical models that describe physical mechanisms of plastic flow in nanocrystalline materials as those associated with transformations of defects (lattice dislocations, grain-boundary dislocations, vacancies) and grain-boundary structures. Following molecular dynamics simulations (Schiotz *et al.*, 1998, 1999; Van Swygenhoven *et al.*, 1999a,b; Van Swygenhoven and Derlet, 2001; Frozeth *et al.*, 2004a,b; Schiotz, 2004), at some critical grain size d_c ($= 10$–20 nm, depending on material), there is a shift in plastic deformation mechanism, from conventional lattice dislocation slip to deformation mechanisms mediated by grain boundaries. For instance, structural transformations of nanocrystalline Ni under mechanical load involve mostly grain-boundary processes at $d < 10$ nm (Van Swygenhoven *et al.*, 1999a,b; Van Swygenhoven and Derlet, 2001). For larger d, both grain-boundary processes and dislocation activity in grain interiors carry plastic deformation. In doing so, lattice dislocation slip in nanoscale grain interiors is conducted by partial and perfect dislocations emitted from grain boundaries and absorbed by them. This deformation regime is called the transition regime (Van Swygenhoven and Derlet, 2001) between grain-boundary-mediated plasticity in nanocrystalline materials with finest grains and conventional lattice dislocation slip which is dominant in coarse-grained polycrystals. Similar results of computer simulation of deformation processes have been reported for nanocrystalline Cu (Schiotz *et al.*, 1998, 1999; Schiotz, 2004) and Al (Yamakov *et al.*, 2003; Wolf *et al.*, 2005).

Parallel with computer analysis of a shift in deformation mechanism, the yield-stress dependence on grain size d has been revealed in simulations of deformation processes in nanocrystalline Cu at constant strain rate (Schiotz and Jacobsen, 2003). This dependence shows an approximately linear Hall–Petch relationship in the range of large grains and inverse Hall–Petch behavior in the range of finest grains with grain size lower than some critical grain size d_c around 15–20 nm. The maximum value of the yield stress calculated in the simulations (Schiotz and Jacobsen, 2003) corresponds to the critical grain size d_c at which the shift in plastic deformation mechanism comes into play. Also, parallel with computer analysis of a shift in deformation mechanism, the strain rate dependence on grain size d has been revealed in simulations of deformation processes in nanocrystalline Al at a constant applied stress (Yamakov *et al.*, 2003). This dependence has the minimum plastic strain rate at critical grain size $d_c = 18$ nm, at which the shift in plastic deformation mechanism comes into play.

In atomistic simulations, grain-boundary-mediated deformation mechanism in nanocrystalline materials is identified as grain-boundary sliding that commonly occurs through atomic shuffling events (see Van Swygenhoven *et al.*, 1999a,b; Van Swygenhoven and Derlet, 2001). In doing so, emission of partial and perfect dislocations from grain boundaries is interpreted as accommodation processes for grain-boundary sliding (Schiotz *et al.*, 1998, 1999; Van Swygenhoven *et al.*, 1999a,b; Van Swygenhoven and Derlet, 2001; Schiotz, 2004; Wolf *et al.*, 2005). Besides, as reported by Van Swygenhoven *et al.* (1999a,b) and Van Swygenhoven and Derlet (2001), grain boundary deformation is accommodated by triple-junction migration, grain-boundary migration and free-volume movement in grain boundaries (see also Section 5.6).

Now let us discuss theoretical models that describe physical mechanisms of plastic flow in nanocrystalline materials and deal with evolution of defects (lattice dislocations, grain-boundary dislocations, vacancies) and transformations of grain boundary structure that occur in mechanically loaded nanocrystalline materials. In the framework of this approach, several models exploit the idea on lattice dislocation motion in grain interiors as the basic deformation mechanism in nanocrystalline materials, which is modified (compared to that in coarse-grained polycrystals) owing to nanoscale effects. At the same time, the generic idea of most models is that the grain-boundary phase provides the effective action of deformation mechanism(s) in nanocrystalline materials, which is (are) different from the lattice dislocation mechanism realized in conventional coarse-grained polycrystals. In context of this generic idea being in agreement with both experimental data (Chapter 4) and results of computer simulations (Schiotz *et al.*, 1998, 1999; Van Swygenhoven *et al.*, 1999a,b; Van Swygenhoven and Derlet, 2001; Frozeth *et al.*, 2004a,b; Schiotz, 2004), the dominant role of the lattice dislocation mechanism causes the grain refinement to strengthen a coarse-grained material, in which case a classical Hall–Petch relationship (5.1) is valid. At the same time, when the deformation mechanisms associated with the active role of grain boundaries effectively come into play, the grain refinement will weaken a specimen; it is the case for nanocrystalline materials with small grains ($d \leq 30$nm). The theoretical models under consideration are distinguished by their identification of the deformation mechanisms inherent to grain boundaries and their description of the competition between these mechanisms and the conventional lattice dislocation slip. The following plastic deformation mechanisms commonly are treated as those associated with the active role of grain boundaries and the nanoscale structure, effectively competing with the lattice-dislocation mechanism in nanocrystalline materials: grain-boundary sliding, grain-boundary diffusional creep, triple-junction diffusional creep and rotational deformation (occurring through transformations of grain-boundary disclinations and dislocations).

In fact, there are many theoretical models of the abnormal Hall–Petch dependence which are based on different deformation mechanisms and, at the same time, are in agreement with the corresponding experimental data from nanocrystalline materials. As a corollary, a comparison of theoretical results with the experimentally observed abnormal Hall–Petch dependence does not allow one to select a theoretical concept from a variety of theoretical models correctly describing the same experimental data, using different ideas. In this situation, validity of a theoretical model should be proved by correspondence of its results with experimental data on the nanoscale phenomena (see points (v)–(x)) that have been detected recently. With selection of the effective models, their representations should be used in explanation of the new macroscale phenomena (iii) and (iv) that are highly interesting from a fundamental viewpoint and of particular importance for wide applications. In this context, in the next sections of this chapter, we will pay special attention to new theoretical models of plastic deformation mechanisms in nanocrystalline materials, that account for the nanoscale phenomena (v)–(x). Also, we will give a brief overview of theoretical concepts and models describing enhanced diffusion in nanocrystalline solids which effectively contributes to plastic and superplastic deformation processes in these materials.

5.4 Nanoscale and interface effects on lattice dislocation slip

In this section, we will consider theoretical models based on the idea of conventional lattice dislocation slip in nanograin interiors as the dominant deformation mechanism in nanocrystalline materials, as with coarse-grained polycrystals. In the framework of this approach, the experimentally documented deviations from the conventional deformation behavior (in particular, the Hall–Petch relationship) in nanocrystalline materials are explained as those related to the influence of the grain-size reduction and high-density ensembles of grain boundaries on the formation of lattice dislocation pile-ups in grain interiors and the penetration of lattice dislocations through grain boundaries in such materials.

Let us start our analysis of this approach with a brief discussion of the models describing the classical Hall–Petch relationship (5.1) in coarse-grained polycrystals. Most of these models can be rationalized in terms of a dislocation pile-up model. These are reviewed in detail by Li and Chou (1970). In deriving the Hall–Petch relation, the role of grain boundaries as a barrier to the dislocation motion is considered in various models. In one type of the models (Armstrong and Head, 1965; Pande and Masumura, 1984), a grain boundary acts as a barrier to pile up the dislocations, causing stresses to concentrate and activating dislocation sources in the neighboring grains, thus initiating the slip from grain to grain. In the other type of the models (Evans and Hirth, 1992; Pande et al., 1993), the grain boundaries are regarded as dislocation barriers limiting the mean free path of the dislocations, thereby increasing strain hardening, resulting in a Hall–Petch-type relation. A review of the various competing theories of strengthening by grain refinement has been discussed by several workers (for a more recent survey, see Lasalmonie and Strudel (1986)). It is clear that a variety of processes, both dislocation and non-dislocation based, could be postulated. It is possible that several of these processes could compete or reinforce the deformation process.

Now let us turn to a discussion of the models focused on the lattice dislocation mechanism of plastic flow in nanocrystalline materials. First, let us consider the dislocation model (Pande and Masumura, 1996). The assumption made in this model is that the classical Hall–Petch dislocation pile-up model is still dominant with the sole exception that the analysis must take into account the fact that in the nanometer-size grains, the number of dislocations within a grain cannot be very large. Further, at still smaller grain sizes, this mechanism should cease when there are only two dislocations in the pile-up.

By considering the conventional Hall–Petch model, Pande and Masumura (1996) showed that a dislocation theory for the Hall–Petch effect gives a linear dependence (5.1) of the yield stress τ on $d^{-1/2}$ only when there is a large number of dislocations in a pile-up and plasticity is not source limited. In this regime, the yield stress increases as grain size d decreases because the pile-ups contain fewer dislocations, the stress concentration at the head falls, and a larger applied stress is required to compensate. When the number of dislocations falls to one, no further increase in the yield stress is possible and it saturates. If, however, sources must operate in each grain, an additional component of the yield stress exists of at least Gb/d (where G is the shear modulus, and b is the Burgers vector magnitude characterizing a perfect lattice dislocation), and the yield stress should rise as $d^{-1/2}$. Thus from these arguments, at a small grain size, either the yield stress should rise faster than $d^{-1/2}$ or it should saturate, but it should not decrease. Chou (1967) considered a pile-up of n straight dislocations with the leading dislocation of Burgers vector mb being locked at a grain boundary, an obstacle for lattice slip. The $(n-1)$ free dislocations of Burgers vector b are pushed toward the locked dislocation by an applied stress. If the number of dislocations n in a pile-up is not too large, the length of the pile-up L is not linear in n but an additional term is necessary. Chou (1967) gives the relation

$$L \cong \frac{A}{2\tau} \left| 4\left(n + m - 1 - 2i_1 \left(\frac{2n}{3}\right)^{1/3}\right)\right|, \qquad (5.2)$$

5.4 Nanoscale and interface effects on lattice dislocation slip

where $i_1 = (6)^{1/3} = 1.85575$, $A = Gb/2\pi$, for edge dislocations, and $A = Gb/2\pi(1-\nu)$, for screw dislocations; G is the shear modulus, and ν is the Poisson ratio. Pande and Masumura (1996) give an improved expression

$$L \cong \frac{A}{2\tau}\left[2(n+m-1)^{1/2} - \frac{i_1+\varepsilon}{12^{1/3}(n+m-1)^{1/6}}i_1\right], \tag{5.3}$$

where ε is a small correction term ($\varepsilon \ll 1$) and can be neglected. They find that for small grain sizes there are additional terms to the Hall–Petch relation

$$\tau = Md^{-1/2} + N(d^{-1/2})^{5/3} + P(d^{-1/2})^{7/3}, \tag{5.4}$$

where M, N and P are constant parameters. This model recovers the classical Hall–Petch at large grain sizes but for smaller grain sizes the τ levels off. This model therefore cannot explain a drop in τ.

Lian et al. (1993) and Nazarov (1996a,b) have developed models similar to that of Pande and Masumura (1996), focusing on the influence of the grain size d on the parameters of lattice dislocation pile-ups in grain interiors. Malygin (1995) has suggested a theory based on a lattice dislocation mechanism, with the effects of grain boundaries taken into account. The dislocation density $\rho(d)$ at any grain size, d, is related in the usual fashion to the square of yield stress and an expression is obtained that connects ρ to the grain size. The expression is based on the assumption that grain boundaries act predominately as sinks for dislocations (just the opposite to that used by Li (1963), who postulated that grain boundaries could be sources for dislocation generation). In Malygin's model, as the grains become finer and finer, more and more dislocations are absorbed by the grain boundaries leading ultimately to a drop in dislocation density and hence in the flow stress since the two are directly related as mentioned above. The model is attractive, and should be considered further. At present, we merely point out two problems with the model. First, it is doubtful if the dislocations play the same role whether the grains are large or small. It is more likely that dislocations in ultrafine grains if present at all, are confined to grain boundaries (Gryaznov et al., 1989, 1991; Evans and Hirth, 1992; Siegel, 1994; Romanov, 1995). Second, in the model (Malygin, 1995), the stress calculated is a work-hardened flow stress rather than a yield stress.

Lu and Sui (1993) assume that both the energy and free volume of grain boundaries decrease with a reduction of the grain size d. This gives rise to an enhancement of lattice dislocation penetration through the grain boundaries and the corresponding softening of nanocrystalline materials. Following Scattergood and Koch (1992), the yield stress of fine-grained materials is controlled by the intersection of mobile lattice dislocations with the dislocation networks at grain boundaries. In this context, with the dislocation line tension assumed to be size dependent, there is a critical grain size that corresponds to a transition from cutting to Orowan bypassing of the dislocation network. This critical grain size characterizes the experimentally detected transition from the conventional to inverse Hall–Petch relationship with reducing the grain size d.

Zaichenko and Glezer (1997) have described the role of triple-junction disclinations – rotational defects formed at triple junctions of grain boundaries – as sinks and sources of the lattice dislocations moving in grain interiors and causing the plastic flow in nanocrystalline materials. Following the model (Zaichenko and Glezer, 1997), the lattice dislocations are emitted and absorbed at the opposite triple junctions of grain boundaries, containing disclinations (formed during fabrication of nanocrystalline materials). In doing so, the yield stress is caused by the interaction between the grain-boundary disclinations and the lattice dislocations moving in grain interiors. The disclinations move in the vicinities of their initial positions (corresponding to the non-deformed state of a material) at the grain boundary junctions when they emit and absorb the lattice dislocations. (This model in many aspects resembles a model (Ovid'ko, 1994) of dislocation–disclination reactions in plastically deformed amorphous solids.)

Despite the good correspondences between theoretically predicted $\tau(d)$ dependences and experimental data, all the above models based on the representations of the lattice dislocation mechanism of plastic flow in nanocrystalline materials meet the question if the lattice dislocations exist and play the same role in nanograin interiors as with conventional coarse grains. As pointed out in papers (Gryaznov *et al.*, 1989, 1991; Evans and Hirth, 1992; Siegel, 1994; Romanov, 1995), the existence of lattice dislocations in either free nanoparticles or nanograins composing nanocrystalline aggregate is energetically unfavorable, if their characteristic size, nanoparticle diameter or grain size, is lower than some critical size which depends on such material characteristics as the shear modulus and the resistance to dislocation motion. The dislocation instability in nanovolumes is related to the effect of the so-called image forces occurring owing to the elastic interaction between dislocations and either the free surface of a nanoparticle or grain boundaries adjacent to a nanograin (Gryaznov *et al.*, 1989, 1991; Nieh and Wadsworth, 1991; Evans and Hirth, 1992; Romanov, 1995). The paucity of mobile dislocations in nanograins has been well documented in electron-microscopy experiments (Chapter 4). The above question arises in the case of $d < 30$ nm, where the abnormal Hall–Petch relationship comes into play. However, the models based on the lattice dislocation slip are effective in explaining the deformation behavior of nanocrystalline materials with grain size $d \approx 30-100$ nm. In any case, grain boundaries serve as dominant sources and sinks of lattice dislocations in nanocrystalline materials. We will discuss this subject in Sections 5.6, 5.7, and 5.8.

5.5 Deformation modes associated with enhanced diffusion along grain boundaries and their triple junctions. Competition between deformation mechanisms. Effect of a distribution of grain size

In this section, the deformation mechanisms associated with enhanced diffusion along grain boundaries and their triple junctions in nanocrystalline materials will be considered, with the combined effects of the competition between various deformation mechanisms (depending on the grain size) and a distribution of grain size on the deformation behavior of nanocrystalline materials taken into account. First, let us consider the model (Chokshi *et al.*, 1989), based on the idea of grain-boundary diffusional creep (Coble creep) as the dominant deformation mode in nanocrystalline solids. Clearly, at sufficiently small grain sizes, the Hall–Petch model based upon the lattice dislocations may not be operative. In this region, Chokshi *et al.* (1989) have proposed the Coble creep to play the crucial role in

5.5 Deformation modes associated with enhanced diffusion

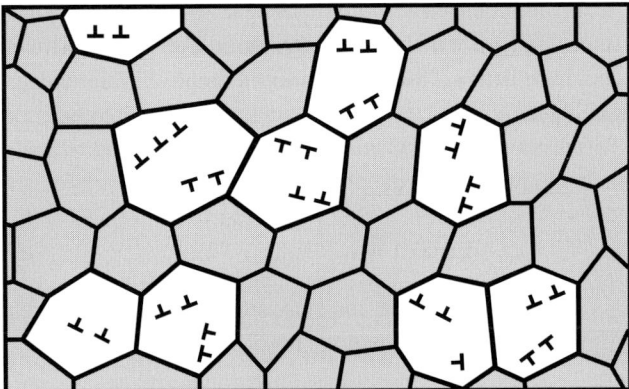

Figure 5.5. Competition between deformation mechanisms in nanocrystalline materials (schematically). Lattice dislocation slip and diffusional deformation modes occur in large grains (with dislocation signs) and small (shaded) grains, respectively.

plastic flow at room temperature. Coble creep occurs via dilatation-stress-driven transport of grain-boundary vacancies along boundaries from regions where tensile stresses exist towards compressed regions. The Coble creep rate $d\varepsilon/dt$ is given as (Coble, 1963; Chokshi and Kottada, 2006):

$$d\varepsilon/dt = 33 D_{gb} a^3 \delta \sigma d^{-3} / k_B T, \qquad (5.5)$$

where D_{gb} is the grain boundary diffusion coefficient, δ is the grain boundary width, a is the crystal lattice parameter, σ is the flow stress, k_B is the Boltzmann constant, T is the absolute temperature, and d is the grain size. In the case of Coble creep, the dependence of the yield stress τ on d is given as (Nieh and Wadsworth, 1991): $\tau \sim d^3 \propto [d^{-1/2}]^{-6}$, i.e. the τ vs. $d^{-1/2}$ curve falls very steeply as $d^{-1/2}$ increases. This exact dependence is not found experimentally, though the trend is correct (Nieh and Wadsworth, 1991).

Even if the Coble creep argument is valid for grain sizes $d < 20$ nm, one still has to explain the behavior in the 20–100 nm range. This is evidently the transition regime between the Hall–Petch and Coble creep-like behavior. The transition regime is effectively described with a distribution of grain size taken into account. It has been done by Masumura et al. (1998). They provided a unified model and developed an analytical expression for τ as a function of the inverse square root of d in a simple and approximate manner that could be compared with experimental data over a wide range of grain sizes. This model, based on the idea of competition between lattice dislocation slip, grain-boundary diffusional creep (Coble creep) and bulk diffusional creep (Figure 5.5), is summarized below. The assumptions in this model are follows:

(1) It is assumed that polycrystals with a relatively large average grain size obey the classical Hall–Petch relation (5.1).
(2) At the other extreme, for very small grain sizes, it is assumed that Coble creep is active, and that the τ vs. d relationship is given by

$$\tau_c = A/d + B d^3, \qquad (5.6)$$

where B is both temperature and strain-rate dependent. The additional term A/d (the threshold term) can be large if d is in the nanometer range. For intermediate grain sizes, both mechanisms might be active if the specimen has a range of grain size distribution.

(3) The statistical nature of the grain sizes in a polycrystal is taken into consideration by using an analysis similar to Kurzydlowski (1990). The volumes (v) of the grains are assumed to be log-normally distributed

$$f(v) = \frac{1}{v\sqrt{2\pi}s_{\ln v}} \exp\left[\frac{-(\ln v - m_{\ln v})}{s_{\ln v}^2}\right], \qquad (5.7)$$

where $m_{\ln v}$ and $s_{\ln v}$ are the mean value and standard deviation of $\ln v$, respectively, and where the m_v is the mean volume of all the grains,

$$m_v = \int_0^\infty v f(v) dv = \exp[m_{\ln v} + (s_{\ln v})^2], \qquad (5.8)$$

and can also be written as $m_v = g\bar{d}^3$, where $\bar{d}$ is mean grain size and with g being a geometrical shape factor considered for this analysis to be equal to 1.

(4) Finally, it is assumed that a grain size d^* exists at which value the classical Hall–Petch mechanism switches to the Coble creep mechanism, $\tau_{hp} = \tau_c$ at $d = d^*$. Using Eqs. (5.1) and (5.6), one has:

$$k(d^*)^{-1/2} = A/d^* + B(d^*)^3 \qquad (5.9)$$

from which d^* can be determined.

Then the yield stress after averaging is given as

$$\langle \tau - \tau_0 \rangle = F_{hp} + F_c, \qquad (5.10)$$

where

$$F_{hp} = \frac{1}{m_v} \int_{v^*}^\infty \tau_{hp} dv, \qquad (5.11)$$

and

$$F_c = \frac{1}{m_v} \int_{v^*}^\infty \tau_c dv. \qquad (5.12)$$

When $\tau_{hp} = \tau_c$, the critical volume is $v^* = k(d^*)^3$.

In the framework of the discussed approach, Masumura *et al.* (1998) have calculated the yield stress τ vs. $d^{-1/2}$ dependence in a wide range of grain size d from nanometers to microns in the exemplary case of NiP. This dependence has been compared with experimental data (MacMahon and Erb, 1989) and shown to be in good agreement.

Thus the model (Masumura *et al.*, 1998) uses conventional Hall–Petch strengthening for larger grains and Coble creep with a threshold stress for smaller grains. In a material with a distribution of grain sizes, a fraction of grains deforms by a lattice dislocation slip process and the rest by vacancy transport (Figure 5.5). As the average grain size decreases, the fraction deforming by slip decreases and the overall response changes from strengthening to softening. The exact form of the curve of yield stress against grain size depends on the relative values of Hall–Petch slope k, the conventional Coble constant B, the threshold constant A, and the width of the grain size distribution β.

The model (Masumura et al., 1998) is supported by results (Yamakov et al., 2002) of computer modeling of plastic deformation processes in nanocrystalline materials, indicating the essential contribution of Coble creep to these processes. Also, it should be noted that Masumura et al. (1998) have suggested a new general approach to the description of mechanical characteristics of deformed nanocrystalline materials, which takes into account a distribution in grain size and suggests simultaneous action of different deformation mechanisms in a mechanically loaded sample. This approach has been exploited in several theoretical works as a method for the description of competing deformation modes, which is more accurate than the "rule-of-mixture" approach.

For instance, with a distribution in grain size, a theoretical model (Fedorov et al., 2002) has recently been suggested describing a contribution of the deformation mechanism associated with triple-junction diffusion to plastic flow in nanocrystalline materials. In recent years, it has definitely been recognized that triple junctions of grain boundaries have structure and properties different from those of the grain boundaries that they adjoin (King, 1999; Rabukhin, 1986; Bokstein et al., 2001). In particular, as has been shown in experiments (Rabukhin, 1986; Bokstein et al., 2001), the triple-junction diffusion coefficient D_{tj} highly (by three or more orders) exceeds the grain-boundary diffusion coefficient D_{gb} in polycrystalline materials. Also, following experimental data (Rabukhin, 1986), creep associated with enhanced diffusion along triple junctions contributes to plastic flow of coarse-grained polycrystalline aluminum.

With these experimental data, Fedorov et al. (2002) suggested that triple-junction diffusion is capable of playing a very important role in plastically deformed nanocrystalline materials with finest grains, where the volume fraction of triple junctions of grain boundaries is extremely high. Actually, triple-junction tubes characterized by high values of diffusion coefficient form a continuous network distributed in all the volume of a nanocrystalline material. When a mechanical load is applied to a nanocrystalline specimen, atoms/vacancies are pumped with a high velocity along the continuous network of triple junctions (Figure 5.6). Atoms/vacancies move from one quadruple point to its neighboring quadruple point along a triple junction and then to a new neighboring quadruple point, and so on. (This process resembles a conventional grain-boundary diffusional creep – Coble creep – which consists of numerous events each being a stress-driven mass transfer from one triple junction to its neighboring junction along a grain-boundary plane bounded by these junctions.) Thus, the directional diffusional creep along the continuous triple-junction network occurs, resulting in a macroscopic plastic deformation of a nanocrystalline specimen. In the illustrative case shown in Figure 5.6, triple-junction tubes provide a diffusional mass transfer from lateral to upper and bottom free surfaces of a nanocrystalline specimen under tensile deformation.

Following Rabukhin (1986), the triple-junction diffusional creep is characterized by rather unusual dependence of plastic strain rate on the grain-size d. More precisely, the grain-size exponent is -4 in this case (Rabukhin, 1986), in contrast to grain-boundary diffusional

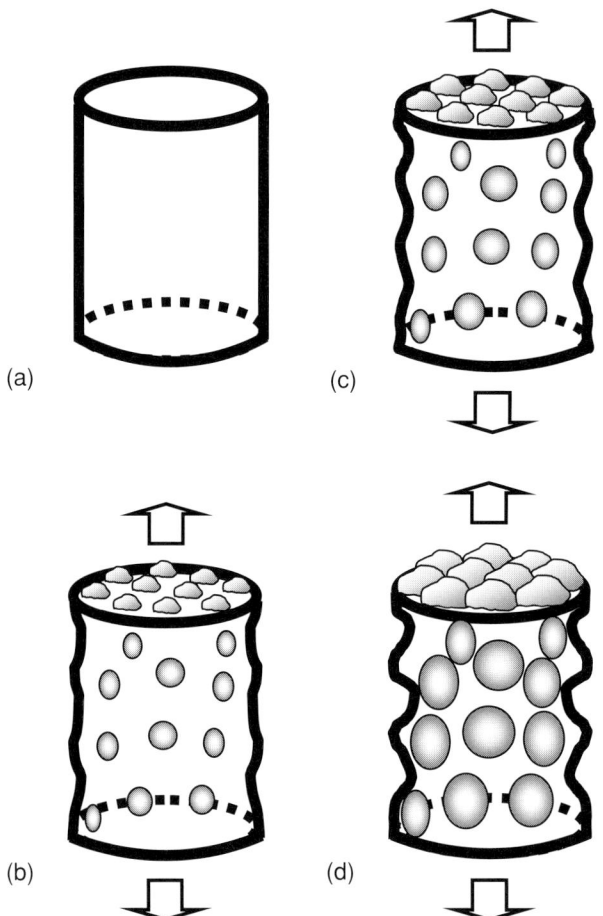

Figure 5.6. Deformation through triple-junction diffusional creep mode. (a) Initial state of a cylindrical nanocrystalline specimen. (b)–(d) Inhomogeneous creep occurs through enhanced diffusion along triple-junction tubes of a nanocrystalline specimen (schematically).

creep (Coble creep) and bulk diffusional creep characterized by grain-size exponents -3 and -2, respectively:

$$\dot{\varepsilon}_{tj} \propto D_{tj} \cdot d^{-4}, \quad \dot{\varepsilon}_{gb} \propto D_{gb} \cdot d^{-3}, \quad \dot{\varepsilon}_{bulk} \propto D_{bulk} \cdot d^{-2}. \quad (5.13)$$

In formula (5.13), ε_{tj}, ε_{gb} and ε_{bulk} are plastic strain rates that correspond to triple-junction diffusional creep, grain-boundary diffusional creep and bulk diffusional creep, respectively. D_{bulk} denotes the bulk self-diffusion coefficient.

Fedorov *et al.* (2002) suggested a theoretical model describing the yield-stress dependence on grain size in fine-grained materials, based upon competition between conventional dislocation slip, grain-boundary diffusional creep and triple-junction diffusional creep. The model (Fedorov *et al.*, 2002) takes into account a distribution in grain size, following the approach (Masumura *et al.*, 1998). As has been shown in (Fedorov *et al.*,

5.6 Grain-boundary sliding in nanocrystalline materials

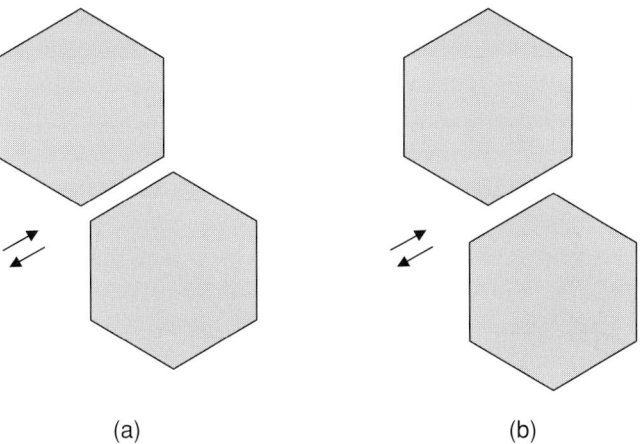

Figure 5.7. Grain-boundary sliding.

2002), the contribution of triple-junction diffusional creep increases with reduction of grain size, causing a negative slope of the Hall–Petch dependence in the range of small grains. These results have been compared with experimental data (Youngdahl *et al.*, 1997; Suryanarayana *et al.*, 1996; Sanders *et al.*, 1996) from copper and shown to be in rather good agreement.

In general, the role of grain-boundary diffusional creep (Coble creep) and triple-junction diffusional creep in plastic flow of nanocrystalline materials increases with reducing grain size and increasing temperature. At the same time, high temperature is capable of inducing grain growth which destroys the nanostructure and thereby masks deformation mechanisms operating in nanocrystalline materials with the finest grains. Besides, with reduction in grain size, alternative deformation mechanisms conducted by grain boundaries can come into play in nanocrystalline materials with the finest grains. We will discuss such deformation mechanisms in the next sections.

5.6 Grain-boundary sliding in nanocrystalline materials

Let us discuss grain-boundary sliding as the deformation mechanism that effectively operates and competes with both the conventional lattice dislocation slip and diffusional deformation modes in nanocrystalline materials with the finest grains. Grain-boundary sliding means a relative shear of neighboring grains that is localized in the boundaries between the grains (Figure 5.7). Commonly grain-boundary sliding occurs in high-angle grain boundaries in coarse-grained and nanocrystalline solids. From a microscopic viewpoint, grain-boundary sliding is treated as occurring by either movement of grain-boundary dislocations (Figure 5.8) or local shear events (Figure 5.9) (Sutton and Balluffi, 1996; Conrad and Narayan, 2000; Van Swygenhoven and Derlet, 2001; Padmanabhan and Gleiter, 2004). In particular, the sliding along grain boundaries with translationally ordered structures can be effectively conducted by movement of gliding grain-boundary dislocations with

230   5 Mechanical properties of structural nanocrystalline materials

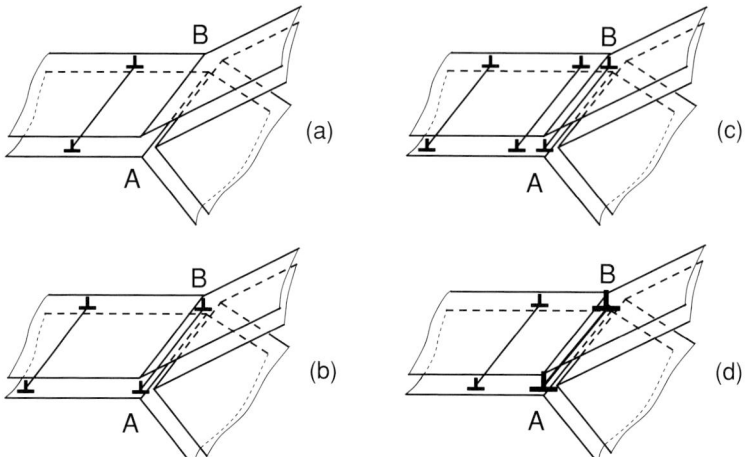

Figure 5.8. Glide of grain-boundary dislocations along a grain-boundary plane. (a)–(c) Grain-boundary dislocations glide along the grain-boundary plane towards triple junction AB. (d) Two head dislocations converge and form a new dislocation at triple junction AB.

Burgers vectors parallel with corresponding boundary planes (Figure 5.8) (Sutton and Balluffi, 1996). Such grain-boundary dislocations in boundaries with ordered structures are defined as defects violating their translation symmetries. They are characterized by Burgers vectors whose magnitudes are small (each magnitude ranges from tentatively $a/5$ to $a/3$, where a is the crystal-lattice parameter) and defined by translational symmetries of grain-boundary structures (Sutton and Balluffi, 1996). Glide of grain-boundary dislocations with Burgers vectors parallel with corresponding boundary planes is driven by the shear stress and carries plastic flow localized in the dislocation cores (Figure 5.8). The dislocation mode of grain boundary sliding is effectively described by means of the dislocation theory in solids; see, e.g., Gutkin and Ovid'ko (2004a).

Besides the dislocation theory (Sutton and Balluffi, 1996; Gutkin and Ovid'ko, 2004a) of grain-boundary sliding, there is an alternative view (Sutton and Balluffi, 1996; Conrad and Narayan, 2000; Van Swygenhoven and Derlet, 2001; Padmanabhan and Gleiter, 2004) that such sliding in coarse-grained and nanocrystalline solids occurs by local shear events thermally activated and driven by the shear stress. (Sometimes, grain-boundary sliding by local shear events is called grain-boundary shear (Conrad, 2003).) Local shear events represent either individual atomic jumps or local transformations of small groups of atoms in the grain-boundary phase. With the definition given by Padmanabhan and Gleiter (2004), local shear events occur in free-volume defects, nanometer-size spheroidal regions where, owing to the presence of an extra free volume, the shear resistance is less than in the rest of the grain boundary (Figure 5.9). This approach, in fact, is based on the theory (Argon, 1979) of thermally activated and shear-stress-driven local shear events carried by free-volume defects in plastically deformed metallic glasses. In this context, sliding by local shear events is expected to be dominant in nonequilibrium grain boundaries with disordered structures and amorphous grain boundaries in single-phase nanocrystalline materials and

5.6 Grain-boundary sliding in nanocrystalline materials

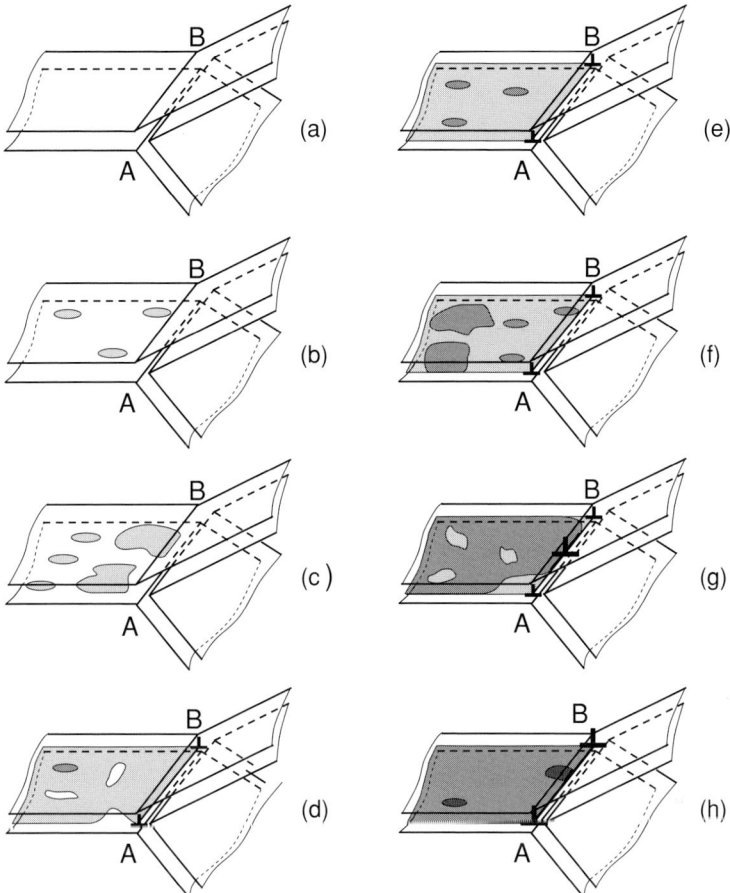

Figure 5.9. Grain-boundary sliding through local shear events. (a)–(h) Local shear events (grey and dark grey ellipses) carry sliding along grain boundary. Shear gradually occurs in all the grain-boundary planes and results in the formation of triple-junction dislocation with Burgers vector varying along triple junction AB.

nanocomposites. In the case of the nanocrystalline materials, the discussed representations on grain-boundary sliding by local shear events are supported by molecular dynamics simulations (Van Swygenhoven and Derlet, 2001) of plastic deformation in nanocrystalline Ni with grain size around 12 nm. Van Swygenhoven and Derlet (2001) examined sliding processes parallel and nonparallel to the grain boundary plane (Figures 5.10 and 5.11, respectively). Their simulations show that grain-boundary sliding occurs by local shear events which are either uncorrelated individual atomic jumps or local transformations of small groups of atoms with a degree of correlation.

With the assumption that grain-boundary sliding results from combined action of thermally activated local shear events, Conrad and Narayan (2000) suggested the following formula for the corresponding macroscopic plastic shear rate:

$$d\gamma/dt = N_v A a \nu \exp[-\Delta G(\tau_e)/k_B T]. \quad (5.14)$$

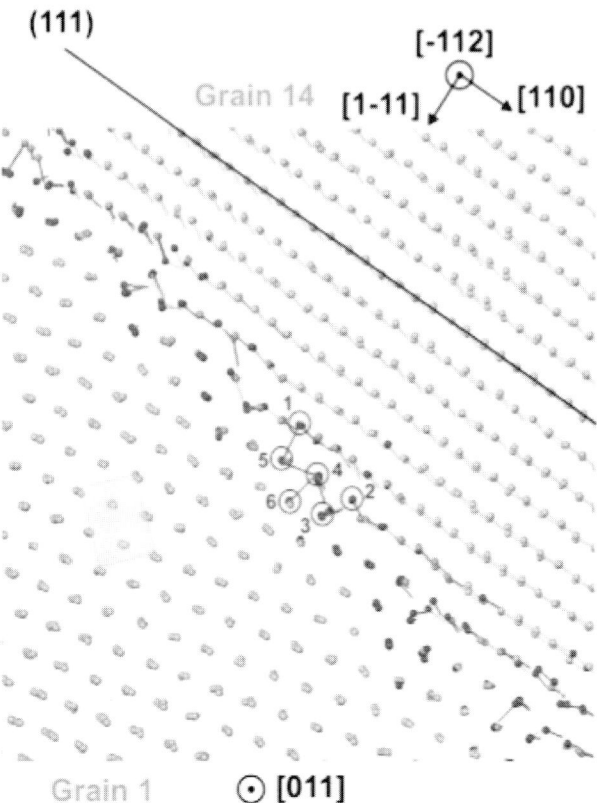

Figure 5.10. A section of the grain boundary between grains 1 and 14. Displacement vectors are shown indicating the change in position between two levels of strain during plastic deformation. Atomic shuffling between the grains can be observed. (Reprinted from *Physical Review B*, Volume 64, Van Swygenhoven, H. and Derlet, P. M., Grain-boundary sliding in nanocrystalline fcc metals, Article 224105, Copyright (2001).)

Here N_v denotes the number of places (free-volume defects) per unit volume where thermally activated shear can occur, A is the area swept out per successive thermal fluctuation, a is the atomic diameter (in terms used in this Chapter, a is the mean interatomic distance in the grain boundary phase), ν is the frequency of vibration, k_B is the Boltzmann constant, T is the absolute temperature, and $\Delta G(\tau_e)$ is the Gibbs free activation energy, which is a decreasing function of the effective shear stress $\tau_e = \tau - \tau_0$, where τ is the applied stress and τ_0 is a back stress or threshold stress. The Gibbs energy is assumed to be given as $\Delta G(\tau_e) = \Delta F - v\tau_e$, where ΔF is the Holmholz free energy (whose value should be approximately that for grain boundary diffusion), and $v = a^3$ is the activation volume. In these circumstances, after expansion of parameters on the right-hand side of Eq. (5.14), Conrad and Narayan (2000) found

$$d\gamma/dt = (6a\nu_D/d)\sinh(v\tau_e/kT)\exp[-\Delta F/kT], \quad (5.15)$$

5.6 Grain-boundary sliding in nanocrystalline materials

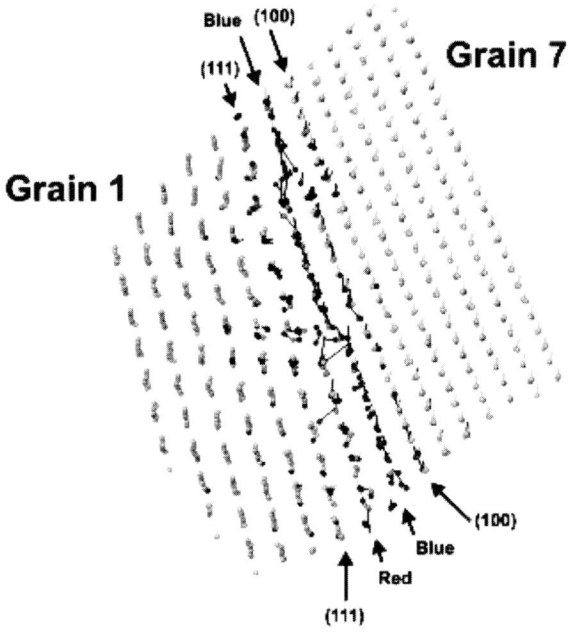

Figure 5.11. A section of the grain boundary between grains 1 and 7. For grain 1, (111) atomic planes are shown and for grain 7, (100) planes are shown. Displacement vectors are shown. The atomic planes indicated by "blue" and "red" constitute the misfit of the grain boundary. (Reprinted from *Physical Review B*, Volume **64**, Van Swygenhoven, H. and Derlet, P. M., Grain-boundary sliding in nanocrystalline fcc metals, Article 224105, Copyright (2001).)

where d is the grain size, and ν_D is the Debye frequency. Using formula (5.15), Conrad and Narayan (2000) calculated the dependence of the yield stress on grain size d in the case where grain-boundary sliding is the dominant deformation mechanism. In particular, in the range of finest grains, the dependence shows a decrease of the yield stress with reducing grain size. This dependence was used in explaining the abnormal Hall–Petch relationship in nanocrystalline materials as that related to competition between the lattice dislocation slip and grain-boundary sliding. It was assumed that polycrystals with a relatively large grain size are deformed by the lattice dislocation slip characterized by the classical Hall–Petch relation (5.1). Nanocrystalline materials with small grain sizes are deformed by grain-boundary sliding characterized by decrease of the yield stress with reducing grain size. Finally, it was assumed that a grain size d^* exists at which value the classical Hall–Petch mechanism switches to the grain boundary sliding mechanism. With these assumptions, Conrad and Narayan (2000) calculated the dependences of the yield stress on grain size d in wide ranges of d, for several metals and alloys. The calculated dependences are shown to be in a good agreement with the corresponding experimental data.

The approach (Conrad and Narayan, 2000), in fact, operates with flat grain boundaries and does not take into account the presence of grain boundary facets, steps and triple junctions where grain-boundary planes change their orientations. However, in real coarse-grained polycrystals and nanocrystalline materials, grain-boundary sliding occurs along grain boundaries that join at triple junctions and form complicated networks with non-trivial geometry. Besides, even individual grain boundaries commonly are not flat, but contain facets and steps. With non-trivial geometry of grain boundaries and their networks, grain-boundary sliding along them is different from "ideal" sliding along a flat grain boundary in a bicrystal. There are geometrical obstacles for grain-boundary sliding along curved grain boundaries. Therefore, grain-boundary sliding always needs to be accommodated in real coarse-grained polycrystals and nanocrystalline materials. In the context discussed, the model (Conrad and Narayan, 2000) is useful in consideration of the yield stress that characterizes the start of plastic deformation in nanocrystalline materials with finest grains. It is because plastic deformation at its starting stage involves the only grain-boundary sliding along straight segments of grain boundaries, in which case accommodation processes are not important. However, the model (Conrad and Narayan, 2000) is hardly relevant in a description of the flow stress that characterizes all the stages of plastic deformation in ductile nanocrystalline materials, because the model ignores accommodation of grain-boundary sliding. At the same time, if grain-boundary sliding essentially contributes to plastic flow and thereby is intensive in a nanocrystalline material, it needs to be effectively accommodated.

Following the classical theory of grain-boundary sliding along non-flat grain boundaries in coarse-grained polycrystals (Sutton and Balluffi, 1996; Zelin and Mukherjee, 1996), the basic mechanisms accommodating such sliding are the elastic accommodation, diffusional accommodation, plastic flow accommodation in the lattice, and local migration of grain boundaries (Figure 5.12). In the case of nanocrystalline materials, especially those with finest grains, grain boundaries are very short, and the amount of their triple junctions is very large. In these materials, grain boundaries change orientations of their planes at mostly triple junctions which, therefore, serve as the key geometrical obstacles for grain-boundary sliding. As a corollary, accommodation of grain-boundary sliding at triple junctions is crucial for its action as a deformation mode in nanocrystalline materials. The accommodation mechanisms operating in nanocrystalline materials are similar to those operating in coarse-grained polycrystals (Figure 5.13). In particular, grain-boundary sliding at triple junctions in nanocrystalline materials is effectively accommodated by diffusional accommodation (Figure 5.13c), plastic flow accommodation in the adjacent nanoscale grains (Figure 5.13d), local migration of grain boundaries and their triple junctions (Figure 5.13e), and rotational deformation (Figure 5.13f). Despite the similarity between the accommodation mechanisms in nanocrystalline materials and coarse-grained polycrystals, there are the specific peculiarities in grain-boundary sliding and its accommodation in nanocrystalline materials. They will be considered below.

To analyse grain-boundary sliding and its accommodation in nanocrystalline materials, first let us discuss the effects of non-accommodated grain-boundary sliding on triple-junction structures. In short, grain-boundary sliding leads to the dislocation storage at triple junctions. For illustration, we consider the dislocation storage at a triple junction owing to grain-boundary sliding along one boundary adjacent to the triple junction (Figures 5.8 and 5.9). The grain-boundary sliding is treated as occurring by either movement of gliding dislocations or thermally activated local shear events. In the former case, when grain-boundary dislocations glide along a boundary plane, they are stopped at triple junction (Figure 5.8a,b

5.6 Grain-boundary sliding in nanocrystalline materials

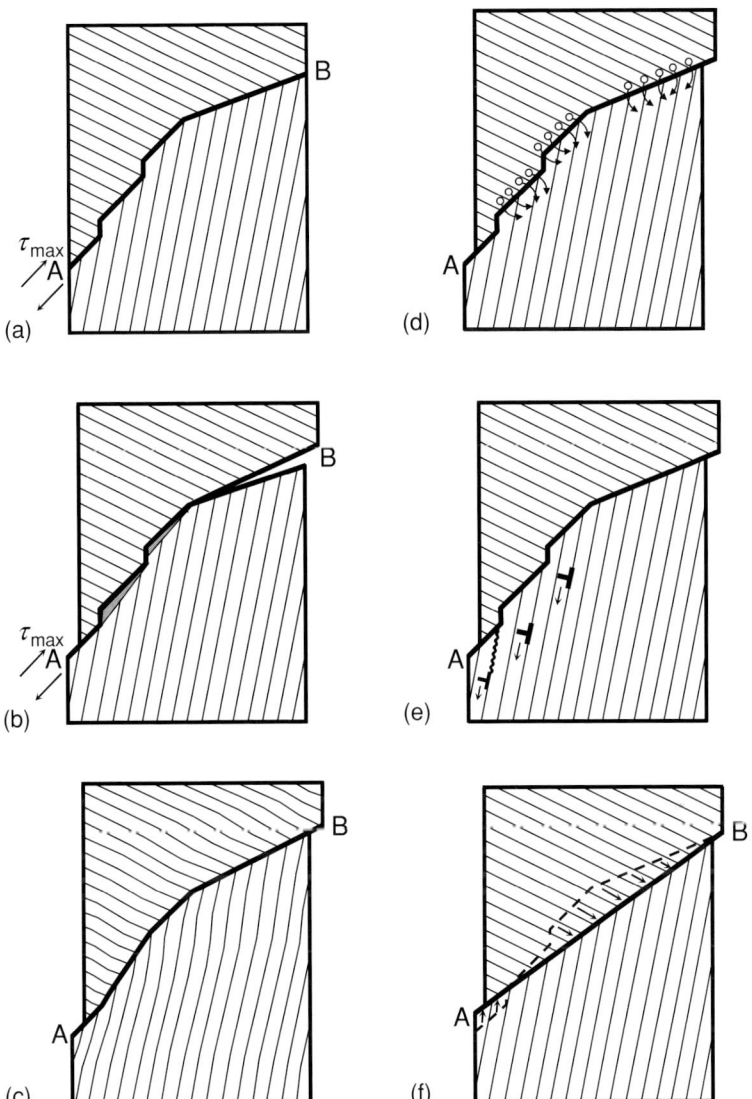

Figure 5.12. Grain-boundary sliding along boundary with steps and facet in a bi-crystal. (a) Initial state. (b) Non-accommodated grain-boundary sliding results in the formation of regions (gray regions) where neighboring grains interpenetrate and regions (white regions) where voids are formed. (c) Elastic accommodation. (d) Diffusional accommodation. (e) Lattice slip accommodation. (f) Accommodating local migration of grain boundaries.

and c). At certain conditions, grain-boundary dislocations can converge at triple junctions which thereby accummulate dislocations with rather large Burgers vectors (with magnitudes exceding the crystal lattice parameter a) (Figure 5.8d).

Let us consider the situation where thermally activated local shear events are driven by the shear stress and carry grain-boundary sliding in a nanocrystalline material (Figure 5.9).

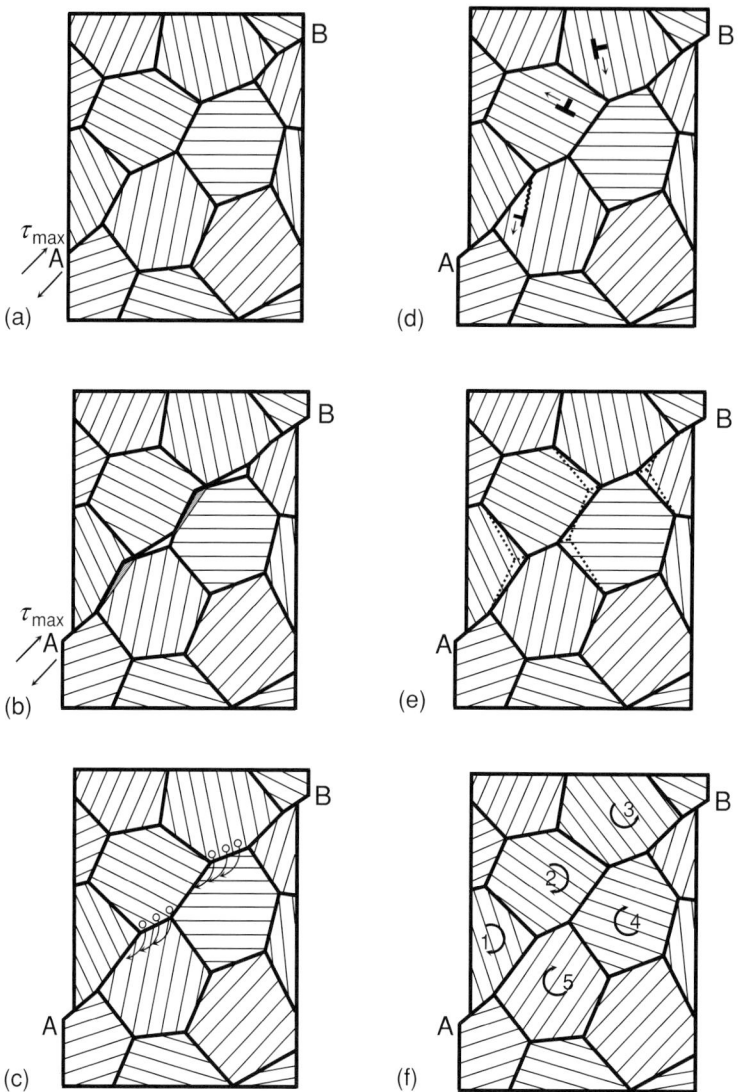

Figure 5.13. Grain-boundary sliding in nanocrystalline solid. (a) Initial state. (b) Non-accommodated grain-boundary sliding results in the formation of regions (gray regions) where neighboring grains interpenetrate and regions (white regions) where voids are formed. (c) Diffusional accommodation. (d) Lattice slip accommodation. (e) Accommodation by local migration of grain boundaries. (f) Accommodation by crystal lattice rotations in nanoscale grains 1, 2, 3, 4 and 5.

As noted previously, grain-boundary sliding by local shear events is very similar to deformation mode occurring by local shear events in plastically deformed metallic glasses (Argon, 1979). In this context, with the interpretation (Argon, 1979) of local shear events as events of the generation of nanoscale dislocation loops, one finds that the grain-boundary sliding by local shear events produces grain-boundary dislocations at triple junctions. Actually, since triple junctions serve as obstacles for grain-boundary sliding, numerous local

5.6 Grain-boundary sliding in nanocrystalline materials

shear events occurring in a grain boundary result in the accumulation of the unfinished shear or, in other words, dislocation charge at opposite sides of the grain-boundary segment (Figure 5.9). That is, numerous local shear events occurring in a grain boundary result in the formation of a defect (Figure 5.9) which is very similar to the conventional grain-boundary dislocation. The difference between the defect (Figure. 5.9h) and the conventional grain-boundary dislocation is in the magnitude of their Burgers vectors. Conventional grain-boundary dislocations are characterized by Burgers vectors whose magnitudes are small (each magnitude ranges from approximately $a/5$ to $a/3$) and defined by translational symmetries of the grain-boundary structures (Sutton and Balluffi, 1996). The defect (Figure 5.9h) resulting from numerous local shear events is characterized by a Burgers vector whose magnitude is arbitrary; it is not restricted by translational symmetries of the grain-boundary structure. Generally speaking, since the defect (Figure. 5.9h) represents the superposition of numerous nanoscale dislocation loops (resulted from thermally activated local shear events (Figure 5.9)), its Burgers vector is not a constant vector along the defect line. The defect (Figure 5.9h) is a Sogmiliana dislocation with a variable Burgers vector whose magnitude can exceed the crystal lattice parameter a.

We have considered the formation of a triple-junction dislocation owing to grain-boundary sliding along one boundary adjacent to the triple junction (Figures 5.8 and 5.9). Its Burgers vector is approximately parallel with the grain-boundary plane that conducts sliding. Similar dislocations are formed at a triple junction owing to grain-boundary sliding along two or three grain boundaries adjacent to the triple junction. In this situation, the Burgers vector of the triple junction dislocation is caused by the combined effects of sliding along the adjacent grain boundaries (for details, see the discussion in the rest of this section).

Thus, triple junctions of grain boundaries, where the boundary planes with various orientations join together, serve as key obstacles for grain-boundary sliding and thereby accumulate dislocations in nanocrystalline materials. The dislocations at triple junctions in nanocrystalline materials are typical defects resulting from grain-boundary sliding conducted by either local shear events or movement of conventional grain-boundary dislocations. In doing so, Burgers vector magnitudes of such grain-boundary dislocation loops can be comparatively large (say, exceed the crystal-lattice parameter a). The aforesaid allows one to effectively use the dislocation theory in a description of grain-boundary sliding, its accommodation and contribution to plastic flow in nanocrystalline materials.

For instance, several theoretical models consider grain-boundary dislocations as carriers of grain-boundary sliding that overcome the obstacles – triple junctions – in nanocrystalline materials. In doing so, the dislocations and grain-boundary structures undergo transformations that accompany and accommodate the grain-boundary sliding. The theoretical models discussed here are distinguished by their interpretation of these transformations occurring in the vicinities of triple junctions and causing a contribution of the grain-boundary sliding to yield and flow stresses.

For illustration, let us consider a theoretical model (Gutkin *et al.*, 2004a) which involves representations on the grain-boundary sliding in a description of the experimentally detected fact (see, e.g., Chapter 4, and Weertman and Sanders (1994), and Volpp *et al.* (1997)) that the $\tau(d)$ relationship in nanocrystalline materials often shows two different behaviors,

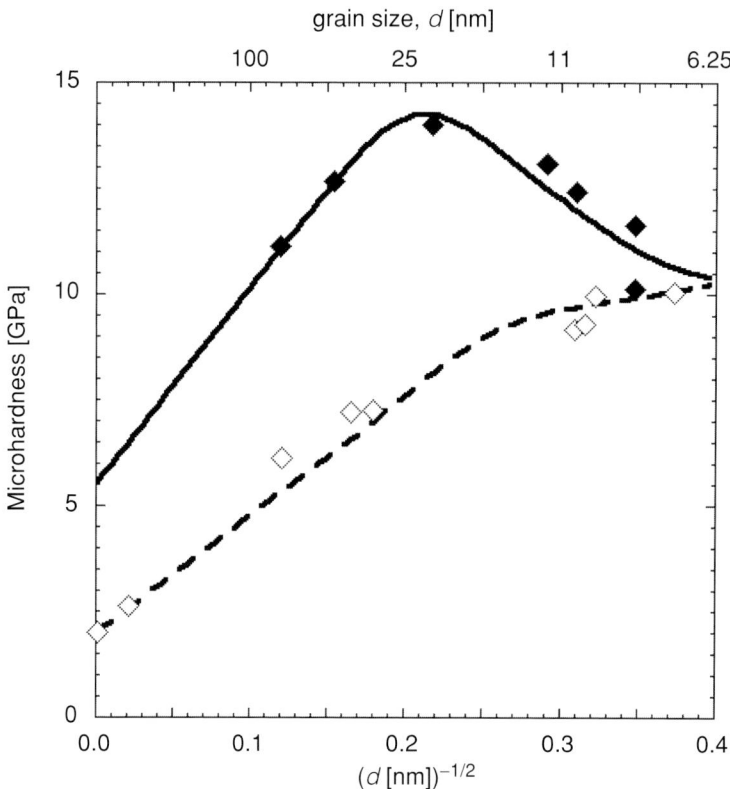

Figure 5.14. Comparison of theoretical predictions (Gutkin *et al.*, 2004a) (solid and dashed curves) with the experimental dependences H_v ($d^{-1/2}$) obtained by Volpp *et al.* (1997) for as-prepared (open boxes) and heat-treated (filled boxes) nanocrystalline materials, respectively.

depending on their processing. In the range of small grain sizes, heat-treated materials exhibit "inverse" Hall–Petch behavior (softening with reduction of grain size), while the yield stress or hardness of as-prepared materials slightly increases or saturates at grain size $d \leq 10$ nm showing little or no "inverse" Hall–Petch behavior (Figure 5.14). Following Gutkin *et al.* (2004a), this difference in the deformation behavior between heat-treated and as-prepared nanocrystalline materials is related to the difference between their defect structures. Different defect structures in heat-treated and as-prepared materials cause the effective action of different deformation modes occurring because of grain refinement. In particular, the contribution of grain-boundary sliding is expected to be high in as-prepared materials commonly characterized by high density of lattice and grain-boundary dislocations which enhance grain-boundary sliding processes. On the other hand, heat treatment is capable of suppressing the grain-boundary sliding, in which case diffusional deformation modes effectively come into play.

Let us consider a nanocrystalline specimen fabricated in highly nonequilibrium conditions. Most of the fabrication routes produce highly defected grain boundaries. In particular, grain boundaries with excess density of grain-boundary dislocations – carriers

5.6 Grain-boundary sliding in nanocrystalline materials

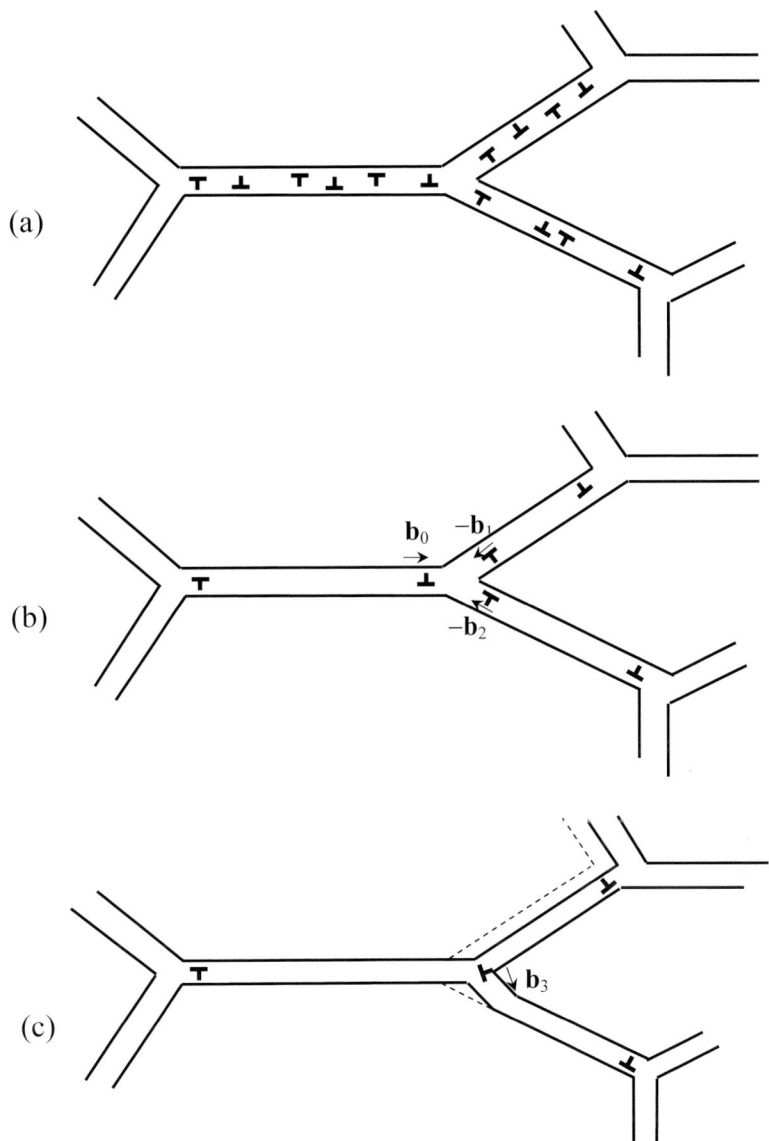

Figure 5.15. Grain-boundary sliding at triple junctions. (a) Grain-boundary dislocation ensemble in as-prepared nanocrystalline specimen. (b) Grain-boundary dislocations are accumulated near triple junctions under action of mechanical load. (c) Dislocation with Burgers vector b_0 moves and comes into reaction with two dislocations (with Burgers vectors $-b_1$ and $-b_2$), resulting in the formation of a grain-boundary dislocation with Burgers vector b_3.

of grain-boundary sliding – often exist in as-fabricated nanocrystalline materials (Valiev et al., 2000; Mohamed and Li, 2001; Valiev, 2004) (Figure 5.15a). When a mechanical load is applied to the specimen, mobile grain-boundary dislocations (with Burgers vectors being parallel to grain-boundary planes) move causing grain-boundary sliding.

Some of moving grain-boundary dislocations anihilate with grain-boundary dislocations having Burgers vectors of opposite sign (Figure 5.15b). Other grain-boundary dislocations are stopped at triple junctions of grain boundaries, that represent effective obstacles for dislocation movement (Figure 5.15b). In general, grain-boundary dislocations stopped near a triple junction are capable of overcoming the junction obstacle by a dislocation reaction when the shear stress reaches some critical value (Figure 5.15c). In nanocrystalline materials with their high-density ensembles of triple junctions, the critical shear stress needed for grain-boundary dislocations to overcome triple junctions specifies the contribution of grain-boundary sliding to the yield stress. With the transformation of grain-boundary dislocations (Figure 5.15) treated as a basic elementary act of plastic deformation at its initial stage in as-prepared nanocrystalline solids, the yield stress is given as follows (Gutkin *et al.*, 2004a):

$$\tau = k_1 + k_2/d, \tag{5.16}$$

where k_1 and k_2 are constants. Grain size d occurs in formula (5.16) as a length scale characterizing elastic interaction between grain-boundary dislocations located near different triple junctions (Figure 5.15b and c) distant by d from each other. Formula (5.23) quantitatively characterizes grain-boundary sliding as a deformation mode (being in competition with other deformation modes) and gives a $\tau(d)$ dependence which is in a rather good agreement with the corresponding experimental data (Volpp *et al.*, 1997) on mechanical characteristics of as-prepared nanocrystalline NiAl materials synthesized by the ball-milling technique (see Figure 5.14).

Heat treatment of nanocrystalline materials gives rise to annihilation of grain-boundary dislocations and thereby suppresses grain-boundary sliding, in which case diffusional deformation mechanisms compete with conventional dislocation slip in heat-treated nanocrystalline materials. The yield stress dependence on grain size d in heat-treated nanocrystalline NiAl materials have been calculated (Gutkin *et al.*, 2004a), taking into account competition between the conventional dislocation slip and Coble creep (treated to be alternative to grain-boundary sliding). The theoretical dependence is in a good agreement with the corresponding experimental data (Volpp *et al.*, 1997); see Figure 5.14.

The model (Gutkin *et al.*, 2004a) suggests just one of the possible explanations of the sensitivity of the $\tau(d)$ relationship on processing. In particular, this model is in conflict with the idea that grain-boundary sliding is always the dominant deformation mechanism responsible for the abnormal Hall–Petch behavior in nanocrystalline materials with finest grains. Besides, the model (Gutkin *et al.*, 2004a) does not take into account the effects of annealing on long-range order in NiAl composition, which in its turn influence mechanical characteristics of NiAl. As a corollary, comparison of the theoretically calculated $\tau(d)$ relationship (Gutkin *et al.*, 2004a) and that measured in experiments (Volpp *et al.*, 1997) with as-fabricated and heat-treated NiAl can be questionable. At the same time, we are unaware of other attempts to theoretically describe the sensitivity of the $\tau(d)$ relationship on processing of nanocrystalline materials. In this context, it is natural to treat the theoretical approach (Gutkin *et al.*, 2004a) as the first approximation model which needs to be

5.6 Grain-boundary sliding in nanocrystalline materials

verified by microstructural experiments and compared to other potential explanations of the sensitivity of the $\tau(d)$ relationship on processing in the future.

In the previous part of this chapter, most attention has been paid to initial stage of plastic deformation, characterized by the yield stress τ and its abnormal Hall–Petch dependence on grain size. That is, we have focused our consideration on the macroscale phenomena (i) and (ii) experimentally detected in mechanically loaded nanocrystalline materials. Now let us turn to a discussion of other experimentally detected phenomena and the role of grain-boundary sliding in these phenomena.

First, let us consider the role of grain-boundary sliding in superplasticity exhibited by some nanocrystalline solids at relatively high strain rates and low temperatures compared to their coarse-grained counterparts (Mayo 1997, 1998; Mishra et al., 1998; McFadden et al., 1999; Islamgaliev et al., 2001; Mishra et al., 2001; Valiev et al., 2001; Mukherjee, 2002a,b; Valiev et al., 2002; Zhan et al., 2003). It is characterized by very high flow stress and strengthening (Mishra et al., 1998; McFadden et al., 1999; Islamgaliev et al., 2001; Mishra et al., 2001; Valiev et al., 2001; Mukherjee, 2002a,b; Valiev et al., 2002) which are the specific features of superplastic nanocrystalline materials, differing in their deformation behavior from that of conventional microcrystalline materials exhibiting superplasticity. These features are the subject of growing fundamental interest motivated by a range of new applications based on the use of superplasticity of nanocrystalline materials. The dominant mode of superplasticity in nanocrystalline materials is viewed to be grain-boundary sliding (Mukherjee, 2002a,b), in which case the unusual strengthening should be related to the specific features of the grain-boundary sliding in nanocrystalline materials. In papers (Gutkin et al., 2003c, 2004b), a theoretical model has been suggested describing the strengthening in nanocrystalline materials exhibiting superplasticity as the phenomenon caused by transformations of grain-boundary dislocations – carriers of grain-boundary sliding – at triple junctions of grain boundaries.

Consider a model configuration of grain-boundary dislocations in a nanocrystalline material, which is formed near a triple junction under the action of shear stress τ (Figure 5.16a). The configuration consists of two grain-boundary dislocations with Burgers vectors b_1 and $-b_2$ (the b_1- and $-b_2$-dislocations, respectively) which are parallel to the corresponding grain-boundary planes adjacent to the triple junction. (This configuration is similar to that shown in Figure 5.15b. However, high-strain-rate superplasticity is exhibited by nanocrystalline materials with comparatively large grain sizes $d \geq 50$ nm. In the case discussed, the interaction between grain-boundary dislocations stopped at different triple junctions is not essential and thereby not taken into account.) The grain-boundary dislocations are stopped in the vicinity of the triple junction when $\tau < \tau_1^{\text{crit}}$. When the shear stress reaches its critical value τ_1^{crit}, the b_1- and $-b_2$-dislocations move over short distances and "meet" at the triple junction (Figure 5.16b). It is accompanied by a dislocation reaction at the triple junction, which involves the b_1- and $-b_2$-dislocations and results in the formation of a sessile grain-boundary dislocation with Burgers vector $b = b_1 - b_2$ (the b-dislocation); see Figure 5.16b. After the dislocation reaction (Figure 5.16a and b) at the triple junction has occurred, two new grain-boundary dislocations are generated and move under the shear stress action towards the triple junction (Figure 5.16c). These moving dislocations elastically interact with the

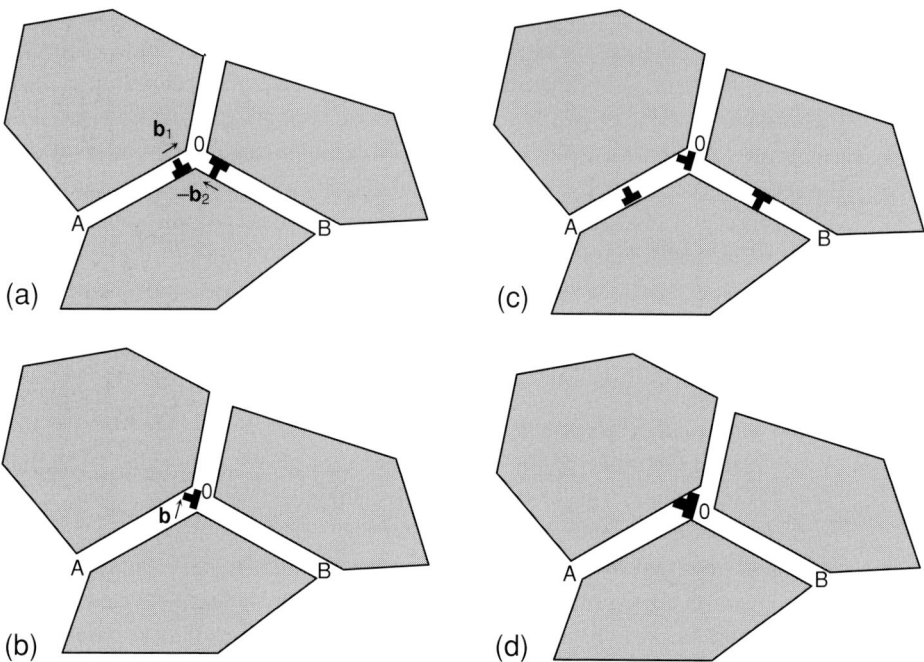

Figure 5.16. Transformations of grain-boundary dislocations near a triple junction. (a) Two gliding grain-boundary dislocations move towards triple junction O. (b) Sessile dislocation with Burgers vector b is formed. (c) Generation of two new gliding grain-boundary dislocations with Burgers vectors b_1 and $-b_2$, that move towards the triple junction. (d) Sessile dislocation with Burgers vector $2b$ is formed.

sessile grain-boundary dislocation which hampers the movement of the new dislocations. At the stress level $\tau_2^{\mathrm{crit}} > \tau_1^{\mathrm{crit}}$, a new elementary act of grain-boundary sliding at the triple junction occurs. With the Burgers vector conservation law, the elementary act results in the formation of a new sessile dislocation (Figure 5.16d). The new and pre-existent sessile dislocations converge resulting in the formation of a new sessile dislocation with the large Burgers vector (Figure 5.16d).

The process under consideration repeatedly occurs and is accompanied by increase of the Burgers vector of the sessile dislocation at each step. Since the sessile dislocation hampers the moving grain-boundary dislocations, the critical shear stress increases with evolution of the defect structure. Following analysis and estimates given by Gutkin et al. (2003c, 2004b), namely this evolution is responsible for the strengthening experimentally detected (Mishra et al., 1998; McFadden et al., 1999; Islamgaliev et al., 2001; Mishra et al., 2001; Valiev et al., 2001; Mukherjee, 2002a,b; Valiev et al., 2002) in nanocrystalline materials at the first extended stage of superplastic deformation.

Now let us discuss the experimentally observed phenomenon of plastic flow localization in nanocrystalline materials. In general, plastic deformation in nanocrystalline materials can be spatially homogeneous or localized in narrow shear bands. Evidence of plastic flow

5.6 Grain-boundary sliding in nanocrystalline materials

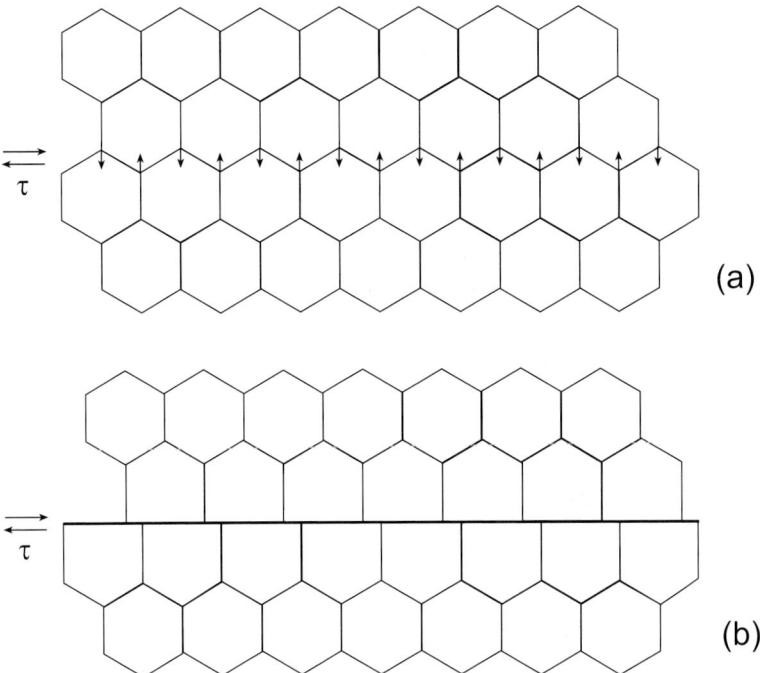

Figure 5.17. Plastic flow localization. (a) Local migration of grain boundaries gives rise to (b) formation of a local zone where grain boundaries are parallel to each other and intensive plastic shear occurs.

localization has been observed in nanocrystalline materials tested in tension (Nieman et al., 1991), fatigue (Witney et al., 1995), compression (Carsley et al., 1995), and microhardness (Andrievskii et al., 1997; Andrievskii, 1998). The model (Hahn et al., 1997; Hahn and Padmanabhan, 1997) describes plastic flow localization as the phenomenon occurring owing to a local migration of grain boundaries that accompanies the grain-boundary sliding. More precisely, the local migration of grain boundaries is considered to provide the formation of an approximately planar ensemble of the grain boundaries (Figure 5.17) along which an intensive plastic shear via the correlated grain-boundary sliding processes occurs (Hahn et al., 1997; Hahn and Padmanabhan, 1997). Triple junctions do not play the role of geometric obstacles at planar grain boundaries along which grain-boundary sliding is thereby enhanced. These macroscopic planar grain-boundary structures are associated with the shear bands where high (super)plastic deformation is localized, resulting in a large macroscopic deformation of a nanocrystalline sample.

There is experimental evidence for the local migration of grain boundaries in superplastically deformed polycrystals (Astanin et al., 1997). Also, recent experimental data (Wei et al., 2002; Jia et al., 2003; Markmann et al., 2003) are indicative of the fact that planar grain-boundary structures are formed in shear bands in nanocrystalline materials with finest and intermediate grains. These data support the model (Hahn et al., 1997; Hahn and

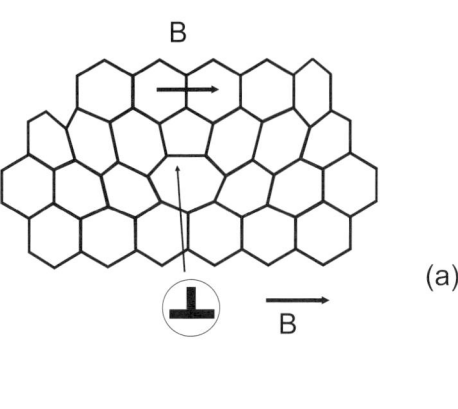

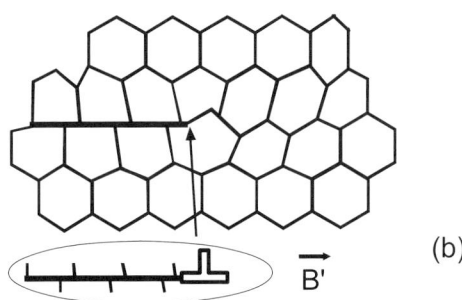

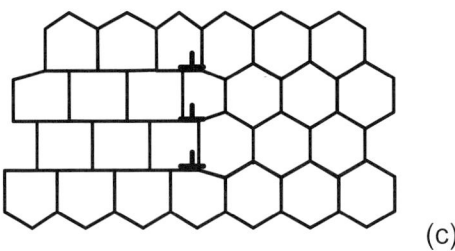

Figure 5.18. Cellular dislocations in nanocrystalline materials. (a) Perfect and (b) non-perfect cellular dislocations in regular array of hexagonal nanograins (cells). (c) Movement of group of non-perfect cellular dislocations causes plastic deformation localized within a shear band.

Padmanabhan, 1997) relating plastic flow localization to migration of grain boundaries and their triple junctions. Also, plastic flow localization at the nanograin level is effectively described as the phenomenon associated with the motion of perfect (Zelin and Mukherjee, 1993, 1995; Zelin et al., 1993, 2001) and partial (Ovid'ko, 2003) cellular dislocations (Figure 5.18), topological defects of arrays of nanograins (cells). In the framework of this approach, grain-boundary sliding, grain-boundary migration and triple-junction migration are elementary processes providing the motion of cellular dislocations (Zelin and Mukherjee, 1993, 1995; Zelin et al., 1993, 2001; Ovid'ko, 2003).

5.7 Interaction between deformation modes in nanocrystalline materials. Emission of lattice dislocations from grain boundaries. Twin deformation mode

At the first stage of theoretical studies of plastic deformation processes in nanocrystalline materials, the "rule-of-mixture" approach has been dominant. In the framework of this approach, grain interiors and grain boundaries (and, sometimes, triple junctions of grain boundaries) are considered as constituent phases characterized by different values of the yield stress (Gryaznov *et al.*, 1993; Gutkin and Ovid'ko, 1993; Carsley *et al.*, 1995; Ovid'ko, 1997; Wang *et al.*, 1995; Kim, 1998; Konstantinidis and Aifantis, 1998; Kim *et al.*, 2000). The yield stress of a nanocrystalline material is calculated as some weighted sum of the yield stresses of grain interiors and grain boundaries. In this approach, the mean grain size (or the volume fraction of the grain-boundary phase) plays the role of parameter controlling the deformation behavior of nanocrystalline materials.

The second approach describes deformation mechanisms operating in nanocrystalline materials at the atomic level in terms of defects of atomic structures, with subsequent extrapolation to a description of their contributions to the macroscopic deformation behavior of a nanocrystalline specimen as a whole (Masumura *et al.*, 1998; Conrad and Narayan, 2000; Fedorov *et al.*, 2002). In doing so, it is commonly assumed that either one deformation mechanism dominates or several mechanisms simultaneously contribute to plastic flow, acting independently from each other. A grain-size distribution in a nanocrystalline specimen is taken into account in calculation of the yield stress as the averaged (over grain sizes) value with different deformation mechanisms independently acting in grains with sizes being in different ranges (Masumura *et al.*, 1998; Fedorov *et al.*, 2002).

However, nanocrystalline materials are aggregates of nanosized grains in which different deformation mechanisms can strongly influence each other. That is, there is a kind of effective interaction between deformation modes in nanocrystalline materials, which definitely should be taken into consideration. In particular, it is of crucial importance in high-strain-rate superplasticity of nanocrystalline materials, which, according to experimental data reviewed by Mukherjee (2002a,b), involves grain-boundary sliding, grain rotations and lattice dislocation slip as the key deformation modes strongly influencing each other. In this context, a third approach in the theory of plastic deformation processes has recently been developed which involves consideration of the combined action of interacting deformation modes that enhance each other. Interaction between deformation modes is sensitive to characteristics of grain-boundary structures, triple-junction geometry and grain-boundary defects, but not only to grain-size distribution. Therefore, this approach is potentially able to effectively explain "preparation–structure–properties" relationships that are of crucial interest from both fundamental and applied viewpoints.

An example of interacting deformation modes is the grain-boundary sliding stimulated by lattice dislocation slip. Lattice dislocation moving in grain interiors come to grain boundaries where they split into grain-boundary dislocations that carry intensive grain-boundary sliding (Valiev *et al.*, 2000) (Figure 5.19). This kind of interaction between deformation modes is well-known in the theory of superplasticity of conventional microcrystalline materials and definitely plays a significant role in nanocrystalline

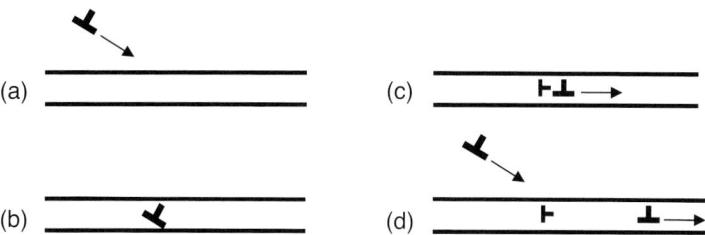

Figure 5.19. Lattice dislocation slip enhances grain-boundary sliding. (a) Lattice dislocation moves in grain interior. (b) Lattice dislocation reaches the grain boundary (c) where it splits into sessile and gliding grain-boundary dislocations. (d) Gliding grain-boundary dislocation carries grain-boundary sliding. New lattice dislocation moves towards the grain boundary.

materials with grain size ranging from 30 nm to 100 nm, in which lattice dislocation slip is intensive.

In nanocrystalline materials characterized by a high volume fraction of the grain-boundary phase, grain boundaries also intensively emit perfect and partial lattice dislocations, but not only absorb. This phenomenon has been observed in experiments (Mukherjee, 2002a,b; Chen et al., 2003; Kumar et al., 2003b; Liao et al., 2003a,b; 2004b). Also, deformation twins associated with partial dislocation activity have been experimentally observed in nanocrystalline materials (Liao et al., 2003a,b; 2004a; Zhu et al., 2005a; Wang et al., 2005). Besides, emission of partial lattice dislocations by grain boundaries has been reported in papers (Schiotz et al., 1998, 1999; Van Swygenhoven et al., 1999a,b, 2002; Van Swygenhoven and Derlet 2001; Yamakov et al., 2001; Frozeth et al., 2004a,b; Hasnaoui et al., 2004; Schiotz, 2004; Shimokawa et al., 2005; Wolf et al., 2005) dealing with molecular dynamics simulations of plastic deformation in nanocrystalline materials.

For instance, Yamakov et al. (2001) performed molecular-dynamics simulations of plastic deformation in nanocrystalline aluminium with columnar grain structure. For columnar grains with a diameter of up to 70 nm, these simulations show the emission of partial and split lattice dislocations from grain boundaries. Yamakov et al. (2001) revealed the two dislocation behavior regimes in nanocrystalline aluminium, depending on grain size d and splitting distance $r_{\text{split}} = Kb^2/(\gamma - b_p m\sigma)$, where $b_p = a/(6)^{1/2}$ denotes the Burgers vector magnitude characterizing a partial dislocation, a is the crystal lattice parameter, γ is the stacking fault energy density (per unit area), m is the Schmid factor, K is a factor depending on the elastic constants of the material, and σ is the external stress. In the situation where $r_{\text{split}} > d$, the typical dislocation behavior is as follows. A partial dislocation is emitted from a grain boundary, propagates through the grain interior and is absorbed by an opposite grain boundary (Figure 5.20a). The stacking fault is formed in the grain interior behind the partial dislocation (Figure 5.20a). For $r_{\text{split}} < d$, two Shockley partial dislocations are consequently emitted from a grain boundary and form a split lattice dislocation that moves towards an opposite grain boundary (Figure 5.20b).

Frozeth et al. (2004a) have modeled by molecular-dynamics methods the structure and deformation behavior of nanocrystalline aluminium with equiaxed grain structure. Following results of their simulations, semi-loops of split dislocations are nucleated at triple

5.7 Interaction between deformation modes

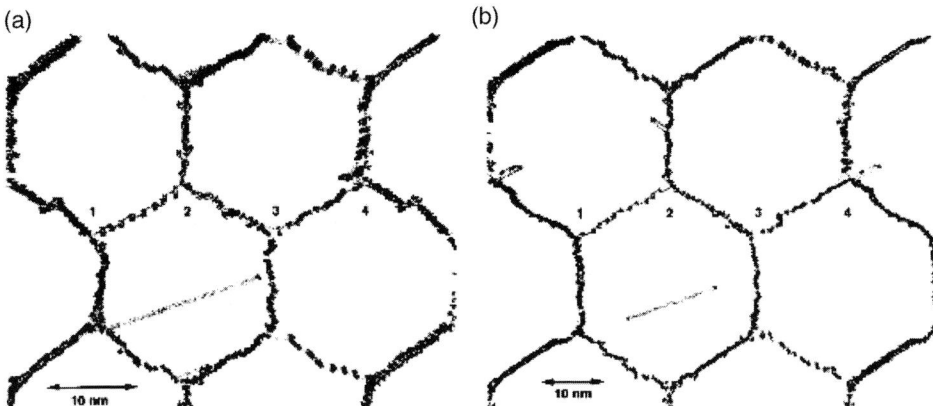

Figure 5.20. Projection of the three-dimensional (3D) simulation cell of columnar microstructure in nanocrystalline aluminium (Yamakov *et al.*, 2001) down the [1$\bar{1}$0] texture axis. The tensile stress of 2.3 GPa was applied in the horizontal direction. (a) Snapshot of the $d = 20$ nm system at 14.6 ps of deformation time; a nucleated partial 1/6 [112] dislocation is observed, leaving a (111) stacking fault behind. (b) Snapshot of a crystallographically identical system, however with a $d = 30$ nm grain size, at 27.6 ps of deformation time. A fully developed extended 1/2 [011] dislocation dissociated into two partials is observed. In both snapshots the applied tensile stress is 2.3 GPa; however, the average internal stress in grains 2 and 4 is 2.77 GPa. The gray dots represent the hcp atoms and the black dots represent disordered atoms, i.e. atoms that are neither fcc nor hcp according to the common-neighbor analysis. (Reprinted from *Acta Materialia*, Volume **49**, Yamakov, V., Wolf, D., Salazar, M., Phillpot, S. R., and Gleiter, H., Length-scale effects in the nucleation of extended dislocations in nanocrystalline Al by molecular-dynamics, pp. 2713–2722. Copyright (2001), with permission from Elsevier.)

junctions of grain boundaries as illustrated in Figure 5.21. Frozeth *et al.* (2004a) questioned the validity of use of the splitting distance in characterizing the nucleation of a split dislocation – configuration consisting of two partial dislocations and a stacking fault between them – at a grain boundary. Their criticism is based on the fact that the nucleation process is highly sensitive to local stress distribution and atomistic mechanism of the nucleation. At the same time, these factors are not taken into account in conventionally used expressions for the splitting distance that characterizes split dislocations in infinite solids. In particular, following Frozeth *et al.* (2004a), once a split-dislocation configuration is emitted from a grain boundary, the splitting distance is much smaller than the distance between the leading partial dislocation and grain boundary prior to nucleation of the trailing partial dislocation.

Schiotz (2004) performed molecular dynamics simulations of plastic deformation in nanocrystalline copper with grain sizes from 5 nm to 50 nm. These simulations show that lattice dislocations are constantly generated at grain boundaries, propagate through grain interiors and are absorbed at opposite grain boundaries. Following Schiotz (2004), the dislocation activity in nanocrystalline copper is qualitatively different from that in coarse-grained polycrystalline materials, because almost all dislocations are generated at grain boundaries in nanocrystalline copper, as the grains are too small to contain a significant number of conventional Frank–Read sources. In doing so, semi-circular dislocation configurations are seen emerging from grain boundaries. Also, simulations (Schiotz, 2004) show the existence

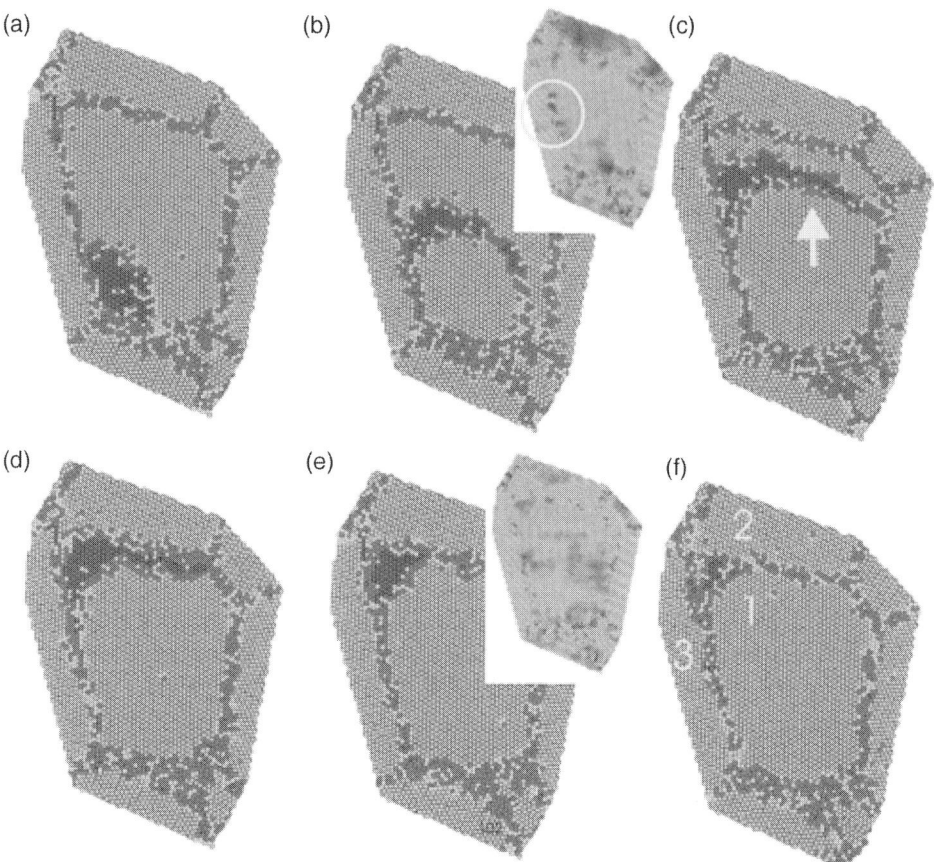

Figure 5.21. Full dislocation in nanocrystalline Al nucleated near a triple junction. (a) The first leading partial has nucleated by 92 ps, (b) the trailing partial has nucleated forming a full dislocation by 94 ps, (c) which then propagates through the grain at 96 ps. (d) At 98 ps, the leading partial is absorbed and (e) at 100 ps the trailing partial is then also absorbed, (f) leaving a small residual stacking fault close to a triple junction that still exists at 110 ps. In (b) and (e) the maximum resolved shear stress distribution is shown, indicating a shear anomaly in the grain boundary that hinders the motion of the full dislocation through the grain, which is removed when the dislocation finally passes by. (Reprinted from *Acta Materialia*, Volume **52**, Frozeth, A. G., Derlet P. M., and Van Swygenhoven, H., Dislocations emitted from nanocrystalline grain boundaries: nucleation and splitting distance, pp. 5863–5870. Copyright (2004), with permission from Elsevier.)

of pile-ups of lattice dislocations (emitted from grain boundaries) in nanocrystalline copper with grain size $d = 50$ nm (Figure 5.22).

Now let us turn to a discussion of analytical models of the role of grain boundaries as lattice dislocation sources. The theoretical model (Fedorov *et al.*, 2003) describes emission of perfect and partial lattice dislocations as a process induced by the preceding grain-boundary sliding. It is a manifestation of interaction between grain-boundary sliding and lattice dislocation slip, which is realized as follows. A pile-up of grain boundary dislocations is generated under the action of mechanical load in a grain boundary in a plastically deformed nanocrystalline sample. Mechanical-load-induced motion of the grain-boundary

5.7 Interaction between deformation modes

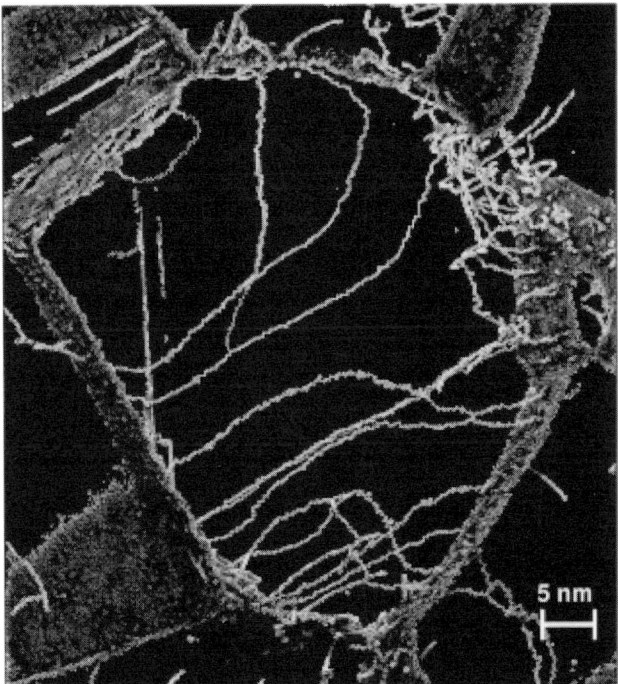

Figure 5.22. A pile-up formed in a grain. The dislocations move from the upper-left to the lower-right corner. For clarity, only a thin slice through the grain is shown, and all atoms in a perfect crystalline neighborhood have been removed. A few dislocations on nearby slip planes, but not part of the actual pile-up, are also seen. Their stress fields perturb the pile-up. (Reprinted from *Scripta Materialia*, Volume **51**, Schiotz, J., Atomic-scale modeling of plastic deformation of nanocrystalline copper, pp. 837–841. Copyright (2004), with permission from Elsevier.)

dislocation pile-up is stopped by a triple junction of grain boundaries (Figure 5.23a). There are several ways of evolution of the grain-boundary dislocation pile-up, including the following (Fedorov *et al.*, 2003). (I) Two (or more) grain-boundary dislocations, each characterized by Burgers vector b, converge into a dislocation with Burgers vector $B = 2b$ having magnitude close to the crystal lattice parameter (Figure 5.23b). Then the resultant dislocation splits into a mobile lattice dislocation with Burgers vector b'_2, which moves to the adjacent grain interior, and an immobile grain boundary dislocation with Burgers vector b'_1, which stays at the triple junction (Figure 5.23c). (II) The head dislocation of the pile-up splits into an immobile grain boundary dislocation (with Burgers vector b_1) that stays at the triple junction and a mobile partial dislocation (with Burgers vector b_2) that moves in the grain interior, in which case a stacking fault is formed behind the moving partial dislocation (Figure 5.23d).

Recently, Gutkin *et al.* (2005a) have extended representations of the model (Fedorov *et al.*, 2003) of the situation where a partial Shockley dislocation is emitted from a dislocated triple junction. In the framework of the model (Gutkin *et al.*, 2005a), triple junctions accumulate dislocations because of grain-boundary sliding across the triple junctions. The dislocation accumulation at triple junctions causes partial Shockley dislocations to be emitted from the

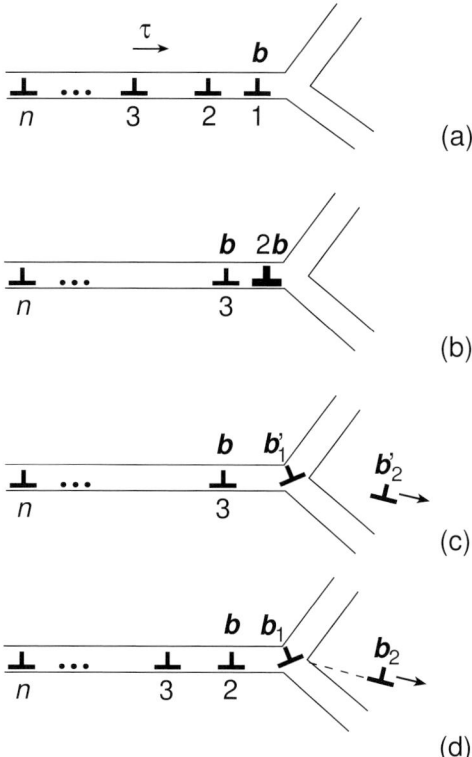

Figure 5.23. Transformations of grain-boundary dislocation pile-ups at the triple junction of grain boundaries. (a) Dislocation pile-up stops at a triple junction. (b) Two dislocations converge into a dislocation with Burgers vector $2b$. (c) Dislocation with Burgers vector $2b$ splits into an immobile grain-boundary dislocation that stays at the triple junction and mobile lattice dislocation that moves in the grain interior. (d) Head dislocation of pile-up splits into an immobile grain-boundary dislocation that stays at the triple junction and a mobile partial dislocation which moves in the grain interior, with a stacking fault (dashed line) formed behind it.

dislocated triple junctions and thus accommodates grain-boundary sliding (Figure 5.24a and b). The dislocation emission process (Figure 5.24a and b) is shown to be energetically favorable in certain ranges of parameters (applied stress, grain size, etc.) in nanocrystalline Al, Cu, and Ni. The dislocation emission process (Figure 5.24a and b) described theoretically by Gutkin *et al.* (2005a) resembles the dislocation emission process (Figure 5.24c and d) revealed in molecular dynamics simulations (Van Swygenhoven *et al.*, 2002) of plastic deformation in nanocrystalline Ni.

The theoretical models (Fedorov *et al.*, 2003; Gutkin *et al.*, 2005a) have described the emission of lattice dislocations from triple junctions of grain boundaries that conduct grain-boundary sliding. These emission processes are initiated by preceeding grain-boundary sliding and serve as a manifestation of the interaction between grain-boundary sliding and lattice dislocation slip. However, in experiments (Mukherjee, 2002a,b; Chen *et al.*, 2003; Kumar *et al.*, 2003b; Liao *et al.*, 2003a,b; 2004b), the emission of lattice dislocations from both triple junctions and flat grain-boundary segments have been observed. In order to

5.7 Interaction between deformation modes 251

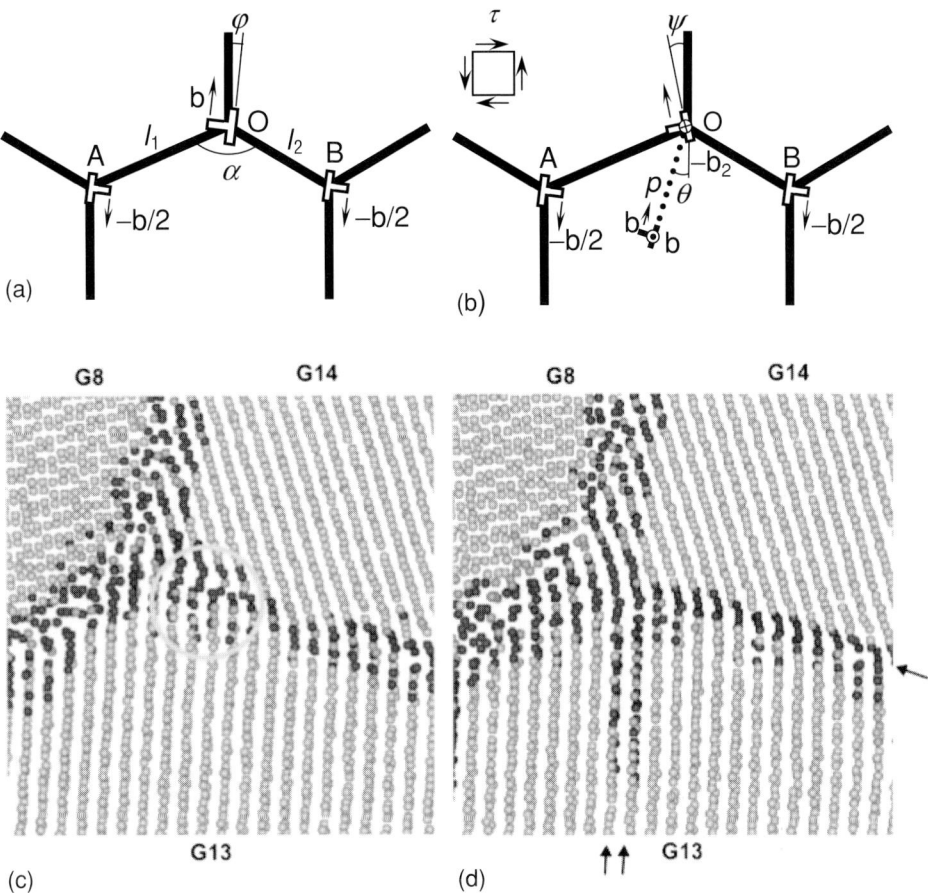

Figure 5.24. Emission of partial Shockley dislocation from triple junction. (a) and (b) Dislocation configuration considered by Gutkin et al. (2005a): (a) Initial configuration with sessile dislocations (resulted from grain-boundary sliding) at triple junctions A, O and B. (b) The sessile dislocation at triple junction O splits into a new sessile dislocation and mobile Shockley partial dislocation which is emitted from the triple junction O (Gutkin *et al.*, 2005a). (c) and (d) Emission of partial Shockley dislocation from triple junction results from the splitting of sessile grain-boundary dislocation at triple junction. A grain-boundary dislocation is highlighted with a circle. (Reprinted from *Physical Review B*, Volume **66**, Van Swygenhoven, H., Derlet, P.M., and Hasnaoui, A., Atomic mechanism for dislocation emission from nanosized grain boundaries, Article 024101. Copyright (2002).) with permission of Elsevier.

account for experimental evidence for the dislocation emission from flat grain-boundary segments, Bobylev *et al.* (2004) have suggested a theoretical model describing the emission of partial dislocations from a high-angle grain boundary containing pre-existent intrinsic dislocations, grain boundary dislocations associated with boundary misorientation misfit. In this situation, the applied stress causes the splitting of a pre-existent intrinsic dislocation into a sessile grain-boundary dislocation and a partial Shockley dislocation which then moves in the adjacent grain interior (Figure 5.25a and b). The dislocation emission

252   5 Mechanical properties of structural nanocrystalline materials

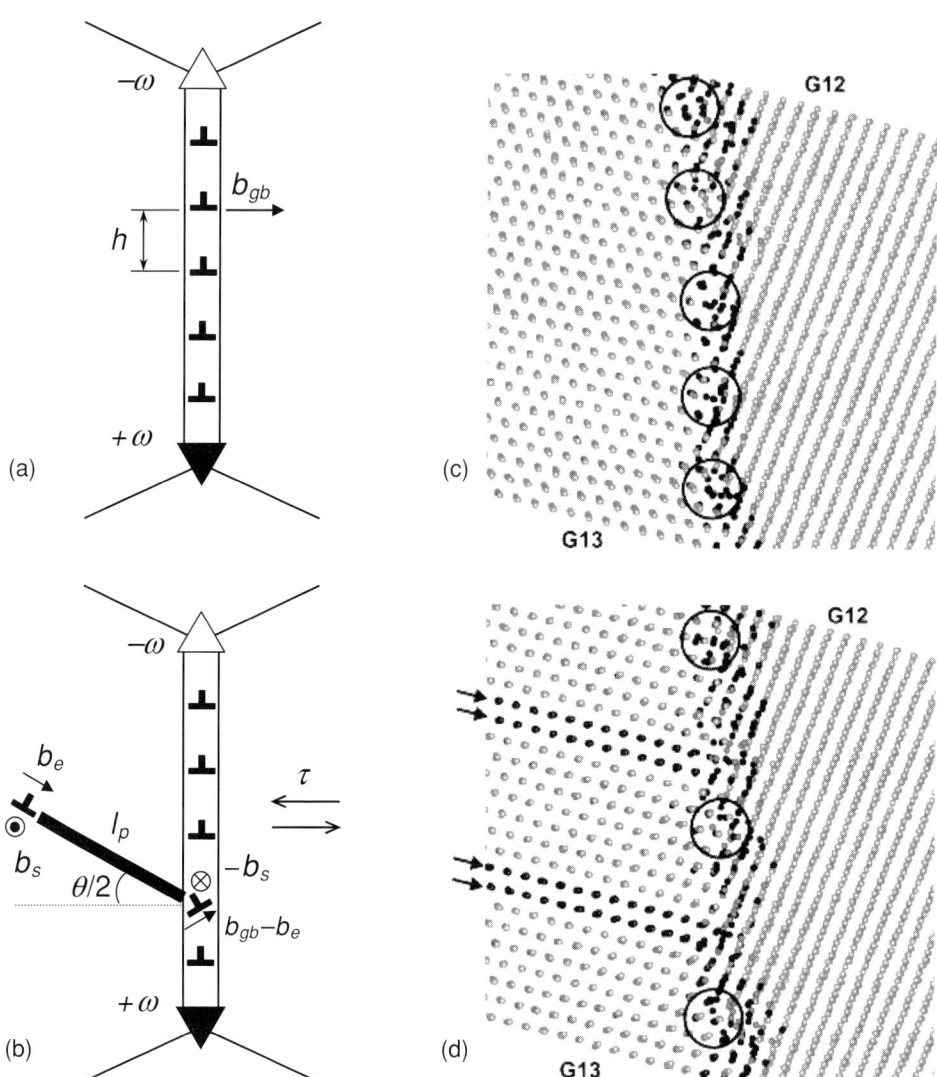

Figure 5.25. Emission of partial Shockley dislocations from a grain boundary. (a) and (b) Dislocation configuration considered by Bobylev *et al.* (2004). (a) Initial configuration with sessile intrinsic dislocations at a high-angle grain boundary. (b) One of the sessile dislocations splits into a new sessile dislocation and mobile Shockley partial dislocation which moves into the grain interior (Bobylev *et al.*, 2004). (c) and (d) Emission of partial Shockley dislocations from a grain boundary results from splittings of sessile grain boundary dislocations at a triple junction. (c) The elastically deformed configuration and (d) the 2.7% strain configuration for the grain boundary 12–13 in the 20 nm sample is shown. For the elastically deformed configuration, five grain-boundary dislocations are identified, which, upon the nucleation of two partial lattice dislocations, reduce to three. (Reprinted figure with permission from *Physical Review B*, Volume **66**, Van Swygenhoven, H., Derlet, P. M., and Hasnaoui, A., Atomic mechanism for dislocation emission from nanosized grain boundaries, Article 024101 2002. Copyright (2002) by the American Physical Society.)

5.7 Interaction between deformation modes

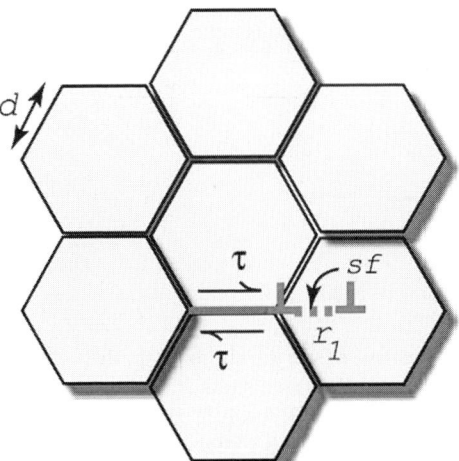

Figure 5.26. A model for stacking fault emission from a grain boundary that acts in a crack-like manner. This has the effect of producing a crack with the size of a typical grain facet. (Reprinted from *Acta Materialia*, Volume **53**, Asaro, R. J., and Suresh, S., Mechanistic models for the activation volume and rate sensitivity in metals with nanocrystalline grains and nano-scale twins, pp. 3369–3382. Copyright (2005), with permission from Elsevier.)

process (Figure 5.25a and b) is shown to be energetically favorable in certain ranges of parameters (applied stress, grain boundary misorientation, etc.) in the exemplary case of nanocrystalline Cu. The model (Bobylev *et al.*, 2004) corresponds to molecular dynamics simulations (Van Swygenhoven *et al.*, 2002) addressing atomistic mechanisms of the partial dislocation emission. Following molecular dynamics simulations (Van Swygenhoven *et al.*, 2002) of plastic deformation in nanocrystalline Ni, partial Shockley dislocations are emitted from a grain boundary because of the splitting transformations of pre-existent intrinsic dislocations (Figure 5.25c and d).

Recently, three-dimensional models (Zhu *et al.*, 2004; Zhu *et al.*, 2005b; Asaro and Suresh 2005; Gutkin and Ovid'ko, 2006) have been suggested describing the emission of dislocation loops and semi-loops from grain boundaries and their triple junctions in deformed nanocrystalline materials. For example, Asaro and Suresh (2005) considered the emission of a leading partial dislocation followed by its trailing partial dislocation from a triple junction serving as stress concentrator due to grain-boundary sliding. Following Asaro and Suresh (2005), the sliding of a grain boundary facet leads to its action as a small crack (Figure 5.26). Figure 5.27 illustrates the emission of a partial dislocation from crack tip, with the possibility of the trailing partial dislocation being emitted at higher levels of stress intensity.

Also, Asaro and Suresh (2005) considered the emission of a leading partial dislocation followed by its trailing partial dislocation from a pre-existent grain-boundary dislocation (Figure 5.28). They calculated the shear stress values characterizing the emission process (Figure 5.28) with various geometric parameters of the partial dislocations. In particular, following Asaro and Suresh (2005), the criterion for emission of a partial dislocation

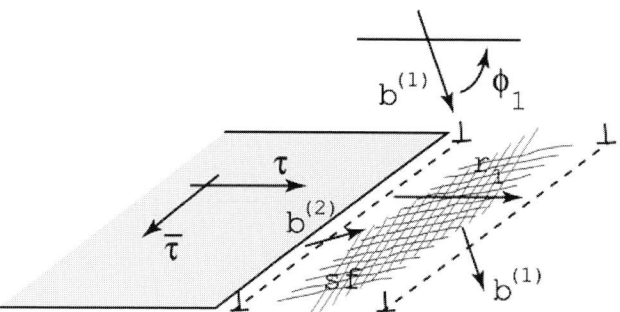

Figure 5.27. A model for partial dislocation emission from a grain boundary that acts in a crack-like manner. The crack is loaded via an in-plane mode II applied stress, τ, and an anti-plane mode III shear stress, $\bar{\tau}$. (Reprinted from *Acta Materialia*, Volume **53**, Asaro, R. J., and Suresh, S., Mechanistic models for the activation volume and rate sensitivity in metals with nanocrystalline grains and nano-scale twins, pp. 3369–3382. Copyright (2005), with permission from Elsevier.)

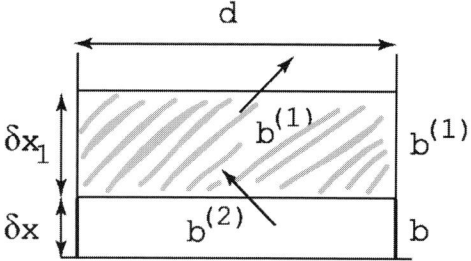

Figure 5.28. Emission of a leading partial dislocation followed by its trailing partial dislocation. The spacing between the two partial dislocations depends on the two stacking energies as described in the text. In some cases, i.e. in materials with low stacking-fault energy, the leading partial dislocation will spread entirely across the grain, whereas in materials with high, or moderately high stacking energies, either full dislocations or complexes containing very large faults will be observed. (Reprinted from *Acta Materialia*, Volume **53**, Asaro, R. J., and Suresh, S., Mechanistic models for the activation volume and rate sensitivity in metals with nanocrystalline grains and nano-scale twins, pp. 3369–3382. Copyright (2005), with permission from Elsevier.)

segment in terms of the shear stress τ resolved along the direction of the lead partial dislocation is given as

$$\tau/G = (\alpha - 1)\Gamma/\alpha + 0.33\, b/d. \tag{5.17}$$

Here G denotes the shear modulus, b is the magnitude of the Burgers vector characterizing a perfect lattice dislocation, d is the grain size, the Γ is reduced stacking-fault energy defined as $\Gamma = \gamma/Gb$, γ is the stacking fault energy density per unit area, $\alpha = d/\delta_{\text{eq}}$, and δ_{eq} is the equilibrium spacing of partial dislocations in the unstressed crystal. A perfect dislocation can be emitted when (Asaro *et al.*, 2003; Asaro and Suresh, 2005)

$$\tau/G = b/d. \tag{5.18}$$

5.7 Interaction between deformation modes

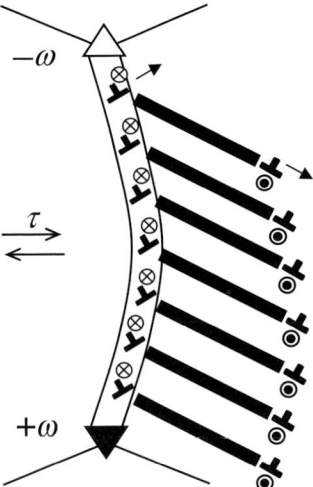

Figure 5.29. Twin deformation is conducted by regularly arranged partial dislocations emitted from a grain boundary.

A similar model has been developed by Zhu *et al.* (2004) in order to account for the experimentally observed existence of split dislocations with anomalously wide stacking faults in deformed nanocrystalline metals. Zhu *et al.* (2004) relate the anomalously wide stacking faults to the nanoscale effect associated with the strong elastic interactions existing between the dislocation segments that form the split dislocation configuration and the residual dislocation at the grain boundary (Figure 5.28). Gutkin and Ovid'ko (2006) have analysed in detail the key modes of the generation of lattice and grain-boundary dislocation loops at pre-existent lattice and grain-boundary dislocation loops in mechanically loaded nanocrystalline materials.

Following experimental data (Liao *et al.*, 2003a,b; 2004a; Zhu *et al.*, 2005a; Wang *et al.*, 2005), a twin-deformation mode has been experimentally observed in mechanically loaded nanocrystalline materials. This phenomenon is closely related to the partial dislocation emission from grain boundaries, because certain regularly arranged groups of partial dislocations are capable of effectively carrying twin deformation; see, e.g., Krujicic and Olson (1998), Mullner and Romanov (2000). In this context, the regularly arranged groups of partial dislocations emitted from grain boundaries can cause twin deformation in nanocrystalline materials (Figure 5.29). Molecular dynamics simulations (Schiotz *et al.*, 1998, 1999; Van Swygenhoven *et al.*, 1999a,b; 2002; Van Swygenhoven and Derlet, 2001; Yamakov *et al.*, 2001; Frozeth *et al.*, 2004a,b; Hasnaoui *et al.*, 2004; Schiotz, 2004; Shimokawa *et al.*, 2005; Wolf *et al.*, 2005) are indicative of deformation twins originating and terminating at grain boundaries in mechanically loaded nanocrystalline metals; see Figure 5.30. Asaro and Suresh (2005) have given a first-approximation analytical examination of the formation of deformation twins at grain boundaries in nanocrystalline materials. However, their analysis is rather approximate and needs further extension in the future.

Thus, emission of lattice dislocations from grain boundaries and twin deformation represent the processes playing the essential role in plastic deformation of nanocrystalline

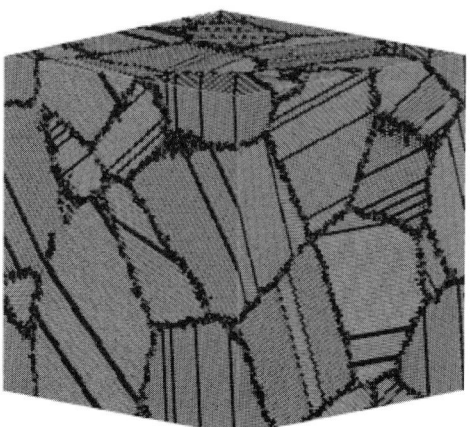

Figure 5.30. The twinned nc-Al sample with an average grain diameter of 12 nm. (Reprinted from *Acta Materialia*, Volume **52**, Frozeth, A. G., Derlet P. M., and Van Swygenhoven, H., The influence of twins on the mechanical properties of nc-Al, pp. 2259–2268. Copyright (2004), with permission from Elsevier.)

materials. This statement is supported by experimental evidence for the emission of lattice dislocations from grain boundaries and twin deformation in nanocrystalline materials (Mukherjee, 2002a,b; Chen *et al.*, 2003; Kumar *et al.*, 2003b; Liao *et al.*, 2003a,b, 2004a,b), results of molecular dynamics simulations (Schiotz *et al.*, 1998, 1999; Van Swygenhoven *et al.*, 1999a,b, 2002; Van Swygenhoven and Derlet, 2001; Yamakov *et al.*, 2001; Frozeth *et al.*, 2004a,b; Hasnaoui *et al.*, 2004; Schiotz, 2004; Shimokawa *et al.*, 2005; Wolf *et al.*, 2005), and analytical theoretical models (Fedorov *et al.*, 2003; Bobylev *et al.*, 2004; Zhu *et al.*, 2004; Zhu *et al.*, 2005b; Asaro and Suresh, 2005; Gutkin and Ovid'ko, 2006) concerning this subject. With the important role of grain boundaries as lattice dislocation sources and deformation twins at the nanoscale level, Cheng *et al.* (2003) and Milligan (2003) suggested a very interesting classification of polycrystalline and nanocrystalline materials. Following Cheng *et al.* (2003) and Milligan (2003), materials are divided into four categories (see also Figure 5.31).

(i) Traditional materials with grain size d larger than approximately 1 μm. In these materials, the lattice dislocation slip is dominant with carriers – perfect lattice dislocations – being generated by mostly dislocation sources (like Frank–Read sources) located in grain interiors.
(ii) Fine-grained materials with grain size d in the range from approximately 30 nm to 1 μm. In these materials, the lattice dislocation slip is dominant with carriers – perfect lattice dislocations – being generated by mostly dislocation sources located at grain boundaries.
(iii) Nano II materials with grain size d in the range from about 10 nm to 30 nm. In these materials, the basic carriers of plastic flow are partial lattice dislocations generated by mostly dislocation sources located at grain boundaries. Since these mobile lattice

5.7 Interaction between deformation modes

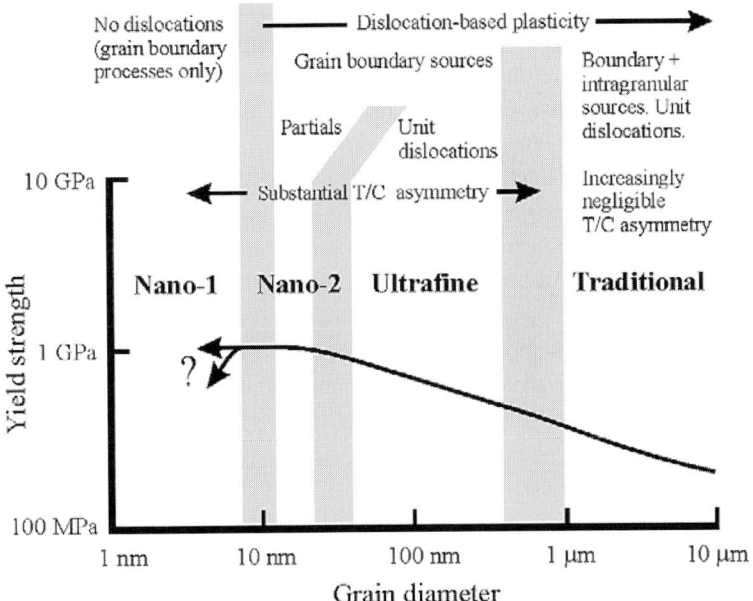

Figure 5.31. Deformation mechanism map for fcc metals. The strength curve is based on copper, which is presented for example purposes. (Reprinted from *Acta Materialia*, Volume **51**, Cheng, S., Spencer, J. A., and Milligan, W. W., Strength and tension/compression asymmetry in nanostructured and ultrafine-grain metals, pp. 4505–4518. Copyright (2003), with permission from Elsevier.)

dislocations are partial, their movement is accompanied by the formation of stacking faults and deformation twins.

(iv) Nano I materials with grain size d lower than approximately 10 nm. In these materials, grain-boundary sliding and other deformation mechanisms conducted by grain boundaries are dominant.

Cheng et al. (2003) and Milligan (2003) emphasized that grain-size ranges used in their classification scheme are very approximate; they can be very different in various materials. In the context of discussion given in this section, notice that the classification (Cheng et al., 2003; Milligan, 2003) indirectly takes into account different forms of the interaction between the lattice dislocation slip and grain-boundary-conducted deformation modes. In this interpretation, grain-boundary-conducted deformation modes provide the generation of either perfect (unit) or partial dislocations from grain boundaries in fine-grained and Nano II solids.

Recently, Zhu et al. (2005b) have suggested a theoretical model of plastic flow in nanocrystalline materials characterized by distributions in grain size. This model is based on the idea of different deformation mechanisms operating in various grain-size ranges, which is very similar to the classification (Cheng et al., 2003; Milligan, 2003) and the model (Asaro and Suresh, 2005). Zhu et al. (2005b) assumed that deformation in materials with grain sizes below, say, 200–300 nm occurs by emission of perfect lattice dislocations from grain boundaries; deformation in materials with grain sizes below, say, 20–30 nm occurs by emission of partial lattice dislocations and twins from grain boundaries; and deformation in materials with even finer grain sizes occurs by grain-boundary sliding. Zhu *et al.* (2005b) involve a grain-size distribution of a specimen in consideration of its plastic flow which, therefore, occurs by simultaneous action of the different deformation mechanisms. The discussed deformation mechanisms – deformation by emission of perfect lattice dislocations from grain boundaries; deformation in materials by emission of partial lattice dislocations and twins from grain boundaries; and deformation by grain-boundary sliding – and their characteristic grain-size ranges are influenced by temperature, crystal lattice parameter, and stacking fault energy. With the relationships between deformation mechanisms, temperature and the material parameters, Zhu *et al.* (2005b) have estimated the strength level of nanocrystalline materials characterized by distributions in grain size. However, the model (Zhu *et al.*, 2005b) uses a lot of adjustable parameters, in which case its predictions are questionable.

5.8 Interaction between deformation modes in nanocrystalline materials. Rotational deformation mode

Now let us turn to a discussion of a rotational deformation mechanism in nanocrystalline materials, that is, plastic deformation accompanied by crystal lattice rotation. Both post deformation and "in situ" electron microscopy experiments are indicative of the essential role of the rotational plastic flow in deformed nanocrystalline materials (Ke *et al.*, 1995; Mukherjee, 2002a,b; Murayama *et al.*, 2002; Shan *et al.*, 2004) (as with conventional coarse-grained materials under high-strain deformation; see reviews (Romanov and Vladimirov, 1992; Klimanek *et al.*, 2001)). Also, molecular dynamics simulations (Schiotz *et al.*, 1999; Shimokawa *et al.*, 2005) have revealed crystal lattice rotations in plastically deformed nanocrystalline materials. For instance, plastic deformation accompanied by crystal lattice rotation (Figure 5.32) has been observed in nanocrystalline Al structures simulated by molecular dynamics methods (Shimokawa *et al.*, 2005).

In general, according to the theory of rotational deformation in conventional polycrystalline solids (Romanov and Vladimirov, 1992; Klimanek *et al.*, 2001), rotational deformation occurs via the motion of grain-boundary disclinations, line defects separating grain-boundary fragments with different misorientation parameters. (The notion of grain-boundary disclinations is effectively exploited in the theory of various grain boundary structures in polycrystalline bulk materials and films (Bollmann, 1989; Bobylev *et al.*, 2001; Gutkin and Ovid'ko, 2001; Seefeldt, 2001). A simple example of the rotational deformation accompanied by crystal lattice rotation in a misorientation band is shown in Figure 5.33. A momentary change of the crystal lattice orientation within the misorientation band – a momentary transfer from the initial state (Figure 5.33a) to the final state (Figure 5.33f) of a deformed crystal – is energetically forbidden. The rotational deformation localized in a misorientation band occurs via consequent stages associated with movement of a dipole of grain-boundary disclinations (Figure 5.33a–f). Movement of a dipole of grain-boundary disclinations (Figure 5.33a–f) with the interspacing h between the disclinations carries

5.8 Rotational deformation mode

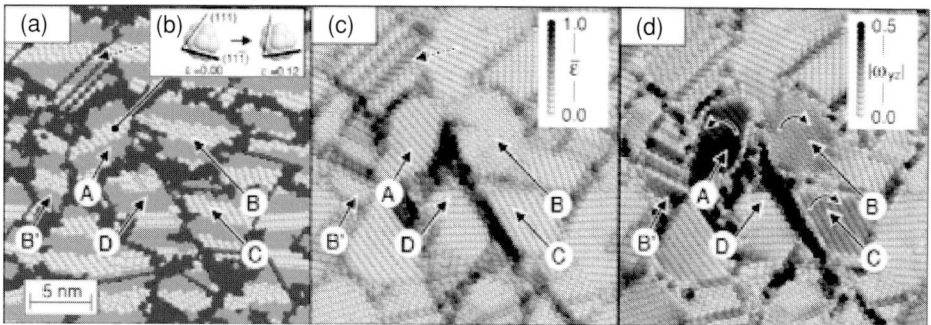

Figure 5.32. Intergranular deformation of nanocrystalline Al in a 5 nm grain when strain $\varepsilon = 0.12$. (a) No dislocation cores exist in the grains. Gaps in the stripes at the grain boundary between grain C and grain D are visible. (b) The crystal orientation of grain A when $\varepsilon = 0.00$ and 0.12. (c) Distribution of the equivalent strain $\bar{\varepsilon}$ from a strain 0.00 and 0.12. Dark-colored atoms are observed at the grain boundary between grain C and grain D. Hence, grain-boundary sliding occurs at this grain boundary. No dark-colored atom exists in grain A, grain B, or grain C. Therefore, no crystal slip occurs in these grains. (d) Distribution of the absolute value of an antisymmetric component of displacement gradient $|\omega_{yz}|$ from $\varepsilon = 0.00$ and 0.12. Grain A, grain B, and grain C consist of dark-colored atoms. Therefore, crystal rotation occurs without crystal slips in these grains. Arrows show the directions of the respective crystals' rotations. (Reprinted figure with permission from *Physical Review B*, Volume **71**, Shimokawa. T., Nakatani, A., and Kitagawa, H., Grain-size dependence of the relationship between intergranular and intragranular deformation of nanocrystalline Al by molecular dynamics simulations, Article 224110. Copyright (2005) by the American Physical Society.)

plastic deformation equivalent to that carried by movement of a superdislocation with Burgers vector magnitude (Romanov and Vladimirov, 1992; Klimanek *et al.*, 2001):

$$B \approx 0.5h \tan(0.5\,\omega). \tag{5.19}$$

Grain-boundary disclination defects and their configurations in coarse-grained polycrystals have been observed in numerous experiments; see the review articles by Romanov and Vladimirov (1992) and Klimanek *et al.* (2001). Grain-boundary disclinations are real structural elements of grain boundaries, causing rotational deformation in coarse-grained polycrystals (Romanov and Vladimirov, 1992; Klimanek *et al.*, 2001; Romanov, 2003). Also, grain-boundary disclinations and their dipole configurations are intensively formed in nanocrystalline materials during their fabrication at highly nonequilibrium conditions, say, ball milling or mechanical compaction (Gryaznov and Trusov, 1993; Murayama *et al.*, 2002). In this context, the notion of grain-boundary disclinations is definitely effective in a description of the rotational deformation mode in nanocrystalline solids (Ovid'ko, 2002; Gutkin *et al.*, 2002; Gutkin and Ovid'ko, 2004a).

Motion of grain-boundary disclinations in plastically deformed solids is commonly treated as that associated with absorption of lattice dislocations (that are generated and move in grains under the action of mechanical load) by grain boundaries (Romanov and Vladimirov, 1992; Valiev and Langdon, 1993; Klimanek *et al.*, 2001). This micromechanism, according to paper (Valiev and Langdon, 1993), is responsible for experimentally observed grain rotations in polycrystalline materials during (super)plastic deformation.

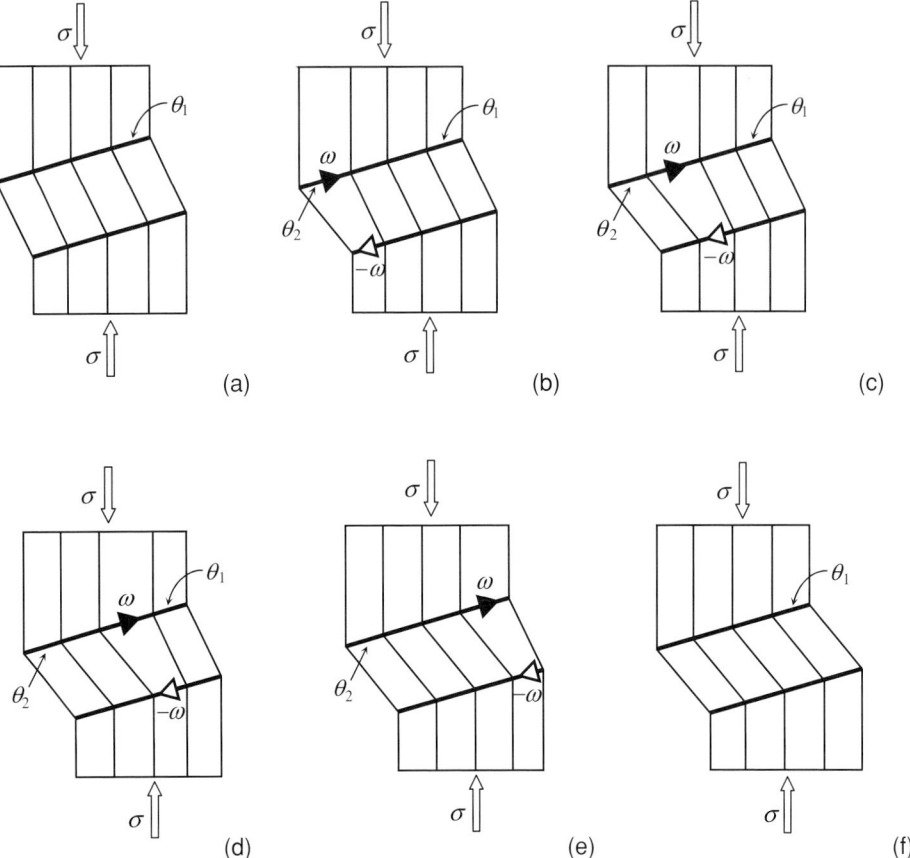

Figure 5.33. Evolution of misorientation band in a mechanically loaded crystal (schematically). Dipole of grain-boundary disclinations (triangles) is formed and moves causing plastic deformation. Movement of disclination characterized by disclination strength $+\omega$ ($-\omega$, respectively) along a grain boundary characterized by misorientation θ_1 ($-\theta_1$, respectively) results in change of grain-boundary misorientation by ω ($-\omega$, respectively).

Recently, a theoretical model (Gutkin *et al.*, 2002, 2003b) has been suggested to describe the rotational deformation in nanocrystalline solids as the processes occurring mostly via the motion of grain-boundary disclinations and their dipoles, associated with the emission of dislocation pairs from grain boundaries into the adjacent grain interiors (Figure 5.34). It is an example of interacting deformation modes in which the rotational deformation and lattice dislocation slip support each other.

Another case of interacting deformation modes is the crossover from grain-boundary sliding to the rotational deformation in nanocrystalline materials (Gutkin *et al.*, 2003a). As has been discussed in Section 5.5, triple junctions of grain boundaries, where grain-boundary planes change their orientations, serve as obstacles for the grain-boundary dislocations that carry grain-boundary sliding. In these circumstances, grain-boundary dislocations stopped at a triple junction are capable of being split into climbing grain-boundary dislocations (Figure 5.35). When this process repeatedly occurs at a triple junction, it results in the

5.8 Rotational deformation mode

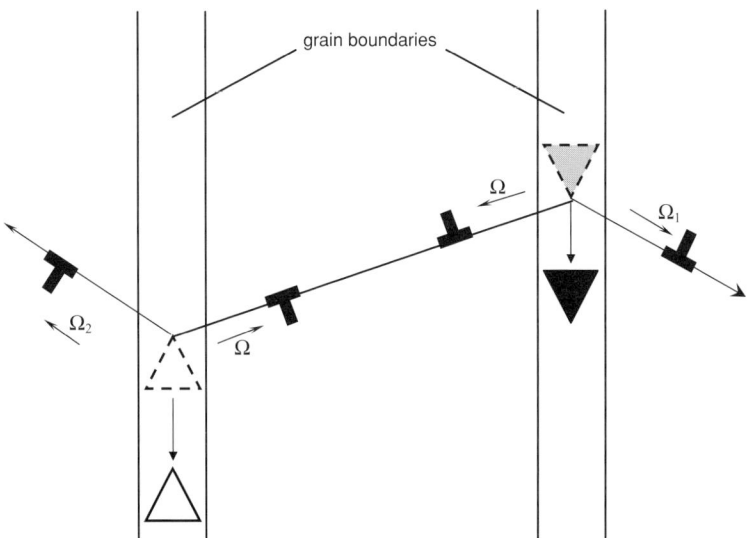

Figure 5.34. Motion of a dipole of grain-boundary disclinations (triangles) is accompanied by emission of lattice dislocations from the grain-boundaries into the adjacent grains.

formation of two walls of dislocations climbing along the grain boundaries adjacent to the triple junction (Figure 5.12). The climbing dislocation walls cause crystal lattice rotation in the grain interior, in which case the repeatedly occurring splitting of gliding grain-boundary dislocations at the triple junction provides the crossover from the grain-boundary sliding to the rotational deformation mode.

Theoretical analysis (Gutkin et al., 2003a) of the energy characteristics of the splitting has indicated that the splitting (Figure 5.35) is energetically favorable in certain ranges of parameters of the system. More precisely, in contrast to the situation considered by Hahn and Padmanabhan (1997), Hahn et al. (1997), and Gutkin et al. (2004b), with grain-boundary sliding which effectively occurs (changing its direction) at triple junctions with low values of the triple-junction angle, the splitting (Figure 5.35) effectively occurs at triple junctions with large values of the triple-junction angle. That is, grain-boundary sliding and rotational mode act as alternative and complementary deformation mechanisms at triple junctions with different geometric parameters. The experimentally detected (Ke et al., 1995; Zelin and Mukherjee, 1996; Mukherjee, 2002a,b; Murayama et al., 2002; Shan et al., 2004) grain rotations in superplastically deformed nano- and microcrystalline materials where grain-boundary sliding is the dominant deformation mechanism, support the theoretical model (Gutkin et al. 2003a).

In general, superplasticity involves also the conventional lattice dislocation slip which is commonly viewed to provide accommodation for grain-boundary sliding and supply lattice dislocations to grain boundaries where the lattice dislocations split into grain-boundary dislocations being carriers of grain-boundary sliding (Valiev and Langdon, 1993; Zelin and Mukherjee, 1996). Following the model of Gutkin et al. (2003a), the lattice dislocations absorbed by grain boundaries also enhance a climb of grain-boundary dislocations that

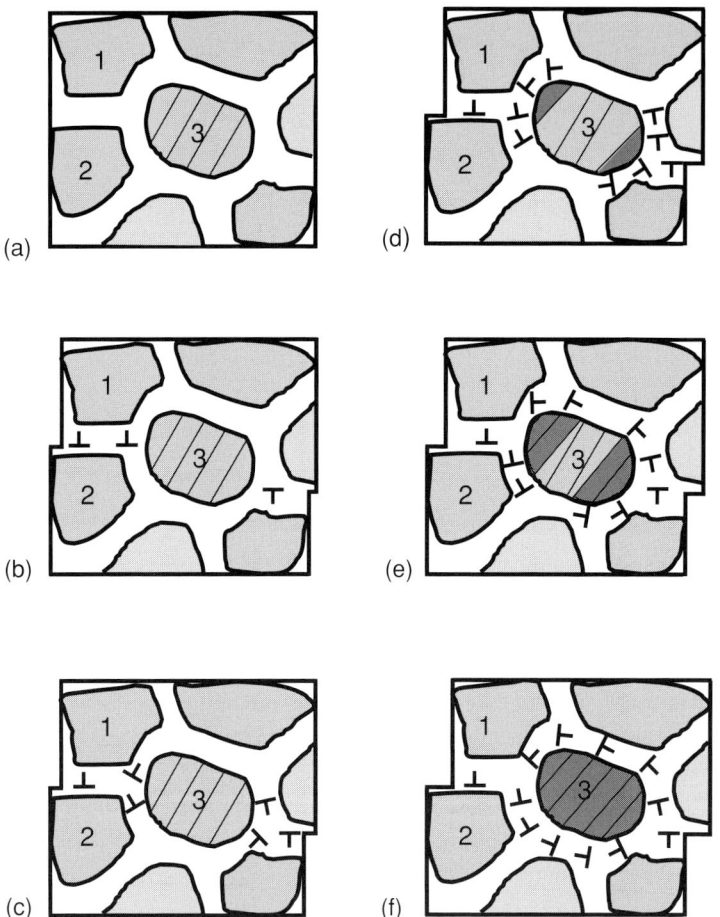

Figure 5.35. Combined action of grain-boundary sliding and rotational deformation mode. (a) Nanocrystalline specimen in a non-deformed state. (b) Grain-boundary sliding occurs via motion of gliding grain-boundary dislocations under shear stress action. (c) Gliding dislocations split at triple junctions of grain boundaries into climbing dislocations. (d), (e) and (f) The splitting of gliding grain-boundary dislocations repeatedly occurs causing the formation of walls of grain-boundary dislocations whose climb is accompanied by crystal lattice rotation in a grain.

carry the rotational deformation. The enhanced grain-boundary dislocation climb stimulated by conventional dislocation slip causes the two following effects crucially important for superplasticity in nanocrystalline materials: (a) plastic deformation is spread in the direction perpendicular to the direction of the maximum shear stress action; (b) triple junctions with large abutting angles, that do not conduct grain-boundary sliding (Gutkin et al., 2003a), effectively conduct the crossover from the grain-boundary sliding to the rotational mode and, therefore, do not serve as stress concentrators enhancing the nucleation of microcracks. As a result of the effects (a) and (b), neither plastic-flow localization nor failure processes occur in a nanocrystalline material which, therefore, is capable of sustaining large plastic deformations. In this context, superplastic deformation in nanocrystalline materials occurs

5.8 Rotational deformation mode

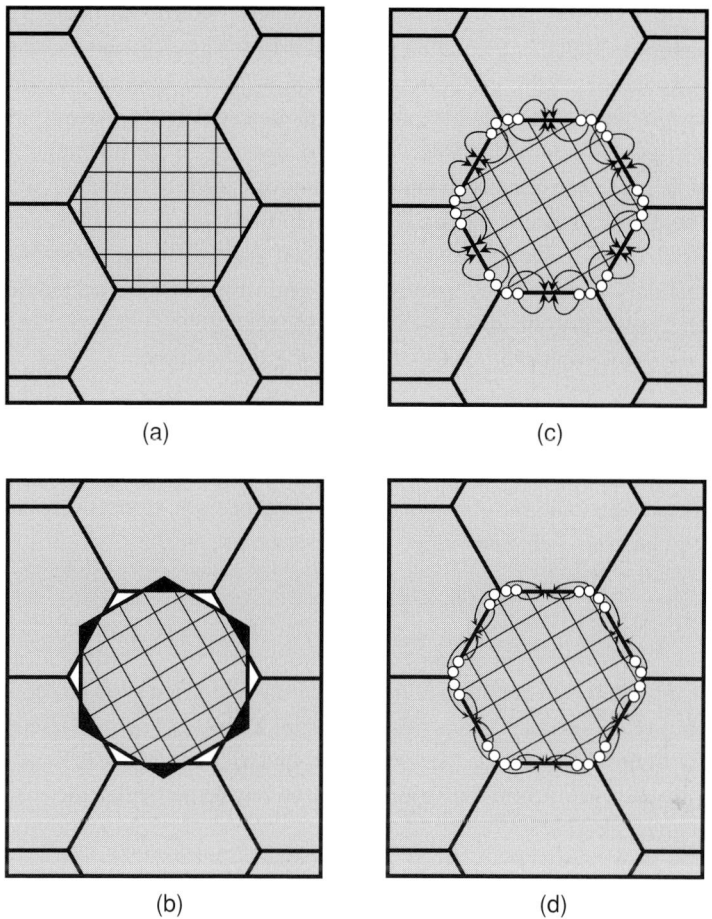

Figure 5.36. Grain rotation in a polycrystalline solid. (a) Initial state. (b) Non-accommodated grain rotation results in the formation of small regions (black regions near boundaries of rotated grain) where neighboring grains interpenetrate and small regions (white regions near boundaries of rotated grain) where voids are formed. (c) Accommodation by bulk diffusion. (d) Accommodation by grain-boundary diffusion.

due to the combined action of grain-boundary sliding and rotational mode enhanced by conventional dislocation slip. The idea of the essential role of interaction between different deformation modes for superplasticity is worth being experimentally tested and theoretically analysed in future studies of nanocrystalline materials exhibiting high-strain-rate superplasticity or good tensile ductility.

Notice that grain rotations can occur in polycrystalline and nanocrystalline bulk materials and films in the absence of mechanical load. Such grain rotations are not related to plastic deformation. They are driven by a release of the grain-boundary energy dependent on the boundary misorientation parameters (Harris *et al.*, 1998; Moldovan *et al.*, 2001) (Figure 5.36). As to details, any grain is bounded by grain boundaries, and energy density $(\gamma_{GB})_i$ of the ith boundary per unit area depends on its misorientation parameter θ_i. The grain tends to rotate, if the sum $\Sigma\, S_i\, d(\gamma_{GB})_i/d\theta_i$ is non-zero, where S_i is the area of the ith grain boundary. In this case, changes in both misorientation parameters of the grain boundaries and thereby crystal lattice orientation of the grain are energetically favorable. Grain rotations in the absence of mechanical load have been experimentally detected in polycrystalline Au films with columnar grains (Harris *et al.*, 1998). Grain rotations under consideration need to be accommodated, because non-accommodated grain rotation leads to formation of voids and regions where neighboring grains overlap (Figure 5.36b).

Moldovan *et al.* (2001) have theoretically described grain rotations driven by a release of the grain-boundary energy. They assumed that grain rotations are accommodated by diffusional mass transfer in grain interiors (Figure 5.36c) and along grain boundaries (Figure 5.36d). Following (Moldovan *et al.*, 2001), the rate of grain rotation

$$\omega = \Omega_L + \Omega_{GB} \tag{5.20}$$

is the sum of the two contributions, Ω_L and Ω_{GB}, associated with accommodation of grain rotation by lattice (bulk) and grain-boundary diffusion processes, respectively. These contributions depend on grain size d as follows (Moldovan *et al.*, 2001):

$$\Omega_L = \Omega_c (d_c/d)^3, \tag{5.21}$$
$$\Omega_{GB} = \Omega_c (d_c/d)^4. \tag{5.22}$$

Here Ω_c and d_c are the characteristic grain rotation rate and grain size, respectively. The characteristic grain rotation rate Ω_c depends on the sum $\Sigma S_i d(\gamma_{GB})_i/d\theta_i$ characterizing the driving force for grain rotation. Also, parameters Ω_c and d_c are sensitive to D_{bulk} and D_{GB}, bulk and grain-boundary diffusion coefficients, respectively. In particular, the characteristic grain size d_c is given as

$$d_c = 3\delta_d D_{GB}/D_{bulk}, \tag{5.23}$$

where δ_d is the diffusion width of grain boundaries along the periphery of the rotated grain. The grain size d_c characterizes crossover from accommodation mechanism controlled by grain-boundary diffusion to that controlled by bulk diffusion (Moldovan *et al.*, 2001). For $d \ll d_c$, grain-boundary diffusion is dominant in accommodation of grain rotation. For $d \gg d_c$, accommodation by bulk diffusion controls the rate of grain rotation. Following estimates (Moldovan *et al.*, 2001) in the model case of Pd at temperature $T = 1400\ °C$, the value of d_c is around 1.8 μm, the characteristic rate $\Omega = \Omega_{GB}$ of grain rotation (controlled by grain-boundary diffusion accommodation) in nanocrystalline Pd with grain size $d = 10$ nm is around 10^7 rad s^{-1}, and the characteristic rate $\Omega = \Omega_L$ of grain rotation (controlled by bulk diffusion accommodation) is around 10^{-8} rad s^{-1} in polycrystalline Pd with grain size $d = 0.1$ mm.

With results of theoretical analysis and estimates given by Moldovan *et al.* (2001), one can conclude that grain rotations in nanocrystalline materials are controlled by grain-boundary diffusion accommodation, and their rate Ω is extremely high. In fact, very high values of grain rotation rate in nanocrystalline materials indicate that crystal lattice orientations in grains of a nanocrystalline specimen very quickly become equilibrium ones corresponding to minimum of the sum boundary energy depending on boundary misorientation parameters.

5.8 Rotational deformation mode

It is worth noting that plastic deformation in nanocrystalline materials causes essential changes in the grain-boundary structures, for instance, because of deformation modes conducted by grain boundaries, absorption of lattice dislocations at grain boundaries and emission of lattice dislocations from grain boundaries (see Sections 5.4–5.7). In these circumstances, plastic deformation in nanocrystalline material essentially changes the energy of grain boundaries and its dependence on boundary misorientation parameters. Owing to the changes in question, grains tend to rotate in nanocrystalline materials during plastic deformation. In doing so, following the theory (Moldovan et al., 2001), the rate of diffusion-accommodated grain rotation increases with diminishing grain size d. This statement is worth being taken into account in explanation of experimental observation (Ke et al., 1995; Zelin and Mukherjee, 1996; Mukherjee, 2002a,b; Murayama et al., 2002; Shan et al., 2004) of grain rotations in deformed nanocrystalline materials.

The accommodation of grain rotations by diffusion (Figure 5.36c and d) is not the sole accommodation mechanism in nanocrystalline materials. Generally speaking, besides the diffusional accommodation (Figure 5.37c) for grain rotations in nanocrystalline materials (Figure 5.37a and b), there exist such accommodation mechanisms as the accommodation by plastic flow in the rotated grain interior (Figure 5.37d), the elastic accommodation resulting in the formation of grain boundary disclinations (Figure 5.37e), the elastic accommodation followed by emission of lattice and partial dislocations into adjacent nanoscale grains (Figure 5.37f). The latter accommodation mechanism, in fact, causes plastic flow in neighboring grains of the rotated grain. These accommodation mechanisms for grain rotations are similar to those for grain-boundary sliding (Figure 5.13) in nanocrystalline materials. However, grain-boundary sliding is also effectively accommodated by local migration of grain boundaries and their triple junctions (Figure 5.13e). In the case of the rotational deformation, local migration of grain boundaries plays an even more significant role as a specific mode of the rotational deformation in nanocrystalline materials (Gutkin and Ovid'ko, 2005). Let us discuss this subject below.

Recent experimental observations (Mukherjee, 2002a; Jin et al., 2004; Soer et al., 2004) and computer simulations (Haslam et al., 2003) indicate that grain-boundary migration and grain-growth processes intensively occur in nanocrystalline materials at plastic and superplastic deformation regimes. However, the nature of these processes and their role in plastic flow of nanocrystalline materials are not understood. Recently, Gutkin and Ovid'ko (2005) have suggested a theoretical model describing stress-induced grain-boundary migration as a new mode of rotational plastic deformation in nanocrystalline materials. At the end of this section, we briefly discuss this mode of rotational plastic deformation as a manifestation of the specific effects of the nanostructure on deformation at the nanoscale level.

Following (Gutkin and Ovid'ko, 2005), let us consider a model arrangement of rectangular grains with low-angle tilt boundaries, including a vertical grain boundary, a finite wall of periodically spaced perfect lattice dislocations, with ends at points A and B as shown in Figure 5.38a and b. Under an external shear stress τ, the vertical grain boundary migrates from initial position AB to a new position A'B'. In this case, both plastic deformation carried by glide of the perfect lattice dislocations (belonging to the vertical boundary) and associated change of crystal lattice orientation occur in the area ABB'A' swept by the migrating

266   5 Mechanical properties of structural nanocrystalline materials

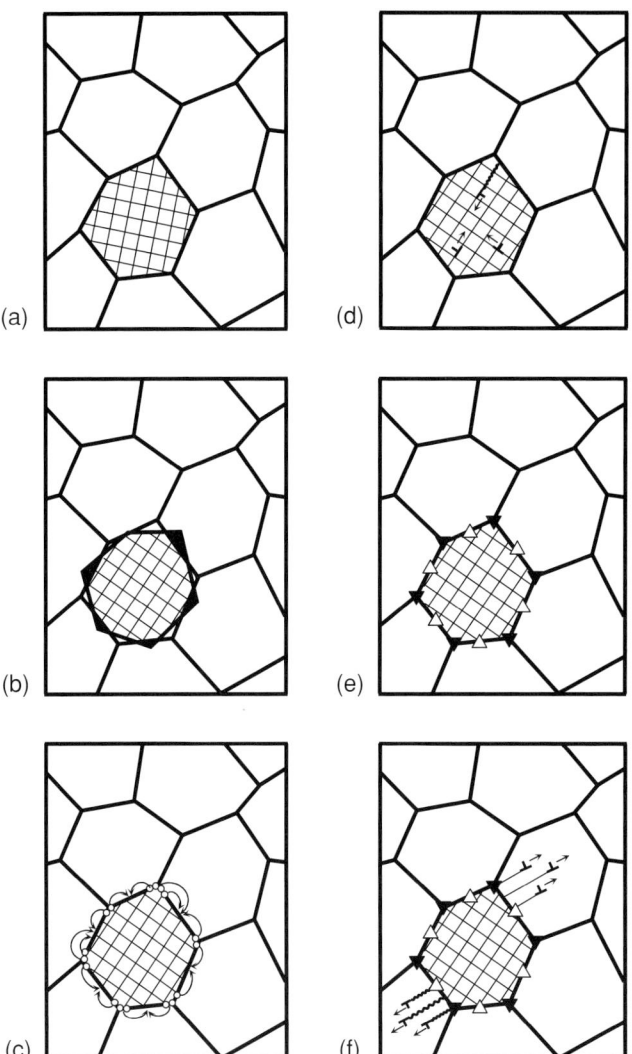

Figure 5.37. Grain rotation in nanocrystalline solid. (a) Initial state. (b) Non-accommodated grain rotation results in the formation of small regions (black regions near boundaries of rotated grain) where neighboring grains interpenetrate and small regions (white regions near boundaries of rotated grain) where voids are formed. (c) Diffusional accommodation. (d) Lattice slip accommodation. (e) Elastic accommodation results in the formation of grain-boundary disclinations (open and black triangles). (f) Elastic accommodation is followed by emission of perfect and partial lattice dislocation emission from disclinated grain boundaries.

boundary (Figure 5.38b). Thus, stress-induced migration of the low-angle tilt grain boundary leads to rotational deformation, plastic deformation accompanied by crystal lattice rotation.

In the initial state (Figure 5.38a and b), the grain boundaries form two triple junctions A and B which are supposed to be geometrically compensated. There are no angle gaps at

5.8 Rotational deformation mode

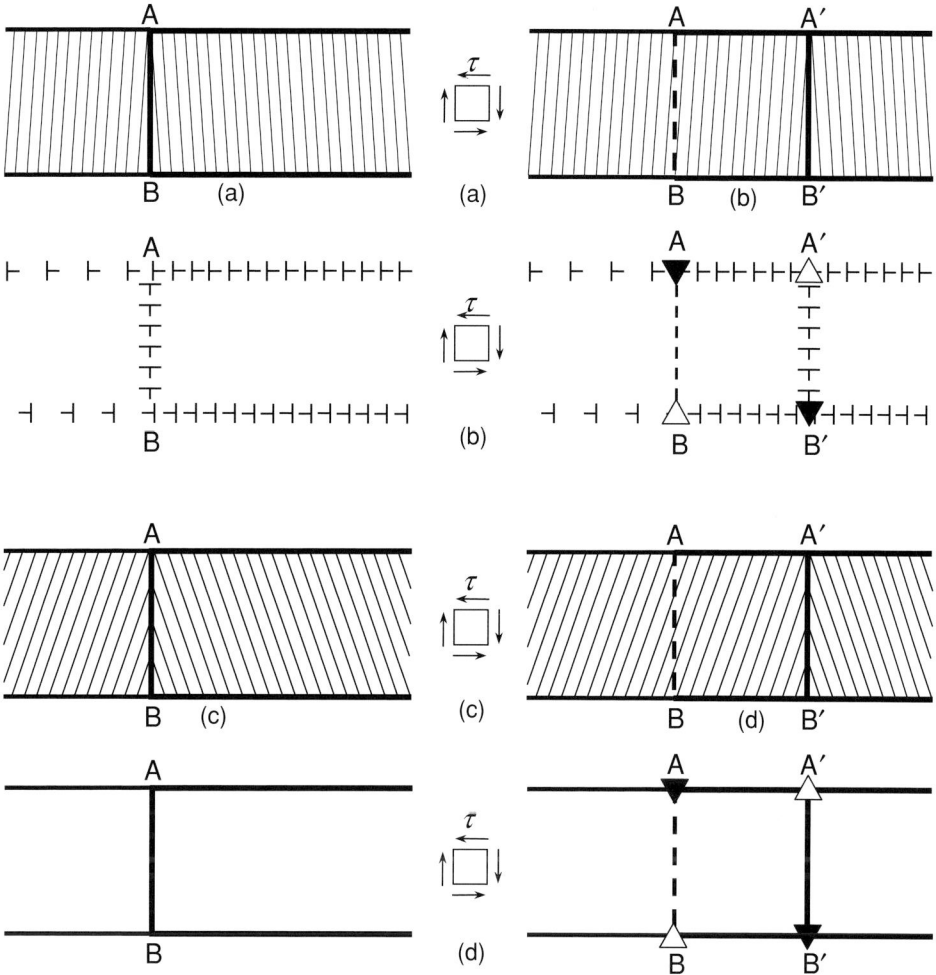

Figure 5.38. Stress-induced migration of a low-angle (a,b) or high-angle (c,d) grain boundary from position AB to position A'B' as a mechanism of rotational deformation realized through the glide of a wall of lattice dislocations (a,b) or motion of a dipole of wedge disclinations (c,d), respectively (see the text for details).

these triple junctions or, in other words, the sum of grain-boundary misorientation angles at each of these junctions is equal to zero. When the vertical grain boundary – a finite wall of lattice dislocations – characterized by the tilt misorientation parameter ω migrates from the position AB to a position A'B' (Figure 5.38a and b), the angle gaps $-\omega$ and $+\omega$ appear at the grain-boundary junctions A' and B', respectively, and two new triple junctions A' and B' are formed. These triple junctions A' and B' are characterized by the angle gaps $+\omega$ and $-\omega$, respectively, carried by the grain boundary. In the theory of defects in solids, straight line defects (junctions) A, B, B', and A' characterized by the angle gaps $\pm\omega$ are defined as partial wedge disclinations (rotational defects) that serve as powerful stress sources characterized by the disclination strength $\pm\omega$ (Romanov and Vladimirov, 1992;

Klimanek *et al.*, 2001). Motion of the disclinations produces rotational plastic deformation (Romanov and Vladimirov, 1992; Klimanek *et al.*, 2001). Thus, stress-induced migration of the low-angle tilt grain boundary carries rotational deformation and results in the formation of a quadrupole of wedge disclinations A, B, B′, and A′ characterized by strength values $\pm\omega$ as shown in Figure 5.38a and b (Gutkin and Ovid'ko, 2005). The same processes occur when the grain boundaries are high-angle tilt boundaries and the grain boundary migrates (Figure 5.38c and d), in which case ω is the tilt misorientation parameter of the high-angle grain boundary.

Gutkin and Ovid'ko (2005) have calculated the critical shear stress $\tau = \tau_{c1}$, the lowest stress at which generation of the disclination configuration AB−A′B′ becomes energetically favorable or, in other words, stable migration of the vertical grain boundary starts to occur. This stress is as follows:

$$\tau_{c1} \approx \frac{D\omega a}{d} \ln \frac{d}{a}. \quad (5.24)$$

Here $D = G/2\pi(1-\nu)$, G denotes the shear modulus, ν is the Poisson ratio, and a is the crystal lattice parameter. It is seen that τ_{c1} strongly depends on the grain size d.

Let us estimate the characteristic values of τ_{c1} for the model case of pure nanocrystalline Al with the elastic moduli $G = 27$ GPa and $\nu = 0.31$, and $a \approx 0.25$ nm. We take two characteristic values for the disclination strength $\omega = 0.085$ ($\approx 5°$) and 0.52 ($\approx 30°$). Then for the grain-size values $2a = 10, 30$ and 100 nm formula (5.21) gives $\tau_{c1} \approx 47$, 23.5 and 7.6 MPa at $\omega = 0.085$ ($\approx 5°$), and $\tau_{c1} \approx 288, 144$ and 46.5 MPa at $\omega = 0.52$ ($\approx 30°$), respectively. These stress values are easily available in real experiments with nanocrystalline Al. Therefore, the grain-boundary migration may give a notable contribution to plastic deformation of nanocrystalline materials where usual dislocation activity is suppressed by small grain size.

5.9 Fracture mechanisms in nanocrystalline materials. Generation, growth and convergence of nanocracks

In order to understand the fundamentals of the unique mechanical properties of nanocrystalline materials, it is very important to adequately describe fracture processes in nanocrystalline materials showing low ductility and analyse mechanisms for fracture suppression in nanocrystalline materials showing good ductility or superplasticity. As noted in Chapter 4, the fracture behavior of nanocrystalline materials is not well defined because of all the questions regarding intrinsic versus extrinsic factors that may affect the ductility and therefore fracture of these materials. Nevertheless, with available experimental data in this area (see Chapter 4), one can distinguish the following tendencies in the fracture behavior of nanocrystalline materials. Both ductile and brittle fracture processes occur in nanocrystaline materials, in which case the fracture mode is sensitive to grain size. In particular, there are several experimentally detected examples of nanocrystalline materials (having mean grain size above 20 nm) showing ductile fracture with preceding neck formation and dimpled structures at fracture surfaces (Kumar *et al.*, 2003a,b; Li and Ebrahimi, 2004;

5.9 Fracture mechanisms in nanocrystalline materials

Youssef *et al.*, 2004, 2005, 2006). In doing so, ductile fracture is viewed to occur through the microvoid coalescence mechanism. At the same time, following (Li and Ebrahimi, 2004, 2005), with reduction in grain size, there is a shift in fracture mode, from ductile mode (for nanocrystalline Ni with mean grain size around 44 nm) to brittle fracture (for nanocrystalline Ni–15%Fe with mean grain size around 9 nm). In the latter case, the main brittle crack is treated to be formed by the generation of multiple intergranular nano/microscale cracks and their convergence. Also, brittle intergranular fracture by the formation of multiple nano/microscale cracks and their convergence often occurs in superhard nanocomposite coatings with finest grains under mechanical load (see Chapter 4).

In general, fracture processes in nanocrystalline materials are crucially influenced by the large amount of the grain-boundary phase. Grain boundaries serve as preferable places for nanocrack nucleation and growth, because the atomic density is low, and interatomic bonds are weak at grain boundaries compared to the bulk phase. In the case of ductile fracture carried by microvoid coalescence, grain boundaries with their high diffusivity enhance the microvoid growth, a process mediated by diffusion. Besides, the extra energy of the grain-boundary phase contributes to the driving force for inergranular fracture with cracks propagating along grain boundaries and releasing the extra elastic energy of grain boundaries, compared to intragranular fracture with cracks propagating through grain interiors. At the same time, grain boundaries are short and curved at numerous triple junctions in nanocrystalline materials. Therefore, if cracks tend to nucleate and grow along grain boundaries, geometry of grain-boundary ensembles causes restrictions on intergranular fracture processes.

In general, with the large amount of the grain-boundary phase in nanocrystalline materials, there is a strong competition between intergranular and intragranular fracture processes. Either one of these processes dominates or they occur in concurrent ways in mechanically loaded nanocrystalline solids, depending on their structural and material parameters as well as conditions of loading. Besides, fracture processes in nanocrystalline materials compete and interact with plastic deformation processes that have the unique peculiarities due to both the nanoscale and interface effects. In particular, as has been discussed in previous sections of this chapter, plastic deformation in nanocrystalline materials effectively occurs by perfect and partial dislocations emitted from and absorbed at grain boundaries, as well as by grain-boundary defects that carry grain-boundary sliding, grain-boundary diffusional creep, triple-junction diffusional creep, and rotational deformation mode. In these circumstances, one expects that brittle and ductile fracture processes in the nanocrystalline matter have the specific peculiarities due to both the structural peculiarities of nanocrystalline materials and the action of the non-conventional plastic deformation mechanisms.

Experiments (Li and Ebrahimi, 2004, 2005) show that reduction in grain size causes a shift in fracture mode, from ductile mode (for nanocrystalline Ni with mean grain sizes around 44 nm) to brittle fracture (for nanocrystalline Ni–15% Fe with mean grain size around 9 nm). The discussed experimental data are indicative of the tendency in the grain-size effect on fracture mode in nanocrystalline materials, but the critical grain size for transition from ductile fracture to intergranular brittle fracture is not exactly determined. It would be interesting to check if the critical grain size for transition in fracture mode is approximately equal to that for transition from the lattice dislocation slip to deformation mechanisms mediated by grain boundaries. In this context, of particular interest are the experiments (Youssef *et al.*, 2004, 2005, 2006) which are indicative of ductile fracture in nanocrystalline Cu and Al–5%Mg alloy with mean grain size being 23 nm and 26 nm, respectively. The mean grain-size values in these materials are very small and close to the critical grain size ($d_c = 10 - 30$ nm) for transition from the lattice dislocation slip to deformation mechanisms mediated by grain boundaries. However, the microstructural characterization (Youssef *et al.*, 2005) of nanocrystalline Cu samples unambiguously shows the lattice dislocation slip to be active. In this situation, with a very high level of the applied stress, void growth by lattice dislocation emission can effectively occur (Lubarda *et al.*, 2004) and provide ductile fracture of nanocrystalline materials.

Besides grain size, other factors strongly affect fracture in nanocrystalline materials. In particular, crack nucleation and propagation can be dramatically enhanced owing to fabrication-produced pores and contaminations. This effect of artifacts has been demonstrated in the experiments (Hugo *et al.*, 2003) dealing with fracture of nanocrystalline Ni films fabricated by d.c. magnetron sputtering and pulsed laser deposition. Following Hugo *et al.* (2003), the nanocrystalline Ni film material fabricated by d.c. magnetron sputtering and characterized by a narrow distribution in grain size (with mean grain size around 19 nm) contains pores and behaves in a brittle manner, with failure occurring via rapid coalescence of intergranular cracks. At the same time, the nanocrystalline Ni film material fabricated by pulsed laser deposition and characterized by a narrow distribution in grain size (with mean grain size around 17 nm) is free from pores and behaves in a ductile manner, with failure occurring via slow ductile crack growth.

To understand the specific fracture behavior of the nanocrystalline matter, results of microstructural experimental characterization of nanocrystalline materials under mechanical load are of crucial importance. Following "in situ" experimental data (Kumar *et al.*, 2003b), nanovoids and nanocracks in ductile nanocrystalline Ni (with the mean grain size around 30–40 nm) commonly nucleate and grow at grain boundaries and their triple junctions during plastic deformation (Figure 5.39). Molecular dynamics simulations (Farkas *et al.*, 2002; Van Swygenhoven *et al.*, 2003) show nanocracks to be generated at triple junctions of grain boundaries near tips of pre-existent large cracks in nanocrystalline Ni with grain size ranging from 5 nm to 12 nm; see Figure 5.40. In the context discussed, with a very high volume fraction of triple junctions of grain boundaries in nanocrystalline materials, nanocracks at triple junctions can be treated as typical elemental carriers of fracture in nanocrystalline materials. This causes particular interest in a description of the mechanisms for nanocrack generation at triple junctions and their correlation with plastic deformation processes conducted by grain boundary defects. In a paper (Ovid'ko and Sheinerman, 2004a), a theoretical model has been suggested describing nucleation and growth of nanocracks at triple junctions as a result of grain-boundary sliding through these junctions in deformed nanocrystalline materials. Below we will consider in detail the basic statements of the model (Ovid'ko and Sheinerman, 2004a), because it is concerned with typical nanocracks, suggests an effective methodology for their theoretical description and relates the specific nanostructural features with fracture mechanisms operating in mechanically loaded nanocrystalline materials.

5.9 Fracture mechanisms in nanocrystalline materials

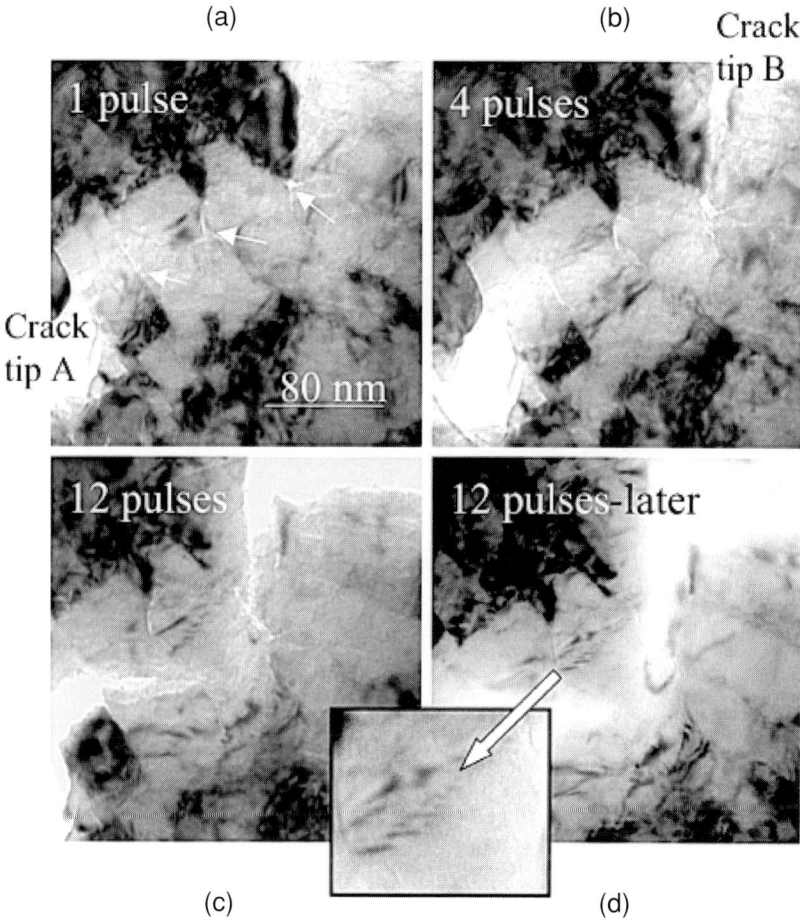

Figure 5.39. A sequence of "freeze-frame" images captured during an in situ deformation test in the TEM of a microtensile specimen. Images (a)–(d) show the microstructural evolution and progression of damage with an increase in the applied displacement pulses. The presence of grain boundary cracks and triple-junction voids (indicated by white arrows in (a)), their growth, and dislocation emission from crack tip B in (b)–(d) in an attempt to relax the stress at the crack tip as a consequence of the applied displacement, can all be seen. The magnified inset in (d) highlights the dislocation activity. (Reprinted from *Acta Materialia*, Volume **51**, Kumar, K. S., Suresh, S., Chisholm, M. F., Horton, J. A., and Wang, P., Deformation of electrodeposited nanocrystalline nickel, pp. 387–405. Copyright (2003), with permission from Elsevier.)

The model (Ovid'ko and Sheinerman, 2004a) describes fracture processes in nanocrystalline materials in which grain boundary sliding essentially contributes to plastic flow. As discussed in Section 5.6, in general, grain-boundary sliding often occurs by local shear events and movement of grain-boundary dislocations in nanocrystalline materials and leads to accumulation of sessile grain-boundary dislocations at triple junctions. In the case of dislocation mode of grain-boundary sliding, mobile grain-boundary dislocations (with the Burgers vector being parallel to grain-boundary planes) move causing the sliding, when

272   5 Mechanical properties of structural nanocrystalline materials

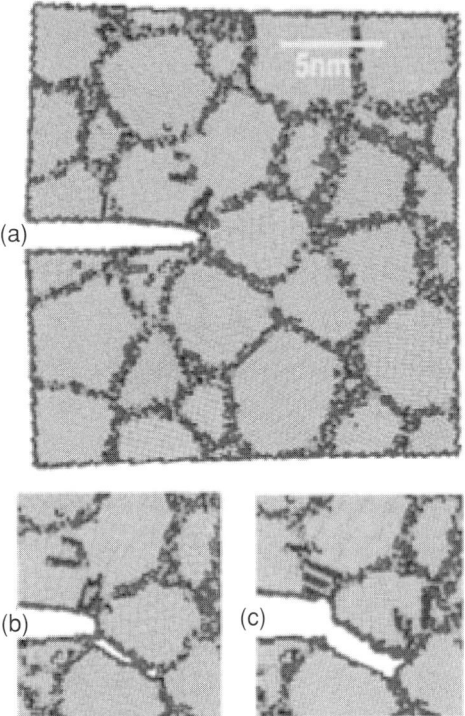

Figure 5.40. Crack-tip configuration in a section of a sample with 5 nm grain size. (a) At the Griffith load. (b) At a load of twice the Griffith value. (c) At a load of 2.5 times the Griffith value. (Reprinted figure with permission from *Physical Review B*, Volume **66**, Farkas, D., Van Swygenhoven, H., and Derlet, P. M., Intergranular fracture in nanocrystalline metals, Article 060101(R). Copyright (2002) by the American Physical Society.)

a mechanical load is applied to the specimen (Figure 5.41a). They are stopped at triple junctions of grain boundaries, where grain-boundary planes are curved and thereby dislocation movement is hampered (Figure 5.41a). At some critical shear stress, grain-boundary dislocations overcome triple junctions, in which case a sessile grain-boundary dislocation is formed at the triple junction (Figure 5.41b) (Gutkin *et al.*, 2003a, 2004b). This process is an elementary act of (super)plastic deformation involving grain-boundary sliding in nanocrystalline materials characterized by a very high volume fraction of triple junctions. The process under consideration repeatedly occurs in a deformed nanocrystalline solid and is accompanied by an increase of the Burgers vector of the sessile dislocation at each step (Figure 5.41c, d and e). Notice that accumulation of sessile dislocations at triple junctions occurs because of grain-boundary sliding by local shear events as well (Figure 5.9). Following Ovid'ko and Sheinerman (2004a), a nanocrack at the triple junction is generated to release the strain energy of the sessile grain-boundary dislocation which resulted from numerous acts of grain-boundary sliding at the triple junction (Figure 5.41f and g). In doing so, the nanocrack may nucleate either in the grain interior (Figure 5.41f) or along a grain boundary adjacent to the triple junction (Figure 5.41g).

5.9 Fracture mechanisms in nanocrystalline materials

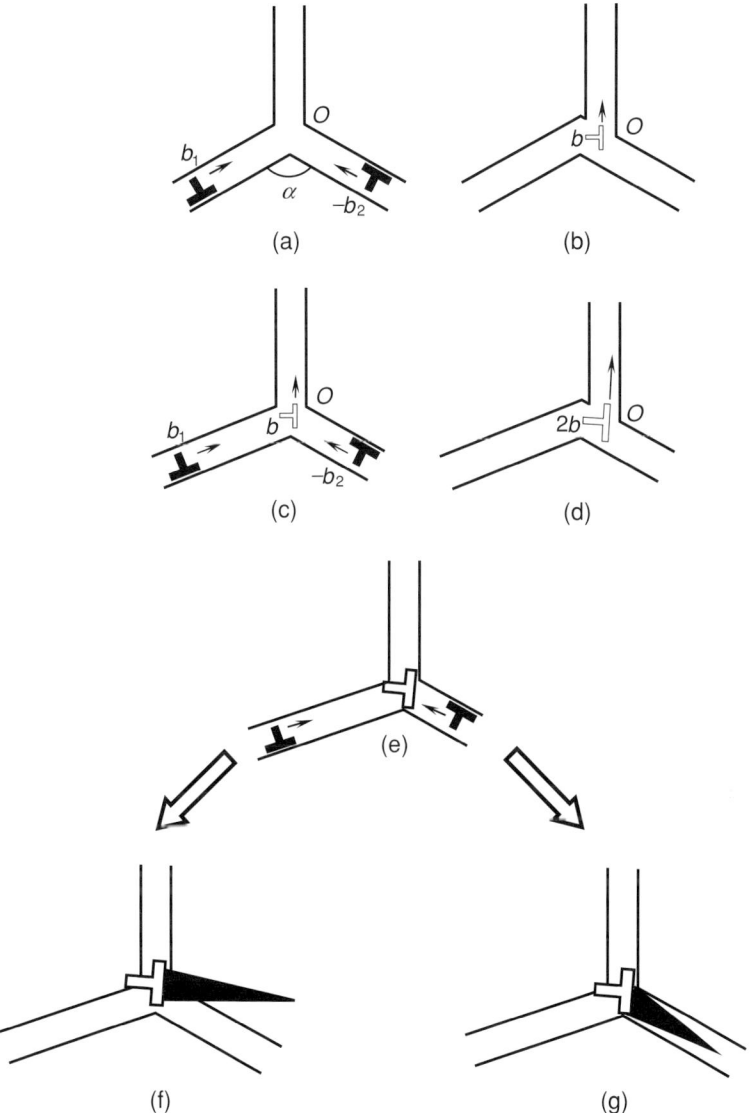

Figure 5.41. Grain-boundary sliding causes (a)–(e) formation of triple junction dislocation whose stress field induces generation of nanocrack either (f) in grain interior or (g) along a grain boundary.

In general, the energetically favorable generation of a crack/nanocrack is characterized in the first approximation by its equilibrium length L_e (see, e.g., Indenbom (1961), Gutkin and Ovid'ko (1994), and Ovid'ko and Sheinerman (2004a)). The equilibrium length L_e of a nanocrack is defined as the nanocrack length corresponding to the maximum or minimum energy of the system. In the first case, a nanocrack tends to rapidly grow (shrink, respectively), if its length L is larger (lower, repectively) than the equilibrium length L_e. That is, the equilibrium state of the nanocrack with length $L = L_e$ is unstable. In the second case (the system energy has a minimum at $L = L_e$), the generation and growth of a nanocrack with length $L < L_e$ are energetically favorable until its length L reaches L_e. That is, the equilibrium state of the nanocrack with length $L = L_e$ is stable. This case includes, in particular, the generation and growth of a triple junction nanocrack (Figure 5.41f and g) which is energetically favorable, if the nanocrack length L is lower than the equilibrium length L_e. This way by which the equilibrium length L_e characterizes triple junction nanocracks (Figure 5.41f and g) generated in the stress field of a sessile grain boundary (super)dislocation is related to the fact that the dislocation stress field rapidly falls with increasing distance from the dislocation line, while the external stress weakly influences the nanocrack generation (Ovid'ko and Sheinerman, 2004a).

In calculation of L_e, Ovid'ko and Sheinerman (2004a) used the configurational force method (Indenbom, 1961) that was effectively exploited in analysis of the generation of plane microcracks in the stress fields of superdislocations (Indenbom, 1961), dislocation pile-ups (Indenbom, 1961), disclination loops (Rybin and Zhukovskii, 1978) and wedge line disclinations (Gutkin and Ovid'ko, 1994). Following the approach (Indenbom, 1961), the configurational force F is defined as the elastic energy released when the crack moves over a unit distance. In the situation with the plane strain state of an elastically isotropic solid with a dislocation, examined in the paper (Ovid'ko and Sheinerman, 2004a), F can be written in its general form as follows (Indenbom, 1961)

$$F = L\pi(1 - v)\left(\tilde{\sigma}_{yy}^2 + \tilde{\sigma}_{xy}^2\right)/4G, \tag{5.25}$$

where G is the shear modulus, v is the Poisson ratio, and $\tilde{\sigma}_{yy}$ and $\tilde{\sigma}_{xy}$ are the mean weighted values of the dislocation stress tensor components σ_{yy} and σ_{xy}, respectively. These mean weighted stress tensor components are calculated using the following formula (Indenbom, 1961)

$$\tilde{\sigma}_{iy} = 2(L\pi)^{-1} \int \sigma_{iy}(x, y = 0)[x/(L - x)]^{1/2}\, dx, \tag{5.26}$$

where $i = x, y$.

The equilibrium length L_e of the nanocrack is derived from the balance $F = 2\gamma_e$ between the release F of the elastic energy and the formation of two new nanocrack surfaces characterized by the surface energy density γ_e per unit area. Here $\gamma_e = \gamma$, the surface energy density per unit area of the free surface, when the nanocrack nucleates in the grain interior (Figure 5.41f), and $\gamma_e = \gamma - \gamma_s/2$, when the nanocrack nucleates along a grain boundary (Figure 5.41g) characterized by the energy density γ_s per unit area of grain boundary plane. The nucleation of the triple junction nanocrack (Figure 5.41f and g) is energetically favorable, if $F > 2\gamma_e$, and unfavorable, if $F < 2\gamma_e$.

Ovid'ko and Sheinerman (2004a) calculated the dependences of F on non-dimensional length L/b of the nanocrack, where b is the Burgers vector magnitude that characterizes mobile grain-boundary dislocations. These dependences of F (written in units of $Gb/[16\pi(1 - v)]$, where G is the shear modulus, and v is the Poisson ratio) are shown in Figure 5.42, for $\beta = 2\pi/3$, $\tau/G = 0.01$, $v = 0.3$ and different values of n, the number

5.9 Fracture mechanisms in nanocrystalline materials

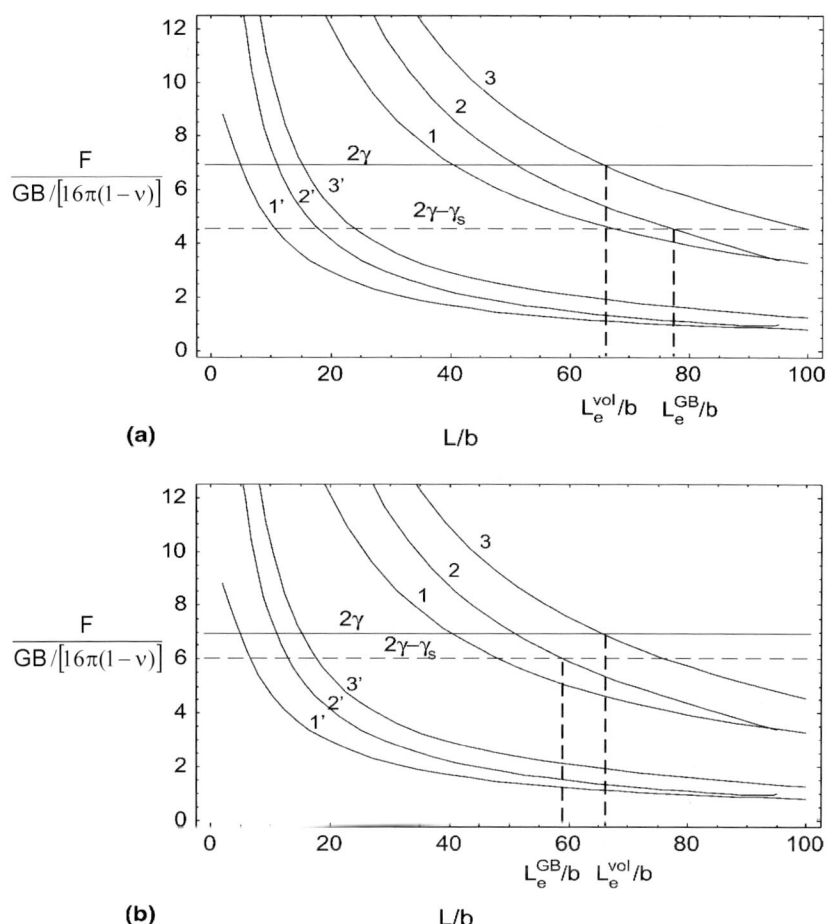

Figure 5.42. Dependence of the configurational force F (in units of $Gb/[16\pi(1-v)]$) on non-dimensional length L/b of the nanocrack, for $\beta = 2\pi/3$, $\tau/G = 0.01$, $l_1 = l_2 = 100b$, $v = 0.3$; $n = 10$ and $\alpha_1 = -\pi/3, \pi/3$ and $2\pi/3$ (see curves 1, 2 and 3, respectively) as well as $n = 5$ and $\alpha_1 = -\pi/3, \pi/3$ and $2\pi/3$ (see curves 1', 2' and 3', respectively). Horizontal lines show values of the effective surface energy density $2\gamma_e$ (in units of $Gb/[16\pi(1-v)]$) in the cases of the nanocrack nucleating in the grain interior (solid line) and along GB plane (dashed line), for $\gamma/(Gb) = 0.1$ and (a) $\gamma_s/(Gb) = 0.07$ and (b) $\gamma_s/(Gb) = 0.03$.

of passes of mobile grain-boundary dislocations through the triple junction, and the angle α_1 characterizing triple junction geometry. The horizontal lines show values of the effective surface energy density $2\gamma_e$ (in units of $Gb/[16\pi(1-v)]$) in the cases of the nanocrack nucleating in the grain interior (solid line) and along a grain-boundary plane (dashed line), shown in Figure 5.41f and g, respectively. The generation of the nanocrack in the grain interior is energetically favorable in the range of parameters where the curve $F(L/b)$ lies above the solid line in Figure 5.42. The generation of the nanocrack along a grain-boundary plane is energetically favorable in the range of parameters where the curve $F(L/b)$ lies

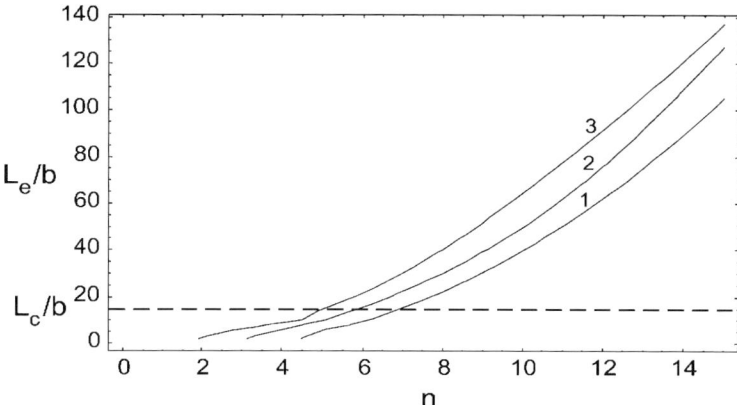

Figure 5.43. Dependencies of non-dimensional equilibrium nanocrack length L_e/b on parameter n, for $\beta = 2\pi/3$, $\tau/G = 0.01$, $l_1 = l_2 = 100b$, $\nu = 0.3$; $\alpha_1 = -\pi/3$, 0 and $2\pi/3$ (see curves 1, 2 and 3, respectively). The dashed horizontal line corresponds to the minimum critical length $L_c = 15b$ at which the nanocrack is generated.

above the dashed line in Figure 5.42. The equilibrium length of the nanocrack corresponds to the point where the curve $F(L/b)$ and the horizontal line (solid line in the case of the nanocrack in the grain interior, and dashed line in the case of the nanocrack nucleating along a grain-boundary plane) intersect. In Figure 5.42, the equilibrium length of the nanocrack in the grain interior and along a grain boundary is denoted as L_e^{vol} and L_e^{GB}, respectively. For $L < L_e$, the nanocrack growth is energetically favorable until its length L reaches L_e.

The notion of a nanocrack (or, more generally, crack) has its sense when the nanocrack length is larger than some critical minimum length L_c being around $5a = 15b$ (with a being the crystal lattice parameter) at which the binding between the atoms of the opposite surfaces of the nanocrack is completely broken. In this context, the generation of a nanocrack is treated to occur, if its equilibrium length $L_e > L_c$. In the opposite case ($L_e < L_c$), the binding between the atoms of the opposite surfaces of the nanocrack causes it to shrink and disappear.

Ovid'ko and Sheinerman (2004a) calculated the dependences of the equilibrium length L_e of a nanocrack at a dislocated triple junction on the parameter n, the number of passes of mobile grain-boundary dislocations through the triple junction. (Also, n characterizes the Burgers vector of the sessile grain-boundary (super)dislocation.) These dependences are shown in Figure 5.43, for different values of the angle α_1. As follows from Figure 5.43, the equilibrium length L_e rapidly increases with rising parameter n in both the cases of the nanocrack growing in the grain interior and along a grain boundary. The dashed horizontal line in Figure 5.43 corresponds to the critical minimum length $L_c = 15b$ of a nanocrack. The points where this horizontal line intersects the curves $L_e(n)$ correspond to the values of n at which the nanocracks are generated. These values are close to 5, indicating that a stable nanocrack is nucleated at a triple junction after just several (about 5) acts of the

5.9 Fracture mechanisms in nanocrystalline materials

grain-boundary dislocation transformation (Figure 5.41) have occurred at the triple junction. Also, L_e grows when the angle β between grain-boundary planes decreases.

Following calculations (Ovid'ko and Sheinerman, 2004a), the equilibrium length L_e rapidly falls with rising free surface energy density γ_e (Figure 5.42). In general, the equilibrium length L_e^{GB} of a nanocrack nucleating along a grain-boundary plane can be either larger or lower than the equilibrium length L_e^{vol} of a nanocrack nucleating in the grain interior. The former case ($L_e^{GB} > L_e^{vol}$) is realized at large values of the grain-boundary energy density γ_s. In doing so, the nucleation of the triple junction nanocrack growing along a grain boundary is energetically preferred compared to the nanocrack growing in the grain interior. The second case ($L_e^{vol} < L_e^{vol}$) is realized at low values of γ_s. In this case, the nucleation of the triple-junction nanocrack growing in the grain interior is preferred.

Thus, the model (Ovid'ko and Sheinerman, 2004a) theoretically describes the nucleation of elemental nanocracks in deformed nanocrystalline materials as a process effectively occurring at triple junctions (Figure 5.41f and g) owing to grain-boundary sliding. The driving force of the nucleation of triple-junction nanocracks is the release of the elastic energy of triple junction (super)dislocations formed because of grain-boundary sliding through triple junctions. The theoretical model (Ovid'ko and Sheinerman, 2004a) accounts for experimental observation (Kumar *et al.*, 2003b) of nanocracks nucleated at triple junctions in deformed nanocrystalline Ni samples exhibiting a substantial ductility. These samples were fabricated by elecrodeposition methods, taking care about absence of nanocracks and nanopores in the as-prepared state. Therefore, triple-junction nanocracks observed by Kumar *et al.* (2003b) in "in-situ" experiments during plastic deformation in nanocrystalline Ni are induced by deformation processes. Notice that the model (Ovid'ko and Sheinerman, 2004a) predicts a certain stability of triple-junction nanocracks. This prediction is in agreement with experiments (Kumar *et al.*, 2003b) in which a good ductility of deformed nanocrystalline materials and a non-catastrophic character of failure processes have been detected.

Now let us discuss the role of diffusion in the generation of triple-junction nanocracks treated as typical elemental carriers of fracture in nanocrystalline materials. Generally speaking, grain-boundary diffusion can be highly enhanced in nanocrystalline materials because of the action of lattice dislocation slip which "supplies" dislocations to grain boundaries where these trapped dislocations climb, split into grain-boundary dislocations and annihilate causing the intensive generation of excess grain-boundary point defects that carry grain-boundary diffusion (see Figure 5.44 and Section 5.11). The enhanced grain-boundary diffusion gives rise to the three following effects responsible for suppression of nucleation of triple-junction nanocracks. (i) The enhanced diffusion provides both intensive movement of vacancies from the local regions where high tensile stresses of the sessile triple-junction dislocations exist and intensive flow of interstitial atoms in the opposite direction (Figure 5.45). In these circumstances, the tensile stresses, in part, are relaxed, and the nucleation of nanocracks is hampered (Ovid'ko and Sheinerman, 2005a). The discussed effect of diffusion gives rise to an increase of the critical plastic strain value at which triple-junction nanocracks (Figure 5.41f and g) are generated. In certain ranges of parameters of a nanocrystalline solid, diffusion is capable of even suppressing the nanocrack generation near

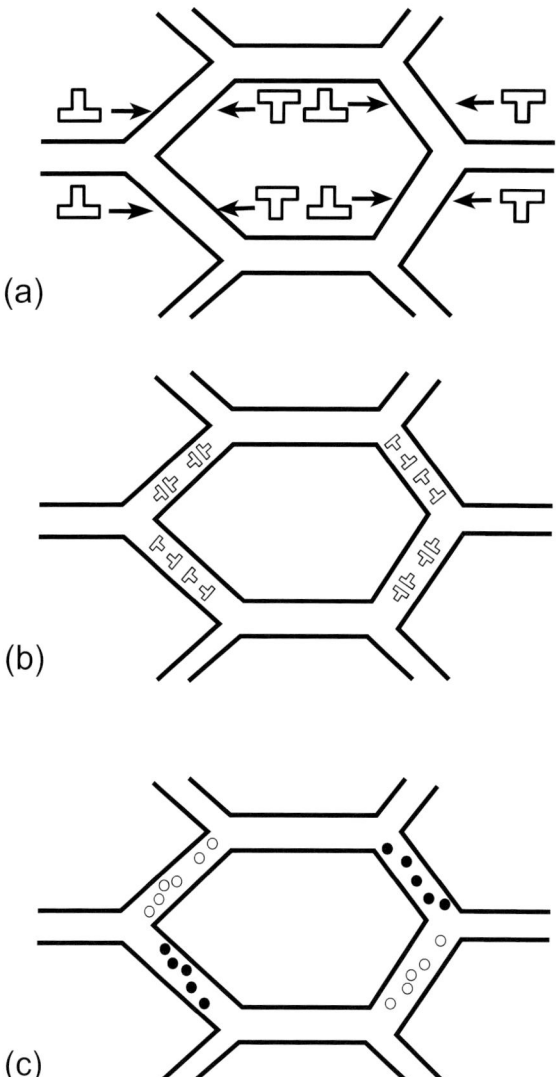

Figure 5.44. Enhancement of grain-boundary diffusion due to absorption of lattice dislocations by grain boundaries (schematically). (a) Lattice dislocations (large open dislocation signs) move under the shear stress action. (b) Climbing dislocations (small open dislocation signs) result from splitting of absorbed lattice dislocations form dipole configurations at grain boundaries. (c) Annihilation of climbing dislocations forming dipole configurations results in the formation of high-density ensembles of nonequilibrium point defects – vacancies (open circles) and interstitials (full circles) – that carry enhanced grain-boundary diffusion.

triple junctions during the extensive stage of plastic deformation (Ovid'ko and Sheinerman, 2005a). (ii) Sessile dislocations formed at triple junctions because of grain-boundary sliding come into reactions with dislocations that intensively climb along grain boundaries (Figure 5.46). These dislocation reactions diminish the Burgers vector magnitudes of the sessile dislocations (Figure 5.46) and thereby decrease the stress concentration at triple junctions.

5.9 Fracture mechanisms in nanocrystalline materials

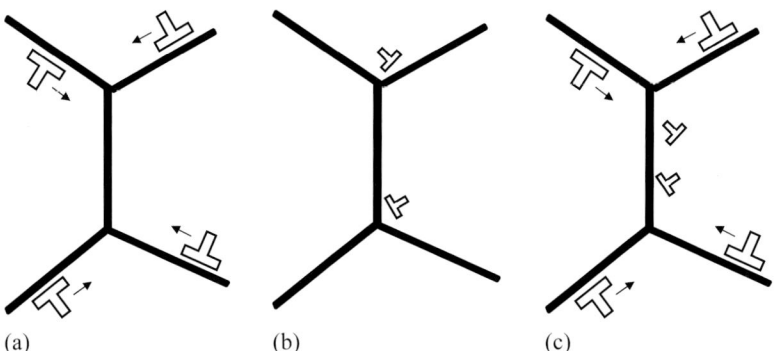

Figure 5.45. Sessile dislocations generated (by grain-boundary sliding) at neighboring triple junctions climb towards each other.

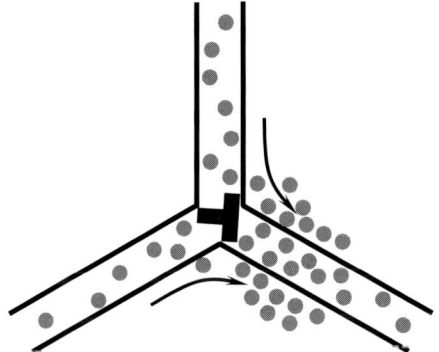

Figure 5.46. Diffusional flow of interstitial atoms causes a partial relaxation of tensile stresses created by triple-junction dislocation.

As a corollary, the grain-boundary dislocation climb (Figure 5.46), whose rate is controlled by grain-boundary diffusion, hampers the nanocrack generation at triple junctions of grain boundaries (Ovid'ko and Sheinerman, 2004b). (iii) The enhanced grain-boundary diffusion provides the effective action of Coble creep (Masumura *et al.*, 1998; Kim *et al.*, 2000; Yamakov *et al.*, 2002) and triple-junction diffusional creep (Fedorov *et al.*, 2002) which thereby effectively compete with grain-boundary sliding. As a corollary, the contribution of grain-boundary sliding to plastic flow decreases, in which case growth of Burgers vectors of the sessile dislocations – nuclei of triple-junction nanocracks (Figure 5.41f and g) – slows down or stops.

It is natural to think that the three effects (i)–(iii) of grain-boundary diffusion (enhanced owing to the action of conventional lattice dislocation slip) give rise to enhanced ductility exhibited by nanocrystalline materials and materials with bimodal (nano- plus micrograined) structures, as well as to high-strain-rate superplasticity exhibited by nanocrystalline materials. In doing so, the combined action of lattice dislocation slip and grain-boundary sliding as dominant deformation modes is crucial. These deformation modes cause mutually consistent plastic flow of both grain interiors and grain boundaries. In addition, the lattice dislocation slip provides "bombardment" of grain boundaries by lattice dislocations which lead to enhancement of grain-boundary diffusion (Figure 5.44). The enhanced diffusion suppresses nucleation of nanocracks induced by grain-boundary sliding in nanocrystalline materials which thereby exhibit a good ductility or even superplasticity (for details, see Section 5.10). In particular, the discussed theoretical representations of the diffusion effects on ductility/superplasticity of nanocrystalline materials are indirectly supported by the experimental data (Islamgaliev *et al.*, 2001) showing degradation of superplastic properties of nanocrystalline materials after a short thermal treatment. More precisely, Islamgaliev *et al.* (2001) reported that nanocrystalline materials fabricated by severe plastic deformation method and then subjected to a short heat treatment do not show superplastic behavior, in contrast to as-fabricated materials. In the experiment (Islamgaliev *et al.*, 2001), heat treatment is not intense enough to cause grain growth but is sufficient to induce annihilation of the non-equilibrium grain-boundary vacancies generated in nanocrystalline materials during their fabrication by severe plastic deformation. In terms of the model (Ovid'ko and Sheinerman, 2005a), after heat treatment, the density of grain-boundary vacancies abruptly decreases and thereby the grain-boundary diffusion (occurring mostly by grain-boundary vacancy transport) becomes a low-intensity process that is not able to suppress the nanocrack nucleation. As a result, though a nanocrystalline specimen in its as-fabricated state shows superplasticity, heat treatment leads to its dramatic degradation.

Though diffusion effectively suppresses the nanocrack nucleation in several nanocrystalline materials showing superplasticity, the effects of diffusion are not sufficient to suppress fracture in most nanocrystalline materials in which nanocracks are intensively generated under mechanical load. Besides, nanocrystalline materials often contain nanocracks formed during their fabrication (see discussion in Chapter 4). The behavior of nanocracks and plastic flow processes under mechanical load are responsible for the fracture mechanism operating in nanocrystalline materials. At the first stage of loading, flat nanocracks are generated along grain boundaries in a nanocrystalline specimen (Figure 5.47a and b). If plastic flow is not intense, these flat nanocracks serve as dangerous stress concentrators inducing new flat nanocracks to be generated in their vicinities (Figure 5.47c, d and e). Then nanocracks located in one specimen section rapidly converge resulting in brittle intergranular fracture through the formation of a catastrophic crack (Figure 5.47f). If plastic flow and diffusion are intense, as-generated flat nanocracks (Figure 5.48a and b) are gradually transformed into nanovoids (Figure 5.48c). Nanovoids grow and are transformed into microvoids (Figure 5.48d and e). Plastic flow is localized and gives rise to the neck formation (Figure 5.48d and e). Then ductile fracture occurs through coalescence of microvoids located in one specimen section of a nanocrystalline specimen (Figure 5.48f). Below we will discuss computer simulations and theoretical models concerning brittle intergranular fracture (Figure 5.47) and ductile fracture mechanism (Figure 5.48) in nanocrystalline materials.

Experiments (Li and Ebrahimi, 2004, 2005) have shown brittle fracture of nanocrystalline Ni–15%Fe with mean grain size around 9 nm. In this situation, the main brittle crack is treated to be formed by the generation of multiple intergranular nano/microscale cracks and their convergence (Figure 5.47). Also, brittle intergranular fracture by the formation of

5.9 Fracture mechanisms in nanocrystalline materials

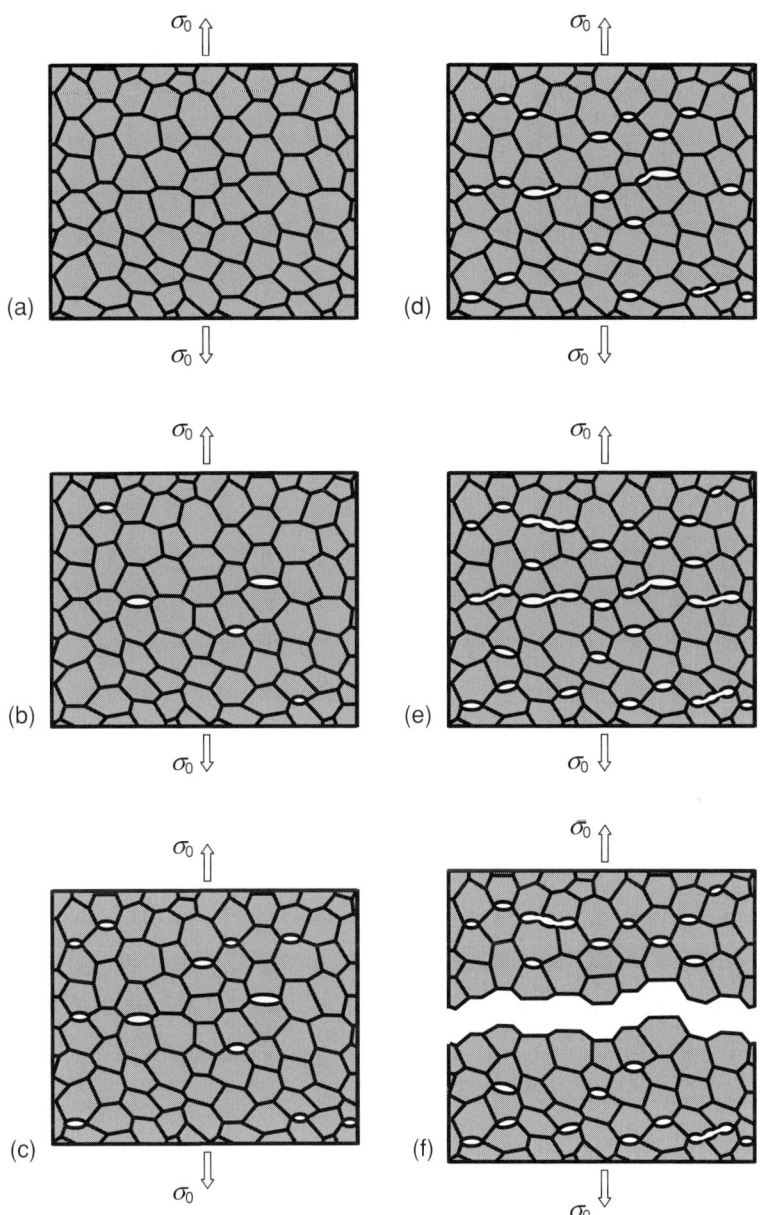

Figure 5.47. Intergranular brittle fracture in nanocrystalline material with finest grains occurs through formation and convergence of nanocracks at grain boundaries.

multiple nano/microscale cracks and their convergence are treated as the dominant fracture mechanism in superhard nanocomposite coatings (see Chapter 4). Computer simulations (Farkas *et al.*, 2002; Van Swygenhoven *et al.*, 2003) of crack growth in nanocrystalline Ni with grain size ranging from 5 nm to 12 nm show the intergranular fracture to be dominant. Intergranular nanocracks at grain boundaries and their triple junctions are found to be

282   5 Mechanical properties of structural nanocrystalline materials

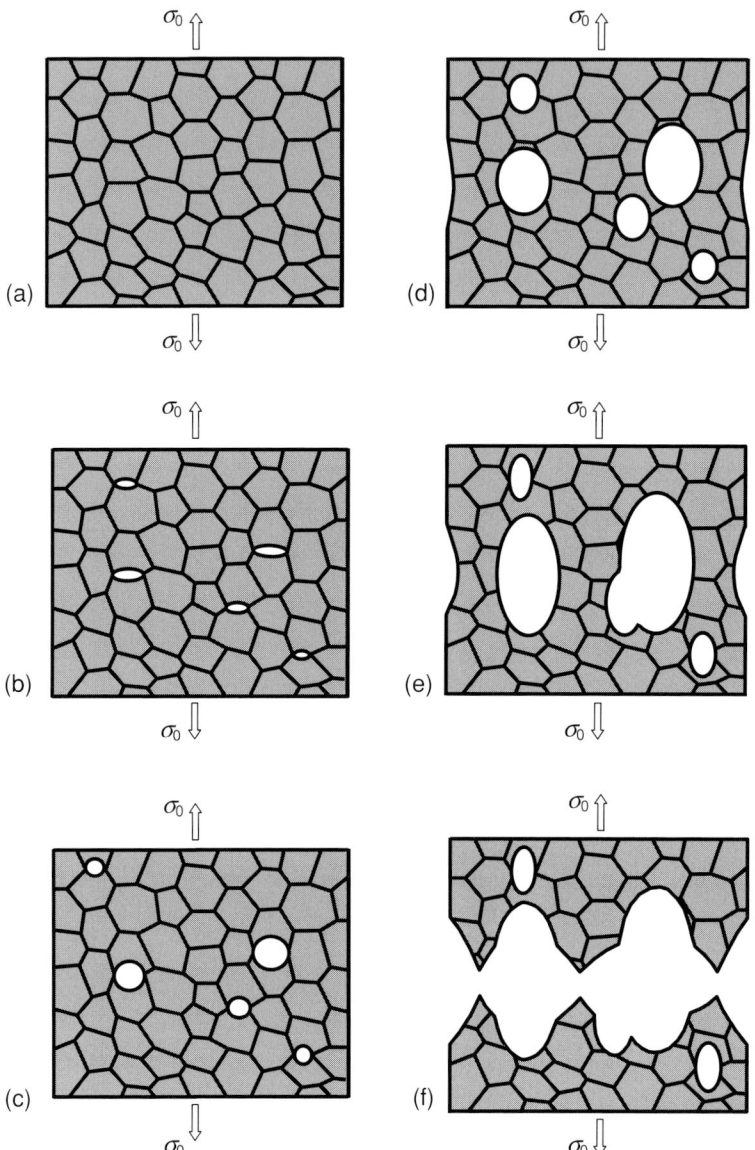

Figure 5.48. Ductile fracture in nanocrystalline material occurs through formation of nanocracks, their transformation into pores, growth of pores and formation of local necks between large pores.

generated near the tip of a pre-existent crack (a mode I semi-infinite crack artificially formed in a nanocrystalline simulation block in its initial state) owing to stress concentration at the tip (Figures 5.40 and 5.49). The crack grows by joining the nanocracks at grain boundaries and their triple junctions in its front (Figures 5.40 and 5.49). In the situation where the pre-existent crack ends at a triple junction in a nanocrystalline Ni sample with the grain size $d = 5$ nm, at the first stage of loading, the crack tip remains at the triple junction, and blunting

5.9 Fracture mechanisms in nanocrystalline materials

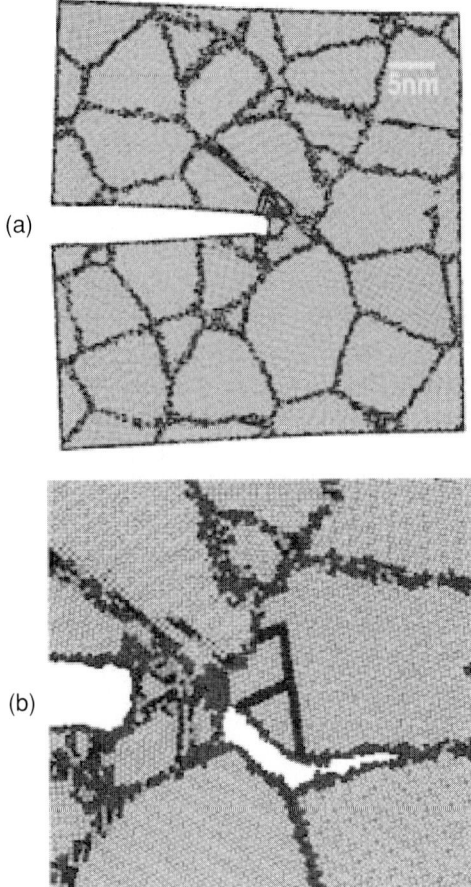

Figure 5.49. Crack-tip configuration in a section of a sample with 10 nm grain size. (a) At a load of twice the Griffith value. (b) At a load of 2.5 times the Griffith value. (Reprinted figure with permission from *Physical Review B*, Volume **66**, Farkas, D., Van Swygenhoven, H., and Derlet, P. M., Intergranular fracture in nanocrystalline metals, Article 060101(R). Copyright (2002) by the American Physical Society.)

is observed (Figure 5.40a) (Farkas *et al.*, 2002). With increase of the applied mechanical stress, two new nanocracks are formed at a grain boundary located in the vicinity of the crack tip (Figure 5.40b). With further increase of the applied mechanical stress, the crack grows (by joining the nanocracks) along a path entirely constituted by grain boundaries (Figure 5.40c). In the situation where the pre-existent crack ends in the grain interior in a nanocrystalline Ni sample with the grain size $d = 10$ nm (Figure 5.49a), the crack emits several partial lattice dislocations that blunt the crack tip region. With increase of an applied load, the stress concentration in the vicinity of the crack tip causes the formation of nanocracks ahead of the crack front at neighboring grain boundaries (Figure 5.49b). At later stages of loading, these cracks join the main crack (Farkas *et al.*, 2002).

Following computer simulations (Farkas *et al.*, 2002; Van Swygenhoven *et al.*, 2003), intergranular fracture processes occurring through growth of a pre-existent crack along grain boundaries are characterized by the energy release rate being approximately three times the expected Griffith value for brittle intergranular fracture. This is indicative of a significant role of plastic flow (realized via emission of partial lattice dislocations and structural transformations in grain boundaries) processes in the intergranular fracture observed in computer simulations (Farkas *et al.*, 2002; Van Swygenhoven *et al.*, 2003) of nanocrystalline metals under mechanical load.

Molecular dynamics simulations (Farkas *et al.*, 2002; Van Swygenhoven *et al.*, 2003), in fact, dealt with toughness of nanocrystalline Ni. Also, toughness of nanocrystalline materials has been theoretically examined in papers (Pozdnyakov, 2003; Pozdnyakov and Glezer, 2005) describing brittle crack growth in such materials. Focus was placed on the competition between intergranular and intragranular fracture modes. In terms of the surface energy density γ_e (per unit area) and grain boundary energy density γ_s (per unit area), intragranular and intergranular fracture modes (Figure 5.50a and b, respectively) in a coarse-grained polycrystal release the specific energies (per unit area of fracture surface) 2γ and $\gamma_e = \eta(2\gamma - \gamma_s)$, respectively, where η is the factor characterizing fracture surface curvature in the case of intergranular fracture. Nanocrystalline materials with the finest grains are specified by very large volume fractions of the grain boundary and triple junction phases. In these circumstances, "pure" intragranular fracture mode can not be realized. When a flat crack grows in such a material (Figure 5.50a), it propagates through regions occupied by grain interiors, grain boundaries and triple junctions. In doing so, a crack releases the energy density (per unit area of fracture surface)

$$\gamma_{\text{flat}} \approx f_b \gamma_e + f_{gb} \gamma_s + f_{tj} \gamma_{tj}, \qquad (5.27)$$

where f_b, f_{gb} and f_{tj} are the volume fractions occupied by the bulk, grain-boundary and triple-junction phases, respectively, and γ_{tj} is the specific energy density of the triple-junction phase. In the context discussed, the toughness of a nanocrystalline sample with a flat crack is characterized by (Pozdnyakov and Glezer, 2005)

$$G_c \approx G_b f_b + G_{gb} f_{gb} + G_{tj} f_{tj}, \qquad (5.28)$$

where G_b, G_{gb} and G_{tj} are the critical energy release rates for the bulk, grain-boundary and triple-junction phases, respectively.

In the case of intergranular fracture, the generation of mode I brittle nanocracks is enhanced along grain boundaries whose planes have a favored orientation, that is, planes at which the action of the external stress is maximum. For any other grain boundary, the external stress action is reduced by a factor (<1) depending on misorientation between the grain-boundary plane and the plane having the favored orientation (for details, see Pozdnyakov, 2003; Pozdnyakov and Glezer, 2005; Morozov *et al.*, 2003; and discussion below).

Nonequilibrium grain boundaries contain high densities of grain-boundary defects, stress sources, and are characterized by large energies compared to conventional, equilibrium grain boundaries. In these circumstances, intergranular fracture processes with cracks growing along grain boundaries are enhanced by the extra energy of the grain boundaries and influenced by stress fields of grain-boundary defects. In the case of nonequilibrium grain boundaries containing disclinations (Figure 5.50c and d), Pozdnyakov and Glezer (2005)

5.9 Fracture mechanisms in nanocrystalline materials

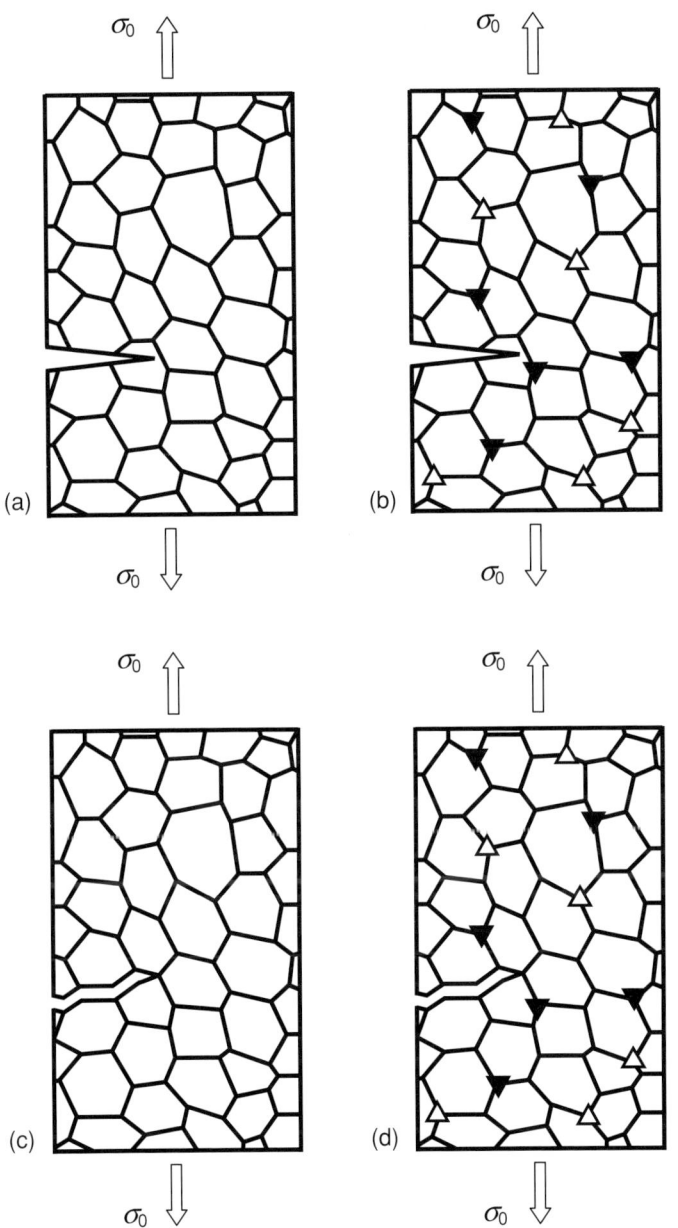

Figure 5.50. Crack-growth modes in nanocrystalline material. (a) Flat crack grows in nanocrystalline material. (b) Flat crack grows in nanocrystalline material with grain-boundary disclinations. (c) Curved crack grows along grain boundaries in nanocrystalline material. (d) Curved crack grows along disclinated grain boundaries in nanocrystalline material.

have found the stress concentration coefficient K_c to be given as

$$K_c = K_c^{eq} + \Delta K, \qquad (5.29)$$

where K_c^{eq} is the stress concentration coefficient in the case of equilibrium grain boundaries, and ΔK characterizes the effects of grain-boundary disclinations on the stress concentration.

Nanocracks – carriers of brittle intergranular fracture – commonly nucleate and grow along grain boundaries. Grain-boundary planes change their orientations at triple junctions whose amount is very large in nanocrystalline materials. In these circumstances, change of nanocrack growth direction at triple junctions represents a critical elemental process of the brittle intergranular fracture, because the crack growth conditions are highly sensitive to crack orientation. To theoretically characterize change of nanocrack growth direction at triple junctions, Gutkin and Ovid'ko (2004b) and Gutkin *et al.* (2005b) proposed theoretical models describing the generation and evolution of curved nanoscale cracks (nanocracks) in single-phase nanocrystalline materials and nanocomposites, respectively. In the framework of the model, the nanocracks are nucleated and grow along grain boundaries, changing the direction of their growth at triple junctions.

We have considered theoretical models and computer simulations of the generation of sole flat and curved nanocracks in the external stress and stress fields of internal stress sources like grain-boundary defects and a pre-existent large crack. In general, besides the generation of individual nanocracks, their convergence crucially influences the fracture behavior of brittle nanocrystalline materials. In a paper (Morozov *et al.*, 2003), a theoretical model was suggested focusing on intergranular brittle fracture associated with evolution of nanocrack ensembles in mechanically loaded nanocrystalline materials. In the framework of this model, events of the nanocrack convergence in a nanocrystalline solid are described as elemental events of a percolation process resulting in the formation of catastrophic macroscale crack.

Following the model (Morozov *et al.*, 2003), let us consider a nanocrystalline solid under tensile stress σ_0 (Figure 5.47a). At some critical values of the external stress, nanocracks – elemental carriers of intergranular brittle fracture – are formed in the mechanically loaded nanocrystalline solid (Figure 5.47b). Following Veprek and Argon (2002), it is assumed that stable nanocracks are formed along grain boundaries and do not penetrate into grain interiors. In doing so, with nanoscopic scales of nanocracks at grain boundaries and large angles made by adjacent grain-boundary planes at triple junctions, the formation of a flat nanocrack at a grain boundary is assumed to be independent of the events of the formation of a nanocrack at any other grain boundaries. As a corollary, when the external stress increases, new stable flat nanocracks at grain boundaries are formed (Figure 5.47c–e). Then the formation of a macroscopic crack – a carrier of the catastrophic failure – occurs (Figure 5.47f), which results from elementary independent events of the formation of flat nanocracks at grain boundaries of a quasistatically loaded nanocrystalline solid.

The macroscopic crack formation under consideration is a partial case of percolation described by the standard mathematical methods of the theory (Ziman, 1979; Stauffer and Aharony, 1992; Sahimi, 1994) of percolation in physical systems. Using these methods, Morozov *et al.* (2003) theoretically described evolution of the nanocrack ensemble in a deformed nanocrystalline solid. In doing so, according to the general representations of the percolation theory (Ziman, 1979; Stauffer and Aharony, 1992; Sahimi, 1994), the macroscopic crack is supposed to be formed when the concentration ρ of stable nanocracks reaches some critical value ρ_c. (In the situation under consideration, the nanocrack concentration ρ is defined as the ratio of the number of grain boundaries at which nanocracks are formed to the total number of grain boundaries.)

5.9 Fracture mechanisms in nanocrystalline materials

A nanocrack of length $2a$ at a grain boundary with length $2a$ and normal **n** to the grain-boundary plane is stable, if $\sigma_{nn} > \sigma_c(d)$. Here σ_{nn} is the stress tensor component at the boundary, and $\sigma_c(a)$ is the critical normal stress characterizing the formation of a stable nanocrack. The critical stress $\sigma_c(a)$ in the first approximation is given by Griffith's formula (Veprek and Argon, 2002):

$$\sigma_c(d) = k(\gamma E/d)^{1/2}, \qquad (5.30)$$

where γ denotes the specific surface energy of the solid, E is the Young modulus, and k is the factor taking into account the nanocrack geometry. Here, for simplicity, one considers the only normal failure mode I, neglecting the shear failure mode II.

For a grain boundary (and, therefore, a nanocrack formed at this boundary) whose plane has the normal **n** making the angle α with axis x, we have: $\sigma_{nn} = \sigma_0(\cos\alpha)^2$. In this situation, the condition $\sigma_{nn} > \sigma_c(d)$ is sensitive to both grain size and grain-boundary orientation characterized by a and α, respectively. With both log-normal distribution in grain size and random distribution in grain-boundary orientation in a nanocrystalline specimen, Morozov et al. (2003) used the percolation theory methods to calculate the stress $\sigma_{\text{catastrophic}}$ at which a catastrophic crack results from independent events of the nucleation of stable nanocracks. The main result of these calculations is that the distributions in grain size and grain-boundary orientation, inherent to nanocrystalline solids, do not essentially influence the macroscopic crack formation in the solid. Values of the stress $\sigma_{\text{catastrophic}}$ were found to be close to the critical normal stress $\sigma_c(\langle d \rangle)$ characterizing the stable state of a nanocrack at a grain boundary with the length equal to the mean grain size $\langle d \rangle$. That is, the crucial effect on the macroscopic crack formation is the result of the mean grain size which causes the stress $\sigma_{\text{catastrophic}}$ at which a catastrophic crack is formed (Morozov et al., 2003). The percolation-theory approach to a theoretical description of fracture processes in nanocrystalline materials is worth developing in the future to take into account both the interaction between nanocracks and plastic flow effects.

We have considered theoretical models of intergranular brittle fracture experimentally observed in nanocrystalline materials (Gan and Zhou, 2001; Li and Ebrahimi, 2004, 2005). However, besides brittle intergranular fracture surfaces, ductile dimples at fracture surfaces are often experimentally observed in nanocrystalline materials (Kumar et al., 2003a, 2003b; Ebrahimi et al., 1998; Mukai et al., 2003; Li and Ebrahimi, 2004, 2005; Youssef et al., 2004, 2005, 2006). They are indicative of the ductile character of fracture occurring in nanocrystalline materials at certain conditions. In particular, in the situation where diffusion and local plastic flow are intense enough, instead of flat nanocracks whose fast nucleation, growth and convergence along grain boundaries carry intergranular brittle fracture in nanocrystalline materials, slowly growing nanopores – carriers of ductile fracture – can be effectively formed whose shape is different from the flat one. Ovid'ko and Sheinerman (2006) theoretically described the nucleation of nanopores of elliptic shape in deformed nanocrystalline and nanocomposite materials, taking into account the presence of high-density ensembles of grain and interphase boundaries that provide both the enhanced diffusion and inhomogeneous plastic flow in these materials. In the framework of the model, elliptic nanopores in nanocrystalline and nanocomposite materials nucleate and grow by vacancy coagulation mechanism at interfaces in the stress fields of interfacial edge dislocations with large

Burgers vectors. In doing so, when elliptic nanopores nucleate, they remove the cores of interfacial dislocations.

As previously discussed in this section, typical interfacial defects that can stimulate nanoscale fracture processes in nanomaterials are interfacial superdislocations, that is, dislocations located at interfaces and characterized by Burgers vector magnitudes being larger than the crystal lattice parameter. Examples of such defect configurations are sessile superdislocations formed at triple junctions (Figure 5.41e) owing to grain-boundary sliding. Following (Ovid'ko and Sheinerman, 2006), let us discuss several scenarios for the formation of other interfacial superdislocations in deformed nanocrystalline and nanocomposite materials.

(i) The lattice dislocation slip is essential or even dominant in nanocrystalline materials with intermediate grains having the grain size in the range from tentatively 30 nm to 100 nm (Kumar *et al.*, 2003a; Zhan *et al.*, 2004; Gutkin and Ovid'ko, 2004a; Wolf *et al.*, 2005; Han *et al.*, 2005; Ovid'ko, 2005a). In such materials, lattice dislocations move under the action of an applied stress from grain interiors towards grain boundaries. In doing so, lattice dislocations emitted by one dislocation source (for instance, a Frank–Read source) in a grain move predominantly along one slip plane. When these dislocations reach a grain boundary, they can converge under the shear stress action and form a superdislocation at a grain boundary step (Figure 5.51a).

(ii) Plastic deformation in nanocomposites consisting of a strong nanocrystalline matrix with large ductile inclusions of the second phase (Figure 5.51b) (He *et al.*, 2003, 2004; Das *et al.*, 2003) can be divided into the two basic stages. At the first stage, plastic deformation of such nanocomposites is concentrated in large ductile inclusions characterized by comparatively low values of the flow stress and surrounded by a strong nanocrystalline matrix. Large ductile inclusions of the second phase are intensively deformed through movement of lattice dislocations which reach interphase boundaries between large inclusions and nanoscale grains and can form superdislocations at these interphase boundaries (Figure 5.51b). At the second stage, the flow stress increases, in which case plastic deformation occurs in both the nanocrystalline matrix and large inclusions. In doing so, plastic deformation in the nanocrystalline matrix is often inhomogeneous with plastic flow being localized in shear bands (Chen *et al.*, 2003; Wang and Ma, 2004a,b). When shear bands propagating through the nanocrystalline matrix meet large inclusions, superdislocations can be formed at interphase boundaries separating large inclusions and nanoscale grains (Figure 5.51c).

(iii) A similar situation comes into play in nanomaterials with a bimodal structure consisting of nanoscale and large grains (Figure 5.51d) (Tellkamp *et al.*, 2001; Sergueeva *et al.*, 2004; Han *et al.*, 2005). At the first stage, plastic deformation of nanomaterials with bimodal structures is concentrated in large ductile grains characterized by comparatively low values of the flow stress and surrounded by a strong nanocrystalline matrix. Large ductile grains are intensively deformed through movement of lattice dislocations which reach grain boundaries of nanoscale grains and can form superdislocations at these grain boundaries (Figure 5.51d). At the second stage, the flow stress increases, in which case plastic deformation occurs in both the nanocrystalline matrix and large

5.9 Fracture mechanisms in nanocrystalline materials

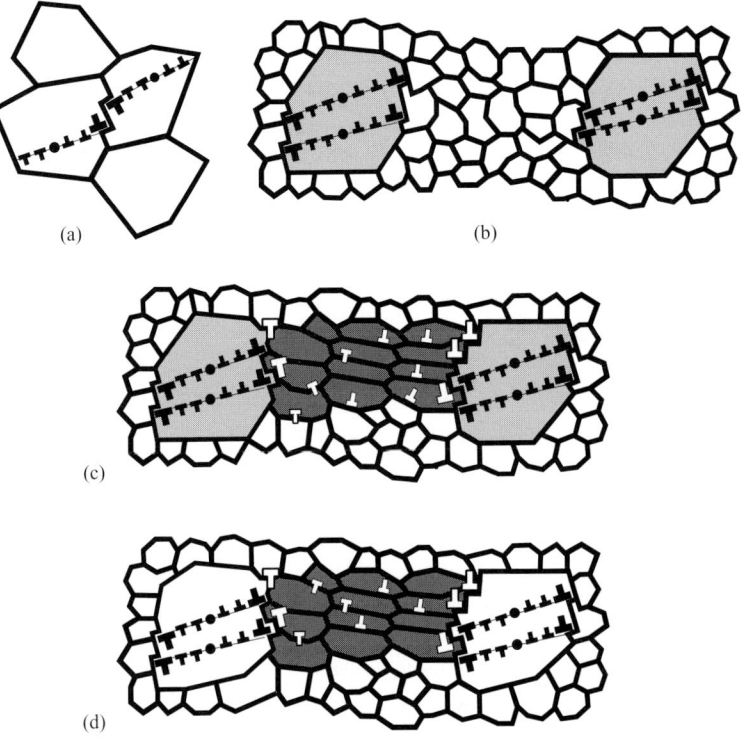

Figure 5.51. Formation of superdislocations at interface steps in deformed nanocrystalline and nanocomposite materials. (a) Lattice dislocations glide along one slip plane towards the grain boundary where they converge and form a superdislocation at a grain-boundary step. (b) Lattice dislocations glide in a large inclusion of the second phase along one slip plane towards the interphase boundary (separating the large inclusion and nanoscale grains in a nanocomposite) where they converge and form a superdislocation at a boundary step. (c) Propagation of shear bands (gray regions) through a nanocrystallline matrix. Shear bands meet interphase boundaries separating large inclusions and the nanocrystalline matrix in a nanocomposite. As a result, superdislocations are formed at the places where shear bands meet interphase boundaries. (d) Shear bands (gray regions) propagate through a nanocrystalline matrix and meet grain boundaries separating large and nanoscale grains in a nanomaterial with bimodal structure. As a result, superdislocations are formed at the places where shear bands meet grain boundaries.

grains. In doing so, plastic deformation in the nanocrystalline matrix is often inhomogeneous, with plastic flow localized in shear bands (Figure 5.51d). When shear bands propagating through the nanocrystalline matrix meet large grains, superdislocations can be formed at grain boundaries separating large and nanoscale grains (Figure 5.51d).

With the results of the theoretical analysis given in the paper (Ovid'ko and Sheinerman, 2006), the nucleation of elliptic nanopores (Figure 5.52) at interface steps containing superdislocations – dislocations with large Burgers vectors – is energetically favorable in deformed nanocrystalline and nanocomposite materials in wide ranges of their parameters. They nucleate in the stress fields of the interfacial dislocations generated due to inhomogeneous plastic deformation in nanomaterials. The size and shape of elliptic nanopores are

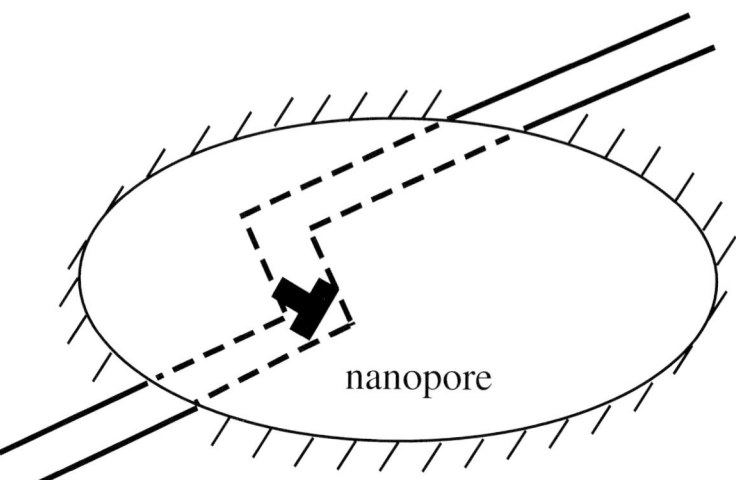

Figure 5.52. Elliptic nanopore nucleates and eliminates the core of interfacial dislocation with large Burgers vector.

sensitive to both geometrical characteristics of the interfaces and the specific energies of the interfaces and nanopore free surface; for details, see (Ovid'ko and Sheinerman, 2006). Elliptic nanopores growing along interfaces represent elemental carriers of ductile fracture processes in deformed nanomaterials. They can serve as nuclei for dimple rupture structures experimentally observed at fracture surfaces in nanomaterials (Kumar et al., 2003a, 2003b; Ebrahimi et al., 1998; Mukai et al., 2003; Li and Ebrahimi, 2004, 2005). In this interpretation, the nucleation, growth and coalescence of elliptic nanopores conduct slow ductile fracture processes in deformed nanomaterials and result in the formation of the mesoscale dimple rupture structures.

Thus, intergranular brittle fracture and ductile fracture are experimentally observed in nanocrystalline materials. Intergranular brittle fracture is expected to occur by the generation of multiple intergranular nano/microscale cracks and their convergence. Ductile fracture is viewed to occur through the microvoid coalescence mechanism. Typical elemental nanocracks in nanocrystalline materials are flat nanocracks nucleated at grain boundaries and their triple junctions in the stress fields of dislocations resulted from grain-boundary sliding across triple junctions (Figure 5.41f and g). Evolution of typical nanocracks under mechanical load causes fracture mechanism operating in a nanocrystalline material and depends on its structural characteristics (first of all, grain size) and material parameters.

In nanocrystalline materials with the finest grains, plastic flow is conducted mostly by grain boundaries and does not provide effective relaxation of stress buildup at triple junctions and elemental flat nanocracks. In these circumstances, new flat nanocracks are easily generated at triple junctions as a result of grain-boundary sliding and stress concentration at neighboring flat nanocracks (Figure 5.47). Thus intergranular brittle fracture occurs which ends at the formation of a macroscopic crack (Figure 5.47f).

In nanocrystalline materials with intermediate grains, plastic flow is conducted by both lattice slip and grain-boundary processes. In this case, plastic flow provides effective relaxation of stress buildup at triple junctions and elemental flat nanocracks which thereby are transformed into ball-like and elliptic pores (Figure 5.48). Because of their shape geometry, stress concentration at ball-like and elliptic nanopores is not so dramatic compared to the situation with flat nanocracks. In these circumstances, ball-like and elliptic pores slowly grow (by vacancy coagulation at pores and lattice dislocation emission from pores) and cause ductile fracture with local necks being formed between pores (Figure 5.48d, e and f).

To summarize, fracture of nanocrystalline materials has the specific peculiarities which are briefly as follows.

(A) Plastic flow in nanocrystalline materials occurs at very large values of flow stress which are close to those needed to induce fracture. Therefore, local nanocracks are easily nucleated in nanocrystalline materials at even low values of plastic strain. This behavior is similar to that of brittle coarse-grained polycrystals (Ashby *et al.*, 1979; Gandhi and Ashby, 1979) in which flow stress values and fracture stress values are close.

(B) Owing to the very large amount of grain boundaries in nanocrystalline materials with the finest grains, the intergranular fracture mechanism is dominant in these materials (Figure 5.47).

(C) Nanocrystalline materials with intermediate grains show plastic flow localization with shear-band formation followed by macroscopic necking. Ductile fracture in these materials occurs through the microvoid coalescence mechanism in neck regions (Figure 5.48).

Finally, notice that theory and computer simulations of fracture processes are in their infancy. We expect rapid growth of research activity in this area in the near future.

5.10 Strain-rate sensitivity, ductility and superplasticity of nanocrystalline materials

Nanocrystalline materials often exhibit superstrength, superhardness and enhanced tribological characteristics, but commonly show low tensile ductility; see Chapter 4 and the literature (Siegel and Fougere, 1995; Hahn and Padmanabhan, 1995; Koch *et al.*, 1999; Gleiter, 2000; Mohamed and Li, 2001; Niederhofer *et al.*, 2001; Padmanabhan, 2001; Veprek and Argon, 2002; Gutkin *et al.*, 2003; Kumar *et al.*, 2003a; Milligan, 2003; Patscheider, 2003; Valiev, 2004; Chokshi and Kottada, 2006; Han *et al.*, 2005; Ovid'ko, 2005a,b; Wolf *et al.*, 2005). Recently, however, several examples of substantial ductility and even superplasticity – ability of a material to undergo large elongations (typically 200% and more) without failure – of superstrong nanocrystalline materials have been reported (Mayo, 1997, 1998; Mishra *et al.*, 1998; McFadden *et al.*, 1999; Islamgaliev *et al.*, 2001; Mishra *et al.*, 2001; Valiev *et al.*, 2001; Mukherjee, 2002a,b; Valiev *et al.*, 2002; Wang *et al.*, 2002; Champion *et al.*, 2003; He *et al.*, 2003; Karimpoor *et al.*, 2003; Kumar *et al.*, 2003b; Ma, 2003; Zhan *et al.*, 2003; He *et al.*, 2004; Wang and Ma, 2004a,b; Youssef *et al.*, 2004, 2005,

2006). Nanocrystalline materials with the unique combination of very high strength and superplasticity (or at least good ductility) represent ideal materials for a wide range of applications in the aerospace and automotive industries, medicine, energy, sports products, etc. In this context, it is very important to understand the underlying mechanisms of the exciting phenomenon of superplasticity in nanocrystalline materials. In this section, a short overview of key theoretical concepts on superplasticity and ductility of nanocrystalline materials is presented. Particular attention will be paid to consideration of both strengthening and strain-rate sensitivity as factors crucially affecting superplasticity and ductility of nanocrystalline materials.

First, let us discuss the effects of grain-size scale on ductility of coarse-grained polycrystalline, ultrafine-grained and nanocrystalline materials (Figure 5.53). As is well-known, the dominant deformation mechanism in conventional coarse-grained polycrystals is the lattice dislocation slip occurring in large grain interiors (Hirth and Lothe, 1982) (Figure 5.53a). Its carriers are perfect lattice dislocations generated and stored in the form of dislocation cells/subgrains in grain interiors during plastic deformation. Grain boundaries serve as obstacles for movement of lattice dislocations. In doing so, grain boundaries influence the level of the yield stress, the stress at which plastic deformation starts to occur. It is described by the well-known classical Hall–Petch relationship (5.1) between the yield stress and grain size. However, in general, the mechanical behavior of coarse-grained polycrystals is crucially affected by evolution of lattice dislocations in grain interiors, but not grain boundaries. For instance, the deformation-induced storage of lattice dislocations in grain interiors in coarse-grained polycrystals is responsible for strengthening. It means that the flow stress increases with rising plastic strain. This standard deformation behavior is exhibited by most coarse-grained polycrystalline materials with grain size d being larger than 300 nm.

With grain refinement, the lattice dislocation slip shows deviations from the standard behavior because of both the nanoscale and interface effects. For illustration purposes, let us consider materials with grain size d in the range from approximately 30 nm to 300 nm (Figure 5.53b). This category of materials includes ultrafine-grained materials (d ranges from 100 nm to 300 nm) and nanocrystalline materials with intermediate grains (d ranges from about 30 nm to 100 nm). The lattice dislocation slip is still dominant in such materials. However, in contrast to the situation with conventional coarse-grained polycrystals, lattice dislocations are not intensively stored in grain interiors. In ultrafine-grained materials and nanocrystalline materials with intermediate grains, the level of the flow stress is crucially affected by grain boundaries operating as active sources and sinks of lattice dislocations (Wang and Ma, 2004a,b). More precisely, grain boundaries serve as the lattice dislocation sinks, in which case lattice dislocations are absorbed at grain boundaries and transformed into "extrinsic" grain-boundary dislocations. Also, grain boundaries serve as the lattice dislocation sources, in which case lattice dislocations are emitted from grain boundaries, in particular, due to transformations of "extrinsic" grain-boundary dislocations (see Section 5.7). In these circumstances, both the storage and annihilation of dislocations at grain boundaries (but not in grain interiors) strongly influences the deformation behavior of ultrafine-grained materials and nanocrystalline materials with intermediate grains (Wang and Ma, 2004a,b).

5.10 Strain-rate sensitivity

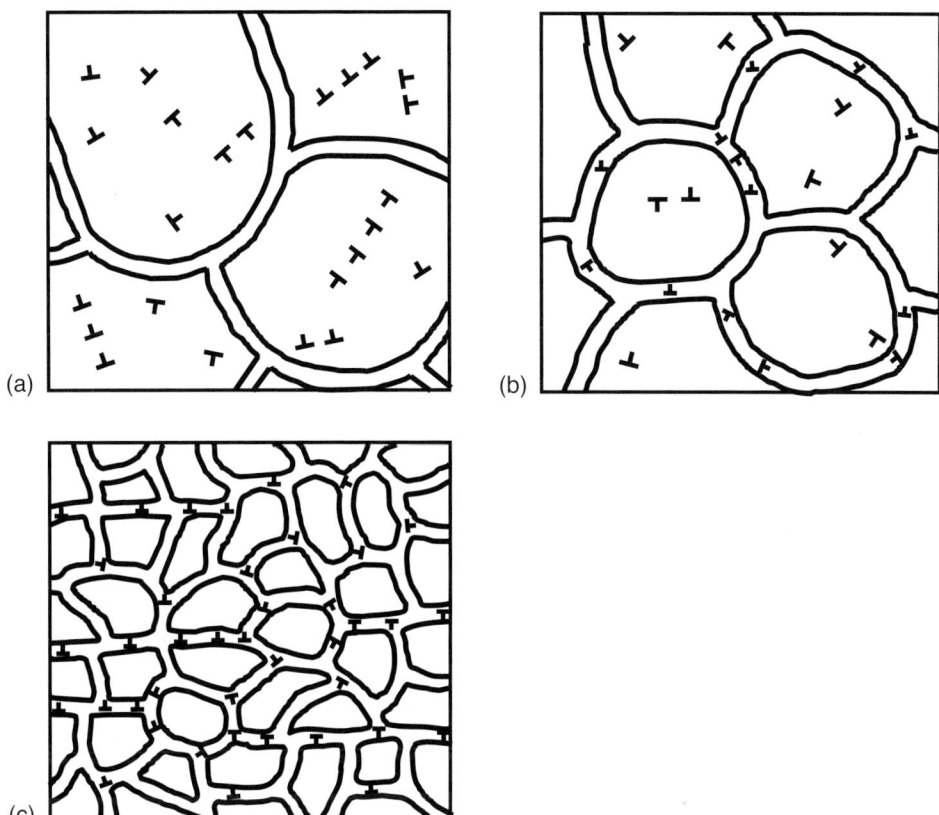

Figure 5.53. Solid-state structures under plastic deformation. (a) Coarse-grained polycrystal. Lattice dislocation slip is dominant which gives rise to dislocation storage in large grain interiors. (b) Nanomaterial with intermediate grains. Lattice dislocation slip is dominant. Its action is controlled by both dislocation storage and annihilation at grain boundaries. (c) Nanomaterial with finest grains. Plastic deformation is conducted by grain boundaries.

The dislocation storage and annihilation at grain boundaries are the two key competing factors crucially affecting the level of the flow stress. The dislocation storage at grain boundaries provides the strengthening of a material during plastic deformation. The dislocation annihilation at grain boundaries provides the softening (recovery) of a material during plastic deformation. Following Wang and Ma (2004a,b), after some initial stage of deformation, the above competing factors reach equilibrium. That is, these factors cause a steady state in which the dislocation storage is completely compensated by the dislocation annihilation. The steady state is characterized by an approximately constant flow stress without any strengthening. When the strengthening is absent, a material under tensile deformation commonly shows plastic flow localization in shear bands (Figure 5.54). Plastic flow localization in materials in the absence of the strengthening results in the macroscopic necking followed by the stress concentration in the neck region. The local stresses acting in the neck region are very high; they become close to the critical stresses causing failure processes.

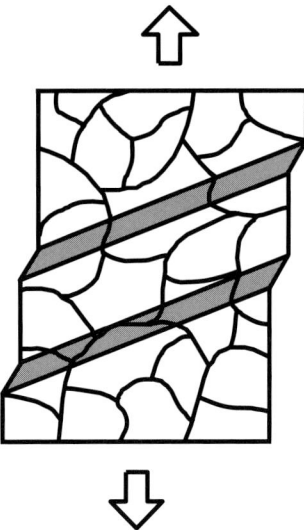

Figure 5.54. Plastic flow localization in shear bands (gray regions) in nanocrystalline material.

In these circumstances, plastic flow localization very quickly leads to intense crack (void) nucleation followed by the macroscopic failure. As shown in a lot of experiments, this deformation behavior – force instability in tension – is typical for most ultrafine-grained materials and nanocrystalline materials with intermediate grains; see Wang and Ma (2004a,b) and references therein.

Nanocrystalline materials with the finest grains having grain size d lower than about 30 nm (Figure 5.53c) under tensile deformation commonly show a brittle behavior with crack nucleation and propagation instability (Figure 5.47), but without a preceding neck formation. The brittle behavior is often the result of fabrication-induced flaws serving as dangerous stress concentrators that cause the enhanced crack nucleation and propagation. With progress in fabrication technologies, it is currently possible to produce flaw-free nanocrystalline materials with the finest grains. However, flaw-free nanocrystalline materials with finest grains also show tendency to brittle behavior owing to their structural peculiarities. The volume fraction occupied by grain boundaries is extremely large in such materials. Therefore, the lattice dislocation slip is suppressed by high-density ensembles of grain boundaries, in which case alternative deformation mechanisms conducted by grain boundaries come into play in nanocrystalline materials with the finest grains (for details, see previous sections) (Figure 5.53c). These deformation mechanisms are characterized by very high values of the flow stress; these values are much higher than those characterizing the lattice dislocation slip in conventional coarse-grained polycrystals and ultrafine-grained materials. At the same time, superstrong nanomaterials with finest grains commonly show low tensile ductility, in particular, owing to the intense formation of nanocracks (Figure 5.47). It is not surprising, because superstrong solids commonly show crack nucleation and/or propagation instability. Cracks are intensively generated at local stress concentrators and

5.10 Strain-rate sensitivity

then converge (see Section 5.9), because high values of the flow stress are close to those needed to induce and sustain fracture processes.

Thus, plastic deformation mediated by grain-boundary processes in nanomaterials with finest grains is characterized by very high values of the flow stress and very quickly leads to intense nanocrack generation and convergence followed by macroscopic fracture (Figure 5.47). Plastic deformation mediated by lattice dislocations in nanomaterials with intermediate grains is characterized by the absence of strengthening and very quickly leads to plastic flow localization (Figure 5.54) followed by ductile fracture (Figure 5.48). The above factors, the absence of the strengthening in nanocrystalline materials with intermediate grains and very high flow stresses in nanocrystalline materials with finest grains, effectively explain numerous experimental data which are indicative of low tensile ductility exhibited by most nanocrystalline materials.

Recently, several examples of substantial tensile ductility and even superplasticity of nanocrystalline materials have been reported (Mayo, 1997, 1998; Mishra *et al.*, 1998; McFadden *et al.*, 1999; Islamgaliev *et al.*, 2001; Mishra *et al.*, 2001; Valiev *et al.*, 2001; Mukherjee, 2002a,b; Valiev *et al.*, 2002; Wang *et al.*, 2002; Champion *et al.*, 2003; He *et al.*, 2003; Karimpoor *et al.*, 2003; Kumar *et al.*, 2003b; Ma, 2003; Zhan *et al.*, 2003; He *et al.*, 2004; Wang and Ma, 2004a,b; Youssef *et al.*, 2004, 2005, 2006). With these experiments, the five basic strategies to achieve ductility in high-strength nanocrystalline materials with intermediate grains have been suggested; see discussion in Section 5.4 and the literature (Tellkamp *et al.*, 2001; Wang *et al.*, 2002, 2005; He *et al.*, 2003, 2004; Koch, 2003; Zhan *et al.*, 2003; Sergueeva *et al.*, 2004; Wang and Ma, 2004a,b; Zhan *et al.*, 2004; Han *et al.*, 2005). These strategies address suppression of plastic flow localization and briefly are as follows.

The first strategy is to suppress plastic flow localization through fabrication of a bimodal single-phase structure composed of nanograins and large grains (Figure 5.55a). *The second strategy* is to suppress plastic flow localization through fabrication of a composite structure consisting of ductile second-phase inclusions embedded into a nanocrystalline matrix (Figure 5.55b). These strategies are based on the large-scale structural organization of nanocrystalline materials and illustrated by the experimental facts that substantial ductility is exhibited by single-phase nanocrystalline materials with bimodal structure (Tellkamp *et al.*, 2001; Wang *et al.*, 2002; Koch, 2003; Sergueeva *et al.*, 2004; Wang and Ma, 2004a,b; Zhan *et al.*, 2004; Han *et al.*, 2005) and nanocomposites with a superstrong nanocrystalline matrix and ductile inclusions of the second phase (He *et al.*, 2003, 2004; Koch, 2003; Zhan *et al.*, 2003). Plastic flow localization is effectively hampered in such bimodal structures and nanocomposites due to the strengthening effects provided by large grains and inclusions of the second phase, respectively. For instance, large grains in bimodal structures (Figure 5.55a) serve as rather ductile structural elements in which plastic deformation is characterized by strengthening. In this case, typical "stress–strain" dependence of a material with the bimodal structure under tensile deformation is characterized by strengthening and obeys the Considere criterion (see Chapter 4) for plastic flow stability during a rather extended deformation stage. As a result, nanomaterials with the bimodal structure often have the same level of ductility as their coarse-grained counterparts and the same level of the flow

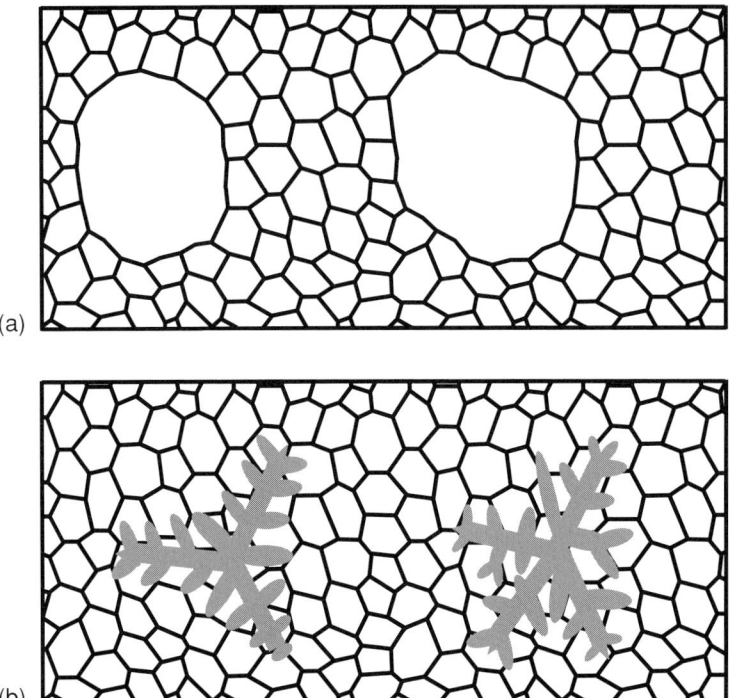

Figure 5.55. Ductile nanomaterials with composite structure. (a) Single-phase nanocrystalline material with bimodal structure consisting of nanoscale and large (microscale) grains. (b) Nanocomposite consisting of nanoscale grains and dendrite-like inclusions of the second phase (schematically).

stress as the yield stress of "pure" nanocrystalline materials. Dendrite-like inclusions (He et al., 2003, 2004) of the second-phase cause similar effects in Ti-based nanocrystalline alloys which therefore show both essential ductility and high strength.

The third strategy is to deform nanocrystalline materials with intermediate grains at low temperatures (Wang and Ma, 2004a,b; Wang et al., 2005). In this case, recovery (the dislocation annihilation) processes at grain boundaries are suppressed and do not compensate the dislocation storage at such boundaries during a rather extended deformation stage. As a result, a nanocrystalline sample under low temperature deformation often shows good ductility due to the strengthening that prevents plastic flow localization.

The fourth strategy is to suppress plastic flow localization through positive strain-rate sensitivity of the flow stress σ (Wang and Ma, 2004a,b; Wei et al., 2004). The strain rate sensitivity is commonly defined as (Wang and Ma, 2004a,b; Wei et al., 2004)

$$m = \{\partial \log \sigma / \partial \ln \dot{\varepsilon}\}_{\varepsilon T}. \tag{5.31}$$

Here ε denotes the plastic strain degree, $\dot{\varepsilon}$ is the plastic strain rate, and T is the temperature. If the strain-rate sensitivity m of a material is positive and sufficiently large, the neck formation is suppressed. It is because a local increase of the plastic strain rate in a nucleus of the neck region leads to a local increase of the flow stress. In the spirit of the theory (Hart, 1967) of

5.10 Strain-rate sensitivity

plastic flow instabilities in solids, the criterion for stable deformation of a nanocrystalline material under tensile load is given as (Wang and Ma, 2004a,b)

$$\sigma^{-1}(\partial\sigma/\partial\varepsilon)_{\dot{\varepsilon}} - 1 + m \leq 0. \tag{5.32}$$

The Hart criterion (5.32) is a generalized version of the Considere criterion ($(\partial\sigma/\partial\varepsilon)_{\dot{\varepsilon}} \leq \sigma$) in the situation with materials having non-zero strain-rate sensitivity. The strain-rate sensitivity is in the following relationship with the activation volume v (Wei et al., 2004):

$$m = k_\text{B} T/\tau v. \tag{5.33}$$

Both m and v represent experimentally measured characteristics that can be useful in identification of deformation mechanisms operating in nanocrystalline materials.

Values of the activation volume in nanocrystalline metals commonly are lower than those in their coarse-grained counterparts. Following Wang and Ma (2004b), Wei et al. (2004) and Asaro and Suresh (2005), typical values of v in nanocrystalline fcc metals are around 3–$10a^3$, where a is the crystal lattice parameter. These values are larger than $v = a^3$, which characterizes Coble creep, and much lower than typical values (around $1000a^3$) of the activation volume, which characterizes thermally activated movement of lattice dislocations intersecting forest dislocations in coarse-grained polycrystals. The activation volume values of around 3–$10a^3$ are assumed (Wang and Ma, 2004b; Wei et al., 2004; Asaro and Suresh, 2005) to be related to emission of partial and perfect lattice dislocations as well as groups of partial dislocations that carry twins from grain boundaries in nanocrystalline fcc metals. In this interpretation, typical values (3–$10a^3$) of the activation volume in nanocrystalline fcc metals are indicative of the action of the specific deformation mechanism by emission of perfect and partial lattice dislocations from grain boundaries.

The fourth strategy based on positive strain-rate sensitivity is intrinsically related to the nanocrystalline state of solids. It is contrasted to the first and second strategies related to macroscale structural organization of materials. Also, it is contrasted to the third strategy related to conditions of mechanical loading. The strategy based on positive strain rate sensitivity, in part, is realized in superplastic deformation of nanocrystalline materials (see below) and represents a promising method for enhancing tensile ductility in the case of "true" nanocrystalline materials with narrow grain-size distributions. Besides, the strain-rate sensitivity is a material property related to specific deformation mechanisms operating in nanocrystalline materials. In this context, an analysis of the strain-rate sensitivity of nanocrystalline materials is important from both applied and fundamental viewpoints.

Notice that all the four strategies under consideration are focused on enhancement of tensile ductility of nanocrystalline materials with intermediate grains and assume that the effects of fabrication-produced pores and contaminations are not significant. In general, however, artifacts can dramatically suppress ductility of nanocrystalline materials; see the discussion in Chapter 4 and the paper (Hugo et al., 2003) reporting on an experimental study of fracture processes in nanocrystalline Ni films fabricated by d.c. magnetron sputtering and pulsed laser deposition. Following Hugo et al. (2003), the nanocrystalline Ni film material fabricated by d.c. magnetron sputtering and characterized by the mean grain size of around 19 nm contains pores and behaves in a brittle manner. At the same time, the nanocrystalline

Ni film material fabricated by pulsed laser deposition and characterized by the mean grain size of around 17 nm is free from pores and behaves in a ductile manner.

With a very important role of fabrication-produced pores and contaminations in plastic flow and fracture processes in nanocrystalline materials, fabrication of nanocrystalline materials free from artifacts represents *the fifth strategy* to reach a good tensile ductility of nanocrystalline materials. Advantages of this strategy were illustrated by the experiments (Youssef *et al.*, 2004, 2005, 2006) in which the in situ consolidation mechanical alloying technique was used to fabricate artifact-free nanocrystalline Cu and Al–5%Mg alloy having mean grain sizes of around 23 nm and 26 nm, respectively. These artifact-free nanocrystalline materials show both ultrahigh strength and good ductility. They are characterized by strengthening and break by ductile fracture mechanism, with ductile dimples at fracture surfaces (Youssef *et al.*, 2004, 2005, 2006). The lattice dislocation slip was reported to be active in artifact-free nanocrystalline Cu and Al–5%Mg alloys. With these data, Youssef *et al.* (2004, 2005, 2006) attributed both the strengthening and thereby good ductility of the materials with lattice dislocation storage in grain interiors. In the framework of this explanaition, it is not clear the reason for lattice dislocation storage in grain interiors, because commonly lattice dislocations in nanoscale grains tend to move towards grain boundaries (where the lattice dislocations are absorbed) owing to the image forces (Gryaznov *et al.*, 1989, 1991; Evans and Hirth, 1992; Siegel, 1994; Romanov, 1995; Chokshi and Kottada, 2006). We think that the lattice dislocations are stored in nanoscale grain interiors in artifact-free nanocrystalline Cu and Al–5%Mg alloy owing to the co-existence of finest and intermediate nanograins in these materials. The mean grain-size values, 23 nm and 26 nm, for the artifact-free nanocrystalline are close to the critical grain size value (10–30 nm) at which transition from the lattice dislocation slip to grain-boundary sliding and other grain-boundary-mediated deformation mechanisms occurs. In these circumstances, the materials under consideration contain both the finest nanograins where grain-boundary sliding and other grain-boundary-mediated deformation mechanisms dominate and relatively large nanoscale grains where the lattice dislocation slip dominates. Owing to the action of grain-boundary sliding, grain boundaries of the finest nanograins are in their highly nonequilibrium state. They contain high-density ensembles of "nonequilibrium" grain-boundary dislocations and other defects created by deformation processes in grain boundaries. (Besides, grain boundaries in their as-prepared state contain many "non-equilibrium" defects, because artifact-free nanocrystalline Cu and Al–5%Mg materials are fabricated by a highly non-equilibrium method, the in situ consolidation mechanical alloying technique.) The "nonequilibrium" grain-boundary dislocations elastically interact with the lattice dislocations moving in intermediate nanograins, prevent their movement towards grain boundaries and thus cause the experimentally observed (Youssef *et al.*, 2005) lattice dislocation storage in intermediate grain interiors. The lattice dislocation storage provides the strengthening and good tensile ductility. It is just a qualitative view whose validity needs further experimental verification and quantitative theoretical analysis.

Another example is artifact-free nanocrystalline Co with hcp crystal lattice and mean grain size being around 12 nm (Karimpoor *et al.*, 2003). The nanocrystalline Co specimens were fabricated by electrodeposition and mechanically tested at three strain rates (10^{-4} s^{-1}, 5×10^{-4} s^{-1}, and 2.5×10^{-3} s^{-1}). Karimpoor *et al.* (2003) reported that

5.10 Strain-rate sensitivity

artifact-free nanocrystalline Co with hcp crystal lattice shows a moderate strengthening, a good ductlility and a decrease of the flow stress with rising strain rate. This deformation behavior was attributed to the dominant role of twin deformation in plastic flow of nanocrystalline Co with finest grains. The discussed experimental data and their interpretation are worth examining in more detail in the future.

Now let us discuss superplasticity in nanocrystalline materials. In general, superplasticity denotes the ability of a solid to undergo large elongations (typically 200% and more) without failure. Following experimental data (Mishra et al., 1998; McFadden et al., 1999; Islamgaliev et al., 2001; Mishra et al., 2001; Valiev et al., 2001; Mukherjee, 2002a,b; Valiev et al., 2002; Zhan et al., 2003), Ni, Ni_3Al, Al- and Ti-based nanocrystalline alloys (with the grain size being in the range from around 50 nm to 100 nm) show superplasticity at lower temperatures and higher strain rates compared to their coarse-grained counterparts. Superplasticity of nanocrystalline materials is characterized by very high values of the flow stress and the strengthening at the first extended stage of deformation; for a review, see (Mukherjee, 2002a,b). At the same time, there is a rather extended second stage of superplastic deformation characterized by the softening.

In general, superplasticity occurs in superstrong nanocrystalline materials, if plastic strain localization and nanocrack nucleation/propagation instabilities are suppressed. In terms of plastic flow mechanisms, superplasticity can be reached because of the mechanisms providing the strengthening at the first stage of superplastic deformation, the positive strain-rate sensitivity of the flow stress at the second stage and the absence of dangerous stress concentrators during both stages of superplastic deformation.

Following experimental data (Islamgaliev et al., 2001; Mishra et al., 2001; Valiev et al., 2001; Mukherjee, 2002a,b; Valiev et al., 2002), grain-boundary sliding crucially contributes to superplastic flow in nanocrystalline materials. Also, lattice dislocation slip and grain rotations have been experimentally observed in nanocrystalline materials under superplastic deformation. With these experimental data, it is suggested (Ovid'ko, 2005b) that superplastic deformation occurs in nanocrystalline materials, if grain-boundary sliding serves as the dominant deformation mechanism whose operation and effective accommodation are provided by lattice dislocation slip, intense diffuson, diffusion-controlled rotational deformation and triple-junction migration.

As to details, grain-boundary sliding is effectively supported by the lattice dislocation slip as follows. Lattice dislocations are effectively generated and move in intermediate grains of deformed nanocrystalline materials. (Following Hirth and Lothe (1982), the critical stress τ_{F-R} for activation of Frank–Read sources of lattice dislocations is given as $\tau_{F-R} \sim Gb/ML$, where b denotes the magnitude of the lattice dislocation Burgers vector, M is the orientation factor, G is the shear modulus, and L is the length of the Frank–Read dislocation segment. In the case of nanocrystalline materials, the length L is around $d/3$, where d is the grain size (Weissmueller and Markmann, 2005). For $b = 0.3$ nm, $M = 0.5$, and the grain-size range d from 50 nm to 100 nm, we have $\tau_{F-R} \sim G/60 - G/30$. These values are close to experimentally measured (Mishra et al., 1998; McFadden et al., 1999; Islamgaliev et al., 2001; Mishra et al., 2001; Valiev et al., 2001; Mukherjee, 2002a,b; Valiev et al., 2002; Zhan et al., 2003) values of the superplastic flow stress in real nanocrystalline materials. Therefore, the flow stress level is high enough to activate Frank–Read sources of lattice dislocations in intermediate grains of nanocrystalline materials.) The lattice dislocations move towards grain boundaries where these dislocations are absorbed and transformed into grain-boundary dislocations (Figure 5.19) that carry grain-boundary sliding. Mobile grain-boundary dislocations – carriers of grain-boundary sliding in nanocrystalline materials under superplastic deformation – glide along boundary planes (Sutton and Balluffi, 1996) and come into dislocation reactions at triple junctions (Figures 5.41a–e and 5.56). As a result of the dislocation reactions, sessile grain boundary dislocations are formed and stored at triple junctions (Figures 5.41a–e and 5.56). The sessile grain boundary dislocations at triple junctions elastically interact with mobile grain boundary dislocations and thereby hamper their glide (Gutkin *et al.*, 2003c, 2004b). Thus, the storage of grain boundary dislocations (Figures 5.41a–e and 5.56) causes the experimentally observed strengthening of nanocrystalline materials at the first stage of superplastic deformation. It is contrasted to coarse-grained polycrystals where the strengthening is provided by the storage of lattice dislocations in grains (Hirth and Lothe, 1982; Seefeldt, 2001).

The storage of grain-boundary dislocations at triple junctions is capable of causing failure processes in deformed nanocrystalline materials, because dislocated triple junctions serve as dangerous stress sources inducing the nucleation of nanocracks (Ovid'ko and Sheinerman 2004a, 2005a) (Figure 5.41f and g). Such nanocracks in the vicinities of triple junctions have been observed by Kumar *et al.* (2003b) in "in situ" experiments with nanocrystalline Ni (Figure 5.39). Thus, the storage of grain-boundary dislocations at triple junctions causes the two competing effects on tensile ductility and superplasticity of nanocrystalline materials. On the one hand, the dislocation storage provides the strengthening and thereby suppresses plastic flow localization. On the other hand, the grain-boundary dislocation storage causes the formation of local stress concentrators and thereby enhances the nanocrack nucleation.

The suppression of the nanocrack nucleation in nanocrystalline materials under superplastic deformation can be reached owing to diffusional accommodation (Figure 5.13c), plastic flow accommodation in the adjacent nanoscale grains (Figure 5.13d), and rotational deformation (Figure 5.13f) (for details, see Section 5.6). All these accommodation processes cause a decrease of the rate of the grain-boundary dislocation storage at triple junctions and thereby suppress the nanocrack nucleation.

The softening of nanocrystalline materials under superplastic deformation is related to triple-junction migration that accommodates grain-boundary sliding (Gutkin *et al.*, 2004b; Padmanabhan and Gleiter, 2004). For instance, as shown in Figure 5.56, grain-boundary sliding – plastic shear of two upper grains relative to the bottom grain along grain boundaries and the triple junction – is accommodated by migration of the triple junction and its adjacent grain boundary. More precisely, the triple-junction migration gradually changes the triple-junction geometry (Figure 5.56) and results in the experimentally observed (Markmann *et al.*, 2003) formation of approximately plane arrays of grain boundaries (Figure 5.56f). In doing so, triple junctions stop being obstacles for grain-boundary sliding along co-planar grain boundaries. Therefore, cooperative grain-boundary sliding along co-planar grain boundaries is enhanced, and the strengthening at some critical plastic strain degree is replaced by the softening (Gutkin *et al.*, 2004b) that characterizes the second stage of superplastic deformation. At the second stage, enhanced cooperative grain-boundary sliding occurs along plane arrays of grain boundaries. These arrays are commonly divided by

5.10 Strain-rate sensitivity

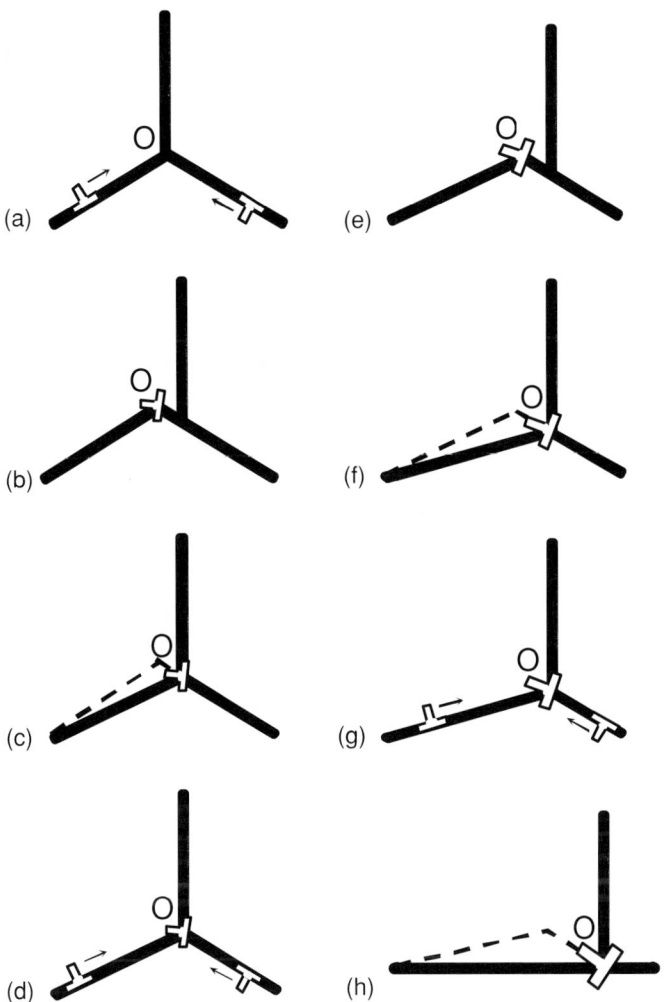

Figure 5.56. Grain-boundary sliding and transformations of defect structures near triple junctions of grain boundaries. (a) Initial state of defect configuration in a deformed nanocrystalline material. Two gliding grain-boundary dislocations move towards the triple junction O. (b) Movement and convergence of two gliding dislocations results in the formation of sessile dislocation, plastic shear of upper grains relative to bottom grain and migration of the triple junction O. (c) Local grain-boundary migration. (d) Movement of two new gliding grain-boundary dislocations towards triple junction O. (e) Movement and convergence of two gliding dislocations results in an increase of Burgers vector magnitude of sessile dislocation, a plastic shear of upper grains relative to bottom grain and migration of the triple junction O. (f) Local grain-boundary migration. (g) and (h) Grain-boundary sliding via numerous acts of transfer of grain-boundary dislocations across the triple junction O results in an increase of Burgers vector magnitude of sessile dislocation and the formation of a planar array of grain boundaries.

grains whose geometry prevents grain-boundary sliding (Figure 5.57). In these grains, the enhanced grain-boundary sliding transforms into a viscous-type deformation mechanism characterized by the positive strain-rate sensitivity. A strong candidate is the slow rotational deformation mediated by diffusion-assisted climb of grain-boundary dislocations

302 5 Mechanical properties of structural nanocrystalline materials

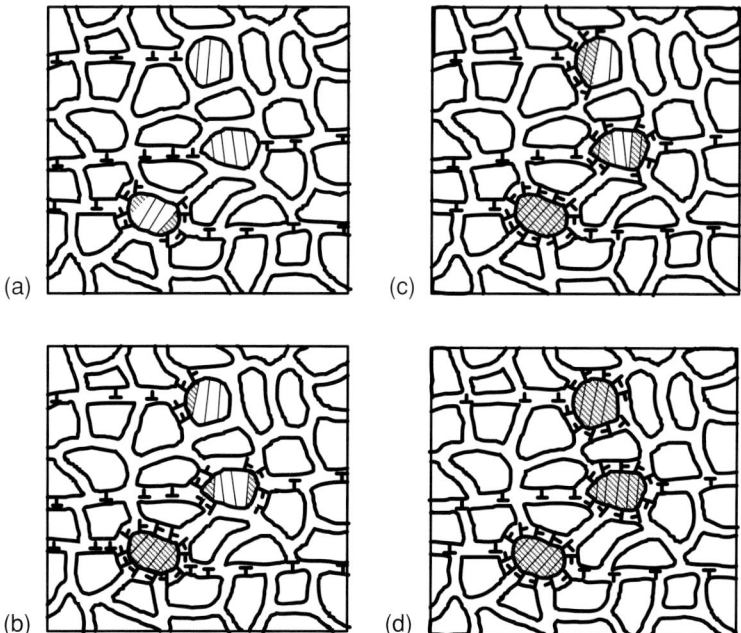

Figure 5.57. Combined action of grain-boundary sliding and the rotational deformation mode in nanocrystalline material under superplastic deformation.

(Figure 5.57) and characterized by the positive strain-rate sensitivity. In the situation discussed, both the enhanced grain-boundary sliding and slow rotational deformation operate at the second stage of superplastic deformation (Figure 5.57). The slow rotational deformation controls the plastic strain rate. With the positive plastic strain sensitivity of the rotational deformation, plastic flow localization is suppressed at the second stage of superplastic deformation.

Thus, with available experimental data and theoretical models concerning superplastic flow mechanisms, it is natural to suggest that superplastic deformation occurs in nanocrystalline materials, if grain-boundary sliding serves as the dominant deformation mechanism whose operation and accommodation are provided by the lattice dislocation slip, intense diffusion, rotational deformation, and triple-junction migration.

In this context, the following specific structural features of nanocrystalline materials can enhance their superplasticity.

(i) Nanocrystalline materials should contain high-angle grain boundaries in their highly defective, non-equilibrium state. High-density ensembles of grain-boundary dislocations at such grain boundaries carry grain-boundary sliding and rotational deformation. High-density ensembles of point defects at nonequilibrium grain boundaries carry intense diffusion processes.

(ii) Nanocrystalline materials should not contain many low-angle boundaries that do not provide grain-boundary sliding and enhanced diffusion processes.

(iii) Nanocrystalline materials should have intermediate grains where Frank–Read sources of lattice dislocations effectively operate. In this case, lattice dislocation slip effectively supports grain-boundary sliding and rotational deformation processes.
(iv) Nanocrystalline materials should contain mobile triple junctions whose migration and structural transformations provide respectively grain-boundary sliding accommodation and transitions between grain-boundary sliding, lattice dislocation slip, and rotational deformation.

These statements are worth being taken into account in further experimental and theoretical studies of superplasticity of superstrong nanomaterials.

5.11 Diffusion in nanocrystalline materials

Let us turn to a discussion of diffusion properties of nanocrystalline materials, which strongly affect the deformation behavior of such materials and, in general, are different from those of conventional coarse-grained polycrystals. Thus, following Horvath *et al.* (1987); Schaefer *et al.* (1995); Kolobov *et al.* (1999, 2000), and Kolobov (2002), nanocrystalline materials exhibit anomalously enhanced diffusion. For instance, the boundary diffusion coefficients in nanocrystalline materials fabricated by high-pressure compaction and severe plastic deformation methods are several orders of magnitude larger than those in conventional polycrystalline materials with the same chemical composition (Horvath *et al.*, 1987; Schaefer *et al.*, 1995; Kolobov *et al.*, 1999, 2000; Kolobov, 2002). With these experimental data, the anomalously fast diffusion is treated as the phenomenon inherent to only nanocrystalline materials and is attributed to their specific structural and behavioral peculiarities (Gleiter, 1989, 1995). At the same time, however, there are experimental data indicating that the boundary diffusion coefficients in dense bulk nanocrystalline materials are lower than those measured in papers (Horvath *et al.*, 1987; Schaefer *et al.*, 1995; Kolobov *et al.*, 1999, 2000; Kolobov, 2002) and similar to the boundary diffusion coefficients in conventional coarse-grained polycrystals or a little higher (Würschum *et al.*, 1996, 1997; Tanimoto *et al.*, 1999, 2000; Schaefer *et al.*, 2000). These data form a basis for the viewpoint that the atomic diffusion in nanocrystalline materials is similar to that in conventional coarse-grained polycrystals. In doing so, the difference in the diffusivities between nanocrystalline and coarse-grained materials is treated as being related only to the difference in the volume fraction of the grain-boundary phase.

Thus, in general, there are controversial experimental data and theoretical representations on diffusion processes in nanocrystalline materials. Nevertheless, mechanically synthesized nanocrystalline bulk materials are definitely recognized to exhibit enhanced diffusion properties compared to those of nanocrystalline materials fabricated by non-mechanical (more equilibrium than mechanical) methods and those of coarse-grained polycrystals (Schaefer *et al.*, 1995; Kolobov *et al.*, 1999, 2000; Kolobov, 2002). The experimentally documented phenomenon in question can be naturally explained as that occurring because of the deformation-induced nonequilibrium (highly defected) state of grain boundaries, in which case the grain-boundary diffusivity increases greatly. In other terms, there is an interaction between the enhanced diffusivity mode and the plastic deformation modes that is capable of affecting properties of the nanocrystalline material.

In this section we will briefly consider theoretical models of diffusion processes in nanocrystalline materials with special attention being paid to the diffusion enhancement occurring in mechanically synthesized nanocrystalline materials. In general, theoretical models of diffusion in nanocrystalline materials can be divided into the following three basic categories. (1) Models that describe grain-boundary diffusion in conventional coarse-grained polycrystals and are directly extended to the situation with nanocrystalline materials, with nanoscale effects neglected. (2) Models focused on the specific features of diffusion in nanocrystalline materials in the general situation, taking into account the nanoscale effects, but neglecting the influence of preparation technologies on the structure and the diffusion properties of nanocrystalline materials. In short, such models deal with the "nanostructure–diffusion" relationship. (3) Models focused on the structural and behavioral peculiarities of grain boundaries in nanocrystalline materials, taking into account the peculiarities of their fabrication. In short, such models are concerned with the "preparation–nanostructure–diffusion" relationship, which is of crucial importance for understanding fundamentals of diffusion-assisted phenomena in nanocrystalline solids and their applications.

Theoretical models of the first type and their applications to a description of diffusion in nanocrystalline materials have been reviewed in detail by Larikov (1995). Such models are, in particular, those (Fischer, 1951; Whipple, 1954; Harrison, 1961; Suzuoka, 1961; Benoist and Martin, 1975) describing spatial distributions of diffusion species within and near grain boundaries in polycrystals, models (Bokshtein, 1978; Bristowe *et al.*, 1980; Balluffi *et al.*, 1981; Brokman *et al.*, 1981; Kwok *et al.*, 1981; Klinger and Gorbunov, 1988; Ma and Balluffi, 1994; Yin *et al.*, 1997; Sørensen *et al.*, 2000) focused on microscopic mechanisms of grain-boundary diffusion, and models (Cocks and Ashby, 1982; Valiev *et al.*, 1980; Pumphrey and Gleiter, 1974; Nazarov *et al.*, 1990; Nazarov, 1996a,b, 1997; Osipov and Ovid'ko, 1992; Ovid'ko and Reizis, 1999; Smirnova and Chuvil'deev, 1999) dealing with diffusion-assisted processes (diffusional creep, diffusion-induced grain-boundary migration, transformations of grain-boundary dislocations and disclinations, etc.) in polycrystals.

Now let us turn to a brief discussion of models of the second type; that is, models describing the enhanced diffusion in nanocrystalline materials in the general situation, with the specific features of their preparation being neglected. In papers (Hoefler *et al.*, 1993; Mishin and Herzig 1995), it has been demonstrated that nanocrystalline materials exhibit a much larger variety of possible kinetic diffusion regimes compared to coarse-grained polycrystals. In particular, together with conventional A-, B- and C-regimes occurring in polycrystals (Harrison, 1961), the additional diffusion kinetic regimes have been distinguished, depending on the grain size, diffusion temperature and time, the grain boundary segregation level in the case of impurity diffusion, and other parameters (Mishin and Herzig, 1995).

Gleiter (1989) considered the anomalously fast diffusion in nanocrystalline materials as the phenomenon related to the following three factors. (1) The diffusion in nanocrystalline materials is essentially enhanced because of the existence of a high-density network of grain-boundary junction tubes which commonly are characterized by more increased diffusion rate than grain boundaries themselves (as is experimentally identified in polycrystals

5.11 Diffusion in nanocrystalline materials

(Rabukhin, 1986; Bokstein *et al.*, 2001). (2) Rigid-body relaxation of grain boundaries that occurs via relative translational displacements of their adjacent crystallites and reduces the boundary free volume is hampered in nanocrystalline materials. This is related to the fact that rigid body relaxations of the various boundaries surrounding every nanocrystallite require its different displacements owing to their different atomic structure. (3) In grain boundaries of nanocrystalline materials the concentration of impurities that often reduce boundary diffusivity is lower than that in grain boundaries in conventional polycrystals.

Let us discuss factors (1)–(3) in terms of the Arrhenius formula for diffusivity in solids. The coefficient D of self-diffusion occurring via transfer of vacancies is characterized by the largest coefficient of diffusion, given by the following Arrhenius relationship (Vladimirov, 1975; Sutton and Balluffi, 1996):

$$D = D_0 \exp[-(\varepsilon_f + \varepsilon_m)/k_B T]. \tag{5.34}$$

Here D_0 denotes the constant dependent on parameters of the ideal crystalline lattice, k_B is the Boltzmann's constant, T is the absolute temperature, and ε_f and ε_m are the energies of formation and migration of vacancies. In nanocrystalline materials (characterized by an extremely high volume fraction of the grain-boundary phase) self-diffusion processes occur mostly via transfer of grain-boundary vacancies. In these circumstances, all the factors (1)–(3) discussed by Gleiter (1989) induce the characteristic boundary free volume to be increased and, as a corollary, both formation and migration of boundary vacancies to be facilitated in nanocrystalline materials as compared to those in conventional polycrystals. In terms of the Arrhenius formula (5.34), the diffusion coefficient D is increased in nanocrystalline materials, because their characteristic energies, ε_f and ε_m, are decreased owing to factors (1) (3) discussed in the paper (Gleiter, 1989).

Now let us discuss theoretical models describing the enhanced diffusion in nanocrystalline materials fabricated at highly nonequilibrium conditions (say, by severe plastic deformation methods). Following extensive work (Masumura and Ovid'ko, 2000; Nazarov, 2000; Ovid'ko and Reizis, 2001; Ovid'ko and Sheinerman, 2003), together with factors (1)–(3) discussed by Gleiter (1989), the existence and transformations of ensembles of grain-boundary dislocations and disclinations strongly enhance diffusion processes in nanocrystalline materials fabricated at highly nonequilibrium conditions. The effects of such defects on diffusion are briefly as follows (Ovid'ko and Reizis, 2001; Ovid'ko and Sheinerman, 2003).

(a) Disordered configurations of grain-boundary dislocations and disclinations are formed in nanocrystalline materials fabricated at highly nonequilibrium conditions. Such dislocations and disclinations re-arrange to annihilate or to form more-ordered, low-energy configurations, in which case they move to new positions. Climb of (grain-boundary) dislocations is accompanied by generation of new point defects (Hirth and Lothe, 1982) – vacancies and interstitial atoms – serving as new carriers of diffusion. In this situation, generation of new vacancies at grain boundaries occurs under the action of the driving force related to a release of the elastic energy of "nonequilibrium" or, in other words, extrinsic grain-boundary dislocations during their transformations. The action of the driving force in question facilitates diffusion processes in nanocrystalline materials. This effect is quantitatively described by formula (5.34) with the sum $\varepsilon_f + \varepsilon_m$

being replaced by $\varepsilon_f - W_v + \varepsilon_m$, where W_v (> 0) is the energy release due to the climb of a grain-boundary dislocation, per one vacancy emitted by the dislocation.
(b) Cores of grain-boundary dislocations and disclinations are characterized by excess free volume, in which case their presence in grain boundaries increases the total free volume that characterizes the grain-boundary phase in nanocrystalline materials. This decreases values of ε_f and ε_m and, therefore, enhances diffusion processes.
(c) Dilatation stress fields of grain-boundary dislocations and disclinations influence migration of vacancies. More precisely, in the spirit of the theory of diffusion in stressed solids (Girifalco and Welch, 1967), the elastic interaction between vacancies and grain-boundary dislocations and disclinations is characterized by the energy:

$$\varepsilon_{int} = -\sigma_{ii} \Delta V / 3, \qquad (5.35)$$

where σ_{ii} denotes the sum of dilatation components (say, components σ_{xx}, σ_{yy} and σ_{zz} written in (x, y, z)-coordinates) of stress fields created by grain boundary defects, and ΔV is the excess free volume associated with a vacancy. To minimize ε_{int}, vacancies migrate to regions where compressive stresses exist. This vacancy migration contributes to the enhanced diffusion in nanocrystalline materials.

According to the theoretical analysis (Ovid'ko and Reizis, 2001; Ovid'ko and Sheinerman, 2003), the most essential contribution to the diffusion enhancement results from the dislocation climb in grain boundaries (see factor (a)). Owing to this factor, the grain-boundary diffusion coefficient in nanocrystalline materials fabricated at highly nonequilibrium conditions is larger by two or more orders than the diffusion coefficient in a nanocrystalline material with a relaxed structure. Relaxation of grain-boundary structures occurs during some time interval in which the diffusion coefficient gradually decreases with time (Nazarov 2003). However, during plastic deformation of a nanocrystalline material, the flow of lattice dislocations to grain boundaries can sustain their enhanced diffusivity. This phenomenon can contribute to enhancement of ductility of nanocrystalline materials in which diffusion suppresses the formation of nanocracks.

5.12 Concluding remarks

In this chapter, we have reviewed theoretical models and computer simulations of plastic flow and fracture processes in nanocrystalline materials with special attention paid to explanation of experimentally detected facts in these materials. With available results of experimental research, theoretical models and computer simulations of mechanical properties of nanocrystalline materials, these properties are highly sensitive to the specific structural features of the nanocrystalline material. In particular, the set of deformation mechanisms in nanocrystalline materials is richer than that in conventional coarse-grained polycrystals. Grain-boundary-mediated deformation mechanisms – grain-boundary sliding, rotational deformation and grain-boundary diffusional creep – effectively operate in nanocrystalline materials with the finest grains (having grain size lower than $d_c = 15$–30 nm). The deformation behavior of nanocrystalline materials with intermediate grains (with grain size ranging

5.12 Concluding remarks

from d_c to 100 nm) is mediated by the lattice dislocation activity controlled by grain boundaries operating as active sources and sinks of perfect and partial lattice dislocations. Also, twin deformation mediated by groups of partial dislocations emitted from grain boundaries contributes to plastic flow of nanomaterials.

Theory and computer simulations of fracture processes in nanocrystalline materials are in their infancy. First results in this area are indicative of the crucial role of grain boundaries and their triple junctions in fracture of nanocrystalline materials. In particular, nucleation of elemental nanocracks is preferred at grain boundaries and their triple junctions owing to plastic flow inhomogeneities provided by the grain-boundary phase in nanocrystalline materials. Evolution of nanocrack ensembles is sensitive to grain size range in nanocrystalline materials. Fracture tends to occur by intergranular brittle fracture mode in nanocrystalline materials with the finest grains and by ductile mode (through microvoid growth and coalescence) in nanocrystalline materials with intermediate grains. It is just a tendency, and the critical grain size for transition from ductile fracture to intergranular brittle fracture is not exactly determined.

Nanocrystalline materials characterized by high strength can show good tensile ductility if they are free from fabrication-induced artifacts and have the nanostructure suppressing neck formation, crack nucleation, and propagation instabilities. Current strategies to optimize ductility and strength of nanocrystalline materials are related to: (i) fabrication of nanocrystalline materials free from pores and contaminations; (ii) fabrication of bimodal and composite nanostructures in which large grains and inclusions of the second phase provide strengthening; (iii) plastic deformation at cryogenic temperatures at which recovery processes are suppressed and thereby dislocation storage provides strengthening; and (iv) plastic deformation by mechanisms characterized by large positive strain rate sensitivity of flow stress. However, further development of these strategies and search for new strategies are of crucial interest for technological applications of structural nanocrystalline materials. Achievement of superplasticity or good ductility in superstrong nanocrystalline materials in a wide range of their compositions and realization of superplastic deformation in nanocrystalline materials at commercially desired high strain rates and room temperature are very important unresolved questions in this hot area of science and technology.

In the future, one can expect rapid progress in molecular dynamics simulations of mechanical properties of nanocrystalline materials as a result of enhancement of computer processor speed and development of parallel computing architectures. In the theory of mechanical properties of nanocrystalline materials, progress is expected in the understanding of critical aspects of competition and interaction between different mechanisms of plastic deformation and fracture operating at various length scales. Current theoretical models are based on widely ranged assumptions and give different explanations of the deformation behavior of nanocrystalline materials, in which case most of them account for the corresponding experimental data. However, it is extremely difficult to experimentally identify the deformation mechanism(s) in nanocrystalline materials because of their very complicated nanoscale structure and its transformations occurring at various length scales during plastic deformation. In addition, the deformation mechanisms may be different in different nanocrystalline materials or even in the same material at different conditions of loading or at different stages of deformation. In these circumstances, further theoretical, computer, and experimental investigations in this area are highly desired for understanding the fundamentals of the outstanding deformation behavior of nanocrystalline materials and development of advanced technologies exploiting their unique mechanical and diffusional properties.

In conclusion, let us outline the key points which are interesting for the future theoretical research and computer simulations of mechanical properties of structural nanocrystalline materials.

(1) Description of deformation mechanisms mediated by grain boundaries, their competition and interaction in structural nanocrystalline materials.
(2) Analysis of specific peculiarities of lattice dislocation slip and twin deformation modes in nanocrystalline materials, with nanoscale and interface effects taken into account.
(3) Description of deformation mechanisms and their geometric coupling in nanocrystalline materials exhibiting high-strain-rate superplasticity and tensile ductility. Determination of structural characteristics and parameters of fabrication and processing that control and optimize superplasticity and ductility of nanocrystalline materials.
(4) With results of (1)–(3) used as input, development of a generalized model that would describe the combined action of competing and interacting deformation mechanisms (lattice dislocation slip, twin deformation, grain-boundary sliding, diffusion deformation mechanisms associated with enhanced diffusion along grain boundaries and their triple junctions, rotational mode, etc.) whose contribution to the plastic flow in nanocrystalline materials is dependent on the material characteristics, grain-size distribution, characteristics of defects and grain-boundary structures, and conditions of fabrication and loading.
(5) Description of the influence of plastic deformation on the structural stability and grain growth as well as the effects of grain growth on plastic flow characteristics of nanocrystalline materials.
(6) Investigation of deformation modes in bulk nanocomposites, nanocrystalline, nanolayered and nanocomposite coatings, with effects of interphase boundaries and different phases (nanocrystalline, crystalline, amorphous) taken into account.
(7) Description of fracture mechanisms operating at various length scales in nanocrystalline materials. Analysis of competition and interaction between fracture and plastic deformation processes in nanocrystalline materials.

References

Andrievskii, R. A., Kalinnikov, G. V., Kobelev, N. P., Soifer, Ya.M., and Shtansky, D. V. (1997). *Phys. Solid State*, **39**, 1661.
Andrievskii, R. A. (1998). *Nanostructured Materials: Science and Technology*. Dordrecht: Kluwer Academic Publishers.
Argon, A. S. (1979). *Acta Mater.*, **27**, 47.
Armstrong, R. W., and Head, A. K. (1965). *Acta Met.*, **13**, 759.
Asaro, R. J., Krysl, P., and Kad, B. (2003). *Phil. Mag. Lett.*, **83**, 733.

Asaro, R. J., and Suresh, S. (2005). *Acta Mater.*, **53**, 3369.
Ashby, M. F., Gandhi, C., and Taplin, D. M. R. (1979). *Acta Mater.*, **27**, 699.
Astanin, V. V., Sisanbaev, A. V., Pshenichnyuk, A. I., and Kaibyshev, O. A. (1997). *Scr. Mater.*, **36**, 117.
Balluffi, R. W., Kwok, T., and Bristowe, P. D. (1981). *Scripta Met.*, **15**, 951.
Benoist, P., and Martin, G. (1975). *Thin Solid Films*, **25**, 181.
Berndt, C. C., Fischer, T., Ovid'ko, I. A., Skandan, G., and Tsakalakos, T. (eds.) (2003). *Nanomaterials for Structural Applications*, Vol. 740, Warrendale: MRS Symp. Proc.
Birringer, R., Gleiter, H., Klein, H. P., and Mazquardt, P. (1984). *Phys. Lett.*, **A 102**, 365.
Bobylev, S. V., Ovid'ko, I. A., and Sheinerman, A. G. (2001). *Phys. Rev.*, **B 64**, 224507.
Bobylev, S. V., Gutkin, M.Yu., and Ovid'ko, I. A. (2004). *Acta Mater.*, **52**, 3793.
Bokshtein, B. S. (1978). *Diffusion in Metals*. Moscow: Metallurgiya (in Russian).
Bokstein, B., Ivanov, V., Oreshina, O., Peteline, A., and Peteline, S. (2001). *Mater. Sci. Eng.*, **A 302**, 151.
Bollmann, W. (1989). *Mater. Sci. Eng.*, **A 113**, 129.
Bristowe, P. D., Brokman, A., Spaepen, T., and Balluffi, R. W. (1980). *Scripta Met.*, **14**, 943.
Brokman, A., Bristowe, P. D., and Balluffi, R. W. (1981). *Appl. Phys.*, **52**, 6116.
Carsley, J. E., Ning, J., Milligan, W. W., Hackney, S. A., and Aifantis, E. C. (1995). *Nanostruct. Mater.*, **5**, 441.
Caro, A., and Van Swygenhoven, H. (2001). *Phys. Rev.*, **B 63**, 134101.
Champion, Y., Langlois, C., Guerin-Mailly, S., Langlois, P., Bonnentier, J.-L., and Hytch, M. (2003). *Science*, **300**, 310.
Chen, M., Ma, E., Hemker, K. J., Sheng, H., Wang, Y., and Cheng, X. (2003). *Science*, **300**, 1275.
Cheng, S., Spencer, J. A., and Milligan, W. W. (2003). *Acta Mater.*, **51**, 4505.
Chokshi, A. H., and Kottada, R. S. (2006). *Transactions of the Indian Institute of Metals*. (in press).
Chokshi, A. H., Rosen, A., Karch, J., and Gleiter, H. (1989). *Scr. Metall.*, **23**, 1679.
Chou, Y. T. (1967). *J. Appl. Phys.*, **38**, 2080.
Chow, G.-M., Ovid'ko, I. A., and Tsakalakos, T. (eds.), (2000). *Nanostructured Films and Coatings, NATO Science Series*. Dordrecht: Kluwer.
Clarke, D. R. (1987). *J. Amer. Ceram. Soc.*, **70**, 15.
Coble R. L. (1963). *J. Appl. Phys.*, **34**, 1679.
Cocks, A. C. F., and Ashby, M. F. (1982). *Progr. Mater. Sci.*, **27**, 189.
Conrad, H. (2003). *Mater. Sci. Eng. A*, **341**, 216.
Conrad, H., and Narayan, J. (2000). *Scr. Mater.*, **42**, 1025.
Das, J., Loeser, W., Kuehn, U., Eckert, J., Roy, S. K., and Schultz, L. (2003). *Appl. Phys. Lett.*, **82**, 4690.
Ebrahimi, F., Zhai, Q., and Kong, D. (1998). *Scr. Mater.*, **39**, 315.
Evans, A. G., and Hirth, J. P. (1992). *Scr. Metall. Mater.*, **26**, 1675.
Farkas, D., Kung, H., Mayo, M., Van Swygenhoven, H., and Weertman, J. (2001). *Structure and Mechanical Properties of Nanophase Materials – Theory and Computer Simulations vs. Experiment*. Warrendale: MRS.
Farkas, D., Van Swygenhoven, H., and Derlet, P. M. (2002). *Phys. Rev.*, **B 66**, 060101 (R).

Fedorov, A. A., Gutkin, M.Yu., and Ovid'ko, I. A. (2002). *Scr. Mater.*, **47**, 51.
Fedorov, A. A., Gutkin, M.Yu., and Ovid'ko, I. A. (2003). *Acta Mater.*, **51**, 887.
Fischer, J. C. (1951). *J. Appl. Phys.*, **22**, 74.
Frozeth, A. G., Derlet P. M., and Van Swygenhoven, H. (2004a). *Acta Mater.*, **52**, 5863.
Frozeth, A. G., Derlet P. M., and Van Swygenhoven, H. (2004b). *Acta Mater.*, **52**, 2259.
Gan, Y., and Zhou, B. (2001). *Scr. Mater.*, **45**, 625.
Gandhi, C., and Ashby, M. F. (1979). *Acta Mater.*, **27**, 1565.
Girifalco, L. A., and Welch, D. O. (1967). *Point Defects and Diffusion in Strained Metals*. Gordon and Breach.
Gleiter, H. (1989). *Progr. Mater. Sci.*, **33**, 223.
Gleiter, H. (1995). *Nanostruct. Mater.*, **6**, 3.
Gleiter, H. (2000). *Acta Mater.*, **48**, 1.
Gottstein, G., King, A. H., and Shvindlerman, L. S. (2000). *Acta Mater.*, **48**, 397.
Gryaznov, V. G., Kaprelov, A. M., and Romanov, A. E. (1989). *Scr. Metall.*, **23**, 1443.
Gryaznov, V. G., Polonsky, I. A., Romanov, A. E., and Trusov, L. I. (1991). *Phys. Rev.*, **B 44**, 42.
Gryaznov, V. G., and Trusov, L. I. (1993). *Progr. Mater. Sci.*, **37**, 289.
Gryaznov, V. G., Gutkin, M.Yu., Romanov, A. E., and Trusov, L. I. (1993). *J. Mater. Sci.*, **28**, 4359.
Gutkin, M.Yu., and Ovid'ko, I. A. (1993). *Nanostruct. Mater.*, **2**, 631.
Gutkin, M.Yu., and Ovid'ko, I. A. (1994). *Phil. Mag.*, **A 70**, 561.
Gutkin, M.Yu., and Ovid'ko, I. A. (2001). *Phys. Rev.*, **B 63**, 064515.
Gutkin, M.Yu., and Ovid'ko, I. A. (2004a). *Plastic Deformation in Nanocrystalline Materials*. Berlin, New York: Springer.
Gutkin, M.Yu., and Ovid'ko, I. A. (2004b). *Phil. Mag. Lett.*, **84**, 655.
Gutkin, M.Yu., and Ovid'ko, I. A. (2005). *Appl. Phys. Lett.*, **87**, 251916.
Gutkin, M.Yu., and Ovid'ko, I. A. (2006). *Phil. Mag.*, **86**, 1483.
Gutkin, M.Yu., Ovid'ko, I. A., and Pande, C. S. (2001). *Rev. Adv. Mater. Sci.*, **2**, 80.
Gutkin, M.Yu., Kolesnikova, A. L., Ovid'ko, I. A., and Skiba, N. V. (2002). Phil. Mag. Lett., **82**, 651.
Gutkin, M.Yu., Ovid'ko, I. A., and Skiba, N. V. (2003a). *Acta Mater.*, **51**, 4059.
Gutkin, M.Yu., Ovid'ko, I. A., and Skiba, N. V. (2003b). *Mater. Sci. Eng.*, **A 339**, 73.
Gutkin, M.Yu., Ovid'ko, I. A., and Skiba, N. V. (2003c). *J. Phys.*, **D 36**, L47.
Gutkin, M.Yu., Ovid'ko, I. A., and Pande, C. S. (2004a). *Phil. Mag.*, **84**, 847.
Gutkin, M.Yu., Ovid'ko, I. A., and Skiba, N. V. (2004b). *Acta Mater.*, **52**, 1711.
Gutkin, M.Yu., Ovid'ko, I. A., and Skiba, N. V. (2005a). *J. Phys.*, **D 38**, 3921.
Gutkin, M.Yu., Ovid'ko, I. A., and Skiba, N. V. (2005b). *Rev. Adv. Mater. Sci.*, **10**, 483.
Hahn, H., and Padmanabhan, K. A. (1995). *Nanostruct. Mater.*, **6**, 191.
Hahn, H., and Padmanabhan, K. A. (1997). *Phil. Mag.*, **B 76**, 559.
Hahn, H., Mondal, P., and Padmanabhan, K. A. (1997). *Nanostruct. Mater.*, **9**, 603.
Han, B. Q., Lavernia, E., and Mohamed, F. A. (2005). *Rev. Adv. Mater. Sci.*, **9**, 1.
Harris, K. E., Singh, V. V., and King, A. H. (1998). *Acta Mater.*, **46**, 2623.
Harrison, G. (1961). *Trans. Faraday Soc.*, **57**, 1191.
Hart, E. W. (1967). *Acta Mater.*, **15**, 351.

Haslam, A. J., Moldovan, D., Yamakov, V., Wolf, D., Phillpot, S. R., and Gleiter, H. (2003). *Acta Mater.*, **51**, 2097.

Hasnaoui, A., Derlet P. M., and Van Swygenhoven, H. (2004). *Acta Mater.*, **52**, 2251.

He, J. H., and Lavernia, E. J. (2001). *J. Mater. Res.*, **16**, 2724.

He, G., Eckert, J., Loeser, W., and Schultz, L. (2003). *Nature Mater.*, **2**, 33.

He, G., Hagiwara, M., Eckert, J., and Loeser, W. (2004). *Phil. Mag. Lett.*, **84**, 365.

Hirth, J. P., and Lothe, J. (1982). *Theory of Dislocations.* New York: McGraw-Hill Publ. Co.

Hoefler, H. J., Hahn, H., and Averback, S. (1993). *Defect and Diffusion Forum*, **68**, 99.

Horvath, J., Birringer, R., and Gleiter, H. (1987). *Solid State Comm.*, **62**, 391.

Hugo, R. C., Kung, H., Weertman, J. R., Mitra, R., Knapp, J. A., and Follstaedt, D. M. (2003). *Acta Mater.*, **51**, 1937.

Indenbom, V. I. (1961). *Sov. Phys. Sol. State* **3**, 1506.

Islamgaliev, R. K., Valiev, R. Z., Mishra, R. S., and Mukherjee, A. K. (2001). *Mater. Sci. Eng.*, **A 304–306**, 206.

Jia, D., Ramesh, K. T., and Ma, E. (2003). *Acta Mater.*, **51**, 3495.

Jin, M., Minor, A. M., Stach, E. A., and Morris Jr, J. W. (2004). *Acta Mater.*, **52**, 5381.

Karimpoor, A. A., Erb, U., Aust, K. T., and Palumbo, G. (2003). *Scr. Mater.*, **49**, 651.

Ke, M., Milligan, W. W., Hackney, S. A., Carsley, J. E., and Aifantis, E. C. (1995). *Nanostruct. Mater.*, **5**, 689.

Kim, H. S. (1998). *Scr. Mater.*, **39**, 1057.

Kim, H. S., Estrin, Y., and Bush, M. B. (2000). *Acta Mater.*, **48**, 493.

King, A. H. (1999). *Interf. Sci.*, **7**, 251.

Klimanek, P., Klemm, V., Romanov, A. E., and Seefeldt, M. (2001). *Adv. Eng. Mater.*, **3**, 877.

Klinger, L. M., and Gorbunov, D. A. (1988). *Structure and Properties of Interfaces in Metals.* Moscow: Nauka (in Russian).

Koch, C. C. (2003). *Scr. Mater.*, **49**, 657.

Koch, C. C., and Narayan, J. (2001). *Mater. Res. Soc. Symp. Proc.*, **634**, B5.1.1.

Koch, C. C., Morris, D. G., Lu, K., and Inoue, A. (1999). *MRS Bullet.*, **24**, 54.

Kolobov, Yu.R. (2002). *Interf. Sci.*, **10**, 31.

Kolobov, Yu.R., Grabovetskaya, G. P., Ratochka, I. V., and Ivanov, K. V. (1999). *Nanostruct. Mater.*, **12**, 1127.

Kolobov, Yu.R., Grabovetskaya, G. P., Ivanov, K. V., Valiev, R. Z., and Lowe, T. C. (2000). In *Investigations and Applications of Severe Plastic Deformation*, ed. Lowe, T. C., and Valiev, R. Z., NATO Science Series. Dordrecht: Kluwer, p. 261.

Komarneni, S., Vaja, R. A., Lu, G. Q., Matsushita, J.-I., and Parker, J. C. (eds.) (2003). *Nanophase and Nanocomposite Materials IV*, vol. 703, Warrendale: MRS Symp. Proc.

Konstantinidis, D. A., and Aifantis, E. C. (1998). *Nanostruct. Maters.*, **10**, 1111.

Krujicic, M., and Olson, G. B. (1998). *Interface Sci.*, **6**, 155.

Kumar, K. S., Suresh, S., and Van Swygenhoven, H. (2003a). *Acta Mater.*, **51**, 5743.

Kumar, K. S., Suresh, S., Chisholm, M. F., Norton, J. A., and Wang, P. (2003b). *Acta Mater.*, **51**, 387.

Kuntz, J. D., Zhan, G.-D., and Mukherjee, A. K. (2004). *MRS Bullet.*, **29**, 22.

Kurzydlowski, K. J. (1990). *Scr. Metall. Mater.*, **24**, 879.
Kwok, T., Ho, P. S., Yip, S., and Balluffi, R. W. (1981). *Phys. Rev. Lett.*, **47**, 1148.
Larikov, L. N. (1995). *Metal. Phys. Appl. Tech.*, **17**, 1.
Lasalmonie, A., and Strudel, J. L. (1986). *J. Mater. Sci.*, **21**, 1837.
Li, J. C. M. (1963). *Trans. TMS-AIME*, **227**, 247.
Li, J. C. M., and Chou, Y. T. (1970). *Met. Trans.*, **1**, 1145.
Li, H., and Ebrahimi, F. (2004). *Appl. Phys. Lett.*, **84**, 4307.
Li, H., and Ebrahimi, F. (2005). *Adv. Mater.*, **17**, 1969.
Lian, J., Baudelet, B., and Nazarov, A. A. (1993). *Mater. Sci. Eng.*, **A 172**, 23.
Liao, X. Z., Zhou, F., Lavernia, E. J., *et al.* (2003a). *Appl. Phys. Lett.*, **83**, 632.
Liao, X. Z., Zhou, F., Lavernia, E. J., He, D. W., and Zhu, Y. T. (2003b). *Appl. Phys. Lett.*, **83**, 5062.
Liao, X. Z., Zhao, Y. H., Srinivasan, S. G., Zhu, Y. T., Valiev, R. Z., and Gunderov, D. V. (2004a). *Appl. Phys. Lett.*, **84**, 592.
Liao, X. Z., Srinivasan, S. G., Zhao, Y. H., Baskes, M. I., Zhu, Y. T., Zhou, F., Lavernia, E. J., and Hu, H. F. (2004b). *Appl. Phys. Lett.*, **84**, 3564.
Lu, K., and Sui, M. L. (1993). *Scr. Metall. Mater.*, **28**, 1465.
Lubarda, V. A., Schneider, M. S., Kalantar, D. H., Remington, B. A., and Meyers, M. A. (2004). *Acta Mater.*, **52**, 1397.
Ma, E. (2003). *Nature Mater.*, **2**, 7.
Ma, Q., and Balluffi, R. W. (1994). *Acta Metall. Mater.*, **42**, 1.
MacHahon, G., and Erb, U. (1989). *Microstruct. Sci.*, **17**, 447.
Markmann, J., Bunzel, P., Roesner, H., Liu, K. W., Padmanabhan, K. W., Birringer, R., Gleiter, H., and Weissmueller, J. (2003). *Scr. Mater.*, **49**, 637.
Malygin, G. A. (1995). *Phys. Solid State*, **37**, 1248.
Masumura, R. A., Hazzledine, P. M., and Pande, C. S. (1998). *Acta Mater.*, **46**, 4527.
Masumura, R. A., and Ovid'ko, I. A. (2000). *Mater. Phys. Mech.*, **1**, 31.
Mayo, M. J. (1997). *Nanostruct. Mater.*, **9**, 717.
Mayo, M. J. (1998). In: *Nanostructured Materials: Science and Technology*, ed. Chow, G.-M., and Noskova, N. I. Dordrecht: Kluwer, p. 361.
McFadden, S. X., Misra, R. S., Valiev, R. Z., Zhilyaev, A. P., and Mukherjee, A. K. (1999). *Nature*, **398**, 684.
Milligan, W. W. (2003). Mechanical behavior of bulk nanocrystalline and ultrafine-grain metals. In *Comprehensive Structural Integrity*, ed. Milne, I., Ritchie, R. O., and Karihaloo, B. Amsterdam: Elsevier, p. 529.
Mishin, Yu., and Herzig, Ch. (1995). *Nanostruct. Mater.*, **6**, 859.
Mishra, R. S., Valiev, R. Z., McFadden, S. X., and Mukherjee, A. K. (1998). *Mater. Sci. Eng.*, **A 252**, 174.
Mishra, R. S., Valiev, R. Z., McFadden, S. X., Islamgaliev, R. K., and Mukherjee, A. K., (2001). *Phil. Mag.*, **A 81**, 37.
Moldovan, D., Wolf, D., and Phillpot, S. R. (2001). *Acta Mater.*, **49**, 3521.
Mohamed, F. A., and Li, Y. (2001). *Mater. Sci. Eng.*, **A 298**, 1.
Morozov, N. F., Ovid'ko, I. A., Petrov, Yu.V., and Sheinerman, A. G. (2003). *Rev. Adv. Mater. Sci.*, **4**, 65.

Mukai, T., Suresh, S., Kita, K., Sasaki, H., Kobayashi, N., Higashi, K., and Inoue, A. (2003). *Acta Mater.*, **51**, 4197.

Mukherjee, A. K. (2002a). *Mater. Sci. Eng.*, **A 322**, 1.

Mukherjee, A. K. (2002b). *Creep Deformation: Fundamentals and Applications*, ed. Mishra, R. S., Earthman, J. C., and Raj, S. V. Warrendale: TMS, p. 3.

Mullner, P., and Romanov, A. E. (2000). *Acta Mater.*, **48**, 2337.

Murayama, M., Howe, J. M., Hidaka, H., and Takaki, S. (2002). *Science*, **295**, 2433.

Nazarov, A. A. (1996a). *Scr. Mater.*, **34**, 697.

Nazarov, A. A. (1996b). *Annales de Chimie (Fr.)*, **21**, 461.

Nazarov, A. A. (1997). *Mater. Sci. Forum*, **243–245**, 31.

Nazarov, A. A. (2000). *Phil. Mag. Lett.*, **80**, 221.

Nazarov, A. A. (2003). *Phys. Sol. State*, **45**, 1166.

Nazarov, A. A., Bachurin, D. V., Shenderova, O. A., and Brenner, D. W. (2003). *Interface Sci.*, **11**, 417.

Nazarov, A. A., Romanov, A. E., and Valiev, R. Z. (1990). *Scripta Metall. Mater.*, **24**, 1929.

Nazarov, A. A., Romanov, A. E., and Valiev, R. Z. (1993). *Acta Met. Mater.*, **41**, 1033.

Niederhofer, A., Bolom, T., Nesadek, P., Moto, K., Eggs, C., Patil, D. S., and Veprek, S. (2001). *Surf. Coat. Technol.*, **146–147**, 183.

Nieh, T. G., and Wadsworth, J. (1991). *Scr. Metall. Mater.*, **25**, 955.

Nieman, G. W., Weertman, J. R., and Siegel, R. W. (1991). *J. Mater. Res.*, **6**, 1012.

Niihara, K., Nakahira, A., and Sekino, T. (1993). In: *Nanophase and Nanocomposite Materials*, ed. Kormaneni, S., Parker, J. C., and Thomas, G. J., MRS Symp. Proc. 286. Pittsburg, p. 405.

Osipov, A. V., and Ovid'ko, I. A. (1992). *Appl. Phys.*, **A 54**, 517.

Ovid'ko, I. A. (1994). *J. Phys. D*, **27**, 999.

Ovid'ko, I. A. (1997). *Nanostruct. Mater.*, **7**, 149.

Ovid'ko, I. A. (2000). *Mater. Sci. Eng.*, **A 280**, 355.

Ovid'ko, I. A. (2002). *Science*, **295**, 2386.

Ovid'ko, I. A. (2003). *Phil. Mag. Lett.*, **83**, 611.

Ovid'ko, I. A. (2004). In: *Encyclopedia on Nanoscience and Nanotechnology*, Vol. 4, ed. Nalwa, H. S. California: American Sci. Publ., p. 249.

Ovid'ko, I. A. (2005a). *Int. Mater. Rev.*, **50**, 65.

Ovid'ko, I. A. (2005b). *Rev. Adv. Mater. Sci.*, **10**, 89.

Ovid'ko, I. A., and Reizis, A. B. (1999). *J. Phys. D*, **32**, 2833.

Ovid'ko, I. A., and Reizis, A. B. (2001). *Phys. Sol. State*, **43**, 35.

Ovid'ko, I.A, and Sheinerman, A. G. (2003). *Phil. Mag.*, **83**, 1551.

Ovid'ko, I. A., and Sheinerman, A. G. (2004a). *Acta Mater.*, **52**, 1201.

Ovid'ko, I. A., and Sheinerman, A. G. (2004b). *Rev. Adv. Mater. Sci.*, **6**, 21.

Ovid'ko, I. A., and Sheinerman, A. G. (2005a). *Acta Mater.*, **53**, 1347.

Ovid'ko, I. A., and Sheinerman, A. G. (2005b). *Rev. Adv. Mater. Sci.*, **9**, 17.

Ovid'ko, I.A, and Sheinerman, A. G. (2006). *Phil. Mag.*, **86**, 1415.

Padmanabhan, K. A. (2001). *Mater. Sci. Eng.*, **A 304–306**, 200.

Padmanabhan, K. A., and Gleiter, H. (2004). *Mater. Sci. Eng.*, **A 381**, 28.

Palumbo, G., and Aust, K. T. (1989). *Mater. Sci. Eng.*, **A 113**, 139.

Pande, C. S., and Masumura, R. A. (1984). In: *Proceedings of Sixth International Conference on Fracture*, p. 857.

Pande, C. S., and Masumura, R. A. (1996). In: *Processing and Properties of Nanocrystalline Materials*, ed. Suryanarayana, C., Singh, J., and Froes, F. H. Warrendale: TMS, p. 387.

Pande, C. S., Masumura, R. A., and Armstrong, R. W. (1993). *Nanostruct. Mater.*, **2**, 323.

Patscheider, J. (2003). *MRS Bulletin*, **28**, 180.

Pozdnyakov, V. A. (2003). *Tech. Phys. Lett.*, **29**, 151.

Pozdnyakov, V. A., and Glezer, A. M. (2005). *Phys. Sol. State*, **47**, 817.

Pumphrey, P. H., and Gleiter, H. (1974). *Philos. Mag.*, **30**, 593.

Rabukhin, V. B. (1986). *Poverkhnost'*, **7**, 126 (in Russian).

Roco, M. C., Williams, R. S., and Alivisatos, P. (eds.), (2000). *Nanotechnology Research Directions*. Dordrecht: Kluwer.

Romanov, A. E., and Vladimirov, V. I. (1992). *Dislocations in Solids*, ed. Nabarro, F. R. N., Vol. 9. Amsterdam: North-Holland.

Romanov, A. E. (1995). *Nanostruct. Mater.*, **6**, 125.

Romanov, A. E. (2003). *European J. Mech.*, **A 22**, 727.

Rybin, V. V., and Zhukovskii, I. M. (1978). *Sov. Phys. Sol. State*, **20**, 1056.

Sahimi, M. (1994). *Applications of Percolation Theory*. London: Taylor and Francis.

Samaras, M., Derlet, P. M., Van Swygenhoven, H., and Victoria, M. (2002). *Phys. Rev. Lett.*, **88**, 125505.

Sanders, P. G., Eastman, J. A., and Weertman, J. R. (1996). In *Processing and Properties of NC Materials*, ed. Suryanarayana, C., Singh, J., and Froes, F. H. Warrendale: TMS, p. 397.

Scattergood, R. O., and Koch, C. C. (1992). *Scr. Mater.*, **27**, 1195.

Schaefer, H.-E., Wurschum, R., Gessmann, T., Stockl, G., Scharwaechter, P., Frank, W., Valiev, R. Z., Fecht, H.-J., and Moelle, C. (1995). *Nanostruct. Mater.*, **6**, 869.

Schaefer, H.-E., Reimann, K., Straub, W., Philipp, F., Tanimoto, H., Brossmann, U., and Würschum, R. (2000). *Mater. Sci. Eng.*, **A 286**, 24.

Schiotz, J. (2004). *Scr. Mater.*, **51**, 837.

Schiotz, J., and Jacobsen, K. W. (2003). *Science*, **301**, 1357.

Schiotz, J., Di Tolla, F. D., and Jakobsen, K. W. (1998). *Nature*, **391**, 561.

Schiotz, J., Vegge, T., Di Tiolla, F. D., and Jakobsen, K. W. (1999). *Phys. Rev.*, **B 60**, 11971.

Seefeldt, M. (2001). *Rev. Adv. Mater. Sci.*, **2**, 44.

Sergueeva, A. V., Mara, N. A., and Mukherjee, A. K. (2004). *Rev. Adv. Mater. Sci.*, **7**, 67.

Shan, Z., Stach, E. A., Wiezorek, J. M. K., Knapp, J. A., Follstaedt, D. M., and Mao, S. X. (2004). *Science*, **305**, 654.

Shimokawa, T., Nakatani, A., and Kitagawa, H. (2005). *Phys. Rev.*, **B 71**, 224110.

Siegel, R. W. (1994). In: *Encyclopedia of Applied Physics*, ed. Trigg, G. L., Vol. 11. Weinheim: VCH, p. 1.

Siegel, R. W., and Fougere, G. E. (1995). *Nanostruct. Mater.*, **6**, 205.

Smirnova, E. S., and Chuvil'deev, V. N. (1999). *Fiz. Met. Metalloved.*, **88**, 74 (in Russian).

Soer, W. A., De Hosson, J. T. M., Minor, A., Moris Jr., J. W., and Stach, E. (2004). *Acta Mater.*, **52**, 5783.

Sørensen, M. R., Mishin, Yu., and Voter, A. F. (2000). *Phys. Rev.*, **B 62**, 3658.

Stauffer, D., and Aharony, A. (1992). *Introduction to Percolation Theory*. London: Taylor and Francis.
Suryanarayana, R., Frey, C. A., Sastry, S. M. L., Waller, B. E., Bates, S. E., and Buhro, W. E. J. (1996). *Mater. Res.*, **11**, 439.
Sutton, A. P., and Balluffi, R. W. (1996). *Interfaces in Crystalline Materials*. Oxford: Oxford Science Publications.
Suzuoka, T. (1961). *Trans. Jap. Inst. Metals*, **2**, 25.
Tanimoto, H., Farber, P., Würschum, R., Valiev, R. Z., and Schaefer, H.-E. (1999). *Nanostruct. Mater.*, **12**, 681.
Tanimoto, H., Pasquini, L., Prümmer, R., Kronmüller, H., and Schaefer, H.-E. (2000). *Scripta Mater.*, **42**, 961.
Tellkamp, V. L., Melmed, A., and Lavernia, E. J. (2001). *Metall. Mater. Trans.*, **A 32**, 2335.
Valiev, R. Z., Gertsman, V.Yu., and Kaibyshev, O. A. (1980). *Phys. Stat. Sol.*, **61**, K95.
Valiev, R. Z., and Langdon, T. G. (1993). *Acta Metall.*, **41**, 949.
Valiev, R. Z., Islamgaliev, R. K., and Alexandrov, I. V. (2000). *Progr. Mater. Sci.*, **45**, 103.
Valiev, R. Z., Song, C., McFadden, S. X., Mukherjee, A. K., and Mishra, R. S. (2001). *Phil. Mag.*, **A. 81**, 25.
Valiev, R. Z., Alexandrov, I. V., Zhu, Y. T., and Lowe, T. C. (2002). *J. Mater. Res.*, **17**, 5.
Valiev, R. Z. (2004). *Nature Mater.*, **3**, 511.
Van Swygenhoven, H., and Derlet, P. M. (2001). *Phys. Rev.*, **B 64**, 224105.
Van Swygenhoven, H., Spaczer, M., and Caro, A. (1999a). *Acta Mater.*, **47**, 3117.
Van Swygenhoven, H., Spavzer, M., Caro, A., and Farkas, D. (1999b). *Phys. Rev.*, **B 60**, 22.
Van Swygenhoven, H., Derlet, P. M., and Hasnaoui, A. (2002). *Phys. Rev.*, **B 66**, 024101.
Van Swygenhoven, H., Derlet P. M., Hasnaoui, A., and Samaras, M. (2003). In *Nanostructures: Synthesis, Functional Properties and Applications*, ed. Tsakalakos, T., Ovid'ko, I. A., and Vasudevan, A. K. Dordrecht: Kluwer, p. 155.
Veprek, S., and Argon, A. S. (2002). *J. Vac. Sci. Technol.*, **20**, 650.
Vladimirov, V. I. (1975). *Einfuhrüng in die Physikalishe Theorie der Plastizität and Festigkeit*. Leipzig: VEB eutscher Verlag für Grundstoffindutrie.
Volpp, T., Göring, E., Kuschke, W.-M., and Arzt, E. (1997). *Nanostruct. Mater.*, **8**, 855.
Wang, N., Wang, Z., Aust, K. T., and Erb, U. (1995). *Acta Metall. Mater.*, **43**, 519.
Wang, Y., Chen, M., Zhou, F., and Ma, E. (2002). *Nature*, **419**, 912.
Wang, Y. M., and Ma, E. (2004a). *Acta Mater.*, **52**, 1699.
Wang, Y. M., and Ma, E. (2004b). *Appl. Phys. Lett.*, **85**, 2750.
Wang, Y. M., Hodge, A. M., Biener, J., Hamza, A. V., Barnes, D. E., Liu, K., and Nieh, T. G. (2005). *Appl. Phys. Lett.*, **86**, 101915.
Weertman, J. R., and Sanders, P. G. (1994). *Solid State Phenom.*, **35–36**, 249.
Wei, Q., Jia, D., Ramesh, K. T., and Ma, E. (2002). *Appl. Phys. Lett.*, **81**, 1240.
Wei, Q., Cheng, S., Ramesh, K. T., and Ma, E. (2004). *Mater. Sci. Eng.*, **A, 381**, 71.
Weissmueller J., and Markmann, J. (2005). *Adv. Eng. Mater.*, **7**, 202.
Witney, A. B., Sanders, P. G., Weertman, J. R., and Eastman, J. A. (1995). *Scr. Metall. Mater.*, **33**, 2025.
Wolf, D., Yamakov, V., Phillpot, S. R., Mukherjee, A. K., and Gleiter, H. (2005). *Acta Mater.*, **53**, 1.

Würschum, R., Kübler, A., Gruß, S., Scharwaechter, P., Frank, W., Valiev, R. Z., Mulyukov, R. R., and Schaefer, H.-E. (1996). *Annales de Chimie (Fr.)*, **21**, 471.

Würschum, R., Reimann, K., Gruß, S., Farber, P., Kübler, A., Scharwaechter, P., Frank, W., Kruse, O., Carstanjen, H. D., and Schaefer, H.-E. (1997). *Phil. Mag.*, **B 76**, 407.

Whipple, R. T. P. (1954). *Phil. Mag.*, **45**, 1225.

Yamakov, V., Wolf, D., Salazar, M., Phillpot, S. R., and Gleiter, H. (2001). *Acta Mater.*, **49**, 2713.

Yamakov, V., Wolf, D., Phillpot, S. R., and Gleiter, H. (2002). *Acta Mater.*, **50**, 61.

Yamakov, V., Wolf, D., Phillpot, S. R., Mukherjee, A. K., and Gleiter, H. (2003). *Phil. Mag. Lett.*, **83**, 385.

Yin, K. M., King, A. H., Hsieh, T. E., Chen, F. R., Kai, J. J., and Chang, L. (1997). *Microscopy and Microanalysis*, **3**, 417.

Youngdahl, C. J., Sanders, P. G., Eastman, J. A., and Weertman, J. R. (1997). *Scr. Mater.*, **37**, 809.

Youssef, K. M., Scattergood, R. O., Murty, K. L., and Koch, C. C. (2004). *Appl. Phys. Lett.*, **85**, 929.

Youssef, K. M., Scattergood, R. O., Murty, K. L., Horton, J. A., and Koch, C. C. (2005). *Appl. Phys. Lett.*, **87**, 091904.

Youssef, K. M., Scattergood, R. O., Murty, K. L., and Koch, C. C. (2006). *Scr. Mater.*, **54**, 251.

Zaichenko, S. G., and Glezer, A. M. (1997). *Phys. Sol. State*, **39**, 1810.

Zelin, M. G., and Mukherjee, A. K. (1993). *Phil. Mag.*, **A 68**, 1183.

Zelin, M. G., and Mukherjee A. K. (1995). *Acta Metall. Mater.*, **43**, 2359.

Zelin, M. G., and Mukherjee, A. K. (1996). *Mater. Sci. Eng.*, **A 208**, 210.

Zelin, M. G., Dunlap, M. R., Rosen, R., and Mukherjee, A. K. (1993). *J. Appl. Phys.*, **74**, 4972.

Zelin, M. G., Guillard, S., and Mukherjee, A. (2001). *Mater. Sci. Eng.*, **A 309–310**, 514.

Zghal, S, Hytch, M. J., Chevalier, J.-P., Twesten, R., Wu, F., and Bellon, P. (2002). *Acta Mater.*, **50**, 4695.

Zhan, G.-D., Kuntz, J. D., Wan, J., and Mukherjee, A. K. (2003). In *Nanomaterials for Structural Applications*, ed. Berndt, C. C., Fisher, T., Ovid'ko, I. A., Skandan, G., and Tsakalakos, T., Vol. 740. Warrendale: MRS Symp. Proc., p. 49.

Zhan, G.-D., Kuntz, J. D., Wan, J., and Mukherjee, A. K. (2004). *MRS Bull.*, **29**, 22.

Zhu, Y. T., Liao, X. Z., Srivansan, S. G., Zhao, Y. H., Baskes, M. I., Zhou, F., and Lavernia, E. J. (2004). *Appl. Phys. Lett.*, **85**, 5049.

Zhu, Y. T., Liao, X. Z., and Valiev, R. Z. (2005a). *Appl. Phys. Lett.*, **86**, 103112.

Zhu, B., Asaro, R. J., Krysl, P., and Bailey, R. (2005b). *Acta Mater.*, **53**, 4825.

Ziman, J. (1979). *Models of Disorder*. Cambridge: Cambridge University Press.

6 Corrosion of structural nanomaterials

6.1 Introduction

Corrosion, degradation, and weathering of advanced materials are problems for which mankind has yet to find a proper solution. From the world's ancient man-made and natural monoliths to today's most modern buildings, bridges, and transportation facilities, the longevities of useful structures are closely regulated by the environment where they are located. Having little control over these aggressive environments, we must carefully select materials that have properties best suited for the conditions to which they are exposed. Even after 100 years of industrial revolution and the development of advanced materials, testing and monitoring procedures, we wonder if our knowledge of today's structural materials and their interactions in the current environment is sufficient to produce long-lasting structures that will benefit society.

For example, the total cost of corrosion and corrosion-related issues in the USA alone is quite significant, amounting to 6% GDP or $500 billion per year (Koch *et al.*, 2002). This cost is distributed over direct costs on materials and structures and indirect costs on loss of productivity. About 90% of the corrosion cost is within iron-based materials. The annual direct cost of corrosion of highway bridges is estimated to be $8.3 billion, of which $3.8 billion is for bridge replacement, and $4 billion is for maintenance. It is estimated that indirect costs to the user due to traffic delays and lost productivity are about $38 billion annually. For more details, see the 2001 report by CC Technologies Laboratories, Inc., FWHA (2001).

In the US electric power industry sector, corrosion-related damage is around $5 billion–$10 billion annually (EPRI estimate). In steam-electric generating plants, corrosion costs exceeded 10% of total power cost. Up to 50% of outages are attributed to corrosion (InTech Magazine, 1998). The corrosion damage in oil and gas pipelines is around $7 billion annually (according to the FWHA report in 2001). One feature that is common to each of the above scenarios is mankind's inability accurately to predict the performance of a particular material in a given environment. Although significant effort is made to determine the mechanical integrity of structures, very little time is spent on determining the causes behind corrosion, and focus on the development of advanced corrosion-resistant materials is lacking.

Thus it is evident that failure of an engineering material is strongly affected by its structure and properties. In many cases, failure starts from the material surface, in processes such as wear and corrosion. In recent years, nanostructured materials have attracted great

6 Corrosion of structural nanomaterials

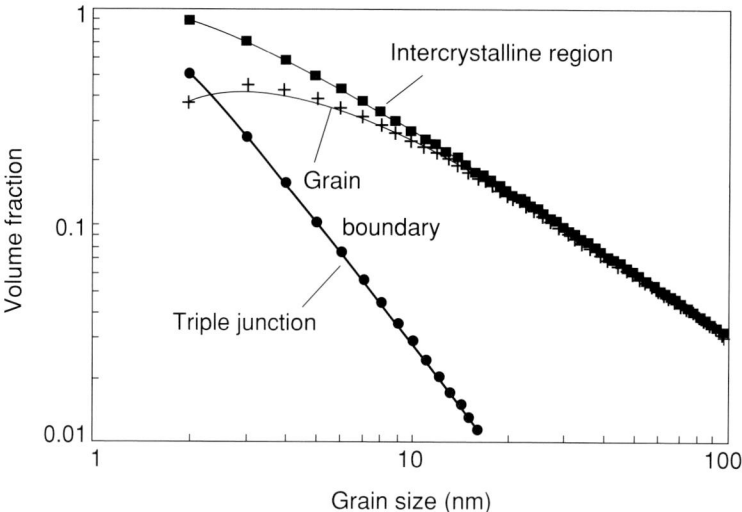

Figure 6.1. Effect of grain size on the calculated volume fractions for intercrystalline regions, grain boundaries and triple junctions (Surayanarayana *et al.*, 1992).

interest because of their high degree of hardness, improved toughness and superior physical properties. It is reported that a nanostructured surface layer of a low carbon steel exhibited greatly improved wear and friction properties (Wang *et al.*, 2001) in selected corrosive environments.

6.2 Effect of defects and grain size

From the point of view of corrosion behavior, a high volume fraction of intergranular defects is often associated with the ultrafine grain size of nanomaterials. The defects are primarily free volumes and microvoids present in the material of interest. These could lead to poor corrosion performance since localized corrosion commonly initiates at surface heterogeneities and weak structural sites. On the other hand, it has been pointed out for alloys with elements that are capable of forming passive films, that the atoms of these elements can diffuse easily through the grain boundaries to the surface of the alloy to form the protective passive layer (Zieger *et al.*, 1995; Tong and Shi, 1995; John *et al.*, 1999).

In the nanocrystalline alloys, the grain-boundary volume represents up to 50% of the total volume. Therefore, these solids are assumed to have a different kind of atomic structure: a crystalline structure with long-range order for all the atoms far from the grain boundaries, and a disordered structure with some short-range order at the interfacial region. It is the presence of two structural components (crystallites and the interfacial component) of comparable volume fractions that characterizes nanocrystalline materials and was claimed to be the origin of novel and improved properties. Figure 6.1 shows the effect of grain size on the calculated volume fractions for intercrystalline regions, grain boundaries and triple

6.4 Nanocrystalline nickel

junctions, assuming a grain boundary thickness of 1 nm (Surayanarayana *et al.*, 1992). The corrosion behavior of these zones in these length scales is unknown and under investigation.

Thus in a medium, where a nanostructured alloy could passivate, the oxide film formed on the alloy surface would be more uniform, and thus more protective, than that on the same microcrystalline alloy. Reviewing the information available on the corrosion behavior of nanocrystalline alloys it seems that is not possible to generalize the effect of a nanoscale structure to validate the corrosion resistance of the alloys (Thorpe *et al.*, 1988; Rofagha *et al.*, 1992; D'Souza *et al.*, 1999; Köster *et al.*, 2001; Alves *et al.*, 2001). In this chapter, we will review the corrosion behavior of nanocrystalline metals and alloys of interest.

6.3 Corrosion of metallic and alloyed nanostructured materials

As described in earlier chapters in this book, electrodeposition is a versatile technique for producing nanocrystalline materials. In a corrosion experiment, it was found that corrosion resistance of a nano-multicomponent $Fe_{32}Ni_{36}Cr_{14}P_{12}B_6$ alloy was greater than its amorphous form (Thorpe *et al.*, 1988), thus establishing the fact that both amorphous and crystalline structures play important roles in corrosion. Nanocrystalline Zn coatings showed improved corrosion resistance compared to electrogalvanized steel in deaerated 0.5 N NaOH solutions (Yousef *et al.*, 2004). In the next few sections, we will discuss the corrosion aspects of several metals and alloys of interest.

6.4 Nanocrystalline nickel

Various reports have cited the corrosion behavior of nanocrystalline Ni, one of the key elements used in many structural materials and components (Rofagha *et al.*, 1991). In a recent study by Mishra and Balasubramaniam (2004), the authors reported the corrosion behavior of electrodeposited nanocrystalline Ni of various grain sizes. Figure 6.2 showed the active–passive polarization behavior of nano-Ni as compared with coarse-grained polycrystalline Ni (Mishra and Balasubramaniam, 2004). There was a progressive shift in the ZCP (zero current potential) towards the noble direction for the nanograins as compared to bulk Ni. The catalysis of hydrogen reduction processes by substantial amounts of crystalline defects at the surface of Ni deposits during electrochemical measurements shifts the ZCP values in the noble direction (Rofagha *et al.*, 1991).

In nanocrystalline Ni, a substantial fraction of atoms lies in the intercrystalline region (Palumbo *et al.*, 1990), hence the positive shift in the ZCP because of changes in the hydrogen reduction process. It is also observed that the nanograin Ni showed higher passive current density than the micron counterpart. This indicates the defective nature of the passive film on the nanocrystalline Ni. Various surface characterization techniques showed the presence of a higher number of defects present in the passive films of the nanocrystalline Ni deposits (Rofagha *et al.*, 1994a,b). This should allow easier diffusion of Ni cations through a more defective Ni film, resulting in higher passive current densities in the passive region of Ni. The imperfect passivation of nano-Ni may also be related to the intrinsic electrochemical

320 6 Corrosion of structural nanomaterials

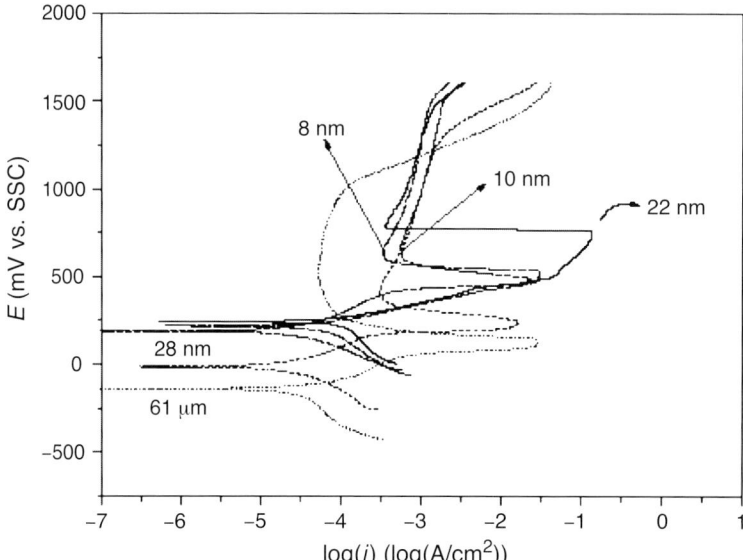

Figure 6.2. Potentiodynamic polarization curves of various nanosized Ni in 1 mol/l H_2SO_4 solution. The experiments were conducted after stabilization of free corrosion potentials (Mishra and Balasubramaniam (2004), reproduced with permission from Elsevier).

behavior associated with the disordered intercrystalline region in Ni (Palumbo and Aust, 1989; Rofagha et al., 1991). It is also known that the compressive strains on the surface lower the corrosion rates when compared to the tensile stresses. The higher compressive microstrain in the 8 nm grain size electrodeposits could also be related to a lower corrosion rate.

6.5 Nanocrystalline cobalt and its alloys

A recent study by Kim et al. (2003) showed that the grain size reduction in Co has little effect on the overall corrosion performance of Co. More recently, Co and nanocrystalline Co have been shown to be potential candidate materials for replacing hexavalent Cr plating in wear-resistant coatings. Recent studies involved investigating the corrosion behavior of electrodeposited nano-Co in 0.25 M sodium sulfate at a pH of 6.5, and subsequently compared this with polycrystalline Co under similar conditions. Figure 6.3a shows typical potentiodynamic polarization curves obtained from mechanically ground as-rolled electrowon Co, annealed electrowon Co, and electrodeposited Co nanocrystalline samples. The overall shape in Figure 6.3 of the potentiodynamic polarization curve for nano-Co is very similar to that for micron Co samples in the E_{corr} values. Although a slightly enhanced dissolution current was observed in the nanocrystalline Co, the overall corrosion behavior of Co was not greatly affected by reducing the average grain size to the nanorange.

In order to study the microstructural effect on the corrosion behavior of Co, electrodeposited Co specimens were annealed at 950 °C for 15 min. This heat treatment resulted in

6.5 Nanocrystalline cobalt and its alloys

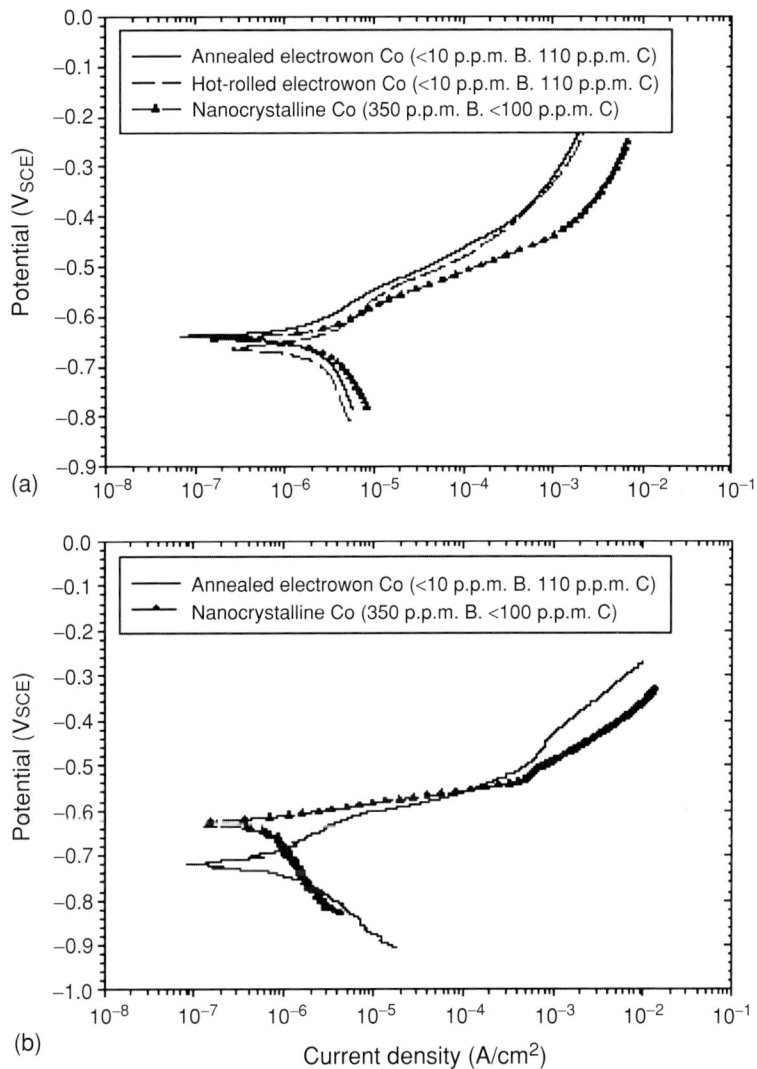

Figure 6.3. Potentiodynamic polarization curves from (a) as ground, (b) electropolished Co surfaces in deaerated 0.25 M Na_2SO_4 at pH 6.5 (Kim et al. (2003), reproduced with permission from Elsevier).

a polycrystalline microstructure with an average grain size of 10 μm. Figure 6.4 represents the potentiodynamic polarization curves of annealed nanocrystalline Co, which is found to be almost identical to that of the as plated nano-Co. However, considerable difference in their corrosion morphologies is noted in Figure 6.5. In the case of as-plated nano-Co (Figure 6.5a), the surface riding marks were smoothened out owing to uniform dissolution of the surface. Uniform metal dissolution can be expected when the surface sulfur impurities are evenly dispersed throughout the nanocrystalline Co film, an observation similar to that made by Waren et al., 2000 in the electrodeposited nano-Ni containing 500 p.p.m. sulfur (S). The surface of the annealed nano-Co (Figure 6.5b) showed considerable preferential

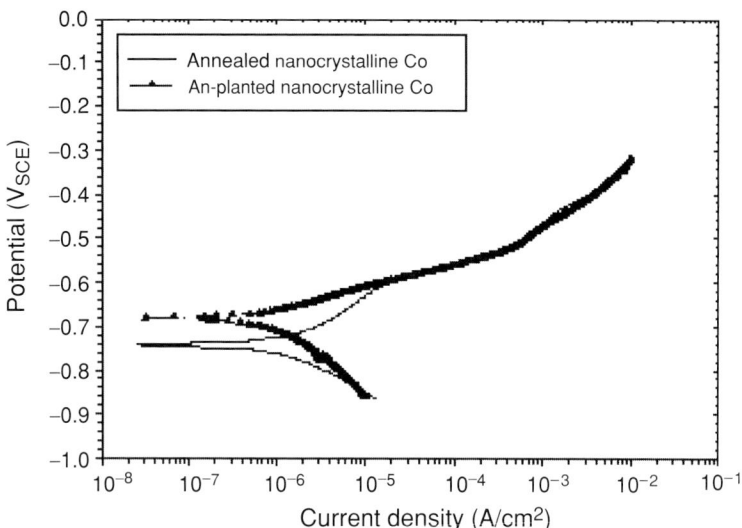

Figure 6.4. Potentiodynamic polarization curves of as-plated nano-Co (containing 440 p.p.m. S and 290 p.p.m. C) and annealed nano-Co at 950 °C for 15 mins (Kim *et al.* (2003), reproduced with permission from Elsevier).

attack along the grain boundaries where segregation of S would be expected after the heat treatment (Klement *et al.*, 1995).

As mentioned earlier, no significant changes were observed in nano- and micro-Co; however, the effect of additives in the electrodeposited nanocrystalline Co can impact its corrosion resistance. A study by Aledresse and Alfantazi (2004) showed that both nano-Co (67 nm) and nano-Co–P (50 nm grain size) did not passivate and the corrosion current densities for nano-Co and Co–P alloys markedly increased by factors of about 2 and 20 times, respectively, when compared to the polycrystalline Co in 0.25 M Na_2SO_4 solution at pH 10.5.

It has been widely accepted that an increase in the P content leads to a transition in the alloy structure from crystalline to amorphous. Research has indicated that, with the addition of P, crystalline and amorphous alloys exhibited better resistance to aqueous corrosion than unalloyed pure metals in acidic media (Krolikowski, 1988; Helfand *et al.*, 1992).

The limited dissolution of P-rich amorphous alloys was mainly associated with an anodic film formation, which further inhibits the anodic process. Figure 6.6 (Jung and Alfantazi, 2005) presents typical corrosion curves of micro-Co, nano-Co, and nano-Co–1.1 P in deaerated 0.1 M H_2SO_4 solution. All samples exhibited active dissolution without any distinctive transition to a passivation up to -0.1 V_{SCE}. In comparison to that of nano-Co, the anodic polarization curve for nano-Co–1.1 P shifted to a more-positive value of potential and decreased the anodic dissolution rate to -0.1 V_{SCE}. The overpotential of H_2 evolution also decreased and the cathodic reaction rates increased very quickly upon increasing the cathodic potential. The effect of P for nano-Co–P on the hydrogen evolution reaction is more significant than grain-size reduction. Considering the nano-Co–1.1 P (10 nm) grain size is smaller than that of nano-Co (20 nm grains), the anodic dissolution kinetics for

6.5 Nanocrystalline cobalt and its alloys

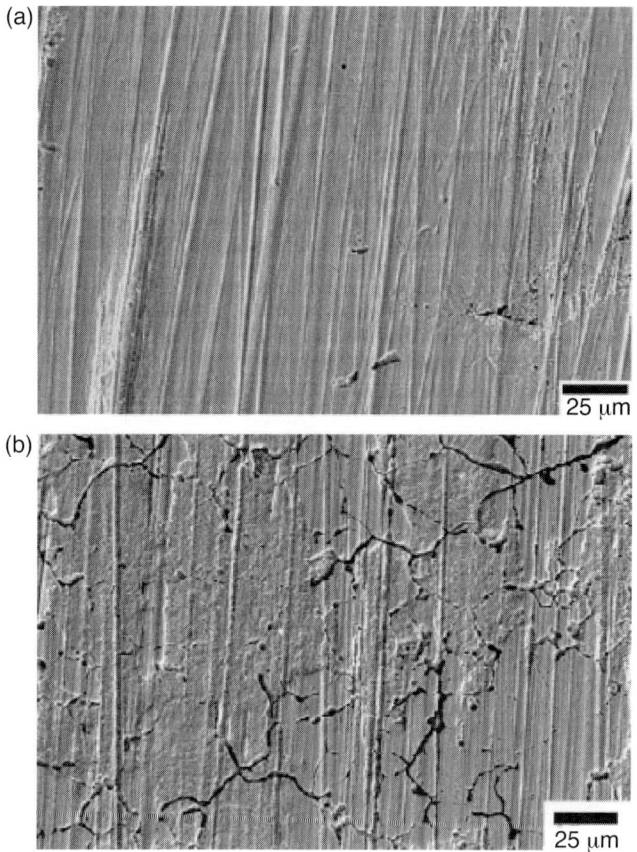

Figure 6.5. SEM images comparing the corrosion morphologies after potentiodynamic tests between (a) as-plated nanocrystalline Co and (b) annealed nanocrystalline Co (Kim *et al.* (2003), reproduced with permission from Elsevier).

the P-containing Co deposits is expected to increase. However, the addition of P content leads to a positive shift of E_{corr} and the anodic dissolution rates are subsequently reduced as compared to that of nano-Co. The beneficial effects in corrosion resistance of alloying Co with small amounts of P have been observed previously (Krolikowski, 1988) in electrodeposited micron-Co–P alloys in H_2SO_4 solutions. As the anodic potential is increased, the differences in the anodic current densities become smaller and eventually similar at higher anodic potentials. This shows that the dissolution kinetics at higher potential is less dependent on crystalline defects.

Figure 6.7 shows SEM micrographs of the corroded surfaces of micro-Co, nano-Co, and nano-Co–1.1 P after potentiodynamic polarization scans in 0.1 M H_2SO_4 solutions. As shown in Figure 6.7a, the micro-Co corroded extensively along the grain boundaries and triple junctions. Nanocrystalline Co (b) exhibits uniform corrosion throughout and the corroded surface still retains the grinding marks. In the case (c) of nano-Co–1.1 P, the surface of the specimen was found to be covered by a non-protective blackish colored film after

324 6 Corrosion of structural nanomaterials

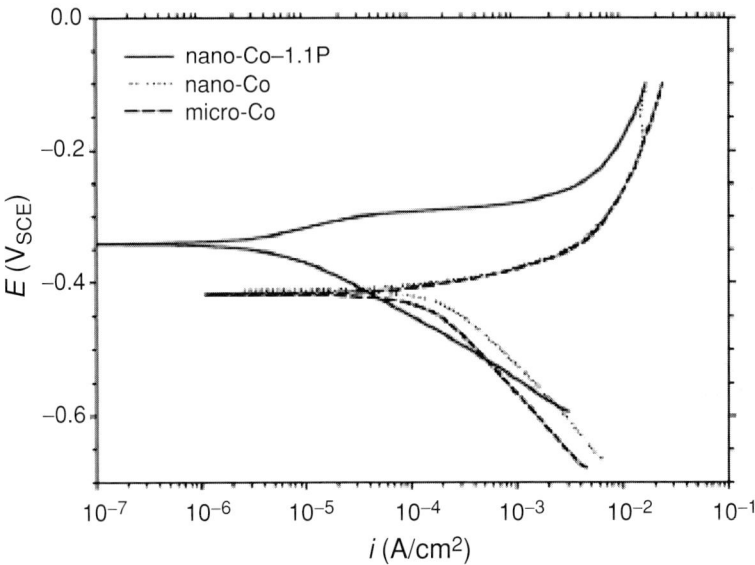

Figure 6.6. Corrosion curves of micro-Co, nano-Co, and nano-Co–1.1 P alloys in deaerated 0.1 M H_2SO_4 solution (Jung and Alfantazi (2005), reproduced with permission from Elsevier).

the corrosion test. As shown in Figure 6.7c, nano-Co–1.1 P exhibited the pitting corrosion with a few micro cracks. The EDS analysis showed intense P presence on the surface. However, no P peak was found in the pits. The metal Co was exposed to solution without the benefit of P alloying in these pits. Therefore it can be speculated that the excellent corrosion resistance of nano-Co–1.1 P near the E_{corr} value deteriorated because of pitting corrosion at high anodic overpotential ranges. This was related to the enrichment of P on the surface.

Electrochemical impedance spectroscopy (EIS) measurements were performed in order to gain a better understanding of the corrosion behavior of nano-Co and Co–1.1 P alloys. As shown in Table 6.1, in the case of nano-Co–1.1 P, the charge transfer resistance was significantly enhanced and the value of R_F was quite high, indicating an effective blocking layer for the anodic dissolution. The measured value of double layer capacitance is $\sim 10^{-4}$ F/cm^2. The n_1 value of less than 1 means that the corroded surface of the sample was not homogenous and may be the result of surface roughness. Results from the EIS measurement showed that the total interfacial impedance of nano-Co–1.1 P was significantly larger than that of the nano-Co at E_{OC}. XPS analysis further revealed that the corroded surface of nano-Co–1.1 P showed an enrichment of P after polarization treatment. Because there are no peaks of the oxidized Co and the oxidized P at E_{OC}, it is safe to say that the elemental P played a major role for enhanced corrosion resistance by hindering the dissolution kinetics of surface Co at E_{OC}. However, at higher anodic overpotential, the superior corrosion resistance did not last because of the formation of a non-protective surface film consisting of mainly hypophosphite and phosphate as well as elemental P.

6.5 Nanocrystalline cobalt and its alloys

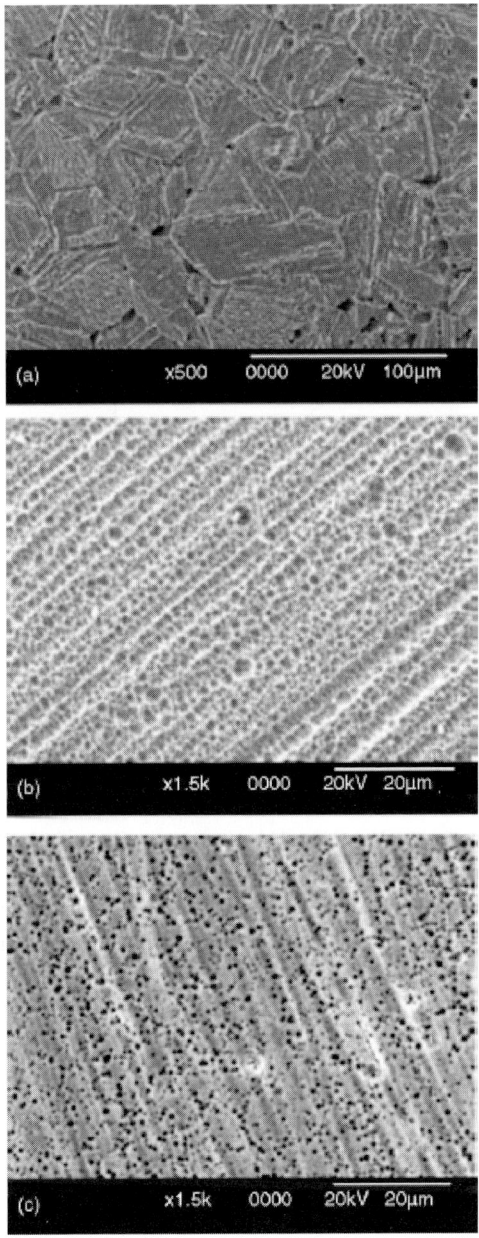

Figure 6.7. SEM micrographs of the corroded surface of (a) micro-Co, (b) nano-Co, and (c) nano-Co–1.1 P after potentiodynamic polarization scans in deaerated 0.1 M H_2SO_4 (Jung and Alfantazi (2005), reproduced with permission from Elsevier).

Table 6.1. *Results of impedance analysis taken at E_{OC} after 30 min of immersion*

Sample	Q_{dl} (F/cm^2)	n_1	R_{ct} (Ωcm^2)	Q_F (F/cm^2)	n_2	R_F (Ωcm^2)
Micro-Co	1.34×10^{-4}	0.88	97			
Nano-Co	1.52×10^{-4}	0.89	70			
Nano-Co–1.1 P	9.93×10^{-5}	0.91	1136	2.92×10^{-4}	0.83	467

6.6 Zirconium and its alloys

Zirconium and its alloys are very useful structural materials for a myriad of applications, especially in the nuclear industry, owing to their low neutron-absorption coefficient. Current trends towards extended burn-up of nuclear fuel in pressurized water reactors have accentuated the demand for Zr-based alloys with higher uniform corrosion resistance under irradiation and lower hydrogen absorption. Zr-based alloys also find major uses in chemical plants. The corrosion of metallic alloys is strongly influenced by alloying elements as well as by the microstructure. Zirconium and crystalline Zr-based alloys are known to exhibit some unique properties, such as chemical as well as electrochemical corrosion resistance, oxidation resistance, and toxicity. During the past few decades there has been an increased interest in the corrosion behavior of Zr-related materials as a function of grain size and crystallinity.

For example, the amorphous and nanoquasicrystalline $Zr_{70}Pd_{30}$ and $Zr_{80}Pt_{20}$ alloys show better corrosion resistance than Zr in NaCl, NaOH, and H_2SO_4 solutions. They have been studied using potentiodynamic polarization (Mondal *et al.*, 2005). Both alloys are susceptible to chloride attack, and pitting has been observed. Complete passivation has been observed in H_2SO_4, while gradual breakdown of the passivating layer occurs in NaOH. In general, the nanoquasicrystalline state in both alloys shows better corrosion resistance than the amorphous state in all solutions. The stabilities of amorphous and nanoquasicrystalline alloys under different environmental conditions are important aspects for their potential applications. In general, amorphous alloys have better corrosion resistance due to their chemically homogeneous structure and the absence of boundaries (Naka *et al.*, 1976).

A number of corrosion investigations have been reported on many multicomponent Zr-based amorphous alloys in different corrosive media (Schroeder *et al.*, 1998; Gebert *et al.*, 2001; Hiromoto *et al.*, 2001; Dutta *et al.*, 1995). Dey *et al.* (1989) carried out investigations on aqueous corrosion of amorphous and crystalline Zr–Ni alloy in different acid media by immersion testing. Another study by Köster *et al.* (2001) shows no significant differences in anodic polarization responses of amorphous and nanocrystalline Zr–Cu–Ni–Al alloys in NaOH solution.

Murty *et al.* (2000, 2001) have demonstrated nanoquasicrystallization of the amorphous phase in binary Zr–Pd and Zr–Pt alloys near their eutectic compositions. In this study, the corrosive behavior of amorphous and nanoquasicrystalline phases in melt-spun Zr–Pd and Zr–Pt alloys in different environments (neutral NaCl, alkaline NaOH, and acidic H_2SO_4 solutions) have been compared with that of Zr (98% purity). The corrosion characteristics

6.6 Zirconium and its alloys

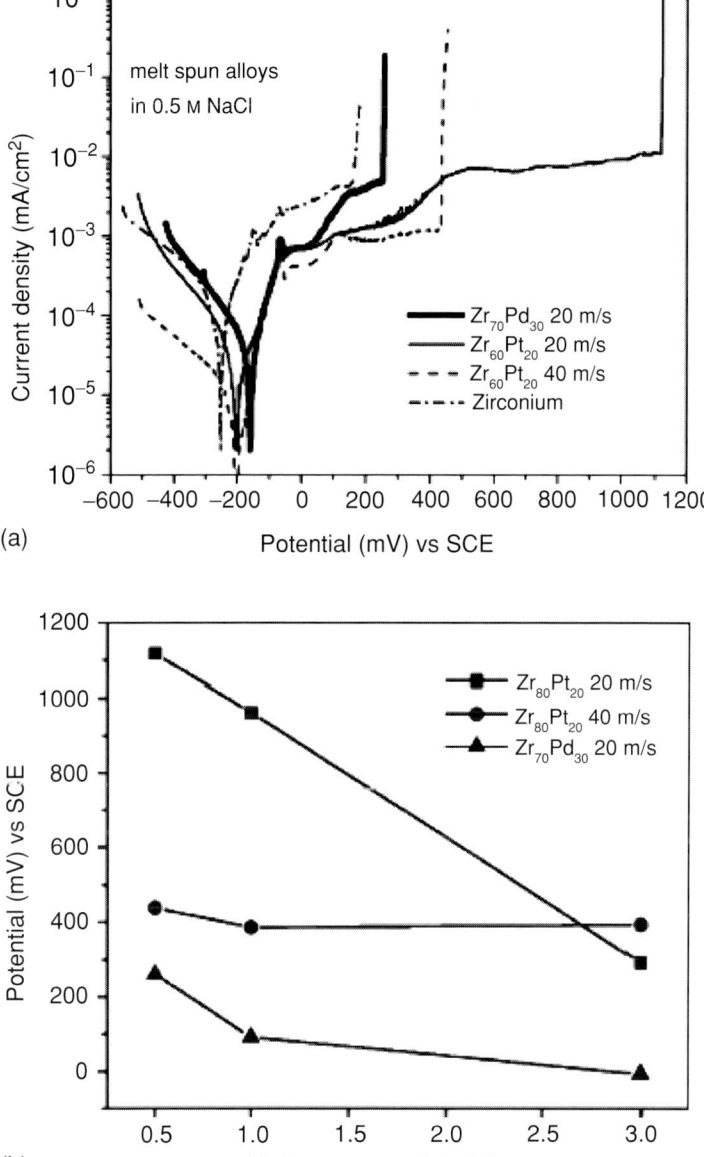

Figure 6.8. (a) Polarization curves of $Zr_{70}Pd_{30}$ (20 m/s), $Zr_{80}Pt_{20}$ (20 m/s and 40 m/s), and Zr metal in 0.5 M NaCl solution; and (b) effect of concentration of NaCl on the E_{BR} of as-melt-spun alloys (Mondal et al. (2005), reproduced with permission from Elsevier).

of the alloys were determined based on their anodic polarization curves obtained by potentiodynamic methods. The samples were tested in 3 M, 1 M, and 0.5 M NaCl, in 0.5 M NaOH, and in 0.25 M H_2SO_4 solutions at room temperature at a rate of 0.5 mV/s.

Figure 6.8a (Mondal et al., 2005) shows the dynamic polarization curves of the three melt-spun alloys in 0.5 M NaCl solution which are compared with Zr. Interestingly, all three

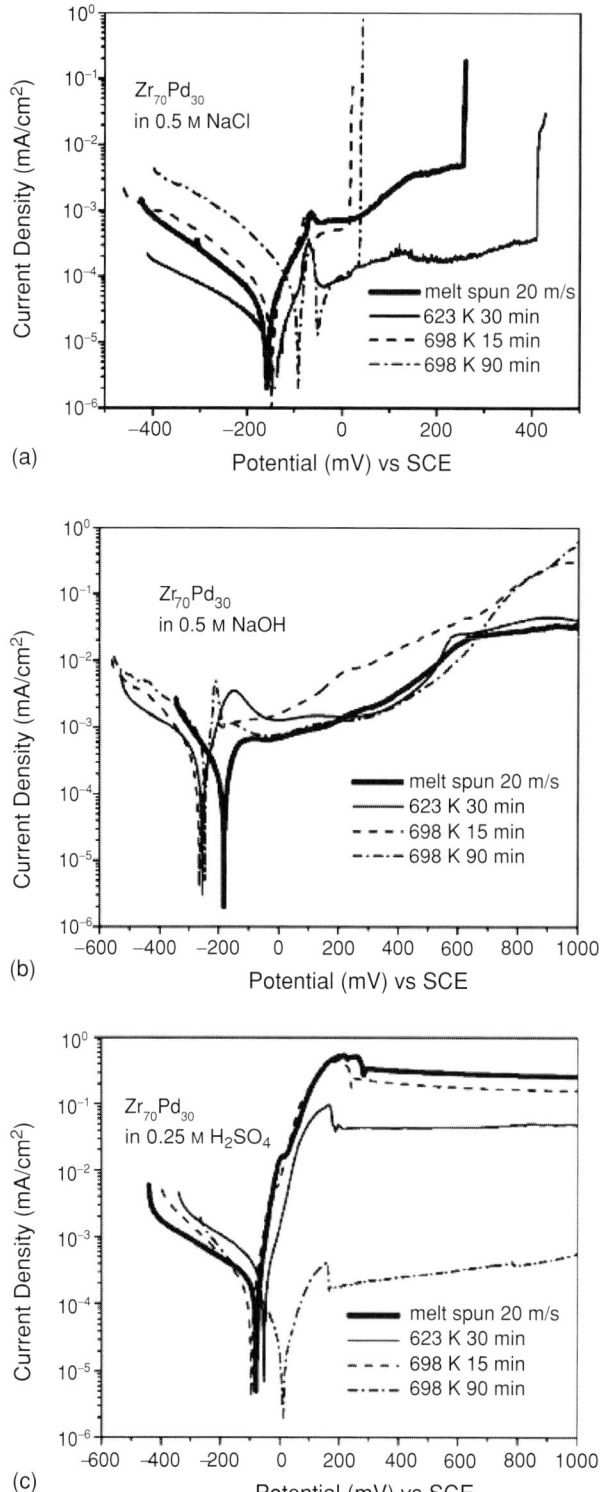

Figure 6.9. Polarization curves of melt-spun and heat-treated $Zr_{70}Pd_{30}$ (20 m/s) in (a) 0.5 M NaCl, (b) 0.5 M NaOH, and (c) 0.25 M H_2SO_4 solution (Mondal et al. (2005), reproduced with permission from Elsevier).

6.6 Zirconium and its alloys

alloys have shown better pitting resistance than that of pure Zr, as indicated by their higher breakdown potential (E_{BR}) compared to Zr.

Figure 6.8b shows the variation of E_{BR} with concentration of NaCl for the three alloys studied. The E_{BR} values of amorphous $Zr_{70}Pd_{30}$ (20 m/s) alloy at all concentrations of NaCl are lower than those of partially nanoquasicrystalline $Zr_{80}Pt_{20}$ alloys, indicating that the latter alloys have better corrosion resistance than the former in NaCl solution. Among the three alloys studied, the Zr–Pt alloy consists of a larger fraction of the nanoquasicrystalline phase. This alloy appears to have the highest corrosion resistance (higher E_{BR}) to NaCl solution up to about 2.5 M.

Figure 6.9a–c shows the polarization curves of melt-spun and heat-treated $Zr_{70}Pd_{30}$ (20 m/s) in 0.5 M NaCl, 0.5 M NaOH, and 0.25 M H_2SO_4 solutions, respectively. It indicates nanoquasicrystallization of the amorphous alloy at 698 K, which brings in an initial marginal increase in the current density in comparison to that annealed at 623 K followed by a significant drop, conveying an increase in the corrosion resistance in all the solutions.

Figure 6.10a–c demonstrates dynamic polarization behaviour of a class of melt-spun $Zr_{80}Pt_{20}$ alloy (both 20 m/s and 40 m/s) and their annealed counterparts in 0.5 M NaCl, 0.5 M NaOH and 0.25 M H_2SO_4 solutions, respectively. In all solutions, the nanoquasicrystalline $Zr_{80}Pt_{20}$ 20 m/s alloy showed a wider passive region in comparison to the annealed alloy (623 K, 30 min).

$Zr_{70}Pd_{30}$ alloy is amorphous when annealed at 623 K for 30 min, suggesting only a possibility of structural relaxation under this condition. The amount of nanoquasicrystalline phase increased with annealing time at 698 K. This is observed in the XRD, which showed extra peaks for nanoquasicrystalline phases at 698 K.

In chloride-containing solution and in oxidizing media, Zr–Pd and Zr–Pt alloys have shown better corrosion resistance compared with pure Zr, as reflected by their lower passive current regions and higher E_{BR} values in NaCl solution (Figure 6.8). The presence of noble metals Pd and Pt has made the alloys more electropositive than pure microcrystalline zirconium (standard reduction electrode potential with respect to normal hydrogen electrode for Pt^{2+}/Pt, Pd^{2+}/Pd and Zr^{4+}/Zr are 1.2 V, 0.987 V, and −1.53 V, respectively (Uhlig, 1971). In addition, the amorphous and nanoquasicrystalline nature of the melt-spun alloys has led to the improvement of their corrosion resistance over the crystalline zirconium.

The corrosion results also indicate that, at the early stages of nanoquasicrystallization, the corrosion resistance to NaCl solution decreases in the case of $Zr_{70}Pd_{30}$ (Figure 6.9a). The initiation of a nanoquasicrystalline phase in an amorphous matrix leads to compositional inhomogeneity in the alloy (Rofagha et al., 1994a,b). In addition, the interface regions between the amorphous and nanoquasicrystalline phases can act as preferential sites for corrosion. However, if the alloy consists of a single nanoquasicrystalline phase, the corrosion resistance increases as observed in the Zr–Pt alloy melt spun at 20 m/s (Figure 6.10a) and in the Zr–Pd alloy for longer annealing at 698 K (Figure 6.9a). When the amount of the nanoquasicrystalline phases increases in the amorphous matrix, the amorphous–nanoquasicrystalline interface is replaced with the nanoquasicrystalline grain boundaries. The latter appear to be more resistant to corrosion than the phase interfaces. This is evident from the results for nanostructured Zr–Pt alloy and those for Zr–Pd alloy annealed for a longer time at 698 K.

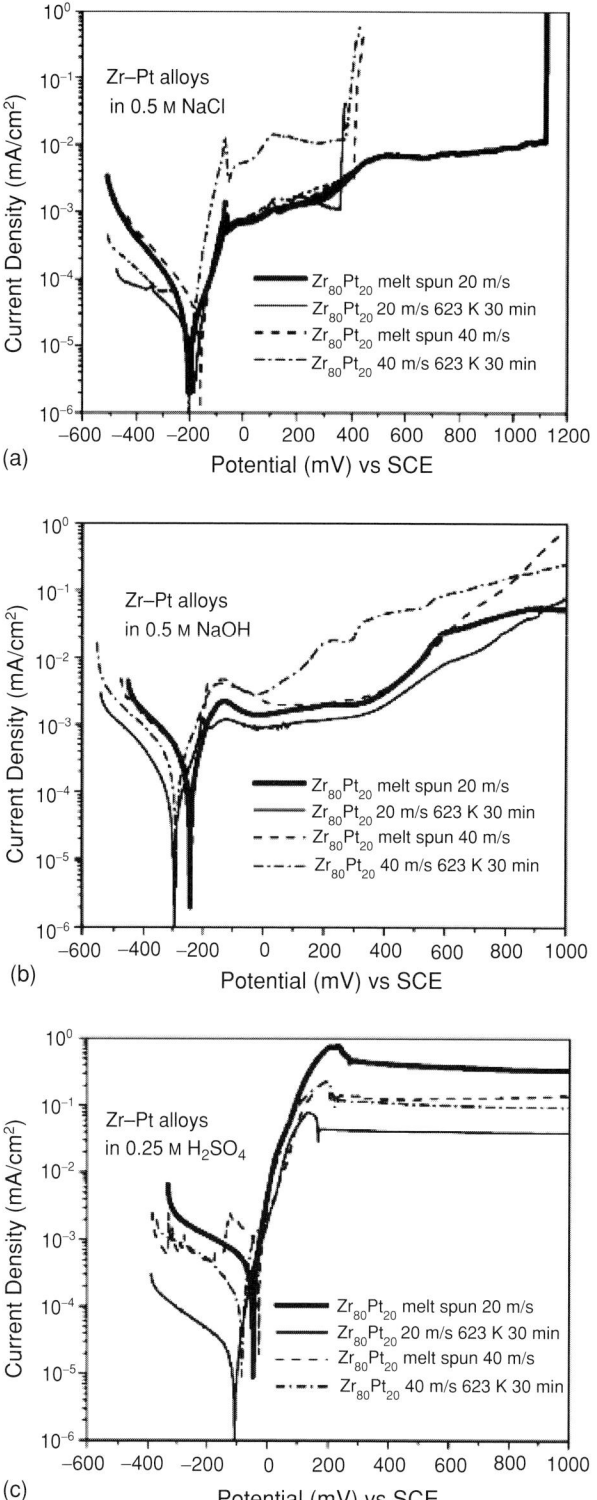

Figure 6.10. Polarization curves of melt-spun and heat-treated $Zr_{80}Pt_{20}$ (20 m/s and 40 m/s) in (a) 0.5 M NaCl, (b) 0.5 M NaOH, and (c) 0.25 M H_2SO_4 solution (Mondal *et al.* (2005), reproduced with permission from Elsevier).

The corrosion of an amorphous alloy generally happens by uniform dissolution throughout the matrix. From the corrosion measurements, it is evident that the fully amorphous alloy sample ($Zr_{70}Pd_{30}$) was found to be more stable than the partially nanoquasicrystalline alloy in the chloride solution. As the crystalline phase appears in the amorphous matrix, the corrosion gradually changes from general uniform dissolution to a preferential dissolution mode at the interface regions. With more nanocrystallization in the matrix, the number of preferential dissolution sites increases. Once the alloy becomes fully nanoquasicrystalline, the defect sites for corrosion are uniformly distributed on an extremely fine scale and hence the alloy attains passivity much earlier than the alloy, which is partially nanoquasicrystalline. It has been observed (Rofagha *et al.*, 1994a,b) that in the case of nanocrystalline alloys, a large number of intercrystalline surface defects is present. This leads to a more defective passive layer. With large defects present in the passive layer, the break down of the passive film will be more uniform, which leads to uniform corrosion throughout the matrix. A uniform distribution of the passive current density along the preferred corrosion sites allows the nanoquasicrystalline alloy to maintain passivity over a large potential range on the noble side and resist localized chloride-induced corrosion. Annealing of melt-spun $Zr_{80}Pt_{20}$ (20 m/s) increases the crystallite size and makes pitting more localized. The breakdown of the passive layer also occurs at a lower potential, lowering the corrosion resistance with increase in grain size of the nanoquasicrystalline phase. The higher corrosion resistance of the nanoquasicrystalline phase over the amorphous phase could be attributed to the presence of ordered regions in the amorphous phase. This will result in localized pitting leading to greater corrosion than the nanoquasicrystalline matrix. Earlier studies (Pardo *et al.*, 2001, 2002) have also shown that the presence of the nanocrystalline phase has better corrosion resistance than the amorphous state under similar conditions.

6.7 304 Austenitic steels: wet corrosion

Stainless steel (SS) is extensively used for high-temperature applications such as superheaters, tube inner liners for exposure to corrosive environments, reheater tubes, and various other components that are subjected to thermal fluctuations in service. In a recent report, Wang *et al.* (2001) showed that a nanostructured surface layer of a low-C steel increases the corrosion resistance significantly. Wang and Li (2003) presented detailed corrosion results of 304 steel, one of the main components of many structural materials. Various sandblasting and annealing treatments are used to create a nanocrystalline surface of a 304SS. A typical microstructure analysis revealed that the sand-blasted surface was heavily deformed and consequently has a high density of dislocations. After annealing treatment, the initially formed dislocation network of fine subgrains (20 nm) changed to nano-sized grains with sharp boundaries. In these nanograins, the dislocations have significantly decreased.

Corrosive immersion tests were carried out on these samples. SEM surface morphologies of as-received, sandblasted, and sandblast-annealed specimens after being immersed in a 3.5% NaCl solution for 16 h are presented in Figure 6.11. As shown in the figures, the sandblasted surface was severely corroded with some red rusts, while the surfaces of as-received and sandblast-annealed specimens kept in fairly good condition.

332 6 Corrosion of structural nanomaterials

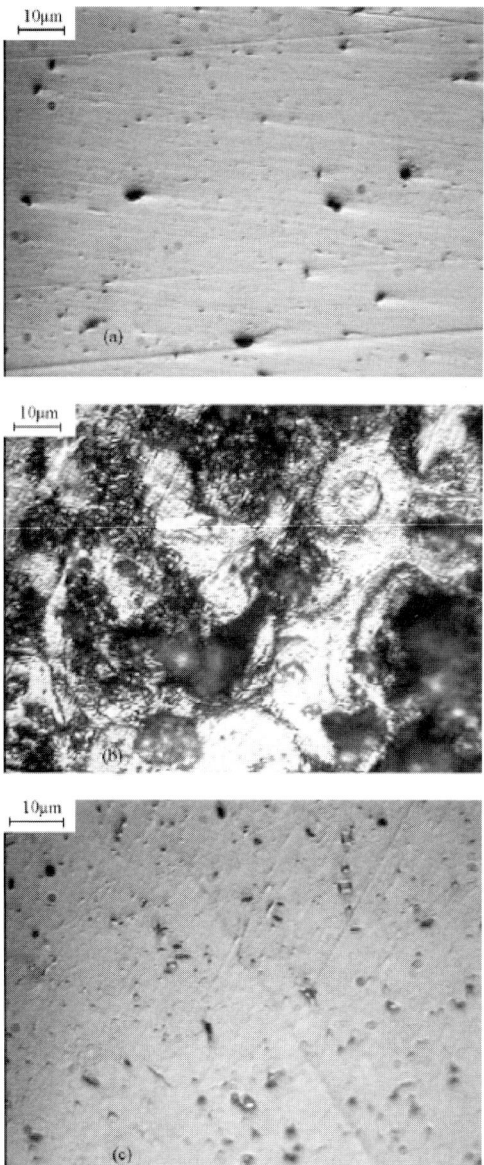

Figure 6.11. Morphologies of different specimens after being exposed to 3.5% NaCl solution for 16 h: (a) 304SS; (b) sandblasted 304SS; and (c) sandblast-annealed 304SS (Wang and Li (2003), reproduced with permission from Elsevier).

Figure 6.12 presents the polarization curves of 304SS, sandblasted 304SS and sandblast-annealed 304SS, respectively. Significant difference in the polarization behavior among the various surface-treated steels was observed. The nanostructure surface considerably improved the polarization behavior of the material. The potentiodynamic polarization curves demonstrated that not only did the sandblast-annealing treatment increase E_{corr}, the free

6.8 304 Austenitic steels: dry corrosion

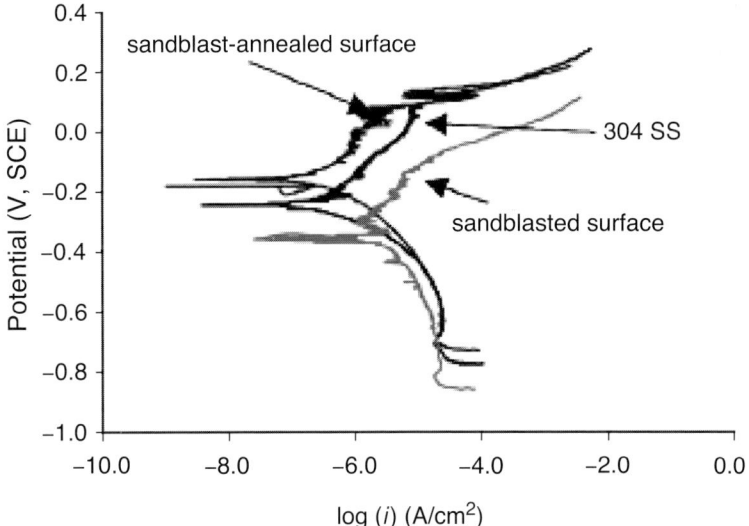

Figure 6.12. Polarization curves of different specimens in a 3.5% NaCl solution. Scanning rate: 20 mV/min (Wang and Li (2003), reproduced with permission from Elsevier).

corrosion potential, but also reduced the passivation-maintaining current, while the sandblasted 304SS showed the worst polarization behavior, resulting from the existence of high-density lattice defect: dislocation.

The differences in the electrochemical behavior among surfaces of the sandblast-annealed, sandblasted, and control steels could result from possible changes in microstructure of the passive films formed. The high corrosion resistance of stainless steel is generally attributed to the formation of a chromium-enriched passive film (Lumsden and Szklarska-Smialowska, 1978), which protects the steel from corrosion attack. When a surface layer is nanocrystalline, high-density grain boundaries could promote the diffusion of Cr to the surface. This results in the formation of a passive film containing more Cr that may increase the corrosion resistance and strength of the film. This might be the reason why the passive film on the sandblast-annealed specimen exhibited superior corrosion resistance and mechanical properties. It should be pointed out that the performance of a thin passive film is also strongly affected by the interfacial bonding between the film and the substrate.

Inturi and Szklarska-Smialowska (1992) have shown superior pitting resistance of sputter-coated nanocrystalline 304SS (grain size 25 nm), compared with conventional 304SS in HCl solution. This is the result of the defective passive film, which provides a uniform distribution of Cl ions on the specimen surface.

6.8 304 Austenitic steels: dry corrosion

The wet corrosion behavior of nanocoatings on 304 steels has been widely reported; however, many of these austenitic class of steels are also used in high temperature environments and undergo severe high-temperature material degradation. In an effort to reduce the high-temperature oxidation of 304 steels various measures have been taken, for example rare earth additions. Addition of rare earth elements such as Ce, Y, Zr, and La or their oxides improve the high-temperature oxidation resistance of the alumina- and chromia-forming alloys because of the reactive element effect (REE). The REE decreases the oxide-scale growth rate with improved resistance to scale spallation by increasing the scale–alloy adhesion. Various researchers have put forward several mechanisms to explain the REE. Antill and Peakall (1967) indicated that the beneficial effect of the rare earth elements was primarily to improve the scale plasticity for accommodating stresses due to the difference in the thermal expansion coefficients between the alloy and the oxide scale.

Application of the reactive element coating on the metals and alloys to improve their high-temperature oxidation resistance is more advantageous than the presence of the rare earths as alloy constituents that usually pose difficulty in alloy processing especially during hot working. Researchers (Hussey *et al.*, 1989; Aguilar *et al.*, 1992; Roure *et al.*, 1994; Seal *et al.*, 1994; Nazeri *et al.*, 1997; Shen *et al.*, 1998; Seal *et al.*, 2000) have suggested that the surface deposition of the reactive element oxide in the case of chromia-forming alloys can harness the beneficial effect of REE. Moon and Bennett (1989) concluded that the scale nucleates at the reactive element oxide particles on the surface, blocks short-circuit diffusion paths by segregating reactive element ions, and reduces the stresses in the oxide scale by altering the microstructure.

Earlier studies (Seal *et al.*, 1994, 2000) have indicated that superficial coating of micron-sized cerium oxide particles is effective in improving the high-temperature oxidation resistance for various grades of stainless steel. However, it was not effective in protecting AISI 304-grade stainless steel from high-temperature oxidation in dry air (Seal *et al.*, 1994). Nanoparticles of ceria are expected to improve the oxidation resistance of chromia- and alumina-forming alloys because of their easy dissolution in the grain boundaries and subsequently their blocking of short-circuit diffusion paths more effectively compared with the micron-sized ceria particles. Preliminary investigations on the improvement of high-temperature oxidation resistance of Ni, Cr, and Ni–Cr super alloys with the application of a nanocrystalline ceria coating were carried out by Czerwinski and co-workers (1993, 1997). Detailed investigations on the use of ceria nanoparticles to improve the high temperature oxidation resistance of chromia- and alumina-forming alloys are yet to be carried out.

Patil *et al.* (2004) reported the comparative oxidation kinetic plots at 1260 K in dry air for 24 h for the bare, micron- and nanosized ceria-coated AISI 304 grade stainless steel (see Figure 6.13). It clearly indicates the effectiveness of the nanocrystalline ceria coating over the micron-sized ceria coating in improving the high-temperature oxidation resistance of the AISI 304 grade stainless steel. Micron-sized ceria-coated stainless steel specimens did exhibit neither a reduction in the overall rate of oxidation nor improvement in the scale adhesion property compared to the bare specimen. However, the micron-sized ceria-coated specimen indicated a smaller mass gain in the initial period of the exposure compared to that of the bare specimen. Such reduction of the kinetics in the initial period can be attributed to the presence of micron-sized ceria particles as discontinuous mechanical barriers at various locations on the surface of the steel substrate. Also such micron-sized ceria particles can provide oxide nucleation sites for a faster and complete coverage of the protective oxide layer on the stainless steel surface compared to the bare alloy. However, as the scale grows,

6.8 304 Austenitic steels: dry corrosion

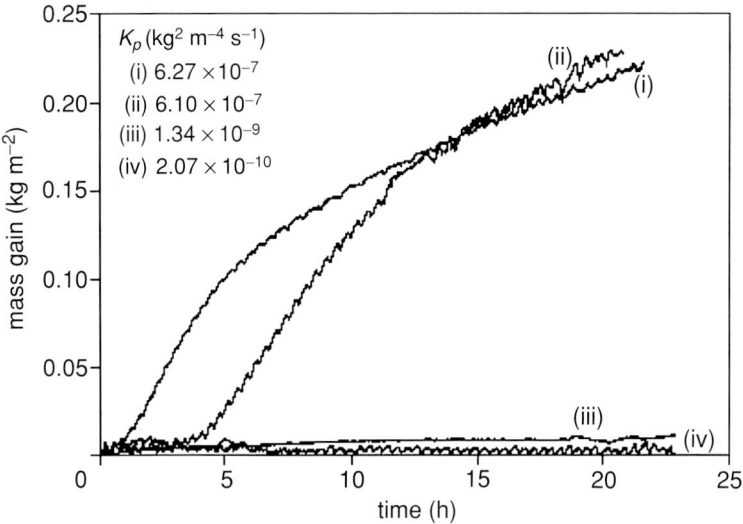

Figure 6.13. Mass gain per unit area versus time plots for isothermal oxidation of AISI 304 stainless steel at 1260 K in dry air for 24 h: (i) bare specimen; (ii) micrometer-sized ceria-coated specimen; (iii) nanosized ceria-coated specimen, first oxidation; and (iv) nanosized ceria-coated specimen, re-oxidization (Patil et al., 2004). Reprinted with permission.

such micron-sized ceria particles probably could not protect from the micro-cracking and spallation of the scale, and hence the overall oxidation rate remained as high as the bare steel. Such easy nucleation of oxide during high-temperature oxidation of stainless steel due to the presence of surface inhomogeneity has been reported earlier (Kuiry et al., 1994).

The ceria nanoparticles provide a large number of heterogeneous nucleation sites per unit surface area on the alloy surface, which has mostly culminated into such fine-grained oxide scale morphology that was expected to exhibit improved scale plasticity as suggested by others (Antill and Peakall, 1967; Stringer, 1970; Whittle and Stringer, 1980; Rhys-Jones et al., 1987; Moon and Bennett, 1989). Consequently, it enhanced the ability of the oxide to absorb thermal and growth-related stresses, thus decreasing the likelihood of stress relief by cracking and subsequent spalling. The presence of Cr on the top surface of the scale indicates the early formation and retention of protective chromia scale in the top scale owing to heterogeneous nucleation sites provided by the ceria nanoparticles.

The researchers further confirmed by XPS analysis (Patil et al., 2004) that the ceria nanoparticles used in the present investigation consisted of both Ce^{+4} and Ce^{+3} oxidation states. The presence of Ce^{+3} ions in the coating layer might also have played a vital role in the improvement of the oxidation resistance of the AISI 304 stainless steel. Although pinpointing the exact nature of such beneficial effect of the presence of Ce^{+3} ions is beyond the scope of the present investigation, the following scenarios may be envisaged. Assuming electrostatic charge neutrality, an oxygen vacancy is created for replacement of each two Ce^{+4} sites by Ce^{+3} ions (Tsunekawa et al., 1999). Oxygen vacancies are mobile and result in faster oxygen diffusion. This behavior can help in changing the oxide-scale growth process from cation outward to oxygen inward.

The ionic radii of Cr^{+3}, Ce^{+3}, and Ce^{+4} are 0.630 Å, 1.283 Å and 1.098 Å, respectively (Shannon, 1976; Pieraggi and Rapp, 1993). The relatively large size difference between Ce ions and Cr ions limits the solubility of the former in the chromia lattice. Accommodating Ce^{+3} and Ce^{+4} ions in the chromia lattice, the lattice strains involved are 104% and 74%, respectively. Hence, incorporation of Ce^{+3} ions in the chromia lattice is less favorable than that of the Ce^{+4} ions. Therefore, the probability of Ce^{+3} ions getting segregated into the grain boundary of oxide is higher than that of Ce^{+4} ions, which perhaps renders the blocking of Cr^{+3} diffusion more effective in the case of the nanosized ceria coating. Moreover, the incorporation of Ce^{+3} ions in the grain boundary of chromia scale is possibly more favorable because of similarity of electrical charge present both in Ce^{+3} and Cr^{+3} ions. Also, the incorporation of Ce^{+3} ions does not require creation of oxygen vacancies to maintain the site relation, unlike that of the Ce^{+4} ion, which might also help reducing the oxygen diffusion along the grain boundaries.

6.9 Magnetic nanocomposites

Magnetic nanocomposites are gaining attention in many industrial sectors. For example, the degree of alignment of the crystallographic c-axis can be tweaked in many multicomponent alloys, which gives a high energy density $(BH)_{max}$ for rapidly growing applications. One such example is the tetragonal $Nd_2Fe_{14}B$ phase. The conventional powder metallurgy route including pre-alignment and sintering (Sagawa et al., 1985) or by hot pressing (Lee, 1985) gives such a texture. The micrograin-sized NdFeB powders can be prepared by melt-spinning (Croat et al., 1984), mechanical alloying (Schultz et al., 1987) and the hydrogen–disproportion–desorption–recombination (HDDR) processes. To obtain optimum magnetic properties and minimum deformation stress, very fine and homogeneous grain sizes in the nanoscale are favorable irrespective of preparation process.

The corrosion behaviour of microcrystalline NdFeB-type (sintered) magnets has been studied extensively in different media (Tokuhara and Hirosawa, 1991; Shimotomai et al., 1990), which showed that the low corrosion resistance can be attributed to the high content of neodymium and their complex multiphase microstructure. The microstructure consists of the ferromagnetic phase matrix ($Nd_2Fe_{14}B$) and the most corrosion sensitive Nd- and B-rich phases in the grain-boundary region (Schultz et al., 1999).

El-Moneim et al. (2002) have reported the production of isotropic nanocrystalline $Nd_{14}Fe_{80}B$ magnets by melt-spinning and subsequent hot pressing at 700 °C. The magnets have grain sizes ranging from 100 nm to 600 nm when annealed at 800 °C for 0.5–6 h and the corrosion resistance of these has been examined in N_2-purged 0.1 M H_2SO_4 electrolyte by in situ inductively coupled plasma solution analysis, gravimetric and electrochemical techniques and hot-extraction [H]-analysis. Potentiodynamic polarization tests were carried out at 0.2 mV/s and 2 mV/s. The electrochemical corrosion behaviour of NdFeB magnets were characterized by galvanostatic polarization technique at various applied anodic and cathodic current densities. It is worthwhile to note that grain growth due to annealing of nanocrystalline melt-spun and hot-pressed magnets results in heterogeneity of the microstructure with a reduction in the fractions of Nd-rich intergranular phase.

6.9 Magnetic nanocomposites

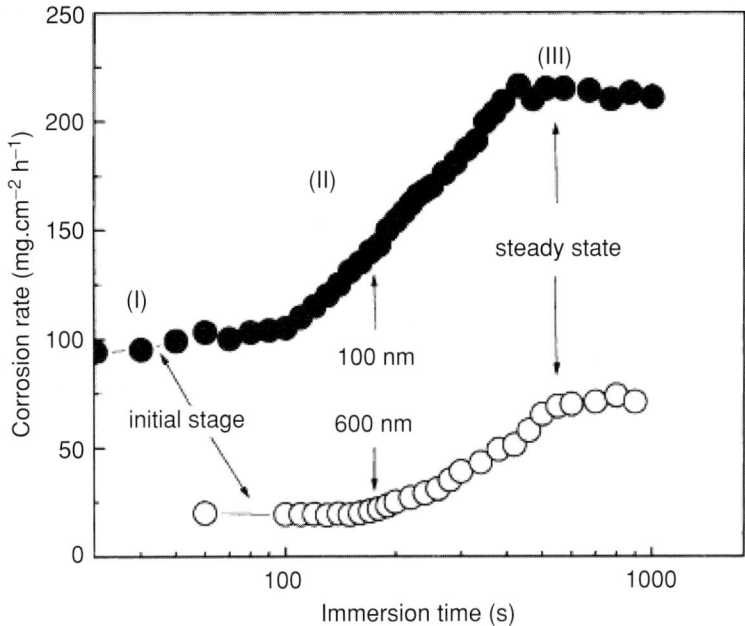

Figure 6.14. Corrosion rates measured by ICP solution analysis for NdFeB magnets with grain sizes of about 100 nm and 600 nm in N_2-purged H_2SO_4 at 25 °C and 720 rpm as a function of immersion time (El-Moneim et al. (2002), reproduced with permission from Elsevier).

Figure 6.14 shows the corrosion rate measured by ICP solution analysis for NdFeB magnets with average grain sizes of 100 nm and 600 nm in N_2-purged 0.1M H_2SO_4 solution at 25 °C and 720 rpm as a function of immersion time. The initial stage (I) of about 1–3 min results from a beginning of preferential attack of Nd-rich intergranular region. The preferential dissolution of Nd-rich intergranular regions and the partial undermining of phase matrix particles cause rapid increase in the corrosion rate in the second stage (II). After 6–7 min, all magnet components dissolve in the same proportion as they appear in the alloy formula and a steady state corrosion rate is attained.

The corrosion rates of NdFeB magnets decrease with an increase in the grain size of the matrix phase. This is attributed to the changes, heterogeneity and the reduction in the volume fraction of Nd-rich intergranular phase, observed in the microstructure upon annealing.

Corrosion behavior under open-circuit conditions showed that nanocrystalline magnets undergo fast corrosion processes in acidic solutions. The processes include selective dissolution of the Nd-rich intergranular phase, hydrogenation of the hard magnet phase and its subsequent mechanical degradation. The large-grain-sized magnets retard fast corrosion processes owing to the heterogeneity and smaller fractions of the Nd-rich intergranular phase in the microstructure.

Figure 6.15 shows potentiodynamic polarization curves of NdFeB magnets with various grain sizes in N_2-purged 0.1M H_2SO_4 solution at 25 °C and 720 rpm using a scan rate of 0.2 mV/s. A decrease in both anodic and cathodic activities of NdFeB magnets occurs with

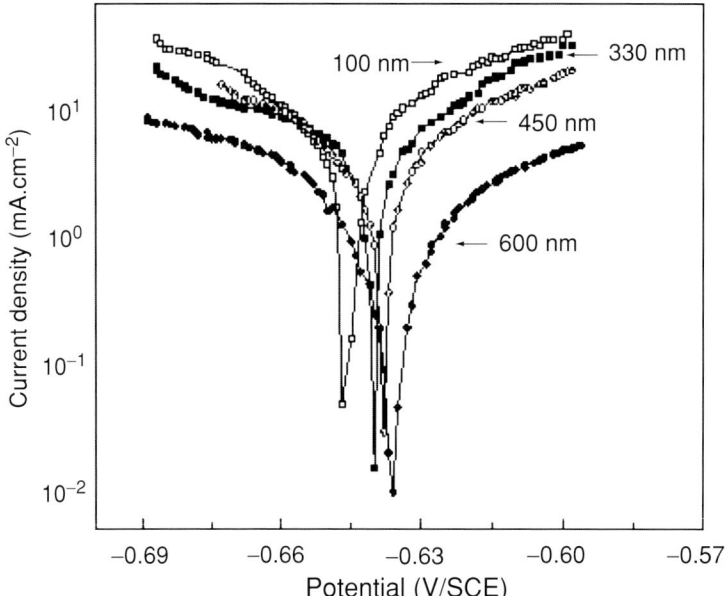

Figure 6.15. Potentiodynamic polarization curves of NdFeB magnets with various grain sizes in N_2-purged H_2SO_4 at 25 °C and 720 rpm using a potential sweep rate of 0.2 mV/s (El-Moneim et al. (2002), reproduced with permission from Elsevier).

an increase in the grain size of the hard magnetic phase. The estimated corrosion current density decreases with increasing grain size.

In conclusion, the increase in the corrosion resistance with average grain size is attributed to the smaller fractions and more heterogeneous distribution of the Nd-rich intergranular phase, along with less hydrogenation of the magnet surface. The abnormal dissolution behavior of magnets with cathodic polarization indicates that applying any kind of cathodic protection can lead to catastrophic corrosion of nanocrystalline NdFeB magnets.

References

Aguilar, C., Colson, J. C., and Larpin, J. P. (1992). *Corros. Sci.*, **89**, 447.
Aledresse, A., and Alfantazi, A. (2004). *J. Mater. Sci.*, **39**, 523.
Alves, H., Ferreria, M. G. S., and Köster, U. (2001). *Corros. Sci*, **45**, 1833.
Antill, J., and Peakall, K. (1967). *J. Iron Steel Inst.*, **205**, 1136.
Croat, J. J., Herbst, J. F., Lee, R. W., and Pinkerton, F. E. (1984). *J. Appl. Phys.*, **55**, 2078.
Czerwinski, F., and Smeltzer, W. W. (1993). *Oxid. Met.*, **40**, 503.
Czerwinski, F., and Szpunar, J. A. (1997). *J. Sol-Gel Sci. Technol.*, **9**, 103.
Corrosion Costs and Preventive Strategies in the United States. Report by CC Technologies Laboratories, Inc., to Federal Highway Administration (FHWA) (2001). Office of Infrastructure Research and Development, Report FHWA-RD-01-156.
Dey, G. K., Savalia, R. T., Sharma, S. K., and Kulkarni, S. D. (1989). *Corros. Sci.*, **29**, 823.
Dutta, R. S., Savalia, R. T., and Dey, G. K. (1995). *Scripta Mater.*, **32**, 207.

D'Souza, C. A., Kuri, S. E., Politti, F. S., May, J. E., and Kiminami, C. S. (1999). *J. Non Cryst. Solids*, **247**, 69.

El-Moneim, A. A., Gebert, A., Schneider, F., Guteisch, O., and Schultz, L. (2002). *Corrosion Science*, **44**, 1097.

Gebert, A., Buchholtz, K., El-Aziz, A. M., and Eckert, E. (2001). *J. Mater. Sci. Eng.*, **A 316**, 60.

Gebert, A., Buchholtz, K., Leonhard, A. A., Mummert, K., Eckert, J., and Schultz, L. (1999). *Mater. Sci. Eng.*, **A 267**, 294.

Helfand, M. A., Sorensen, N. R., and Nelson, G. C. (1992). *J. Electrochem. Soc.*, **133**(9), 2121.

Hiromoto, S., Tsai, A. P., Sumita, M., and Hanawa, T. (2001). *Mater. Trans., JIM*, **42**, 656.

Hussey, R. J., Papaiacovou, P., Shen, J., Mitchell, D. F., and Graham, M. J. (1989). *Mater. Sci. Eng.*, **A120**, 147.

Inturi, R. B., and Szklarska-Smialowska, Z. (1992). *Corrosion*, **48**, 398.

InTech Magazine Online (1998). Published at www.isa.org.

John, A., Zeiger, W., Scharnweber, D., Worch, H., and Oswald, S. J. (1999). *Anal. Chem.*, **365**, 136.

Jung, H., and Alfantazi, A. (2005). *Electrochemical Acta* (Online DOI: 10.1016/j.electacta.2005.06.037).

Kim, S. H., Aust, K. T., Erb, U., Gonzalez, F., and Palumbo, G. (2003). *Scripta Mater.*, **48**, 1379.

Koch, G. H., Brongers, M. P. H., Thompson, N. G., Virmani, Y. P., and Payer, J. H. (2002). *Federal Highway Adminstration Technical Report No. FHWA-RD-01-156*, March 2002 (www.corrosioncost.com), 773.

Klement, U., Erb, U., El-Sherik, A. M., and Aust, K. T. (1995). *Mater. Sci.*, **177**, A203.

Köster, U., Zander, D., Triwikantoro, A., Rüdiger, A., and Jastrow, L. (2001). *Scripta Mater.*, **44**, 1649.

Krolikowski, A. (1988). *Key. Engr. Mater.*, **20**, 1169.

Kuiry, S. C., Seal, S., Bose, S. K., and Roy, S. K. (1994). *Iron Steel Inst. Japan International*, **34**, 599.

Lee, B. W. (1985). *Appl. Phys. Lett.*, **46**, 790.

Lumsden, J. B., and Szklarska-Smialowska, Z. (1978). *Corrosion*, **34**, 169.

Mishra, R., and Balasubramaniam, R. (2004). *Corrosion Science*, **46**, 3019.

Mondal, K., Murty, B. S., and Chatterjee, U. K. (2005). *Corrosion Science*, **47**, 2619.

Moon, D. P., and Bennett, M. J. (1989). *Mater. Sci. Forum*, **43**, 269.

Murty, B. S., Ping, D. H., and Hono, K. (2000). *Appl. Phys. Lett.*, **77**, 1102.

Murty, B. S., Ping, D. H., Ohnuma, M., and Hono, K. (2001). *Acta Mater.*, **49**, 3453.

Naka, M., Hashimoto, K., and Masumoto, T. (1976). *Corrosion*, **32**, 146.

Nazeri, A., Trzaskoma-Paulette, P. P., and Bauer, D. (1997). *J. Sol-Gel Sci. Technol.*, **10**, 317.

Palumbo, G., and Aust. K. T. (1989). *Mat. Sci. Eng A.*, **113**, 139.

Palumbo, G., Thorpe, S. J., and Aust, K. T. (1990). *Scripta Metall.*, **24**, 1347.

Pardo, A., Otero, E., Merino, M. C., Lopez, M. D., Vazquez, M., and Agudo, P. (2001). *Corros. Sci.*, **43**, 689.

Pardo, A., Otero, E., Merino, M. C., Lopez, M. D., Vazquez, M., Agudo, P., and Hich, A. M. (2002). *Corrosion*, **58**, 987.
Patil *et al.* (2004). *Proc. R. Soc. Lond. A.*, **460**, 3569.
Pieraggi, B., and Rapp, R. A. (1993). *J. Electrochem. Soc.*, **140**, 2844.
Rhys-Jones *et al.* (1987). *Corr. Sci.*, **27**, 49.
Roure, S., Czerwinski, F., and Petric, A. (1994). *Oxid. Met.*, **42**, 75.
Rofagha, R., Langer, R., El-Sherik, A. M., Erb, U., Palumbo, G., and Aust, K. T. (1991). *Scripta Mat.*, **25**, 2867.
Rofagha, R., Langer, R., El-Sherik, A. M., Erb, U., Palumbo, G., and Aust, K. T. (1992). *Mater. Res. Res. Soc. Symp.*, **283**, 751.
Rofagha, R., Splinter, S. J., Erb, U., and Mc Intyre, N. S. (1994a). *Nanostructured Materials*, **4**, 69.
Rofagha, R., Plinter, S. J., Erb, U., McIntyre, N. S. (1994b). *Nanostruct. Mater.*, **4**, 69.
Sagawa, M., Fujimura, S., Togawa, M., Yamamoto, H., and Mastura, Y. J. (1985). *J. Appl. Phys.*, **55**, 2083.
Schroeder, V., Gilbert, C. J., and Ritchie, R. O. (1998). *Scripta Mater.*, **38**, 1481.
Schultz, L., Wecker, J., and Hellstern, E. J. (1987). *Appl. Phys.*, **61**, 3583.
Schultz, L., El-Aziz, A. M., Barkleit, G., and Mummert, K. (1999). *Mater. Sci. Eng.* A, **267**, 307.
Seal, S., Bose, S. K., and Roy, S. K. (1994). *Oxid. Met.* **41**, 139.
Seal, S., Roy, S. K., Bose, S. K., and Kuiry, S. C. (2000). *J. Mater. (Electronic)*, **52**(1), 1.
Shannon, R. D. (1976). *Acta Crystall. Sect. A: Cryst. Phys., Diffr. Theor., Gen. Crystallogr.*, **A32**, 751.
Shen, J., Zhou, L., and Li, T. (1998). *J. Mater. Sci.*, **33**, 5815.
Shimotomai, M., Fukufa, Y., Fujita, A., and Ozaki, Y. (1990). *IEEE Trans. Mag.*, **26**, 1939.
Stringer, J. (1970). *Corros. Sci.*, **10**, 513.
Surayanarayana, C., Mukhopadhayay, D., Patanker, S. N., and Froes, F. H. (1992). *J. Mater. Res.*, **7**, 2114.
Thorpe, S. J., Ramaswami, B., and Aust, K. T. (1988). *J. Electrochem. Soc*, **135**, 2162.
Tong, H. Y., and Shi, F. G. (1995). *Scripta Metall. Mater.*, **32**, 511.
Tokuhara, K., and Hirosawa, S. (1991). *J. Appl. Phys*, **69**, 5521.
Tsunekawa, S., Sivamohan, R., Ito, S., Kasuya, A., and Fukuda, T. (1999). *Nanostructured Materials*, **11**, 141.
Uhlig, H. H. (1971). *Corrosion and Corrosion Control: An Introduction to Corrosion Science and Engineering*. New York: John Wiley & Sons, Inc.
Wang, Z. B., Yong, X. P., Tao, N. R., and Li, S. (2001). *Acta Metall. Sinica*, **37**, 1251.
Wang, X. Y., and Li, D. Y. (2003). *Wear*, **255**, 836.
Waren, P. J., Thuvander, M., Abraham, M., Lane, H., Cerezo, A., and Smith, G. D. W. (2000). *Mater. Sci. Forum*, **701**, 343.
Whittle, D. P., and Stringer, J. (1980). *Phil. Trans. R. Soc. Lond.*, A **295**, 309.
Yousef, Kh. M. S., Koch, C. C., and Fedkiw, P. S. (2004). *Corros. Sci.*, **41**, 51.
Zieger, W., Schneider, M., Scharnweber, D., and Worch, H. (1995). *Nanostructured Materials*, **6**, 1013.

7 Applications of structural nanomaterials

7.1 Introduction

The very early application of nanomaterials is utilized in systems where nanopowders are used in their free form, without consolidation or blending. For a simple example, nanoscale titanium dioxide and zinc oxide powders are now commonly used by cosmetics manufacturers for facial base creams and sunscreen lotions for UV protection. Nanoscale iron oxide powder is now being used as a base material for rouge and lipstick. Paints with reflective properties are also being manufactured using nanoscale titanium dioxide particles. Figure 7.1 lists some of the common current and future applications of nanomaterials. The list is not complete, however; it is a current summary and the view of the present authors.

Nanostructured wear-resistant coatings for cutting tools and engineering components have been in use for several years. Nanostructured cemented carbide coatings are used on some Navy ships for their increased durability. Recently, more sophisticated uses of nanoscale materials have been realized. Nanostructured materials are in wide use in the area of information technology, integrated into complex products such as the hard disk drives that provide faster communication in today's world.

Many uses of nanoscale particles have already appeared in specialty markets, such as defense applications, and in markets for scientific and technical equipment. Producers of optical materials and electronics substrates such as silicon and gallium arsenide have embraced the use of nanosize particles for chemomechanical polishing of these substrates for chip manufacturing. Nanosize particles of silicon carbide, diamond, and boron carbide are used as lapping compounds to reduce the waviness of finished surfaces from corner to corner and produce surface finishes to 1–2 nm smoothness. The ability to produce such high-quality components is significant for scientific applications and could become even more important as electronic devices shrink and optical communications systems become a larger part of the nation's communications infrastructure.

Several nanoscale technologies appear to be 4–5 years away from producing practical products. For example, quantum dots of different sizes can be attached to the different molecules in a biological reaction, allowing researchers to follow all the molecules simultaneously during biological processes with only one screening tool. Advances in the feeding of nanopowders into commercial sprayer systems are also promising. It should soon be possible to coat plastics with nanopowders for improved wear and corrosion resistance. Now,

7 Applications of structural nanomaterials

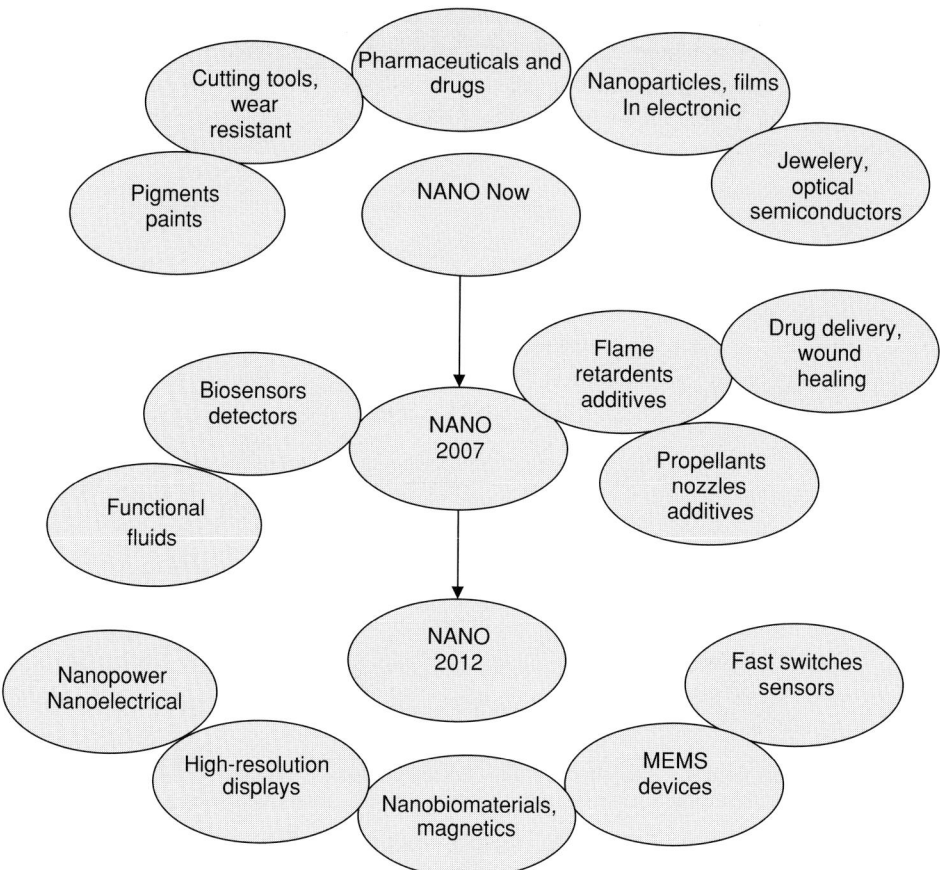

Figure 7.1. Current applications in nanotechnology and a timeline of events for future direction – author view.

one can imagine scenarios in which plastic parts replace heavier ceramic or metal pieces in weight-sensitive applications. The automotive industry is researching the use of nanosized powders in so-called nanocomposite materials.

Several companies have demonstrated injection-molded parts or composite parts with increased impact strength. Full-scale prototypes of such parts are now in field evaluation, and use in the vehicle fleet is possible within 3–5 years or even sooner. Several aerospace companies have programs under way for the use of nanosized particles of aluminum or hafnium for rocket-propulsion applications. The improved burn and speed of ignition of such particles are significant factors for this market for further technological advancement.

A number of other near-term potential applications are also emerging. The use of nanomaterials for coating surfaces to give improved corrosion and wear resistance is being examined on different functional substrates. Several manufacturers have plans to use nanomaterials in the surfaces of catalysts. The ability of nanomaterials such as titania and zirconia to facilitate the trapping of heavy metals and their ability to attract biorganisms make them excellent candidates for filters that can be used in liquid separations for industrial processes or waste-stream purification. Similarly, new ceramic nanomaterials can be used for water-jet nozzles, injectors, armor tiles, lasers, lightweight mirrors for telescopes, and anodes and cathodes in energy-related equipment. A few applications of structural nanomaterials are presented in the following sections.

7.2 Ceramic nanocomposites for load-bearing applications

Researchers have observed unusual behavior, in the form of high contact-damage resistance with no corresponding improvement in toughness, in Al_2O_3/nanotube composites (Wang *et al.*, 2004). Researchers from the University of Connecticut and the National Institute for Materials Science in Tsukuba, Japan, used the spark-plasma sintering (SPS) method to produce the Al_2O_3/single-walled nanotube (SWNT) composites and observed their behavior under Vickers (sharp) and Hertzian (blunt) indentation tests. These composites are not tough in the classical sense, but are resistant to indentation (contact) damage. Such a combination is uncommon in ceramics. The researchers observed similar behavior in Al_2O_3/graphite composites. The reason for this is the unique way in which SWNTs go to the grain boundaries in alumina and provide shear weakness under indentation. The unusual properties of Al_2O_3/SWNT composites could be used in engineering and biomedical applications where contact loading, rather than toughness, is important.

7.3 Bulk nanoribbons can tie up photonic circuits

Researchers at the University of California, Berkeley, and Lawrence Berkeley National Laboratory have fabricated SnO_2 nanoribbons that could be used as optical waveguides in photonic circuits (Law *et al.*, 2004). Researchers synthesized ribbon-shaped SnO_2 nanowires up to 1500 μm long. The single-crystal nanoribbons have rectangular cross-sections that are 100–400 nm in width and thickness. The nanoribbons can act as waveguides much like conventional optical fibers. If laser light is focused onto one end of the ribbon, the generated photoluminescence is strongly guided along the structure to emanate at the opposite end. Visible and ultraviolet light can be channeled into the sub-wavelength structures with little optical loss.

The SnO_2 nanowires are also sufficiently large and strong to be manipulated under an optical microscope. Twists and bends can be introduced with radii as small as 1 μm without disrupting the waveguiding ability. Individual waveguides can be coupled together by overlapping their ends, or assembled into functional geometries such as Y-junctions, branch networks, or ring oscillators. The SnO_2 nanoribbons can also be linked to ZnO nanowire light sources and detectors. Assembling nanowire building blocks with different functions, such as light creation, routing, and detection, is one way of achieving photonic circuits that can manipulate light pulses within submicrometer volumes.

7 Applications of structural nanomaterials

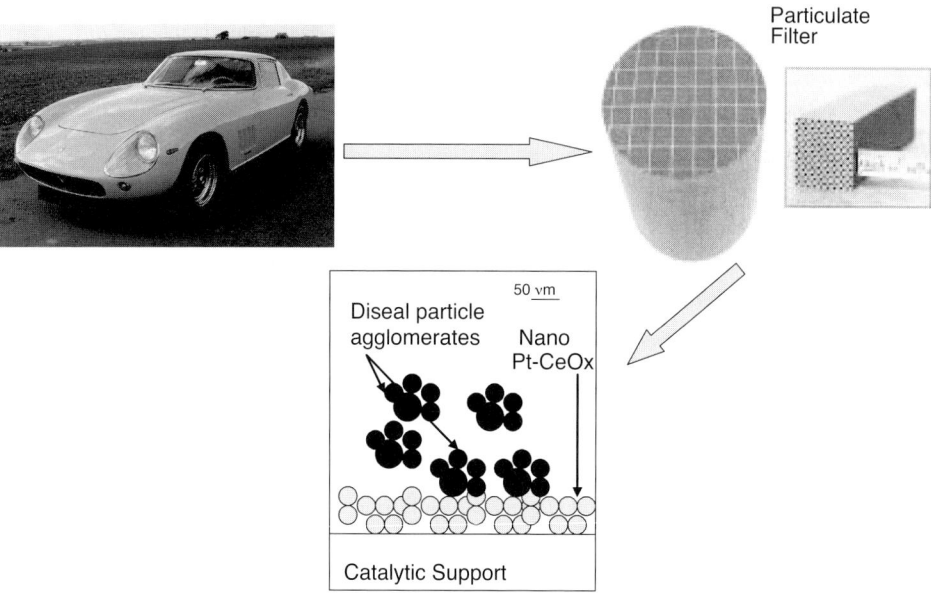

Figure 7.2. Use of nanoparticles in a future diseal particulate filter.

7.4 Functionally gradient nanoparticles

The size of nanoparticles is well known to produce electronic properties different from the bulk. But structural differences between nanoparticles and bulk material are less well understood. If you make a nanoparticle, say a small piece of ZnS, the idea is that in the middle it is basically bulk-like with maybe a bit of relaxation on the surface. Gilbert *et al.* (2004) have taken an experimental and theoretical approach to determine the crystallinity and disorder in ZnS nanoparticles. The nanoparticles show bondlength contraction, random atomic displacements, and complex strain distribution. In other words, the nanoparticle structure is stiffer than the bulk structure. The results have implications for understanding and modeling nanoparticle behavior, particularly in composites and nanoelectromechanical systems.

7.5 Nanotechnology in automotive applications

Diseal particulate-based filters for cars are being developed using a nanotechnology approach. Nano Pt-CeO$_x$ particles are developed to optimize the contact between the particle and diesel soot. Because of the lower particle size a relatively lower filter regeneration temperature is achieved (less than 450° C). Figure 7.2 shows a schematic of such a potential nanofilter for an automobile.

Nanocomposites are currently used in Toyota or GM cars instead of regular plastic. In addition to being stronger and longer-lasting than conventional plastics, nanocomposites

7.5 Nanotechnology in automotive applications

(a)

(b)

Figure 7.3. (a) The step-assist on the 2002 GMC Safari represents a first for the exterior application of a TPO-based nanocomposite (Current nanotechnology applications, 2003). (b) Use of nanocomposites in a Hummer vehicle at different parts.

also have a very important advantage: they are lighter. It is important for a car to be as light as possible. Lighter cars use less fuel, which means that they are less expensive for people to own.

The new bumpers that Toyota has been making for its cars are 60% lighter than the old bumpers and twice as hard to scratch or dent (Current nanotechnology applications, 2003). According to GM, the nanocomposite step-assists are 7%–8% lighter than the old step-assists and have more strength and better surface quality. GM is now testing how nanocomposites might be used in other car parts, like in the panels that make up the body of the car or in the panels that support the doors and passenger compartment (Buchholz, 2003).

The 2002 Chevrolet Astro and GMC Safari midsize vans are the first vehicles to use an advanced thermoplastic olefin nanocomposite on an exterior application, according to General Motors Corp., but not the last. About the vehicles' optional step-assist – a running board designed to help people enter and exit a vehicle. The step-assist on the 2002 GMC Safari represents a first for the exterior application of a TPO-based nanocomposite (see Figure 7.3).

GM and three supplier partners – Basell, Southern Clay Products, and Blackhawk Automotive Plastics, Inc. – worked to bring the TPO-based nanocomposite to production. Basell, the world's largest producer of polypropylene resins for plastics, provides the resin and processing technology for the new nanocomposite.

The technology breakthrough deals with how the very thin flakes of a clay filler are peeled apart. The greater the extent of exfoliation, the thinner the particles are, and a greater surface area is available for interaction with the polymer matrix, resulting in better performance. When properly exfoliated, these layered materials have dimensions on the order of 1 nm (0.04 μin) thick by 100–1000 nm (3.94–39.4 μin) long.

When compared with a conventional TPO, the new nanocomposite can provide up to a 20% weight saving (between 7.5% and 8% on the step-assist), a similar stiffness – 1000–1200 MPa (145 000–174 000 lbf/in^2) – and more ductile strength at low temperature. The new nanocomposite is also TPO cost-equivalent, has a class A surface, is more recyclable (because of less additive material), and does not require new tooling.

Filler content for the nanocomposite is based on a class of clays known as smectites and represents 2.5%–3% of the part (the figure for TPO filler content is typically 15%–20%). In addition to improved throughput, the new nanocomposite provides improved paint adhesion and quality when compared to a talc-filled TPO.

In another venture Dow Chemical and Magma International of America have established a joint Advanced Technology Program (ATP) by the National Institute of Science and Technology (NIST) to develop practical synthesis and manufacturing technologies to enable the use of new high-performance, low-weight nanocomposite materials in automobiles. Proposed potential application would save 15 billion liters of gasoline and reduce CO_2 emission by more than 5 billion kilograms over the life of one year's production of vehicles by the American automotive industry.

These nanomaterials are also likely to find use in non-automotive applications such as pipes, fittings for the building and construction industry, refrigerator lines, medical, consumer equipment housing, and recreational vehicles.

7.6 Nanoclay–polymer composites for structural applications

Emerging nanotechnologies offer the potential for revolutionary new polymer materials with enhanced physical features: reduced thermal-expansion coefficients, increased stiffness and strength, barrier properties, and heat resistance, without loss of impact strength. Nanocomposites, which contain nanometer-scale particles that are homogeneously dispersed throughout traditional polymers, can provide stiffness and strength approaching that of metals, but with significant reductions in weight.

While the most cost-effective nanosized reinforcing particles are clays, they require surface modification with surfactants to achieve homogeneous dispersion at the nanoscale. Even at a cost of $3–7/lb, commercially available nanoclays are only truly compatible with polar polymers like nylon. The proper design of surfactant coatings used to make nanoclays remains a critical technical/economic obstacle to the commercial production of engineered plastics for high-volume applications like automobile body panels.

Research efforts at Argonne National Laboratory have focused on the development of an integrated approach to achieve good exfoliation of nanosized clays in a broad range of polymers, including polyolefins. The Argonne approach begins with a patented clay purification process that selectively separates and recovers exfoliated nanoclays. The surface modification of the mineral surface begins during the early stages of the purification process and utilizes a range of unique surface chemistries with enhanced thermal stability and unusually

7.7 Nanotechnology in the consumer world

Technology/Application	Estimated Market Size (by 2009)
Polymer-Clay Nanocomposites:	Over 1 billion pounds:
Packaging	367 million pounds
Automotive	345 million pounds
Building & Construction	151 million pounds
Coatings	63 million pounds
Industrial	48 million pounds
Other	67 million pounds

Figure 7.4. A potential nano polymer-clay composite market. (Source: Principia partners, *Nanocomposite polymer technology for the next century*).

high affinity for nonpolar polymers like polypropylene. These self-activated nanoclays are fully dispersible without expensive compatiblizing agents.

This translates into high performance at a low cost. For example, oxygen barrier properties of Argonne composites are more than 200 000 times better than oriented polypropylene and more than 2000 times better than Nylon-6. The Argonne technology can be used to produce transparent nanoclay dispersions in a variety of thermoplastic polymers (e.g. polyolefins, elastomers, ionomers) at clay loadings of 20 wt% or higher. Utilizing recently developed Argonne technology, the amount of costly organic modifiers can be reduced by a factor of more than ten, while improving the properties of the nanocomposites.

The Argonne technology offers the possibility of nanoclays at a cost of between $0.30–0.70/lb. Argonne is applying for several key patents on its unique approach for making clay/polymer nanocomposite materials. Figure 7.4 shows a potential market for the nanocomposite polymer technology

7.7 Nanotechnology in the consumer world

The consumer world is exploding with "nanotechnology enhanced" products. Consumer products is an area where the experts are saying the most immediate nanotechnology impacts will be made and recognized by the majority of people in the world. Currently there are numerous products on the market that are the result of nanotechnology.

For the sporting enthusiast, we have tennis balls that last longer, tennis rackets that are stronger, golf balls that fly straighter, nano ski wax that is easier to apply and more effective than standard wax, and bowling balls that are harder; and these products are just scratching the surface of the market. These products use nanostructured materials to give an enhanced performance. For example, the Double Core balls are more impressive because, unlike other balls that were built to last beyond the standard set-and-a-half of playing time, these entries are not quite as heavy nor as spin-resistant as their predecessors.

Wilson has "nanocoated" the balls with a butyl-based barrier called Air D-Fense by InMat Inc. which the company says inhibits air permeation by 200% (Figure 7.5). The balls did not fuzz up quickly after taking a pounding on California's hard courts, nor did they lose their spring deep in the second set after some long, brutal rallies from the baseline.

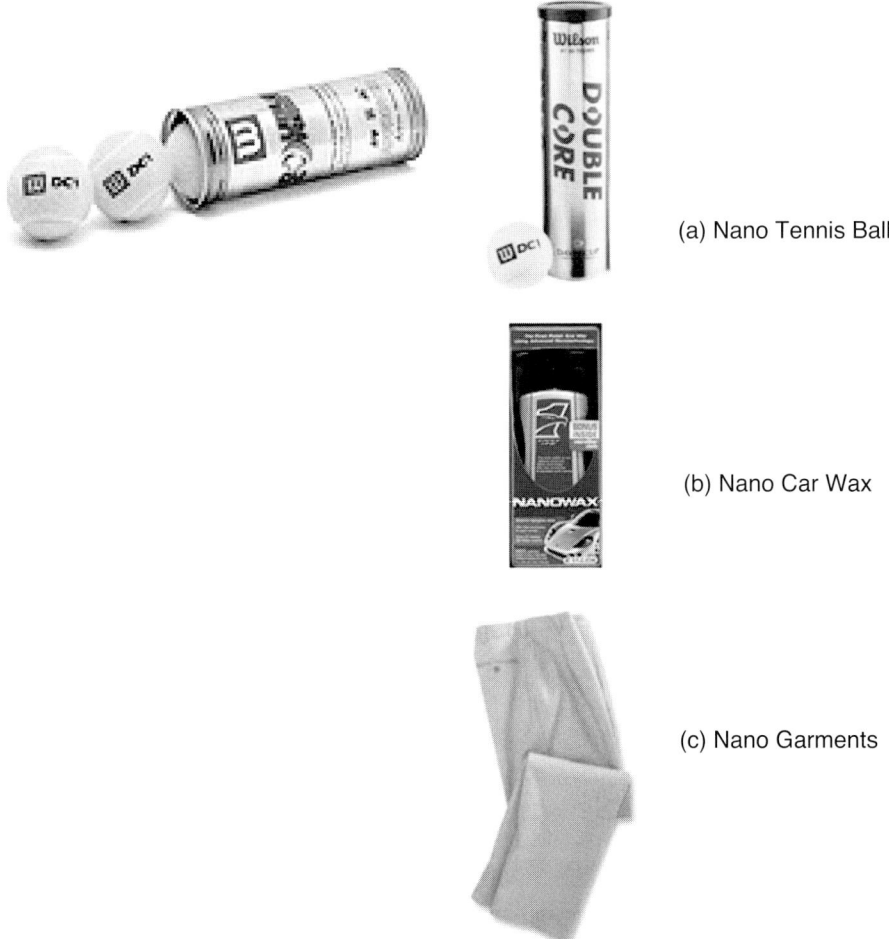

Figure 7.5. Nanoparticles in consumer good (a) tennis ball, (b) car wax, (c) garments.

It has already been approved by one of the sport's leading bodies – the International Tennis Federation – and has been used in the men's professional international team competition, the Davis Cup. It seems the only potential problem the Double Core faces in becoming a permanent part of the tennis landscape is its price (a case of 24 costs $141.54 on Amazon.com) and whether Wilson continues to produce standard balls to compete with the nanocoated variety.

Babolat's VS NCT (Nano Carbon Technology) Drive is a superlight racket at 9.0 ounces with a large, 110-inch head. Babolat joined with Nanoledge SA to obtain the nanotubes, which they say are ten times stiffer than conventional graphite and are placed around the racket head.

The Babolat VS NCT Control is a medium-powered racket that is characterized by its tear-drop head shape and maneuverable control. The VS NCT Control will find favor among 3.0–4.5 intermediate level players who like mobility but still require a stable feel. Although

7.7 Nanotechnology in the consumer world

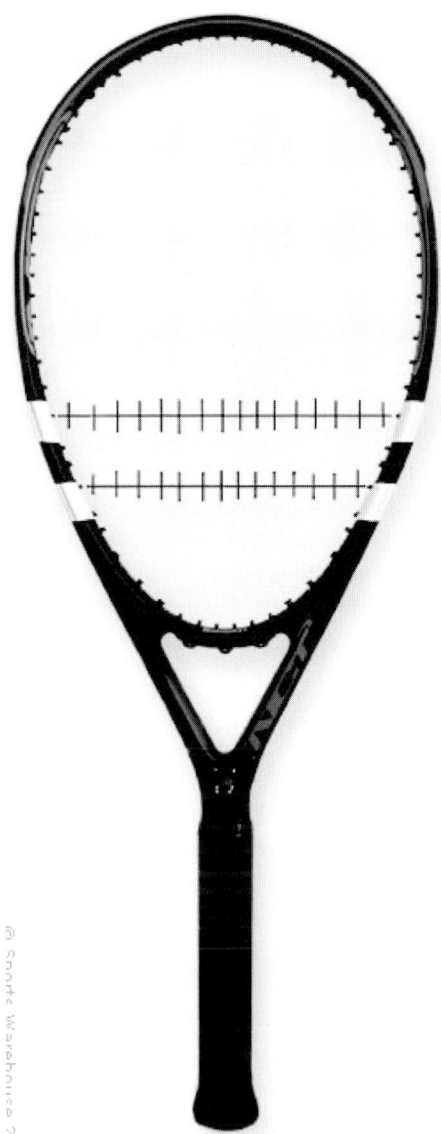

Figure 7.6. Tennis racket made out of reinforced carbon nanotubes.

designed primarily for players with medium-length swings, we found enough control for faster-swinging spin players too. The racket's maneuverability was a plus when it came to quick exchanges at the net or when hitting groundstrokes on the run.

NCT (Nano Carbon Technology) rackets are constructed with two "stabilizers," rigidified by carbon nanotubes, that are located at each side of the racket center giving it a unique shape and texture. According to Babolat, the result should be 20% more rigidity than a standard rectangular structure. Figure 7.6 shows a nanotube-reinforced superlight racket.

Speaking of scratching the surface, we also have nano car wax (Figure 7.5) that fills in those tiny cracks more effectively and gives you a shinier vehicle. Current wax products utilize particles that are too large to effectively fill and remove many hairline scratches, thus causing swirl marks to appear on the paint surface, leaving a surface that reflects light unevenly. By using nanotechnology, Eagle One has developed a wax that conceals swirl marks and hairline scratches. The polishing agent particles used in this formula are nano-sized and are much more efficient at filling and sealing scratches, creating a more even surface for the microemulsion Carnauba wax. This provides a smoother, more-consistent coating, allowing light to reflect more evenly off the paint surface. This smoother coating also minimizes smearing after you have applied the wax and allows easy removal. And because nanoparticles are so tiny, NanoWax eliminates the white powder residue that other waxes leave behind.

In the clothing world, we have pants that repel water (Figure 7.5) and won't stain, shirts and shoe inserts that keep you cool in the summer and warm in the winter, and nano socks that don't "stink" owing to the inclusion of nanotech materials (nanosized silver particles). Nanoceramic coatings are being utilized on photo-quality picture paper to deliver sharper, higher-quality "homemade" digital photo reproductions on your ink jet printer. How about that DVD you watched last night? Any idea how big the features on that now ubiquitous product are? DVD "bumps" to store information are 320 nm wide.

7.8 Nanobelts for actuator applications

Zinc oxide nanoscale ribbons were made by Professor Wang at Georgia Institute of Technology and this breakthrough was published in the news section of *Advanced Materials Process (AMP)*, in 2004. Called nanosprings, the structures have piezoelectric and electrostatic polarization properties that could make them useful in small-scale sensing and microsystem applications. The structures are several millimeters long but only 10–60 nm wide and 5–20 nm thick. These nanostructures are important because they could be transducers and actuators for future-generation nanoscale devices. In addition, the piezoelectric properties of these structures could make them useful in detecting and measuring very small fluid flows, tiny stress/strain forces, high-frequency acoustic waves, and even small air flows.

7.9 Nanosteel for high wear, toughness and hardness applications

While not producing nanoscale microstructures, conventional steel technology is based on manipulating a solid–solid state transformation called a eutectoid transformation. The first step is to heat up the alloy into the single-phase region (austenite). The alloy is then cooled or quenched at various cooling rates to form multiphase structures (i.e. ferrite + cementite). Depending on the rate of cooling, a wide variety of microstructures (i.e. pearlite, bainite, and martensite) can be obtained with a wide range of properties. This manipulation of the transformation has resulted in the wide variety of engineering steels now available.

The NanoSteel Company (http://www.nanosteelco.com/) is developing advanced, nano-structured materials solutions for industry, including upstream and downstream oil and gas industry applications where durability, anti-corrosion, strength, and toughness are sought-out necessities or enhancements.

The NanoSteel Company (Choudhary, 2003) utilizes an entirely analogous but different solid-state processing route, called glass devitrification, to produce steels with bulk nanoscale microstructures. In this case, the supersaturated solid solution precursor material is a supercooled liquid, called a metallic glass. The glass structure itself is a very-high-strength structure since entire classes of defects including one-dimensional dislocations and two-dimensional grain and phase boundaries are eliminated. Unfortunately, the glass is not a defect-free material, since it contains a large fraction of free volume defects, so the full strength of the iron atomic bond is not realized. Upon subsequent heating, the metallic glass precursor transforms into multiple solid phases through devitrification. In glass-forming steel systems, depending on the specific composition, the crystallization temperature usually is in the range from 450 °C to 750 °C and the enthalpy of the glass to crystalline transformation varies typically from −75 J/g to −200 J/g. Since the glass-forming steels commonly melt from 1000 °C to 1300 °C, this means that the glass devitrification occurs at low fractions of the melting temperature (typically 0.4–0.7 T_m) where diffusion is limited and where the driving force, owing to the metastable nature of the glass state, is extremely high. Thus, during devitrification a very high nucleation frequency occurs with limited time for grain growth before impingement between neighboring grains.

Analogous to the characteristic microstructures developed from the eutectoid transformation, the devitrified steels form specific characteristic microstructures which are nanoscale. By this route it is possible to develop very stable nanostructures that resist coarsening at elevated temperatures. It is this type of novel approach that has resulted in the development of amorphous/nanocomposite steels which exhibit strength levels which exceed the existing paradox to a significant extent (45% of theoretical). The result of this basic approach was the development of a family of amorphous/nanocomposite steel alloys which have been dubbed "Super Hard Steel."

TNC produces a range of thermal spray and weld overlay hardfacing products in powder and wire form to address wear, abrasion, erosion, cavitation, and corrosion problems in industry. TNC is focused on bringing new developments in nanomaterials from basic discovery to mainstream technology in an ever-increasing array of industrial processing techniques using feedstock that is conventional in form and readily usable by existing application systems. Incorporation of nanotech coating solutions into existing material systems is expected to vastly extend lifetimes, reduce total ownership costs, and spawn entirely new material technologies which are able to perform in new and demanding environments and in ways previously not possible.

7.10 Copper–carbon nanotube composite for high heat applications

A new copper/CNT composite with potentially lower cost and greater thermal conductivity than diamond composites, as well as excellent thermal expansion properties, has been developed by Omega Piezo Technologies, Inc., State College, PA (Pickrell, 2005). This nanocomposite was designed to manage heat in high-power radar systems and other heat-generating electronic devices. The researchers are engaged in developing a Cu–CNT material that will have thermal conductivity of 1300 watts/meter Kelvin (W/m·K), compared with 600–1200 W/m·K for competing diamond/copper composites. This new composite is attractive to equipment makers who want to pack more chips into products operating at high power for enhanced performance.

7.11 Metal matrix nanocomposites for structural applications

Metal matrix composites (MMCs) such as continuous carbon or boron fiber reinforced aluminum and magnesium, and silicon carbide reinforced aluminum, have been used for aerospace applications owing to their lightweight and tailorable properties. There is much interest in producing metal matrix nanocomposites that incorporate nanoparticles and nanotubes for structural applications, as these materials exhibit even greater improvements in their physical, mechanical and tribological properties as compared to composites with micron-sized reinforcements (Takagi et al., 2001; Bhattacharya and Chattopadhyay, 2001; Ferkel and Mordike, 2001; Dong et al., 2001; Kuzumaki et al., 2000). The incorporation of carbon nanotubes in particular, which have much higher strength, stiffness, and electrical conductivity compared with metals, can significantly increase these properties of metal matrix composites. Currently metal matrix nanocomposites are being explored for structural applications in the defense, aerospace and automotive sectors.

Concurrent with the interest in producing novel nanocomposite materials is the need to develop low-cost means to produce these materials. Most of the prior work in synthesizing nanocomposites involved the use of powder metallurgy techniques, which are not only high cost, but also result in the presence of porosity and contamination (Xu et al., 1999; Ying and Zhang, 2000). Solidification processing methods, such as stir mixing, squeeze casting, and pressure infiltration are advantageous over other processes in rapidly and inexpensively producing large and complex near-net-shape components; however, this area remains relatively unexplored in the synthesis of nanocomposites. Stir mixing techniques, widely utilized to mix micron-size particles in metallic melts (El-Eskandarany, 1998; Rohatgi and Asthana, 1986), have recently been modified for dispersing small volume percentages of nanosize reinforcement particles in metallic matrices. Although there are some difficulties in mixing nanosize particles in metallic melts resulting from their tendency to agglomerate, a research team in Japan has published research on dispersing nanosize particles in aluminum alloys using a stir mixing technique (Kawabe et al., 1999) and researchers at the Polish Academy of Science (Liu et al., 1997) have recently demonstrated the incorporation of greater than 80 vol% nanoparticles in metals using high-pressure infiltration with pressures in the GPa range. Composites produced by this method possess the unique properties of nanosize metallic grains and are certainly useful in structural nanocomposites.

7.12 Application of ferrofluids with magnetic nanoparticles

A ferrofluid is a simple liquid medium containing a colloidal suspension of ferromagnetic nanoparticles. Ferrofluids exhibit remarkable properties in the presence of a magnetic field.

7.12 Application of ferrofluids with magnetic nanoparticles

Typical ferrofluids consist of 10 nm particles of magnetite suspended within an appropriate solvent. Most commercial ferrofluids utilize an oil-based liquid solvent. The nanoparticles in the ferrofluids are individual permanent magnets; when placed in suspension, the net magnetization of the ferrofluid is zero until a magnetic field is applied. What distinguish ferrofluids from other fluids are the body and surface forces and torques that arise in the ferrofluids when magnetic fields are applied to them, which in turn give rise to unusual fluid-mechanical response.

Present-day applications require very small volumes of ferrofluids. Current applications that have an estimated total market size of ~ $30–60 million include the following.

- Contaminant exclusion seals on almost every PC disk drive, in the medical field MRI and CAT scan equipment.
- Vacuum seals for high-speed, high-vacuum, motorized spindles.
- Use as viscous dampers in the air gaps of stepper motors used in aircraft and various other machines.

Various other applications are envisioned in the next few years. The success lies in the advance of the science of ferrofluids subjected to various time-dependent magnetic fields for both heated and isothermal scenarios. For all applications good knowledge of the thermophysical properties of the ferrofluids is essential. The following are examples.

- Enhanced cooling and electrical insulation of power transformers.
- Magnetic separation of ores in mining and scrap metal separation.

For example, the potential market (in terms of revenue) for ferrofluid transformers lies in the range of $0.5–1.0 billion; in addition, there will be substantial savings to the utilities. Researchers are looking at electrorheological (ER) and magnetorheological (MR) fluid dampers for stability augmentation of hingeless and bearingless rotors in automobiles and in bridge suspensions (the latter for storm and earthquake protection). ER and MR fluids are being studied because of their ability to undergo changes in yield stress with the application of an electric or magnetic field, which leads to dampers with smart capabilities. It is shown that, for a constant field, as the force increases, the plug thickness decreases and the equivalent viscous damping constant decreases. Also for a constant force, as the field increases the plug thickness increases and the equivalent viscous damping constant increases. (Hurt and Wereley, 1996). Figure 7.7 shows the use of ferrofluids in the tunable suspension in automobiles.

One of the most exciting new applications for magnetorheological fluid technology is that of real-time controlled dampers for use in advanced prosthetic devices. In such systems a small magnetorheological fluid damper is used to control, in real-time, the motion of an artificial limb based on inputs from a group of sensors. A "smart" prosthetic knee system based on a controllable magnetorheological fluid damper was commercially introduced to the orthopedics and prosthetics market in 2000. The benefit of such an artificial knee is a more natural gait that automatically adapts to changing gait conditions. Figure 7.8 shows such an application (Carlson et al., 2001).

Application of MR fluids in the long term is very promising. NASA's center of microgravity research is investigating how ferrofluids can play an important role in space, where

Figure 7.7. Tunable suspension systems using ferrofluids in various sections of an automobile.

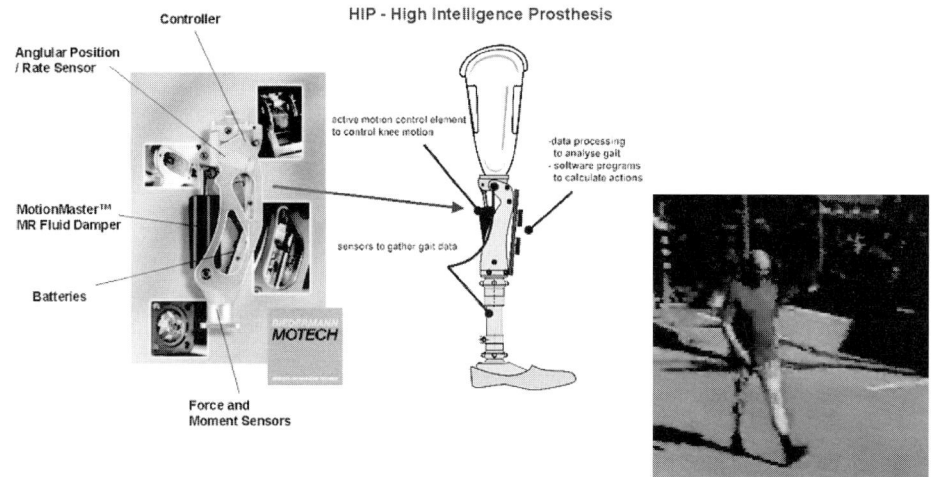

Figure 7.8. Ferrofluids in advanced prosthetics (Carlson *et al.*, 2001).

gravity is essentially absent and free convection dependent processes can be sustained by replacing the gravitational forces with a magnetic force. Long-term applications include the following.

- Development of ferrofluid-cooled and insulated power equipment for extended human space missions.
- Development of nanoscale bearings for levitation.
- Magnetically controlled heat conductivity for precision temperature control in small electronic devices and components.

7.13 Industrial applications of nanocomposite coatings

The industrialization of the "Ti–Al–Si–N" coatings with Si dissolved in the $(Ti_{1-x}Al_x)N$ metastable solid solution was pioneered by Tanaka *et al.* (2001), whereas Jilek and Holubar were developing industrial applications of the nc-$(Ti_{1-x}Al_x)N/a$-Si_3N_4 nanocomposites (Holubar *et al.*, 1999, 2000; Jilek *et al.*, 2003, 2004). Already the "Ti–Al–Si–N" metastable solution coatings showed an enhanced hardness, improved oxidation resistance and a better cutting performance as compared to the state-of-the art $(Ti_{1-x}Al_x)N$ ones (Tanaka *et al.*, 2001). However, as pointed out by Tanaka later, the nc-$(Ti_{1-x}Al_x)N/a$-Si_3N_4 nanocomposites showed an even better cutting performance (Tanaka *et al.*, 2004). The "Ti–Si–N" solid-solution coatings with the fcc structure of TiN were recently studied in some detail by Flink *et al.* (2005). These authors have shown that, in this solid solution, the hardness increases almost linearly with silicon content and reaches about 45 GPa for Si content of 14 at.%. Upon annealing, the hardness remains nearly constant up to about 900 °C and decreases afterward reaching about 26 GPa after annealing to 1100 °C. Compared with the results shown in Figure 3.13 it is clear that the nanocomposites possess a higher thermal stability.

Hitachi company recently also announced the development of a new coating material based on the "Ti–Si–N" system (Hitachi, 2005), and much development at other places is on the way. According to the available information (Hitachi, 2005) the material of their coatings consists of nanocomposites with a Ti-rich TiSiN crystallite phase and Si rich TiSiN amorphous phase. Much higher hardness and higher oxidation resistance than in a conventional solid solution hardened coating, such as TiAlN, was reported. Tools with these coatings are suitable for high-speed cutting and highly efficient machining of hardened die steels for molding. Because the majority of the current development in the companies is proprietary, little information is available. Therefore we shall limit our discussion to several examples which are well documented in the publications or accessible via internet.

When understanding the generic design principle (Veprek and Reiprich, 1995; Veprek *et al.*, 2005a, 2006) it is obvious that the nanocomposites with a high thermal stability cannot be easily deposited in conventional PVD industrial coating systems which usually operate at a low deposition temperature of ≤ 350 °C and low partial pressure of nitrogen. It is also usually difficult, in the large industrial coating systems which have to operate economically,

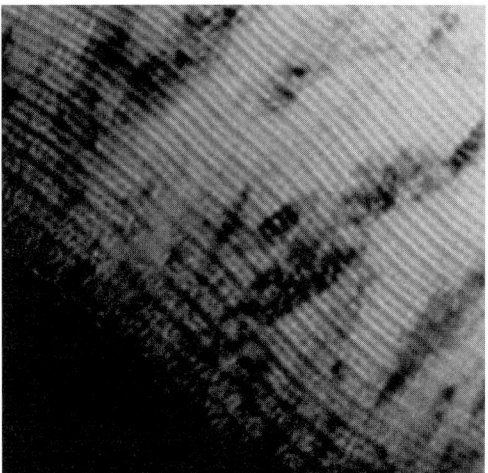

Figure 7.9. Transmission electron micrograph of nc-($Ti_{1-x}Al_x$)N/a-Si_3N_4 nanocomposites deposited by the LARC® technology with composition modulation of about 8 nm (Cselle et al., 2003).

to keep the oxygen impurity level in the coatings below about 0.3–0.5 at.% (see Figure 4.40). Therefore it was necessary to develop a new coating technology. One example is the LAteral Rotating ARC Cathodes, LARC®, technology in combination with a pre-cleaning of the cathodes by means of Virtual Shutter® (Jilek et al., 2003, 2004; Cselle et al., 2003). We refer the reader to the quoted papers for further details and, in the following part of this section, we concentrate on several illustrative examples of the applications which can be found in the published papers.

The asymmetric arrangement of the cathodes (e.g. Ti and Al/Si cathodes), together with the planetary rotation of the tools being coated, results in the deposition of nc-($Ti_{1-x}Al_x$)N/a-Si_3N_4 nanocomposites with compositional modulation at the nanoscale, as illustrated in Figure 7.9 (Cselle et al., 2003). This modulation enhances the resistance of the coating against crack propagation in a similar manner as in multilayered coatings. Moreover, by adjusting the speed of the rotation to the deposition rate, the modulation period can be adjusted in such a way as to reach the maximum hardness as reported for heterostructures. This is illustrated in Figure 7.10 (Cselle et al., 2003).

Because the deposition of multicomponent coating by PVD is usually done from several cathodes where the substrates rotate with respect to them, such nanolayers can be in principle obtained also by other PVD techniques, as reported by Münz et al. for the deposition of heterostructures by reactive sputtering (Münz et al., 2001; Münz, 2003). These researchers also reported a significantly enhanced cutting performance of the heterostructures with a period in the range of several nanometers. Combining the heterostructures with superhard nanocomposites brings combined advantages.

Figure 7.11 shows an example of the drilling performance of the nanolayered, nanocomposites coatings deposited as multilayers in comparison to a variety of conventional (TiN, ($Ti_{1-x}Al_x$)N and advanced coatings. It is seen that already the pre-cleaning of the cathodes by means of the virtual shutter results in a significant improvement of the

7.13 Industrial applications of nanocomposite coatings

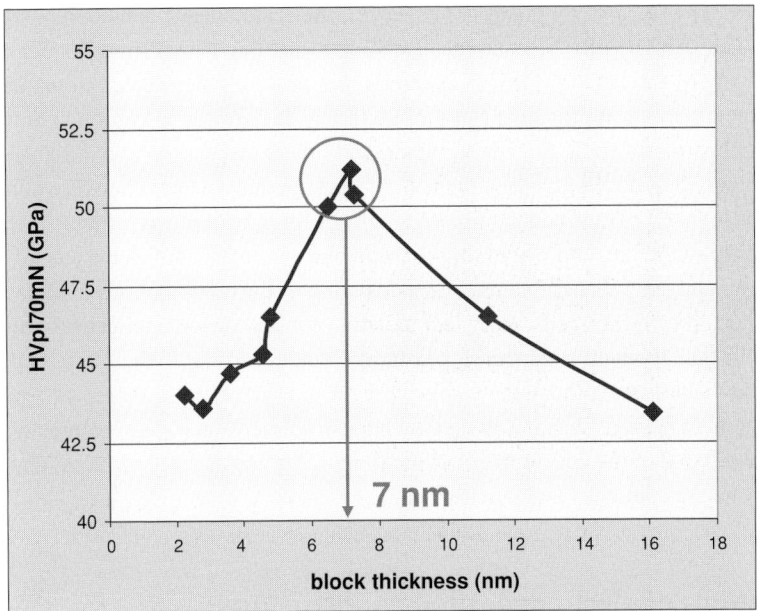

Figure 7.10. Example of the adjustment of the modulation period of the nanolayered, nanocomposite coatings via the adjustment of the rotation speed to the deposition rate in order to obtain the highest hardness (Zindulka, 2003; Cselle *et al.*, 2003).

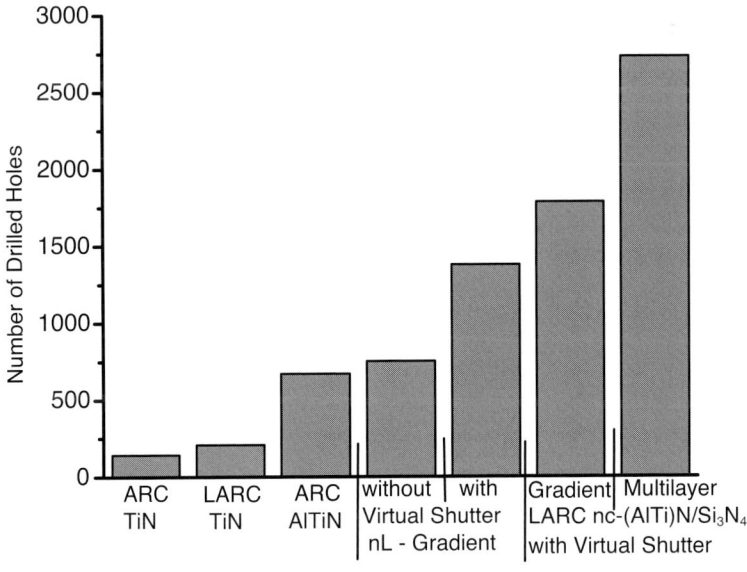

Figure 7.11. Comparison of the cutting performance of conventional, state-of-the-art TiN and AlTiN coatings with the new ones prepared by means of the LARC coating technology (see text). Dry drilling of steel with 5 mm diameter drill made of cemented carbide with corner wear = 200 μm. The numbers indicate the total number of drilled holes (Cselle *et al.*, 2003; Jilek *et al.*, 2004).

state-of-the art $(Ti_{1-x}Al_x)N$ coatings. However, the nanolayered, nanocomposites multi-layer nc-$(Ti_{1-x}Al_x)$N/a-Si_3N_4 coatings reach a life time that is four times longer compared with the state-of-the art $(Ti_{1-x}Al_x)N$ coatings that presently dominate the market for dry machining. A similar improvement of the life time was reported also for a variety of other machining operations such as milling, taping and others (see Cselle *et al.*, 2003).

The most recent development in the area of the superhard nanocomposites concentrates on the development of new nanocomposite coatings for machining of difficult-to-cut alloys, such as nickel-based superalloys for high-temperature turbines of airplane jets, of titanium-based alloys and the "Glare" – laminates (consisting of the combination of Al-based alloys and graphite-fiber reinforced carbon composites) for aerospace applications, and others. Most of this development is of proprietary nature but, according to the limited information available to the present authors, the results of the R & D achieved so far are very promising. Therefore, there is hardly any doubt that the nanostructured superhard coatings will bring a new prospective into the development in the field of dry machining, forming, and also into new areas where coatings have failed so far.

7.14 Applications of electrodeposited nanostructures

Electrodeposition, as outlined in Chapter 2, is a one-step process for production of nano-structured materials. It is based on an old technology, but with refinements such as pulse plating much work has been done in recent years to develop nanocrystalline materials by electrodeposition. As shown in Chapter 4, electrodeposited nanocrystalline metals can exhibit both high strength and good ductility. Electrodeposited nanocrystalline materials can be in the form of thick coatings, and free-standing foils, plates, or tubes, so can be considered to be bulk structural materials. Palumbo *et al.* (2004) have reviewed the applications for electrodeposited nanostructures. These authors list a wide variety of applications, some already in commercial use, some potential. One example of a large-scale structural application of a bulk nanostructured material was the use of an electrodeposited nickel microalloy as an *in situ* repair technique for nuclear steam generator tubing (Palumbo *et al.*, 1997). Electrodeposited bulk nanostructured Ni and Co alloys have been recently developed as environmentally benign replacements of hard Cr electroplating (Palumbo *et al.*, 2002). Microelectromechanical structures (MEMS) have been produced by an electrodeposition molding technique (through-mask electroplating or LIGA, originally developed in Germany). The LIGA method uses photoresist technology to fabricate high-aspect-ratio components. However, conventional MEMS components made by LIGA suffer from the anisotropy of properties owing to the large columnar grains formed by the conventional electrodeposition. Severe reliability problems can occur because of the anisotropic mechanical properties such as elastic modulus and strength that occur for different grain orientations when only a few grains make up the MEMS component. It has been shown that nanostructured MEMS components exhibit considerably improved cross-sectional properties both in terms of uniformity (hardness and elastic modulus) and overall hardness level.

7.15 Potential military applications

Topic	Current	1–5 years	6–10 years	10–50 years
Structural/heavy use sector	Nanoparticles for pigments, paints, self-cleaning windows for commercial use	Smart/responsive nanocoatings for food packaging	Lab on a chip nanotechnology for screening synthesis/catalysis for environmental control	Ultralight materials for cheaper construction and transportation high-strength, light-weight, structures
Energy	Nano-catalyst enhanced fuels for better efficiency	Nanomaterials for fuel cells/batteries	Efficient solar cells using nanotechnology	Nanomateirals for hydrogen storage, fuel cells, alternative energy sources
Health care/medicine	Sun screens	Bio nanosensors for diagnostics	Artificial muscle lab on a chip technology for more efficient drug discovery	Nanomachines for in vivo treatment nanopumps/valves for tissue engineering/artificial organs
Security		Nano bar coding and tagging nanotubes for thermal protection	Remote detection	
Electronics			Carbon nanotube electronic components	Nanomaterials in light-emitting diodes/fsts and PV devices, single electron/molecule devices
Communication			Flat-panel flexible displays using nanotechnology high density data storage	Fast processing using quantum computing DNA computers

Figure 7.12. Potential applications and future markets for nanoscience and nanotechnology.

7.15 Potential military applications

Nanocrystalline materials can be tough, because of the possible combination of high strength and good ductility. It is claimed that Ni–Fe armor plating can be twice as tough as the required specifications for military vehicles when made in nanocrystalline form (Zhong, 2004).

Electromagnetic launchers, rail guns, use electrical-magnetic energy to propel penetrators/projectiles at velocities up to 2.5 km/s. This increase in velocity over conventional explosives can deliver projectiles with an impact velocity of Mach 5 to targets at ranges of 250 miles. Since a rail gun operates on electrical energy, the rails need to be very good conductors of electricity. They also need to be strong and rigid so that the rail gun does not sag while firing and buckle under its own weight. While a good electrical conductor like Cu might be the choice for the rails, Cu is too weak and does not have sufficient wear resistance or high temperature strength. Therefore, nanocomposites of tungsten, copper, and titanium diboride are being studied which may provide the combination of good electrical conductivity and strength required (see the Nanomat website, www.nanomat.com/index2.html).

Kinetic-energy penetrators made from nanocrystalline bcc metals such as tungsten heavy-metal alloys are being evaluated for possible replacement of depleted uranium projectiles. Plastic flow localization has been studied in bulk tungsten with an ultrafine grain structure (Wei *et al.*, 2005). This shear localization is not found in conventional-grain-size tungsten, and is responsible, along with crack formation, to discard material in the "self-sharpening" penetration that is desired.

7.16 Concluding remarks

The future applications of nanotechnology could be immense. It is, however, crucial that expectations of the opportunities afforded by nanotechnology are both upbeat and realistic. Over-blown claims for nanotechnology will not help in communicating the reality of scientific progress and in differentiating between science and science-fiction. Some present and possible future applications for nanotechnology with a suggested time-frame are given in Figure 7.12. It must be recognized that the real benefits of nanotechnology are likely to be many years away. It should also be noted that not all of the technological challenges we face will be solvable by simply reducing the physical dimensions of existing materials.

References

Bhattacharya, V., and Chattopadhyay, K. (2001). *Scripta Mater.*, **44**, 1677.
Buchholz, K. (2003). Nanocomposite debuts on GM vehicles. Automotive Engineering International Online, April 30, 2003. http://www.sae.org/automag/material/10-2001/index.htm
Carlson, J. D., Matthis, W., and Toscano, J. R. (2001). Industrial and commercial applications of smart structures technologies (Lord Corporation), ed. Anna-Maria R. McGowan. *Proceedings of SPIE*, **4332**, 308.
Cselle, T., Morstein, M., Coddet, O., Geisser, L., Holubar, P., Julik, M., Sima, M., and Janak, M. (2003). Werkzeug Technik, **77** (March 2003), 1. (Accessible at www.shm-cz.cz and www.platit.com)
Choudhary, P. (2003). *Advanced Materials and Processes*, ASM International Journal, July 2003, **8**.
Current nanotechnology applications. (2003). *Nanotechnology Now*, May 1, 2003. http://nanotech-now.com/current-uses.htm
Dong, S. R., Tu, J. P., and Zhang, X. B. (2001). *Mater. Sci. Eng.*, **A313**, 83.
Ferkel, H., and Mordike, B. L. (2001). *Mater. Sci. Eng.*, **A298**, 193.
El-Eskandarany, S. M. (1998). *J. Alloys Compd.*, **279**, 263.
Flink, A., Larsson, T., Sjölén, J., Karlsson, L., and Hultman, L. (2006). *Surf. Coat. Technol.*, **200**, 1535.
Gilbert, B., Feng, H., Zhang, H., Waychunas, G. A., and Banfield, J. F., (2004). *Science*, **305**, 651.
Hitachi Tool Engineering, Ltd., Narita Works (2005). http://www.hitachi-tool.co.jp
Holubar, P., Jilek, M., and Sima, M. (1999). *Surf. Coat. Technol.*, **120–121**, 184.
Holubar, P., Jilek, M., and Sima, M. (2000). *Surf. Coat. Technol.*, **133–134**, 145.
Hurt, M. K., and Wereley, N. M. (1996). AIAA-96–1294-CP, AIAA/ASME/AHS Adaptive Structures Forum, Salt Lake City, UT, Apr. 18, 19, 1996, Technical Papers (A96–27071 06–39), 247.

Jilek, M., Holubar, P., Veprek-Heijman, M. G. J., and Veprek, S. (2003). *Mater. Res. Soc. Symp. Proc.*, **697**, 393.

Jilek, M., Cselle, T., Holubar, P., Morstein, M., Veprek-Heijman, M. G. J., and Veprek, S. (2004). *Plasma Chem. Plasma Process.*, **24**, 493.

Kawabe, A., et al. (1999). *J. Jpn. Inst. Met.,* 149.

Kuzumaki, T., Ujiie, O., Ichinose, H., and Ito, K. (2000). *Adv. Eng. Mater.*, **2**, 416.

Law, M., Sirbuly, D. J., Johnnson, J. C., Goldberger, J., Saykally, R. J., and Yang, P. (2004). *Science*, **305**, 1269.

Liu, H., et al. (1997). *Mater. Manuf. Processes.*, **12**, 831.

Münz, W.-D. (2003). *MRS Bulletin*, **28**, 173.

Münz, W.-D., Lewis D. B., Hovsepian, P. Eh., Schönjahn, C., Ehiasarian, A., and Smith, I. J. (2001). *Surf. Engineering*, **17**, 15.

Palumbo, G., Gonzalez, F., Brennenstuhl, A. M., Erb, U., Shmayda, W., and Lichtenberger, P. C. (1997). *NanoStructured Mater.*, **9**, 737.

Palumbo, G., Erb, U., McCrea, J. L., Hibbard, G. D., Brooks, I., and Gonzalez, F. (2002). *AESF SUR/Fin Proc. Q*, 204.

Palumbo, G., McCrea, J. L., and Erb, U. (2004). In *Encyclopedia of Nanoscience and Nanotechnology*, ed. Nalwa, H. S. American Scientific, Publishing, pp. 89–99.

Pickrell, D. (2005), *Advanced Materials and Processes, ASM International Journal*, **March issue**, 11.

Rohatgi, P. K., and Asthana, R. (1986). *Int. Mat. Rev.* **31**, 115.

Takagi, M., Ohta, H., Imura, T., Kawamura, Y., and Inoue, A. (2001), *Scripta Mater.*, **44**, 2145.

Tanaka, Y., Ichimiya, N., Onischi, Y., and Yamada, Y. (2001). *Surf. Coat. Technol.*, **146–147**, 215.

Tanaka, Y., Kondo, A., and Maeda, K. (2004). Invited lecture at the International Conference on Metallurgical Coatings and Thin Films, Session B, San Diego, April/May 2003 (unpublished).

Veprek, S., Veprek-Heijman, G. M. J., Karvankova, P., and Prochazka, J. (2005a). *Thin Solid Films*, **476**, 1–29.

Veprek, S., Männling, H.-D., Karvankova, P., and Prochazka, J. (2006). *Surf. Coat. Technol.*, **200**, 3876.

Veprek, S., and Reiprich, S. (1995). *Thin Solid Films*, **268**, 64–71.

Wang, X., Padture, N. P., and Tanaka, H. (2004). *Nat. Mater.*, **3**, 539.

Wei, Q., Ramesh, K. T., Ma, E., Keskes, L. J., Dowding, R. J., Kazykanov, V. U., and Valiev, R. Z. (2005) *Appl. Phys. Lett.*, **86**, 101907.

Xu, C. L., Wei, B. Q., Ma, R. Z., Liang, J., Ma, X. K., and Wu, D. H. (1999). *Carbon*, **37**, 855.

Ying, D. Y., and Zhang, D. L. (2000). *Mater. Sci. Eng.*, **286**, 152.

Zhong, L. W. (2004). *Advanced Materials and Processes, ASM International Journal*, **Jan. issue**, 22.

Zindulka, O. (2003). Unpublished.

Index

aerogel 10
anelastic properties of nanostructured materials 136
   elastic energy dissipation coefficient 136
   internal friction 136, 138
Arrhenius formula 305
artifact-free nanocrystalline 298, 299
atomic force microscopy (AFM) 97

ball milling, *see* mechanical attrition
bimodal structure 155, 156, 209, 217, 279, 288, 295
Bragg's law 95
brick-like grain structure 218
bulk nanostructures 5

carbon nanotubes 16
ceramic
   nanocrystalline ceramic 177
   nanocomposite 343
chemical synthesis of nanopowders 36
   agglomeration 37, 42, 44
   intermetallics 58
   precursors 37, 44, 176
   reducing agents 58
   sonochemical decomposition 58, 61
   steel powders 61, 62
   ultrasonic irradiation 58
chemical vapor deposition (CVD) 72
   thermal CVD 74
   plasma-induced CVD 76, 82, 120
coatings
   decomposition 122, 125
   friction 123
   preparation 72
   hardness 118, 121, 122
   self-hardening 121, 122, 123, 125
   spinodal decomposition 126, 127, 128
colloids, stability 40
   attractive potential 40
   dispersion forces 40, 41
   electrostatic stabilization 40
   Hamaker constant 40
   steric stabilization 41
concave grain boundary 102
configurational force method 274
Considere criterion 152
convex grain boundary 102

corrosion
   austenitic steels wet corrosion 331
   behavior of nanocrystalline materials 318, 319, 336
   behavior of nanocrystalline nickel 319
   behavior of nanocrystalline cobalt 320, 324
   behavior of zirconium 326
   dry corrosion 333
   resistance 319, 322, 323, 324, 326, 329, 331, 333, 336, 338, 341
   wet corrosion 333
creep
   Coble creep 137, 140, 167, 173, 174, 175, 176, 224, 225, 226, 227, 228, 229, 240, 279, 297
   dislocation creep 173, 176
   Nabarro–Herring creep 173
crystal lattice rotation 218, 258, 261, 266
cutting tools 10

deformation localization 32
dendrite-like inclusions 217, 296
differential scanning calorimetry (DSC) 98
dislocation activity in nanograins 162, 163, 165
dislocation climb 103, 261
drag effect of triple junctions 106, 108
ductility 1, 151, 155, 156

elastic properties of nanostructured materials 1, 134
   shear modulus 135
   Young's modulus 134, 137
electrodeposition 70, 72, 97, 110, 151, 153, 174, 175, 176, 178, 298, 319, 358
elliptic pores 291
elongation 152, 155, 157
energetic ion bombardment 182, 192
extrinsic defects 216
extrinsic grain-boundary dislocations 292, 305

fabrication produced pores 297, 298
fatigue
   behavior of nanocrystalline materials 177, 178
   crack growth 178, 180, 196
   failure 178
   high cycle fatigue 178
   low cycle fatigue 178
   performance 178
ferrofluids 352, 353, 355

Index

ferromagnetic nanoparticles 352
forest dislocations 297
fracture
   behavior of nanocrystalline materials 180, 268
   brittle fracture 268, 269, 280, 287, 290
   ductile fracture 268, 269, 270, 280, 287, 290, 291, 295, 307
   intergranular fracture 269, 280, 281, 283, 284, 286, 287, 290, 291, 307
   intragranular fracture 269, 284
   stress 151
   toughness 151, 180, 181, 192, 193, 196

glass transition temperature 34
gradient theory 206
grain boundaries
   amorphous grain boundaries 210, 211, 230
   diffusion 107, 114, 117
   diffusion coefficient 303
   emission of dislocation 162, 163
   high-angle 210, 211
   in nanocrystalline materials 107
   grain-boundary defects 107
   low-angle 210, 211
   migration 101, 103, 108, 109
   mobility 110, 112
   nonequilibrium 210, 211, 213, 216, 230, 284
   pinning 66, 110, 149
   rotation 33, 162, 164
   sliding 33, 162, 164
grain growth 93, 99
   abnormal growth 99, 109, 118
   normal growth 99
grain growth in coarse-grained polycrystals 99, 101, 103, 107, 108
   2D models 102
   driving forces 100, 101
   Euler equation 102
   grain-boundary migration 101, 103, 108
   grain-boundary sweeping 104
   hampering forces 100, 108
   kinetic equation 101, 104
   mean-field theories 103, 108
   parabolic law 101, 103, 105
   statistical theory 104
   stochastic theory 105, 108
   structural gradient 104
   topology 101
   triple junctions mobility 105, 106, 108
   von Neumann–Mullins relationship 103
grain growth in nanocrystalline materials 98, 107, 108, 111, 116
   activation energy 114, 116
   crystal lattice rotation 108
   driving forces 108, 109
   hampering forces 108
   isothermal grain growth kinetics 114, 117
grain-growth inhibition 109
   chemical ordering 112
   Gibbs equation 113
   grain-size stabilization 112
   kinetic approach 109

   porosity drag 109
   second phase drag 110
   solute drag 111
   thermodynamic approach 109, 112
Griffith criterion 148, 187

Hall–Petch behavior 139, 145, 147, 161, 217, 219, 221, 222, 223, 225, 226, 233, 292
Hall–Petch mechanism 188, 233
hardness
   of ceramics 148
   of intermetallic compounds 145, 147
   of metals and alloys 1, 138, 141, 144, 157
   of multiphase materials 148, 149, 150
   strain rate effect 167, 168, 169
heat treatment 240, 280
hybrid deposition techniques 84

image forces 224, 298
in situ consolidation 31, 32, 154, 158
inert gas condensation 27, 29, 109, 110, 117, 173, 175, 177, 178
instrumental broadening 94, 96
interfacial enthalpy 98
intrinsic defects 216
intrinsic dislocation 251, 253
inverse Hall–Petch effect 140, 145, 147, 219, 221, 224, 233, 240, 241

Kissinger method 98

Langevin equation 105
local shear event 230, 231, 234, 235, 236, 237, 271, 272

macroscopic crack 286, 287, 290
magnetic nanomaterials 8
mechanical attrition 29, 32, 65, 240, 259
   cryogenic milling 31, 32, 35, 172, 209
   in polymers 34
   mechanical alloying 29, 33, 34
   mechanical milling 29, 33
   minimum grain size 29
   temperature effect 30
metastable grain size 113
microemulsion technology 46
   critical micelle concentration (CMC) 48
   emulsification 47
   emulsifier, see surfactant
   free energy of microemulsion 50
   interfacial energy 47, 52
   hydrophile–lipophile balance (HLB) 47
   precipitation 52
   reaction kinetics 52
   surfactant 47, 48, 54
   thermodynamics 49
microhardness 217, 243
microvoid 180, 269, 280, 291, 307, 318
misorientation band 258
molecular dynamics simulation 188, 206, 207, 216, 219, 231, 246, 247, 250, 253, 255, 256, 258, 270, 284, 307

model
    analytical model 206, 248, 256
    continuum model 205, 206
    geometric model 205

nanobelts 350
nanosteel 350
nanoclay 346, 347
nanocomposites
    copper–carbon nanotube composites 351
    magnetic nanocomposites 336
    metal matrix nanocomposites 352
    superhard nanocomposites 4, 72, 74, 77, 120, 127, 184, 188, 189, 193, 269, 281, 356, 358
    ultrahard nanocomposites 189, 193
nanopowders 25, 341
    compaction 28
    consolidation 26, 31, 32, 64, 151
    contamination 26, 29, 35, 65
    densification 64, 65
    porosity 28, 109, 134, 151
    sintering 26, 64, 65
nanoribbons 343
nanoscale and interface effects 204, 205, 207, 220, 221, 269, 292, 308
nanoscale dislocation loop 236, 237
necking 152, 154, 280, 294, 296, 307

Orowan bypassing 223
overpotential 71
oxidation resistance 334, 335, 355
oxygen impurity 190, 192

partial dislocation 246
    leading 247, 253
    trailing 247, 253
Peierls stress 167
phonon confinement model 98
pinning pressure 110
plastic flow localization 242, 243, 244, 262, 287, 289, 291, 293, 294, 295, 296, 300, 302
polymer nanocomposites 11

quantum dots 341

Raman spectroscopy 97
reactive element effect (REE) 334
rule of mixture 219, 227, 245

scanning electron microscopy (SEM) 96, 97
scanning tunneling microscopy (STM) 97
severe plastic deformation (SPD) methods 67, 153, 177, 178, 280, 303
    accumulative roll-bonding 68
    equal channel angular pressing (ECAP) 68
    high-pressure torsion (HPT) 67
    multiple forging 68
shear bands 32, 144
Sherrer equation 94

Sherrer formula 95
Shockley dislocation 166, 246, 249, 251, 253
Sogmiliana dislocation 237
sol–gel synthesis 37, 45
    alcoholic solutions 37
    alkoxide 37
    condensation 38
    diffusion growth 39, 40
    emulsifiers, see surfactants
    hydrolysis 38
    mononuclear growth 38, 40
    polynuclear growth 39, 40
    surfactants 42
solid solution hardening 141
specific structural features of
    composite nanocrystalline materials 208, 213
    single-phase nanocrystalline materials 208, 211
sputtering 78, 120
    background pressure 81
    magnetron sputtering 79, 121
stacking fault, anomalously wide 255
strain hardening 154, 161, 162
stress–strain curve 155, 156, 157, 158, 160, 161
structural nanomaterials
    definition 25
    fabrication 27
superhard coatings 182, 190, 358
superhard heterostructures 183
superlattice 2, 3
superplasticity 1, 169, 170, 171, 172, 195, 218, 241, 245, 263, 279, 299, 308
    equation for elevated temperature plasticity 170
sweep constant 104

Tabor approximation 142
temperature effect on deformation 167
thermal plasmas deposition 78
thermal stability of nanostructures 93, 110, 120, 125, 127, 156
topological classes of grains 104
transmission electron microscopy (TEM) 96, 162
tube-like regions 208
twinning 157, 161, 165, 166, 255, 256, 257

vacuum arc evaporation 82
    "PLATIT" technique 82
    LARC technology 83
voids 162

Warren–Averbach method 95, 96
Williamson–Hall method 95, 96

X-ray line-broadening analysis 94, 109

yield strength 142, 146, 151, 155, 157, 167
    asymmetry in tension and compression 144

zero current potential (ZCP) 319